Sustainable Food Packaging Technology

Sustainable Food Packaging Technology

Edited by

Athanassia Athanassiou

WILEY-VCH

Editor

Dr. Athanassia Athanassiou
Italian Institute of Technology
Smart Materials
Via Morego, 30
16163 Genova
Italy

Cover Images
© Andrei Mayatnik / Shutterstock,
© Nelli Syrotynska / Shutterstock

Library of Congress Card No.:
applied for

British Library Cataloguing-in-Publication Data
A catalogue record for this book is available from the British Library.

Bibliographic information published by the Deutsche Nationalbibliothek
The Deutsche Nationalbibliothek lists this publication in the Deutsche Nationalbibliografie; detailed bibliographic data are available on the Internet at <http://dnb.d-nb.de>.

Print ISBN: 978-3-527-34556-4
ePDF ISBN: 978-3-527-82006-1
ePub ISBN: 978-3-527-82008-5
oBook ISBN: 978-3-527-82007-8

Typesetting SPi Global, Chennai, India
Printing and Binding Printed and bound by CPI Group (UK) Ltd, Croydon, CR0 4YY

Printed on acid-free paper

10 9 8 7 6 5 4 3 2 1

Contents

Preface *xiii*

Part I Review on Biopolymers for Food Protection *1*

1 **Emerging Trends in Biopolymers for Food Packaging** *3*
Sergio Torres-Giner, Kelly J. Figueroa-Lopez, Beatriz Melendez-Rodriguez, Cristina Prieto, Maria Pardo-Figuerez, and Jose M. Lagaron
1.1 Introduction to Polymers in Packaging *3*
1.2 Classification of Biopolymers *4*
1.3 Food Packaging Materials Based on Biopolymers *7*
1.3.1 Polylactide *7*
1.3.2 Polyhydroxyalkanoates *8*
1.3.3 Poly(butylene adipate-*co*-terephthalate) *9*
1.3.4 Polybutylene Succinate *10*
1.3.5 Bio-based Polyethylene *11*
1.3.6 Bio-based Polyethylene Terephthalate *13*
1.3.7 Poly(ethylene furanoate) *14*
1.3.8 Poly(ε-caprolactone) *15*
1.3.9 Thermoplastic Starch *15*
1.3.10 Cellulose and Derivatives *17*
1.3.11 Proteins *17*
1.3.11.1 Gelatin *18*
1.3.11.2 Wheat Gluten *18*
1.3.11.3 Soy Protein *20*
1.3.11.4 Corn Zein *20*
1.3.11.5 Milk Proteins *21*
1.4 Concluding Remarks *21*
References *24*

2 Biopolymers Derived from Marine Sources for Food Packaging Applications *35*
Jone Uranga, Iratxe Zarandona, Mireia Andonegi, Pedro Guerrero, and Koro de la Caba
2.1 Introduction *35*
2.2 Fish Gelatin Films and Coating *37*
2.2.1 Collagen and Gelatin Extraction *37*
2.2.2 Preparation and Characterization of Fish Gelatin Films and Coatings *39*
2.2.3 Food Shelf Life Extension Using Fish Gelatin Films and Coatings *40*
2.3 Chitosan Films and Coatings *42*
2.3.1 Chitin and Chitosan Extraction *42*
2.3.2 Preparation and Characterization of Chitosan Films and Coatings *43*
2.3.3 Food Shelf Life Extension Using Chitosan Films and Coatings *44*
2.4 Future Perspectives and Concluding Remarks *46*
References *46*

3 Edible Biopolymers for Food Preservation *57*
Elisabetta Ruggeri, Silvia Farè, Luigi De Nardo, and Benedetto Marelli
3.1 Introduction *57*
3.2 Polysaccharides *61*
3.2.1 Alginate *63*
3.2.2 Carrageenans *63*
3.2.3 Cellulose *67*
3.2.4 Chitosan *69*
3.2.5 Pectin *70*
3.2.6 Pullulan *71*
3.2.7 Starch *71*
3.3 Proteins *72*
3.3.1 Casein *73*
3.3.2 Collagen *74*
3.3.3 Gelatin *74*
3.3.4 Wheat Gluten *75*
3.3.5 Whey Protein *75*
3.3.6 Silk Fibroin *76*
3.3.7 Zein *77*
3.4 Lipids *78*
3.4.1 Beeswax *80*
3.4.2 Candelilla Wax *80*
3.4.3 Carnauba Wax *81*
3.4.4 Shellac *81*
3.5 Edible Composite Materials *82*
3.6 Active Coatings *85*
3.6.1 Antimicrobial Agents *85*
3.6.2 Antioxidant Agents *85*

3.7 Materials Selection and Application *86*
3.8 Conclusions *87*
References *88*

Part II Food Packaging Based on Individual Biopolymers and their Composites *107*

4 Polylactic Acid (PLA) and Its Composites: An Eco-friendly Solution for Packaging *109*
Swati Sharma
4.1 Introduction *109*
4.2 Synthesis of PLA and Its Properties *110*
4.3 Properties Required for Food Packaging *111*
4.3.1 Barrier Properties *111*
4.3.2 Optical Properties *113*
4.3.3 Mechanical Properties *114*
4.3.4 Thermal Properties *114*
4.3.5 Antibacterial Properties *115*
4.4 General Reinforcements for PLA *116*
4.4.1 Natural Fibers *116*
4.4.2 Synthetic Fibers *121*
4.4.3 Functional Fillers *122*
4.4.3.1 Clay/PLA Composites *122*
4.4.3.2 Metal-oxide/PLA Composites *123*
4.5 Biodegradability of PLA *123*
4.6 Conclusions and Future Prospects *124*
References *124*

5 Green and Sustainable Packaging Materials Using Thermoplastic Starch *133*
Anshu A. Singh and Maria E. Genovese
5.1 Sustainability and Packaging: Toward a Greener Future *134*
5.1.1 The Plastic Threat *134*
5.1.2 The Call for Sustainability *135*
5.1.3 Biomaterials for Sustainable Packaging *135*
5.2 Thermoplastic Starch *137*
5.2.1 Starch: Physicochemical Properties, Processing, Applications *137*
5.2.2 From Starch to Thermoplastic Starch *141*
5.2.3 Plasticizers of Starch *142*
5.2.4 Processing of Thermoplastic Starch *143*
5.3 Thermoplastic Starch-Based Materials in Packaging *145*
5.3.1 Technical and Legal Requirements for Packaging Materials *145*
5.3.2 Composites of TPS with Fillers *146*
5.3.3 Composites of Thermoplastic Starch with Polysaccharides *147*
5.3.4 Composites of Thermoplastic Starch with Polyesters *149*

5.3.5 Composite of TPS Based on Chemical Modification *152*
5.3.6 Commercial Packaging Materials Based on Thermoplastic Starch *152*
5.4 Conclusions *153*
References *155*

6 Cutin-Inspired Polymers and Plant Cuticle-like Composites as Sustainable Food Packaging Materials *161*
Susana Guzmán-Puyol, Antonio Heredia, José A. Heredia-Guerrero, and José J. Benítez
6.1 Introduction *161*
6.1.1 Bioplastics as Realistic Alternatives to Petroleum-Based Plastics *161*
6.1.2 The Plant Cuticle and Cutin: The Natural Food Packaging of the Plant Kingdom *166*
6.1.3 A Comparison of Cutin with Commercial Plastics and Bioplastics *169*
6.1.4 Tomato Pomace is the Main and Most Sustainable Cutin Renewable Resource *172*
6.1.5 Toward a Sustainable Industrial Production of Cutin-Inspired Commodities *173*
6.2 Synthesis of Cutin-Inspired Polyesters *173*
6.2.1 The Influence of the Monomer Architecture in the Physical and Chemical Properties of Cutin-Inspired Polyhydroxyesters *173*
6.2.2 The Effect of Oxidation in the Structure and Properties of Cutin-Inspired Fatty Polyhydroxyesters *177*
6.2.3 Surface vs. Bulk Properties *180*
6.3 Cutin-Based and Cutin-like Coatings and Composites *183*
6.3.1 Cutin-Inspired Coatings on Metal Substrates *183*
6.3.2 Plant Cuticle-like Film Composites *186*
6.4 Concluding Remarks *188*
Acknowledgments *189*
References *189*

7 Zein in Food Packaging *199*
Ilker S. Bayer
7.1 Introduction *199*
7.2 Solvent Cast Zein Films *202*
7.3 Chemical Characteristics of Solvent-Cast Zein Films *206*
7.4 Extrusion of Zein *209*
7.5 Zein Laminates with Various Packaging Films *212*
7.6 Zein Blend Films with Other Biopolymers *214*
7.7 Outlook and Future Directions *217*
7.8 Conclusions *219*
References *220*

Part III Biocomposites of Cellulose and Biopolymers in Food Packaging *225*

8 Cellulose-Reinforced Biocomposites Based on PHB and PHBV for Food Packaging Applications *227*
Estefania L. Sanchez-Safont, Luis Cabedo, and Jose Gamez-Perez
8.1 Introduction to Bioplastics *227*
8.2 PHB and PHBV: a SWOT (Strength, Weakness, Opportunity, and Threat) Analysis *229*
8.2.1 Polyhydroxyalkanoates (PHA): Poly-3-hydroxybutyrate (PHB) and Poly-3-hydroxybutyrate-*co*-3-hydroxyvalerate (PHBV) *229*
8.2.2 PHB and PHBV: Strengths *231*
8.2.3 PHB and PHBV: Weaknesses *232*
8.2.4 PHB and PHBV: Opportunities *235*
8.2.5 PHB and PHBV: Threats *236*
8.3 Cellulose Biocomposites *236*
8.3.1 Structure, Composition, and General Properties of Lignocellulosic fibers *237*
8.3.2 Lignocellulosic Fibers in Polymer Composites *240*
8.3.2.1 Fiber Modification *241*
8.3.2.2 Fiber-matrix Chemical Anchor *242*
8.4 PHA/Fiber Composites *242*
8.4.1 PHB and PHBV/Cellulose Composites: Achievements and Limitations *242*
8.4.2 New Trends in PHB and PHBV/Cellulose-Reinforced Biocomposites *245*
8.4.3 The Potential Use of PHA-Based Composites in the Food Packaging Sector *247*
8.5 Conclusions *248*
References *250*

9 Poly-Paper: Cellulosic-Filled Eco-composite Material with Innovative Properties for Packaging *263*
Romina Santi, Silvia Farè, Alberto Cigada, and Barbara Del Curto
9.1 Introduction *263*
9.2 Materials *265*
9.2.1 Matrix *265*
9.2.2 Reinforcement *266*
9.2.3 Composite Formulations *266*
9.2.4 Extrusion Process *267*
9.3 Mechanical Properties *268*
9.4 Suitable Processes for Poly-Paper *268*
9.4.1 Injection Molding *269*
9.4.2 Thermoforming *270*
9.4.3 Poly-Paper Expansion *270*
9.5 Additional Properties of Poly-Paper *272*

9.5.1 Shape Memory Forming 272
9.5.2 Self-Healing by Water 273
9.6 End-of-Life 275
9.7 Conclusions 277
References 278

10 Paper and Cardboard Reinforcement by Impregnation with Environmentally Friendly High-Performance Polymers for Food Packaging Applications 281
Uttam C. Paul and José A. Heredia-Guerrero
10.1 Introduction 281
10.2 Improving the Barrier Properties of Paper and Cardboard by Impregnation in Capstone and ECA Solutions 282
10.2.1 Preparation of the Samples 283
10.2.2 Morphological Characterization 283
10.2.3 Chemical Characterization 285
10.2.4 Barrier Properties, Wettability, and Water Uptake 285
10.2.5 Mechanical Characterization 291
10.3 Water, Oil and Grease Resistance of Biocompatible Cellulose Food Containers 292
10.3.1 Preparation of the Samples 294
10.3.2 Morphological Analysis 295
10.3.3 Water and Oil Resistance Properties 296
10.3.4 Mechanical, Grease Resistance, and Barrier Properties of Treated Paper 296
10.4 Conclusions 300
References 300

11 Nanocellulose-Based Multidimensional Structures for Food Packaging Technology 305
Saumya Chaturvedi, Sadaf Afrin, Mohd S. Ansari, and Zoheb Karim
11.1 Introduction 305
11.2 Necessities in Food Packaging Industry 307
11.3 An Overview of NC 308
11.4 Cellulose Fibrils and Crystalline Cellulose 308
11.5 Why NC for Packaging? 310
11.6 Effect on NCs on Networking 310
11.7 Migration Process of Molecules Through NC Dimensional Film 312
11.8 Processing Routes of NC-based Multidimensional Structures for Packaging 312
11.9 CNFs for Barrier Application 314
11.10 CNCs for Barrier Application 315
11.11 Conclusion 316
References 317

Part IV Natural Principles in Active and Intelligent Food Packaging for Enhanced Protection and Indication of Food Spoilange or Pollutant Presence *323*

12 Sustainable Antimicrobial Packaging Technologies *325*
Selçuk Yildirim and Bettina Röcker
12.1 Introduction *325*
12.2 Antimicrobial Food Packaging *326*
12.3 Natural Antimicrobial Agents *328*
12.3.1 Plant Extracts *328*
12.3.2 Organic Acids, Their Salts and Anhydrides *335*
12.3.3 Bacteriocins *336*
12.3.4 Enzymes *337*
12.3.5 Chitosan *338*
12.4 Conclusions and Perspectives *340*
References *341*

13 Active Antioxidant Additives in Sustainable Food Packaging *349*
Thi-Nga Tran
13.1 Introduction *349*
13.2 Antioxidant Capacities of Plant-Based Food Packaging Materials *352*
13.2.1 Antioxidant Natural Extracts in Food Packaging *353*
13.2.2 Antioxidant Raw Materials Derived from Food Wastes and Agro-Industry by-Products *359*
13.3 Conclusions and Future Perspectives *361*
References *363*

14 Natural and Biocompatible Optical Indicators for Food Spoilage Detection *369*
Maria E. Genovese, Jasim Zia, and Despina Fragouli
14.1 Food Spoilage *370*
14.1.1 Food Spoilage: A Never-ending Challenge *370*
14.1.2 Microbial Spoilage *370*
14.1.3 Physical and Chemical Spoilage *372*
14.1.4 Factors Determining Food Spoilage *372*
14.2 Food Spoilage Detection *372*
14.2.1 Conventional Methods and Technologies for the Detection of Food Spoilage *372*
14.2.2 On Package and on Site Sensing Technologies: A New Strategy for Food Spoilage Detection *373*
14.3 Natural and Biocompatible Optical Indicators for Food Spoilage *379*
14.3.1 Optical and Colorimetric Detection *379*
14.3.2 Natural and Biocompatible Indicators *379*
14.3.3 Detection of pH, Acids, and Amines *380*
14.3.4 Detection of Oxygen *386*

14.3.5 Detection of Carbon Dioxide *387*
14.3.6 Detection of Bacteria *388*
14.4 Concluding Remarks and Future Perspectives *388*
References *389*

Part V Technological Developments in the Engineering of Biocomposite Materials for Food Packaging Applications *395*

15 Biopolymers in Multilayer Films for Long-Lasting Protective Food Packaging: A Review *397*
Ilker S. Bayer
15.1 Introduction *397*
15.2 Biopolymer Coatings and Laminates on Common Oil-Derived Packaging Polymers *399*
15.3 Multilayer Films Based on Proteins *405*
15.4 Multilayer Films Based on Polysaccharides *409*
15.5 Coatings on Biopolyesters *415*
15.6 Summary and Outlook *418*
References *420*

Index *427*

Preface

There could not be a clearer and at the same time shocking demonstration of our non-sustainable way of living than the recent COVID-19 pandemic, which started in late 2019 in Wuhan, China, expanding all over the world in just few months. The occurrence and the extremely rapid expansion are connected to human dietary shifts toward consumption of animal products never used before due to increasing nutrition demand, extraordinary population densities, and unprecedented environmental pollution. The pandemic, its fast spread, and its consequences all over the planet in a very short period made evident that we are living in a closed system, interconnected in ways that are out of our control and we have to face the global problems with common strategies. Sustainable living has become essential, and a global sustainable consciousness must be formed and immediate decisions and actions need to be taken toward this direction.

Sustainability should be radically established in our lifestyle, habits, and actions. The massive use of plastics and their uncontrolled disposal in the last four decades are habits that need to be changed immediately. The uncontrolled production and use of plastic have brought the planet's pollution to levels never seen before. Only in 2018, 359 million metric tons of plastic were produced globally, while a total of about 9.2 billion metric tons were produced between 1950 and 2017. From all this plastic ever produced, it is estimated that about 9% has been recycled, 12% incinerated, and the remaining 79% has ended up in landfills or the environment. For example, it is estimated that 4.8 to 12.7 million metric tons of plastic enter the marine environment every year [1]. Photographs of animals and fish suffocating in their habitats due to dumped plastics, or of extended areas full of plastic garbage, especially from developing countries that have become the waste disposal fields for the developed ones, as well as studies on how toxic chemicals released from wrongly disposed plastics compromise our health are reaching us daily.

The most important source of badly disposed plastic waste is packaging. In 2017, around 15 million metric tons of plastic packaging waste was generated only in the European Union. In the general packaging sector, food packaging has the most important plastic demand. Plastic food packaging production in Europe is 8.2 million tons per year, included in the 20.5 million tons per year production for the general packaging sector and in the 51.2 million tons per year of the total European plastic demand [2]. Its short lifetime and frequent contamination from food makes it the most voluminous, wrongly disposed, plastic waste. For this reason, the introduction of biodegradable-compostable plastic packaging, either from petrochemical sources or preferably from natural renewable resources, has become mandatory and attracts a great deal of research and industrial interest. This, in combination with the various governmental stringent requirements and incentives related to plastic reduction throughout the planet, makes sustainable food packaging an emerging application area that expectantly will find its way to the market substituting the currently used recalcitrant plastic packaging solutions.

This book deals exactly with this rapidly emerging research and application field of Sustainable Food Packaging. It starts with Part I "Review on Biopolymers for Food Protection." This part of the book presents review chapters 1, 2, and 3 on the most relevant biopolymers that slowly find their way to the food packaging market, but also on biopolymers that are not yet industrialized (either due to high costs of extraction and transformation in packaging materials or due to lack of investment), but have a great potential due to their unique properties. In particular, Chapter 1 "Emerging Trends in Biopolymers for Food Packaging," by Sergio Torres-Giner et al., starts with a detailed and comprehensive introduction to the different types of biopolymers, and their classification according to their origin and biodegradability characteristics. The chapter continues with the presentation of the most important biopolymers that are currently available and describes their origin, chemistry, synthesis/extraction, and/or chemical modification methods. It also positions these biopolymers in the current plastic market and describes their prospects, advantages, and disadvantages in the sector of food packaging. In their concluding remarks, the authors give an expert point of view on where the bioplastic efforts for food packaging should be directed in order to have an important positive future environmental impact. Chapter 2 "Biopolymers Derived from Marine Sources for Food Packaging Applications," by Jone Uranga et al., presents the two most important biopolymers for food packaging, originating from marine biomass, fish gelatin and chitosan. Regarding gelatin, the chapter describes the extraction methods of collagen by fish waste biomass, and the subsequent production gelatin by partial hydrolysis of collagen. The authors continue with a presentation of the methods of development of gelatin coatings and films as food packaging and their impact on the food shelf life extension. Regarding chitosan, the authors first analyze the extraction methods of chitin by marine biomass, such as crustacean shells and squid pens, before its transformation to chitosan by deacetylation. Finally, the development, properties, and effect on the packaged food life extension are analyzed for the various chitosan coatings and films presented in the literature as food packaging solutions. Chapter 3 "Edible Biopolymers for Food Conservation," by Elisabetta Ruggeri et al., describes the innovative idea of natural polymeric protective coatings or films for food preservation and freshness extension that can be consumed together with the food, accompanied by the various regulations that would cover such use. The authors analyze the various biopolymers that can be transformed into edible packaging, classifying them as polysaccharides, proteins, lipids, and their mixtures. They present the various ways of development of the films for wrapping or of the coatings applied directly onto the food, their properties, and the possibility to act as matrices for functional additives, like antimicrobial and antioxidant agents. Finally, the authors provide information on the possible limitations and on the future perspectives of natural edible food packaging.

The book continues with Part II "Food Packaging-Based on Individual Biopolymers and Their Composites," where the most promising biopolymers and their composites for the food packaging sector are presented separately. Part II describes both the biopolymers that have already found their way to the market together with their future challenges, and the biopolymers that are not yet in the market but due to their unique properties have a good potentiality. Indeed, Chapter 4 "Polylactic acid (PLA) and Its Composites: An Eco-friendly Solution for Packaging," by Swati Sharma, describes research advancement related to food packaging based on PLA and its composites, indisputably the most available and promising biopolymer currently present in the market. The chapter starts with the synthetic routes of PLA and continues with the description of its physical properties, emphasizing the ones essential for food packaging materials. Subsequently, the author presents a review on the various fillers, i.e. synthetic and natural fibers or nanoparticles, that have been used to enhance the relevant properties of PLA, and closes the chapter with the current uses in the market and the future perspectives of this exceptional biopolymer. Chapter 5 "Green and Sustainable Packaging Materials using Thermoplastic Starch," by Anshu Anjali Singh and Maria Erminia Genovese, is dedicated to biocomposites based on thermoplastic starch, a biopolymer with great potentiality for food packaging due to its abundance, biodegradability, and low price, that has already found its way to the market. Although the chapter is dedicated to a specific biopolymer, the authors make an analytical introduction to the plastic threat that has made vital the need of sustainable polymers especially in the food packaging sector, and present the various categories of biopolymers available for this purpose emphasizing on starch. They continue with the presentation of research studies on thermoplastic starch composites developed for packaging applications underlining their most relevant properties, such as mechanical properties, gas and vapor permeability and biodegradability, their processing methods, their possible drawbacks, while they also present the commercially available packaging solutions. Finally, they conclude with the challenges to be addressed and the future developments needed for starch, its composites, and its derivatives in food packaging. Chapter 6 "Cutin-inspired Polymers and Plant Cuticle-like Composites as Sustainable Food Packaging Materials," by Susana Guzman-Puyol et al., presents the unique biopolymer cutin, the main component of cuticle, which constitutes the outer surfaces of plants and serves as a protective layer from the environment. The authors introduce the dramatic effect of plastics to the environment and the alternative bio-based and biodegradable biopolymers. They continue with the description of plant cuticle, its natural role, its structure and its composition, where the principle component cutin is introduced. A quantitative comparison of the physical properties of cutin and various biodegradable biopolymers reveals the strong and weak points of the former, and the ways to obtain cutin from its main resource, the tomato pomace, are described. The authors also present scalable techniques for the synthesis of cutin-inspired polyesters and methods to tune their properties. Finally, the methods to fabricate cutin-inspired coatings and cuticle-like composites as protective packaging for food are presented. Chapter 7 "Zein in Food Packaging," by Ilker S. Bayer, gives an exhaustive presentation of the protein biopolymer, zein, highly promising for food packaging mainly due to its film-forming capability and hydrophobic properties. Its origin, molecular structure, and general properties are first introduced. The focus of the chapter is on the fabrication methods of zein-based films and their characterization as food packaging materials. The presented methods are solvent casting, melt extrusion, solvent or melt blending with biopolymers, and lamination, either with other bio-polymers or petroleum-based polymeric films. The author closes the chapter with the future perspectives and his point of view on the directions that zein development should follow in order to be a valid candidate for the food packaging industry.

Part III of the book is dedicated to the most abundant natural polymer of our planet, the cellulose, and is entitled "Biocomposites of Cellulose and Biopolymers in Food Packaging." It presents sustainable composites of biopolymers with cellulose in various forms, with exceptional properties, competitive to the conventional nonbiodegradable plastics, for the food packaging market. In particular, Chapter 8 "Cellulose-reinforced Biocomposites Based on polyhydroxybutyrate (PHB) and poly(3-hydroxybutyrate-co-3-hydroxyvalerate) PHBV for Food Packaging Applications," by Estefania Lidon Sanchez-Safont et al., deals with the microbial polymers polyhydroxyalkanoates (PHAs) and discusses how cellulose fibers can improve their properties to make them competitive for the food packaging market. Specifically, the authors analyze separately, first the strengths, weaknesses, opportunities, and threats (SWOT) of two bacterial biopolymers, PHB and PHBV, and subsequently the properties of lignocellulose fibers that make them ideal candidates as reinforcing fillers in composites. Next, they review the research that has been done so far on the combination of PHB and PHBV with various lignocellulose fibers and they present the great potentiality of such composites as food packaging materials, indicating the weak points to be overcome. Chapter 9 "Poly-Paper: Cellulosic-filled Eco-composite Material with Innovative Properties for Packaging," by Romina Santi et al., focuses on a very interesting patented material, made of cellulose microfibers in the sustainable, water-soluble poly(vinyl alcohol) (PVA) matrix, with the name Poly-Paper. Poly-paper can be very versatile in terms of manufacturing processes and final properties, and as such, can find many ways to enter into the packaging market. Indeed, it can be processed by extrusion, injection molding, and thermoforming, while at the end of life can reenter in the paper recycling chain. Chapter 10 "Paper and Cardboard Reinforcement by Impregnation with Environmentally Friendly High-performance Polymers for Food Packaging Applications," by Uttam C. Paul and José A. Heredia-Guerrero, describes their research on a different approach to cellulose-based food packaging. Cellulose is not present as micro or nanofiller in a polymer, but in the form of an intact substrate, like paper or cardboard, impregnated with environmentally safe, biocompatible polymers and composites, in order to attain properties that can expand and reinforce its positioning in the food packaging market, without compromising its biodegradable nature. The treatment offers to the cellulose substrates water and oil resistance, resistance to moisture, and mechanical reinforcement, all properties that cellulose lacks intrinsically, so it can be a viable solution to increase its request in the sustainable food packaging market. The book continues with Chapter 11 "Nanocellulose-Based Multidimensional Structures for Food Packaging Technology," by Saumya Chaturvedi et al., which deals with nanocellulose-based food packaging solutions. The authors present an overview of the different kinds of nanocellulose, which include fibrils or crystallites with at least one dimension in the nanoscale range, and their properties depending on the origin, i.e. plants or bacteria, and the isolation methods. The chapter proceeds with the ways that nanocellulose, alone or combined with polymers, forms various compact structures that can be used for packaging. Finally, the authors focus on two types on nanocellulose, the cellulose nanofibers and the cellulose nanocrystals, emphasizing their differences and reporting the research that was done so far on their freestanding films either alone or as polymer fillers, comparing them with conversional plastics films used in the food packaging sector.

A book on Sustainable Food packaging could not be complete without Part IV dedicated to "Natural Principles in Active and Intelligent Food Packaging for Enhanced Protection and Indication of Food Spoilage or Pollutant Presence," since sustainability is closely related to the extension of the shelf life of food and the prevention of food waste. Wasting food is economically nonviable, not ethical, and drains the already very limited natural resources. Chapter 12 "Sustainable Antimicrobial Packaging Technologies," by Yildirim Selçuk and Bettina Röcker, presents advancements in the use of bioactive substances combined within biopolymers from renewable resources as protective food packaging with antimicrobial action against foodborne pathogens and spoilage microorganisms. The authors introduce the concept of active packaging, its different categories, and specific actions, with a dedicated section on the antimicrobial active packaging, its classes, research advances, and regulations. Then, they analyze separately the most studied natural antimicrobial agents, i.e. essential oils and phenolic compounds; organic acids, their salts and anhydrides; bacteriocins and enzymes; and the antimicrobial polymer chitosan, with references of their use as active additives in packaging of food systems. The authors conclude their chapter with the strategies needed for a successful and rapid introduction of active sustainable antibacterial packaging in the food packaging industry. Chapter 13 "Active Antioxidant Additives in Sustainable Food Packaging," by Thi Nga Tran, deals also with active packaging but in this case with antioxidant activity. The author starts with an introduction to the urgent need of a significant reduction of food losses and wastes, and how protection from oxidation, using packaging systems of biopolymers combined with natural antioxidant substances, could help. The chapter continues with a detailed analysis of the various antioxidant molecules extracted by plants, their combination with biopolymers into active food packaging, and the properties of the obtained packaging systems, including, of course, their antioxidant activity. A particular mention is made to the possibility of using raw dried plants powders, even from agricultural by-products, as antioxidant fillers into biopolymers for the development of active sustainable food packaging, avoiding the extraction costs. Part IV of the book ends with Chapter 14 "Natural and Biocompatible Optical Indicators for Food Spoilage Detection," by Maria E. Genovese et al., which presents another very interesting approach in food waste prevention. The authors describe packaging materials with incorporated natural or biocompatible molecules that change their molecular structure, and thus their optical properties, in the presence of food spoilage. Consequently, when a specific food spoilage by-product is present, the active packaging changes one or more optical properties (i.e. color, spectral absorption, fluorescence) enabling a real-time and direct naked eye spoilage detection. The authors introduce the factors determining food spoilage, and analyze thoroughly the conventional methods, as well as the most recent portable technologies for on-site and on-package detection of the spoilage, together with the functioning principles of these technologies. Then, the authors focus on the description of the various functional components used for the optical and colorimetric spoilage indication usually embedded in a polymeric, most of times natural renewable, support, as well as the specific spoilage by-product they can detect. A particular emphasis is given on the sensing potential of natural dyes and pigments extracted from plants, i.e. curcumin and anthocyanins, as well as their synthetic counterparts, due to their eco-friendly nature.

The book closes with Part V "Technological Developments in the Engineering of Biocomposite Materials for Food Packaging Applications," where Chapter 15 "Biopolymers in Multilayer Films for Long Lasting Protective Food Packaging: A Review," by Ilker S. Bayer, presents the possibilities that technology provides to take advantage of the various biopolymers and composites combining them in unique solutions for food packaging. Apart from melt extrusion, injection molding, blow molding, and thermoforming, all techniques used broadly in the plastic industry and mentioned in the various chapters of this book, Chapter 15 describes the ways of making multilayer films that can combine the unique properties of the various biopolymer layers into one material. The chapter reviews both multilayer laminates of biopolymers with conventional oil-derived polymers and all sustainable laminates, based on proteins, polysaccharides, or biopolyesters. The author concludes that multilayer laminates of carefully chosen biopolymers and biocomposites could be the ideal materials for food packaging since they combine sustainability with optimized desired properties due to their unique construction.

Genova, Italy
29 September 2020

Athanassia Athanassiou

References

1 Jambeck, J.R., Geyer, R., Wilcox, C. et al. (2015). Plastic waste inputs from land into the ocean. *Science*: 768–771.
2 Data for the year 2018 From ING Economics Department and https://www.statista.com/statistics/282732/global-production-of-plastics-since-1950.

Part I

Review on Biopolymers for Food Protection

1

Emerging Trends in Biopolymers for Food Packaging

Sergio Torres-Giner, Kelly J. Figueroa-Lopez, Beatriz Melendez-Rodriguez, Cristina Prieto, Maria Pardo-Figuerez, and Jose M. Lagaron

Novel Materials and Nanotechnology Group, Food Safety and Preservation Department, Institute of Agrochemistry and Food Technology (IATA), Spanish Council for Scientific Research (CSIC), Calle Catedrático Agustín Escardino Benlloch 7, Paterna 46980, Spain

1.1 Introduction to Polymers in Packaging

According to the Food and Agriculture Organization of the United Nations (FAO), approximately one-third of all food produced globally is lost or wasted [1]. Food waste is produced throughout the whole food value chain, from the household to manufacturing, distribution, retail, and food service activities. Taking into consideration the limited natural resources available, it is more effective to reduce food waste than to increase food production. For this reason, several efforts have been put for the development of more effective food packaging strategies [2, 3]. Packaging items have become essential to protect food from different environmental conditions. Depending on the type of food, the packaging article can be customized to prevent or inhibit microbial growth, avoid food decomposition by removing the entrance of light, oxygen, and moisture, or even to prevent spoilage from small insects. Additionally, novel packaging items can be monitored to give information about the quality of the packaged food, ultimately diminishing food waste during distribution and transport [4].

Common materials utilized for food packaging include glass, paper, metal, and plastic. The latter are nowadays more frequently used since they have a large availability at a relatively low cost and can display good characteristics for packaging items, such as mechanical strength, barrier properties, and transparency [4, 5]. The most commonly used petrochemical materials for packaging applications can be divided into various families:

- Polyolefins and substitutes of olefins, such as low-density polyethylene (LDPE) and linear low-density polyethylene (LLDPE), polypropylene (PP), polystyrene (PS), oriented polystyrene (OPS), polyvinyl alcohol (PVOH), polyvinyl chloride (PVC), and polyvinylidene chloride (PVDC). Polyolefins are frequently used in reusable bags, paper cups, and stand-up pouches, while substitutes of olefins such as PVC are popularly used in cling films and in some prepackaged meals.

Sustainable Food Packaging Technology, First Edition. Edited by Athanassia Athanassiou.

- Copolymers of ethylene, such as ethylene-vinyl acetate (EVA) and ethylene-vinyl alcohol (EVOH), are typically used to make lid films for trays and barrier interlayers.
- Polyesters, such as polyethylene terephthalate (PET) and other aliphatic and aromatic polyesters, are mainly used to make water bottles.
- Polyamides (PAs) are commonly employed in films or trays for food products that are very sensitive to oxygen.

Most of these materials are made by condensation or addition polymerization of monomers of hydrocarbon or hydrocarbon-like raw materials, which means that due to their fossil-based nature and high chemical stability, they are not biodegradable and will accumulate in landfills over the years, causing a negative impact on the environment. Although several recycling strategies are currently being carried out, packaging materials are often contaminated with leftover food, making recycling economically inconvenient and thus unviable. In 2010, primary plastic production was 270 million tons, yet plastic waste was 275 million tons since plastics produced in previous years entered the waste stream, where the sector of packaging was the highest producer of plastics, that is, 146 million in 2015. This has led to an increase in the number of campaigns requesting the removal of single-use plastics, with the European parliament aiming to ban single-use plastic cutlery, cotton buds, straws, and stirrers by 2021. Such environmental awareness and implementation of stringent environmental regulations are leading to research for alternatives to food packaging materials and, thus, efforts are being directed, at both academic and industrial levels, at the use of bioplastics in a variety of consumer products.

1.2 Classification of Biopolymers

The above-described environmental issues, together with the scarcity of oil sources, are the main drivers behind the interest for the development of new materials for food packaging applications. Although bioplastics only account for 1% of the approximately 350 million tons of plastics produced annually, being mostly applied as packaging materials [6, 7], including high-performance thermoplastic materials and foams, they represent an important part of the Bioeconomy and will undoubtedly shape the future of the plastic industry [8]. As a result, the use of biopolymers in packaging has increased considerably over the past few years due to their sustainable feedstock, biodegradability, and similar processing characteristics as existing thermoplastics [9].

Biopolymers comprise of a whole family of materials with different properties and applications. They include polymers with a "bio-based" origin and "biodegradable" polymers or polymers featuring both properties. Bio-based polymers refer to any kind of polymer that is produced from renewable resources, which include both naturally occurring polymers and synthetic polymers produced by means of monomers obtained from biological sources [10]. Naturally occurring polymers are biomacromolecules, that is, molecules of large molecular weights (M_W) produced in nature by living organisms and

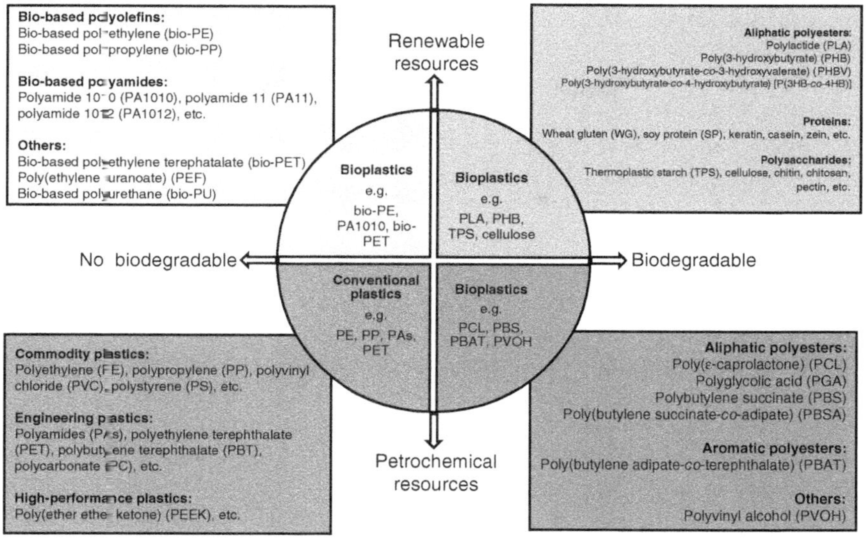

Figure 1.1 Classification of biopolymers widely used in packaging.

plants. Biodegradable polymers are defined as those polymer materials whose physical and chemical properties undergo deterioration and completely degrade, when exposed to the enzymatic action of microorganisms, to carbon dioxide (aerobic process), methane (anaerobic process), water (aerobic and anaerobic processes), inorganic compounds, and biomass [11]. Bio-based polymers can be biodegradable but not all biodegradable polymers are bio-based. Additionally, some synthetic biodegradable polymers could be in a near future partially or fully developed from bio-based monomers. Bio-based polymers offer the advantage of conservation of fossil resources by using biomass that regenerates (annually) and the unique potential of carbon neutrality whereas biodegradability is an add-on property of certain types of polymers that offers additional means of recovery at the end of a product's life [12, 13].

Figure 1.1 summarizes the classification of biopolymers grouped according to their origin and biodegradability characteristics. On the top right of the figure, bio-based and biodegradable polymers are gathered. Nature produces over 170 billion metric tons per year of biomass, yet only 3–4% of this material is being used by humans for food and nonfood purposes [14]. Biomass derived carbohydrates are the most abundant renewable resources available, representing approximately 75% of this biomass, which are currently regarded as the basis for the green chemistry of the future. Most of these biopolymers are mainly made from carbohydrate-rich plants such as corn or sugarcane, that is, the so-called food crops, which are also currently referred to as the "first generation feedstock" of bioplastics. This currently represents the most efficient feedstock for the production of bioplastics, as it requires the least amount of land to grow and produce the highest yields. Biomass derived polyesters such as polylactide (PLA) and thermoplastic starch (TPS) are among the most promising biodegradable polymers and they contribute to up to 65% of the family of this type of biopolymers.

The same durability properties that have made traditional petroleum derived plastics ideal for many applications, such as those found in packaging, are leading to terrible waste-disposal problems as these materials are resistant to microbial degradation and plastics accumulate in the environment. For this reason, some biodegradable polymers, yet based on petrochemical polymers, have been developed in the last few years. The most frequently studied polymers are included in the bottom-right group, which are aliphatic or aliphatic–aromatic polyesters since neat aromatic polyesters based on terephthalic acid are generally insensitive to hydrolytic degradation and to enzymatic or microbial attack due to their high stability. Indeed, the biodegradation rate increases rapidly when the concentration of terephthalic acid becomes lower than 55 mol%. These petrochemical biodegradable polymers can find several uses in both flexible and rigid packaging applications.

Bio-based but not biodegradable polymers, which are shown at the top-left of the figure, currently offer important contributions by reducing the dependence on fossil fuels and through the related positive environmental impact, that is, reduced carbon dioxide emissions. New approaches go toward the complete or partial substitution of conventional plastics by renewable resources such as biomass [15]. Conventional polymers from feedstock routes are being explored for well-known applications, including the packaging industry [16]. These are generally based on monomers derived from agricultural and food-based resources such as corn, potatoes, and other carbohydrate feedstock. The new branch of these "green polymers" reflects the "biorefinery" concept [17]. The monomers to produce these bio-based polymers can be obtained from natural resources, for example catalytic dehydration of bioethanol obtained by microbial fermentation. Although these biopolymers are not biodegradable, they have the same processing and performance as conventional polymers made from natural gas or oil feedstocks. Such developments have recently led to the new paradigm for sustainable food packaging: "Bio-based but not biodegradable" [18]. This is further evidenced by the recent development of fully bio-based polyethylene terephthalate (bio-PET), where the ethylene glycol and the terephthalic acid are both derived from plant-based sugars and agricultural residues.

The discussion about the use of biomass for industrial purposes is still often linked to the question about whether the conversion of potential food and feed into materials is ethically justifiable. Although the surface required to grow sufficient feedstock for current bioplastics production is only about 0.01% of the global agricultural area of a total of 5 billion hectares (bioplastics), the bioplastics industry is also researching the use of nonfood crops and agricultural residues, the so-called "second generation feedstock," with a view to its further use. Innovative technologies are focusing on nonedible by-products as the source for bioplastics, which includes large amounts of cellulosic by-products and wastes such as straw, corn stover, or bagasse. This leaves significant potential for using biotechnological processes to create platform chemicals for industrial purposes, among them the production of bioplastics. Therefore, the trend for the development of next generation of bioplastics is currently led by the emergence of conventional polymers made from renewable and nonfood sources.

1.3 Food Packaging Materials Based on Biopolymers

The development of different kinds of packaging materials from biopolymers has seen an increase in the last few years [19]. The use of biopolymers in food packaging can provide physical protection during storage and transportation, and create proper physicochemical conditions for maintaining quality and safety and for extending the shelf life of food [20]. All subsections in Section 1.3 summarize the most important trends in biopolymers for food packaging applications.

1.3.1 Polylactide

The commercialization of PLA began in 1990 though it was known since 1845. PLA is mostly obtained from bacterial fermentation of carbohydrates that can come from renewable resources such as corn or sugar. Fermentation turns the sugar into lactic acid, which is the building block for PLA, but most industrial applications make use of its dimer lactide. NatureWorks LLC (previously Cargill Dow LLC) and, more recently, Corbion (former Purac Bioplastics) are the major suppliers of PLA with a production capacity of over 100 ktons per year, in which this biopolymer is produced in a continuous efficient synthetic process via ring-opening polymerization (ROP) of the lactide dimer [21]. There are two optical forms: L-lactide, the natural isomer, and D-lactide, the synthetic one [22]. The production of the different isomers depends on the microbial strain used during the fermentation process. The polymer crystallinity and the properties of PLA can significantly vary depending on the ratio and stereochemical nature of the monomer [23].

Nowadays, PLA is one of the most researched and commercialized biopolymers and it is seen as a potential substitute for conventional polymers as packaging materials since it is bio-based and compostable [24]. It has similar properties to traditional polymers such as PET, PS, and polycarbonate (PC) [25]. The most relevant characteristics of this biopolymer are high rigidity, good transparency, heat sealability, printability, and melt processability. Also, it can be processed on large-scale production lines such as injection molding, blow molding, thermoforming, and extrusion [26]. It is classified as generally recognized as safe (GRAS) by the United States Food and Drug Administration (FDA) and is safe for all food packaging applications [27]. However, it also has limiting properties for its use in packaging applications. For example, its low glass transition temperature (T_g) limits its utilization above 55 °C and it also shows low toughness and ductility [28]. Nevertheless, these narrow circumstances can be improved by varying the ratio of L/D isomers, modifying its stereochemistry, or mixing with other polymers and fillers to improve the mechanical and thermomechanical properties [29].

In the field of food packaging, PLA is ideal for fresh products and those that do not require protection against oxygen, but also it can be used in food trays, bottles, candy wraps, and cups [30]. Accordingly, its high permeation to water makes it suitable for some packaging applications such as extension of the life period of fresh fruit and bread. Some coatings are used as a kind of barrier layer

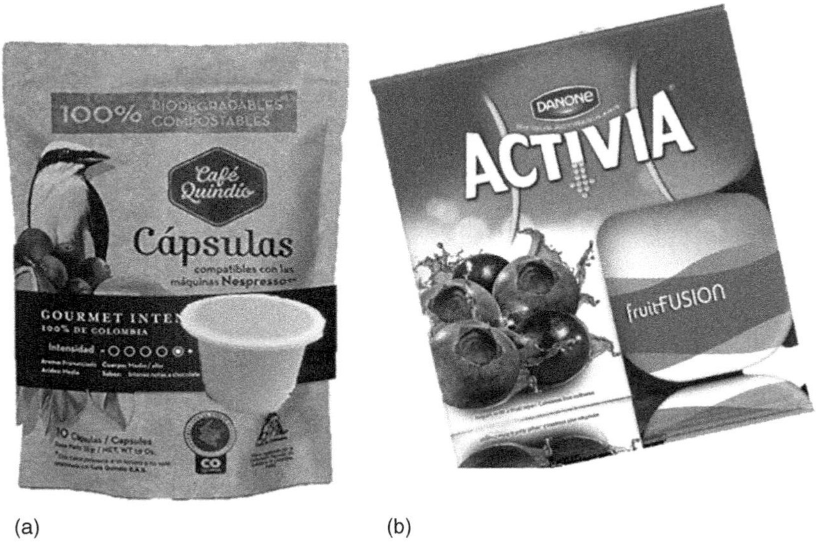

Figure 1.2 Commercial food packaging articles made of polylactide (PLA): (a) Coffee capsules. (b) Yogurt cups. Source: Courtesy of Danone (Paris, France).

to reduce permeability of this biopolymer [31, 32]. Figure 1.2 shows some commercial packaging articles made of PLA to contain food products.

1.3.2 Polyhydroxyalkanoates

Polyhydroxyalkanoates (PHAs) currently represent one of the most important alternatives to fossil derived polymers in the frame of the Circular Economy [33], showing the highest potential to replace polyolefins in packaging applications due to their biocompatibility and physical properties [34]. PHAs are a family of biopolyesters synthesized by a wide range of microorganisms as carbon storage material. Although they have suitable characteristics such as biodegradability, thermoplasticity, and similar mechanical strength and water resistance to other polymers such as PP and PS [35], their production is expensive due to the high costs of the fermentation and downstream processes. The use of industrial by-products and waste or mixed microbial cultures (MMCs) represents a viable option to reduce the production costs of PHAs [36].

Among PHAs, the most widely studied and first bacterial member of this family identified was poly(3-hydroxybutyrate) (PHB). This isotactic homopolyester is biodegradable not only in composting conditions but also in other environments such as marine water and it presents similar thermal and mechanical properties with some petrochemical polymers. However, its use is limited due to its poor impact-strength resistance and a narrow processing temperature window [37]. To improve these shortcomings, its copolymers such as poly(3-hydroxybutyrate-*co*-3-hydroxyvalerate) (PHBV) and poly(3-hydroxybutyrate-*co*-4-hydroxybutyrate) [P(3HB-*co*-4HB)] have been explored. Particularly, PHBV is a good candidate since it has a much lower crystallinity and melting temperature, decreased stiffness and brittleness, and

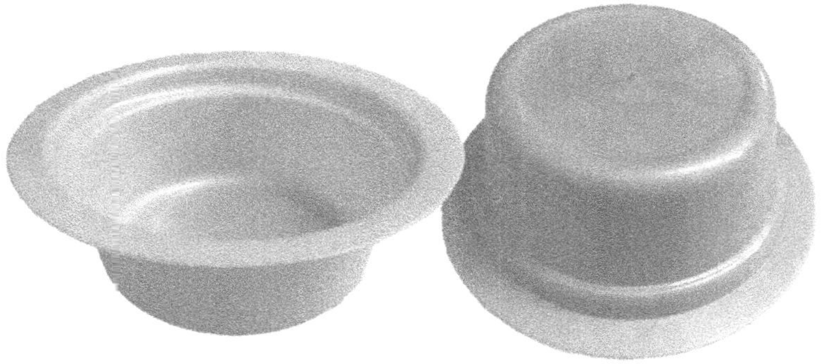

Figure 1.3 Biodegradable food tray made of poly(3-hydroxybutyrate) (PHB) obtained by injection molding.

higher ductility. The T_g of these PHA copolymers varies from −40 to 5 °C, and the melting temperature (T_m) also range from 50 to 180 °C, depending on their chemical composition [38].

Biomer (Krailling, Germany) and Tianan Biologic Materials (Ningbo, China) are the main producers of microbial PHB and PHBV, respectively. PHAs can be processed by common methods such as extrusion, injection molding, thermoforming, film blowing, and so on [39]. These materials are suitable for very different areas of food and cosmetic packaging, for instance blow-molded bottles, milk cartons, cosmetic containers, feminine hygiene products, adhesives, paper coatings, waxes, paints, and so on [40]. Figure 1.3 depicts, as an example, a tray made of PHA. Moreover, the use of nanofillers or active substances, such as antimicrobial and/or antioxidant substances, incorporated into a PHA-based packaging material can change the packed food condition extending the shelf life and improving the protection and/or sensory properties, adding an extra value to the final product [41, 42].

1.3.3 Poly(butylene adipate-*co*-terephthalate)

Poly(butylene adipate-*co*-terephtalate) (PBAT) is a biodegradable aliphatic–aromatic copolymer, synthetized by the reactions of the monomer 1,4-butanediol, adipic acid, and terephthalic acid. It is biocompatible and biodegradable, being degraded in a few weeks by lipases from *Pseudomonas cepacia* and *Candida cylindracea* [43]. Also, it is highly amorphous, it has high ductility and thermomechanical properties comparable with LDPE, and is compostable [44, 45]. Despite these suitable characteristics, PBAT shows a high permeability to water, that is, a poor water barrier property, and fails to achieve the mechanical strength required for some applications [46]. Different approaches have been considered to overcome these disadvantages, such as surface modifications, use of polymer blends, and reinforcement by fillers [47]. For instance, blends of PBAT with PLA can result in a biodegradable article with balanced mechanical properties, though the use of a chain extender or a cross-linking agent to enhance the interfacial adhesion is usually necessary [48]. Also, the use of nanocomposites based on PBAT and layered silicates have yielded materials with improved mechanical properties [49].

PBAT films can be prepared using cast film extrusion, blown film extrusion, thermo-compression, and solvent casting methods [50] and they can be used for agriculture, food packaging, waste and compost bags, among others [51]. PBAT is sold commercially by the trade names of Ecoflex® by BASF (Ludwigshafen, Germany), Easter Bio® from Eastman Chemical (Kingsport, USA), and Origo-Bi® from Novamont (Novara, Italy). It is allowed for food packaging applications by the FDA [52, 53] and the incorporation of antimicrobial substances for active food packaging has recently shown very promising results to enhance food safety [54].

1.3.4 Polybutylene Succinate

Polybutylene succinate (PBS) is a biodegradable and compostable aliphatic polyester produced by polycondensation of succinic acid and 1,4-butanediol. PBS was exclusively derived from petroleum-based monomers, but since more recently the monomers can also be obtained by the bacterial fermentation route to produce fully bio-based polybutylene succinate (bio-PBS) [29, 55]. So far succinic acid has been mainly produced by electrochemical synthesis due to the high yield, low cost, high purity of the final product, and very low or no waste formation [56]. However, the production of succinic acid by bacterial fermentation uses renewable resources and consumes less energy compared to chemical process. For this reason, companies such as Corbion (Geleen, the Netherlands) and BASF are working on the scaling up of an economically feasible bio-based succinate production process, despite the fact that these processes have traditionally suffered from poor productivity and high downstream processing costs. Other examples are the development of a biomass-derived succinic acid production by Mitsubishi Chemical (Tokyo, Japan) in collaboration with Ajinomoto (Tokyo, Japan) to commercialize bio-PBS or the development of a commercially feasible fermentation process for the production of succinic acid, 1,4-butanediol, and the subsequent production of PBS by DSM (Heerlen, the Netherlands) and Roquette (Lestrem, France). Myriant (Quincy, USA) and Bioamber (Plymouth, USA) have also developed a fermentation technology to produce the monomers [57, 58]. Thus, in 2015, the annual production capacity of bio-based succinic acid reached 200 000 tons [59]. In the case of 1,4-butanediol, conventional production processes use fossil fuel feedstocks, such as acetylene and formaldehyde. Nevertheless, the bio-based process to obtain the diol involves the use of glucose from renewable resources to produce succinic acid followed by a chemical reduction to produce butanediol [29]. PBS with excellent mechanical properties and processing capabilities can be then produced from the renewable monomers by transesterification, direct polymerization, and condensation polymerization reactions followed by chain extension and lipase-catalyzed synthesis.

PBS is a semicrystalline aliphatic polyester with a good melt processability and balanced mechanical properties, closely comparable to those of PP. It is tougher than PLA and it shows similar thermal behavior than LDPE and a melting point lower than that of PLA [29, 60]. Its thermal and mechanical properties highly depend on the crystal structure and the degree of crystallinity [61]. The T_g and T_m are approximately −32 and 115 °C, respectively. In terms of mechanical properties, PBS has a good tensile and impact strength with moderate rigidity and hardness [29]. PBS has a wide processing window, which makes the resin suitable for extrusion, injection molding, thermoforming, fiber spinning, and film blowing. PBS has been employed as film, in foaming, and in food packaging containers [60]. However, the relatively poor mechanical flexibility of PBS limits the applications of 100% PBS-based products. This issue can be overcome by blending PBS with other biopolymers and fillers to improve the mechanical properties to suit the required application and biodegradation rate [62–65]. The development of PBS copolymers can also lead to biopolymers with a decreased degree of crystallinity, depressed heat distortion temperature, and improved elongation [56, 66]. Copolymerization is achieved by adding a third monomer such as sebacic acid, adipic acid, terephthalic acid, succinic acid with substituted side groups, 1,3-propanediol, and other substituted glycols, which can be also produced from renewable resources. Poly(butylene succinate-*co*-terephthalate) (PBST) and poly(butylene succinate-*co*-adipate) (PBSA) are the most employed copolymers, PBSA being the most common for flexible packaging applications. Additionally, the flexibility of the PBS backbone and the presence of readily hydrolyzable ester bonds, which are prone to catalytic degradation by microorganisms or enzymes, promote rapid degradation process of the PBS copolyesters [67]. Some studies have evaluated the biodegradability of biodegradable polymers in the mature compost soil and found that the biodegradability of PBS was slower than that of PBSA in compost soil [68]. However, the results of enzymatic hydrolysis as well as environmental degradation of PBS are highly dependent on the environment, the composition of the PBS grade, and its compound [55].

1.3.5 Bio-based Polyethylene

Bio-based polyethylene (bio-PE), also called "microbial" or "green" polyethylene, can be produced by the catalytic dehydration of bioethanol obtained by microbial fermentation, followed by normal polymerization to produce polyethylene (PE) [69, 70]. Bio-PE is not biodegradable and has the same properties as PE made from natural gas or oil feedstocks [29]. Conventional PE is manufactured by polymerization of ethylene under pressure, temperature, in the presence of a catalyst. Traditionally, ethylene is produced through steam cracking of naphtha or heavy oils or ethanol dehydration. However, the concept of producing PE from bioethanol is not particularly new. In the 1980s, Braskem (São Paulo, Brazil) had already made bio-PE from bioethanol, however, the limitations of the biotechnology processes made the technology unattractive at that time [71]. Currently, bio-PE produced on an industrial scale from bioethanol is derived from renewable feedstocks, including sugarcane and beet, starch crops such as maize, wood, wheat, corn, and other plant wastes through microbial strain and biological fermentation process. In a typical process, extracted sugarcane juice with high sucrose content is anaerobically fermented to produce ethanol. At the end of the fermentation process, ethanol is distilled in order to remove water

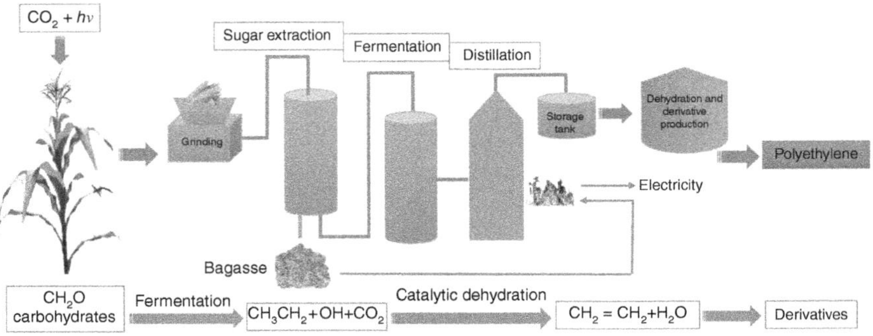

Figure 1.4 Schematic flow diagram of the production of bio-based polyethylene (bio-PE) from sugarcane via fermentation into ethanol and subsequent dehydration into ethylene. Source: From Koopmans [73]. © 2013, John Wiley & Sons.

and to yield an azeotropic mixture of hydrous ethanol. Ethanol is then dehydrated at high temperatures over a solid catalyst to produce ethylene and, subsequently, bio-PE [69, 72]. Figure 1.4 shows a schematic flow diagram of the bio-PE production.

Braskem is the largest producer of bio-PE, mainly bio-based high-density polyethylene (bio-HDPE), with 52% market share, with an annual production capacity of 200 000 tons per year made from ethanol obtained from sugarcane [71] and this is the first certified bio-PE in the world. Similarly, Braskem is developing other bio-based polymers such as bio-based polyvinyl chloride (bio-PVP), bio-based polypropylene (bio-PP), and their copolymers with similar industrial technologies. Braskem's current bio-based PE grades are mainly targeted toward food packaging, cosmetics, personal care, automotive parts, and toys. Dow Chemical (Midland, USA) in cooperation with Crystalsev (São Paulo, Brazil) is the second largest producer of bio-PE, having 12% market share. Solvay (Brussels, Belgium), another producer of bio-PE, has 10% share in the current market. However, Solvay is a leader in the production of bio-PVC with similar industrial technologies. China Petrochemical Corporation (Pekin, China) also plans to set up production facilities in China to produce bio-PE from bioethanol [74]. LyondellBasell (Rotterdam, The Netherlands) and Neste (Espoo, Finland) have recently announced the first parallel production of bio-PP and bio-based low-density polyethylene (bio-LDPE) at a commercial scale, being marketed under the trade names Circulen and Circulen Plus [75].

Bio-PE can replace all the packaging applications of current fossil derived PE because of its low price, good lifetime performance, and especially recyclability [75]. The price of bio-PE is currently about 50% higher as compared with petrochemical PE, but it will take advantage from the scale-economy. Current upcoming applications by multinationals include yogurt cups produced by Danone (Paris, France), fruit juice bottles by Odwalla (Atlanta, USA), and plastic caps and closures for aseptic paperboard cartons by Tetra Pak (Lund, Sweden) [76].

1.3.6 Bio-based Polyethylene Terephthalate

PET is a copolymer of monoethylene glycol (MEG) and terephthalic acid, and one of the most studied polymers to be transformed in commercial bio-based plastics, which can be derived from plant-based sugars and agricultural residues. This interest gave life to a technological collaboration between several companies, such as Coca-Cola (Atlanta, USA), Ford (Detroit, USA), Heinz (Sharpsburg, USA), Nike (Eugene, USA), Danone and Procter & Gamble (Cincinnati, USA), Avantium (Geleen, the Netherlands), and Micromidas (San Jose, CA), to develop commercial processes for the production of bio-PET. The source of petrochemical terephthalic acid is primarily from the oxidation of *p*-xylene, obtained from the catalytic reforming of naphtha. More than 98% of the *p*-xylene produced globally is converted to terephthalic acid and the global demand for purified terephthalic acid is expected to exceed 60 million tons by 2020, being most of it used for PET production. As a result of this cost-competitive demand, bio-based terephthalic acid made at equal purity and cost as the petroleum derived terephthalic acid would have a clear market advantage as well as a lower price volatility due to the non-dependence on petrochemical *p*-xylene [77]. However, bio-PET is currently produced from plant-based monoethylene glycol (bio-MEG) whereas terephthalic acid is still derived from petroleum. Toyota Tsusho (Nagoya, Japan) and China Manmade Fibers Corporation (Taipei, China) jointly founded a company in November 2010 that manufactures bio-MEG made from plant derived bioethanol [78]. Other companies such as SCG Chemicals (Bangkok, Thailand) also produce bio-MEG from residues of agricultural activities including molasses, hay, and bagasse [79]. The first partially bio-PET product was commercialized by The Coca-Cola Company (Atlanta, USA) under the trade name of PlantBottle™, where a 30 wt% of bio-MEG was used for the production of bio-PET (Company). At the moment, some researchers are working on the synthesis of bio-based terephthalic acid to obtain fully bio-PET. This process is based on an integrated method to convert forest residues to isobutanol [80], which can be processed into *p*-xylene [81], the precursor of terephthalic acid. Although the 100% bio-based bottle was released in Milan in June 2015, the so-called PlantBottle™, shown in Figure 1.5, has not reached yet the price parity to equal the price of producing current Coca-Cola PET bottles [82]. PepsiCo (New York, USA) also announced the use of a PET bottle made entirely with renewable resources coming from waste carbohydrate biomass obtained from the food industry such as orange peels, oat hulls, corn husks, and potato scraps [83].

Similar to bio-PE, bio-PET is not biodegradable but it has the same properties as conventional PET made from natural gas or oil feedstocks. Current partially bio-PET is used to make a number of products including drinking water and soda bottles, making them environmentally friendly and a new packaging alternative. Products made from bio-PET have the same qualities as regular PET in its distinctive functions, weight, appearance, and it can also be recycled and reused [79]. This material can be recycled, incinerated, or landfilled, but it can also be intended for disposal by composting, where it undergoes soil degradation to CO_2 and water [9]. Enzymatic hydrolysis of aromatic/aliphatic polyesters was first demonstrated in the 1990s for PET with several esterases, lipases, and especially

Figure 1.5 Image of the PlantBottle™ made up to 30% from biomass and 100% recyclable. Source: Courtesy of the Coca-Cola Company (Atlanta, USA).

cutinases [84, 85]. Nevertheless, PET hydrolysis by enzymes is a relatively slow process, since the biocatalysts are specialized to attack natural polyesters such as cutin and were not designed by nature for degrading manmade synthetic polyesters in the first instance [85].

1.3.7 Poly(ethylene furanoate)

A potential green substitute for terephthalic acid is 2,5-furandicarboxylic acid (FDCA), which is a bio-based building block that can be polymerized with bio-MEG to form a new 100% bio-based polyester called poly(ethylene furanoate) (PEF) [85]. PEF can be synthesized by polycondensation, ROP, and solid-state polymerization. Polycondensation is the most commercially relevant method but it results in long exposure times to high processing temperatures, around 200 °C, which increases the production cost and, even more importantly, leads to thermal degradation and discoloration of the biopolymer. Solid-state polymerization (SSP) is a milder process, though a bottle-grade PEF has not been achieved yet [86]. SSP involves heating of the starting partially crystalline polyester at a temperature between its T_g and T_m, which is used mainly for PET manufacturing to get over its relatively low M_W [87]. The company Avantium is developing PEF using bio-MEG and FDCA coming from the dehydration of carbohydrates [77]. In 2016, it was announced a new technology involving a highly efficient separation technology and catalyst that would result in economically feasible production of FDCA starting from 2016. The planned industrial production capacity is estimated to exceed 300 000 tons per year while the company has established collaborations with major endusers from the food and beverage industry [84].

PEF shows excellent thermal properties and lower T_m values than PET because of which it can be processed at lower temperatures. It has the ability to withstand high temperature due to its higher T_g as well as thermal stability up to 320 °C. PEF outperforms the barrier properties of PET. Specifically, PEF's O_2 barrier is more than ten times higher than that of PET, the CO_2 barrier is four times higher, and the water vapor barrier is two times higher [88]. Other attractive properties of PEF include excellent mechanical strength, reduced carbon footprint, and ability to formulate in films, fibers, and almost replace PET in water bottles [87]. PEF can be recycled in very similar ways to PET [88]. Recently, the enzymatic synthesis of FDCA-based polyesters was studied by the groups of Loos and Boeriu with *Candida antarctica* lipase B [89, 90]. However, apart from numerous reports around the emerging topic of PEF production and application, there is no information on enzymatic hydrolysis of PEF. Nevertheless, especially for bio-based PEF, biotechnological processing for functionalization and chemical recycling of monomers could have a large potential to replace harsh chemicals [84].

1.3.8 Poly(ε-caprolactone)

Poly(ε-caprolactone) (PCL) is a biodegradable aliphatic polyester derived from the chemical synthesis of crude petroleum [91]. It is obtained from ROP of ε-caprolactone in presence of metal alkoxides and also through the polycondensation of ϵ-hydroxyhexanoic acid. PCL is commercially available under the trade names CAPA (Solvay, Belgium), Tone (Union Carbide, USA), Celgreen (Daicel, Japan), and many others [92].

Some features of PCL include biodegradability, good solubility, flexibility, low T_m (approximately 60 °C) and a T_g of around −60 °C, and easy processing. PCL is used in flexible packaging materials in the form of films or coatings for extending the shelf life of food products. However, due to its high price and long biodegradability cycles, PCL is commonly blended with other biopolymers, such as chitosan and starch [93]. Among them, starch has been proposed as a reinforcing agent to improve the mechanical strength of PCL [94]. There are also several studies on the hydrolysis and biodegradability of PCL. The degradation process of PCL takes place through hydrolysis, thereby leading to molecular fragmentation or chain scission [94]. Moreover, enzymes and fungi easily biodegrade PCL. However, to improve the degradation rate, several copolymers with lactide or glycoside have been developed [94].

1.3.9 Thermoplastic Starch

Starch is a natural polysaccharide that can be obtained from a great variety of crops such as cassava and corn. It is considered as one of the most promising biopolymers for food packaging due to its availability, biodegradability, and low price [95]. Starch is composed of a mixture of two polymers, namely amylose and amylopectin. Amylose is a linear polysaccharide composed entirely of D-glucose units joined by α-1,4-glycosidic linkages. Amylopectin is a branched-chain polysaccharide composed of glucose units linked primarily by α-1,4-glycosidic bonds but with occasional α-1,6-glycosidic bonds, which are responsible for the branching. Relative percentages of amylose and amylopectin in starch are in the range 10–30% amylose and 70–90% amylopectin. The starch having a crystallinity between 20% and 40% is termed as semicrystalline in which the amorphous region of starch contains amylose and the branching points of amylopectin [96]. Native dry starch has a limited range of applications because of its high brittleness, high viscosity, retrogradation, insolubility in cold water, and poor melt processability because its T_g is probably above the decomposition point so it does not soften and flow [97]. BIOTEC® (Emmerich, Germany) has three product lines of TPS that include Bioplast® granules for injection molding, Bioflex® for film applications, and Biopur® as foamed starch [98]. In any case, films developed from native starch typically show moderate oxygen barrier properties but poor moisture barrier and mechanical properties, which limit their applications in food packaging [99].

Fortunately, starch can be modified by processing with plasticizers, grafting with vinyl monomers, and blending with other polymers. Plasticizers increase both the flexibility and processability of starch, which exhibits thermoplastic behavior and takes the name of TPS when plasticized by relatively low levels, in the 15–30 wt% range, of molecules that are capable of hydrogen bonding with the starch hydroxyl groups, mainly water, glycerol, and sorbitol. TPS can readily flow at elevated temperature and pressure and it can be extruded to give both foams or shaped into solid articles by injection molding. TPS products with different viscosity, water solubility, and water absorption have been prepared by altering the moisture/plasticizer content, amylose/amylopectin ratio of the raw material, and the temperature and pressure in the extruder [98]. A great deal of research has been performed on the plasticization of TPS using glycerol [100], sorbitol [101], urea or formamide [102], dimethyl sulfoxide [103], and low-M_W sugars [104]. Figure 1.6 shows different packaging articles obtained from starch, ranging from food trays to cups.

Unfortunately, the properties of neat TPS materials are not still good enough for most packaging applications. For example, the properties of films made of water- and glycerol-plasticized TPS are poor at high humidity, present poor dimensional stability, and become brittle as water is lost. Fortunately, TPS can be blended with other polymers and fillers to improve the mechanical properties and also attained water resistance [105]. For instance, TPS/PLA blends show chemical resistance, improved flexibility and toughness, and low cost [106]. The

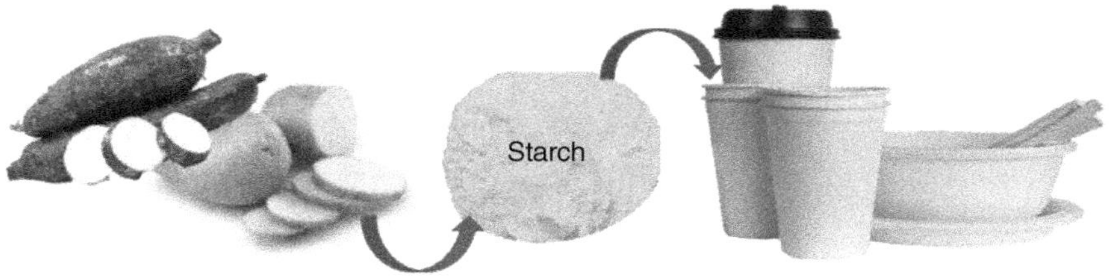

Figure 1.6 Biodegradable packaging articles based on starch.

use of TPS in biopolymer blends is particularly relevant to obtain materials with high elongation at break properties in food packaging [107].

1.3.10 Cellulose and Derivatives

Cellulose is the most abundant natural polymer on earth. The major source of cellulose is certainly wood, which contains 40–50 wt%, being the fundamental component of the cell walls of plants and natural fibers. Cellulose is a linear naturally occurring polymer composed of 1,4-linked-β-D-anhydroglucopyranose units that are covalently linked via acetal functions between the equatorial –OH group of C_4 and the C_1 carbon atom [108]. Neat cellulose is, however, unsuitable for film production because it is highly crystalline and also insoluble in water due to the strong intra- and intermolecular hydrogen bonding between the individual chains and its highly crystalline structure [109]. Therefore, cellulose is usually dissolved in a mixture of sodium hydroxide and carbon disulfide and recast into sulfuric acid. This chemical treatment results in the production of the so-called cellophane film, which has good mechanical properties. However, it is often coated with nitrocellulose wax or PVDC to improve its moisture sensitiveness. Coated cellophane is then used for baked goods, fresh products, processed meat, cheese, and candy though it is not heat sealable due to its non-thermoplastic nature [110].

Alternatively, cellulose can be chemically modified to produce water-soluble cellulose ester or ether derivatives by either esterification or etherification, respectively, of individual hydroxyl groups on the polysaccharide backbone. Commercial derivatives of cellulose include cellulose acetate, ethyl cellulose, hydroxylethylcellulose, and hydroxyl-propyl cellulose, among others [111]. Figure 1.7 summarizes some of the hydrophilic and hydrophobic cellulose derivatives categorized according to their pH-responsive behavior and chemistry. Steps involved in making these thermoplastic materials include, first, making cellulose derivatives biopolymers in powder form and, thereafter, extruding them in the presence of different additives and plasticizers such as citrates. Although the gas and moisture barrier properties of cellulose acetate are not optimal for food packaging, this film is excellent for products demanding high moisture as it allows respiration and reduces fogging [110]. Mazzucchelli (Castiglione Olona, Italy) and Planet Polymer (California, USA) manufacture biodegradable plastics under the trade names of BIOCETA® and EnviroPlastic® Z, respectively, based on cellulose acetate. BIOCETA® has been applied for the manufacture of biodegradable packaging films, retractable films, and tubes [98].

1.3.11 Proteins

Protein-based films have lately become a hot research topic due to their film-forming capacity and cohesiveness, low cost, and biodegradability features. Proteins present good barrier against oxygen and aroma, among others gases. However, they also show high water vapor permeability due to their hydrophilic

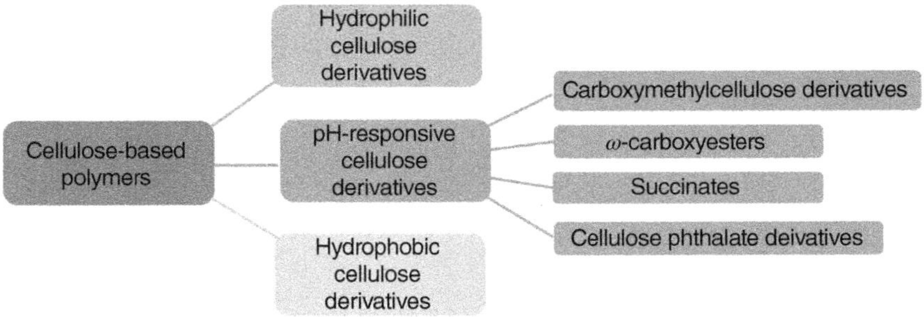

Figure 1.7 Cellulose derivatives categorized based on their pH-responsive behavior and chemistry.

nature [112, 113]. Protein films have been developed from gelatin, corn zein, wheat gluten (WG), soy protein (SP), casein, and whey protein [114–116].

1.3.11.1 Gelatin

Gelatin is a water-soluble protein that is prepared by thermal denaturation of collagen in the presence of dilute acid (gelatin type A) or alkali (gelatin type B). Collagen is found in animal skins and bones such as connective tissues, skin, and bones (see Figure 1.8) [117]. Its structure is triple helix, being stabilized by the formation of hydrogen bridges between the chains through amino and carboxyl groups. When collagen is denatured, the triple helix breaks and the polypeptide chains adopt a random configuration, forming gelatin composed mainly of glycine, proline, and 4-hydroxyproline residues [118–120]. Then, gelatin results in a heterogeneous mixture of single or multi-stranded polypeptides, each with extended left-handed proline helix conformations and containing between 300 and 4000 amino acids. Gelatin is primarily used as a gelling agent forming transparent elastic thermoreversible gels on cooling below 35 °C. It can be used as a valuable biopolymer in tissue engineering applications. Moreover, the gelification properties of gelatin, its capacity of forming and stabilizing emulsions, and its adhesive properties and dissolution behavior make this biopolymer a potential polymer for the manufacture of bio-based films [121]. However, its poor mechanical properties, especially in the wet conditions, limit its application as a packaging material, [122, 123]. Many techniques, including vapor cross-linking, orientation, and use of fillers such as hydroxyapatite nanoparticles (nHAs) and tricalcium phosphate (TCP), have been developed to reinforce gelatin-based films [124].

1.3.11.2 Wheat Gluten

WG is a protein complex having two major components, known as glutenin and gliadin. It is composed of proteins containing water-insoluble and ethanol-soluble prolamins and water- and ethanol-insoluble glutelins (70–80 wt%) in combination with small amounts of wheat oils, starch, and insoluble hemicellulose [125]. The gliadins are mainly monomeric single chain polypeptides, whereas the glutenins are polymeric and disulfide linked polymeric chains [126]. These components are responsible for the physical and chemical properties of WG and confer it with higher viscoelastic properties compared

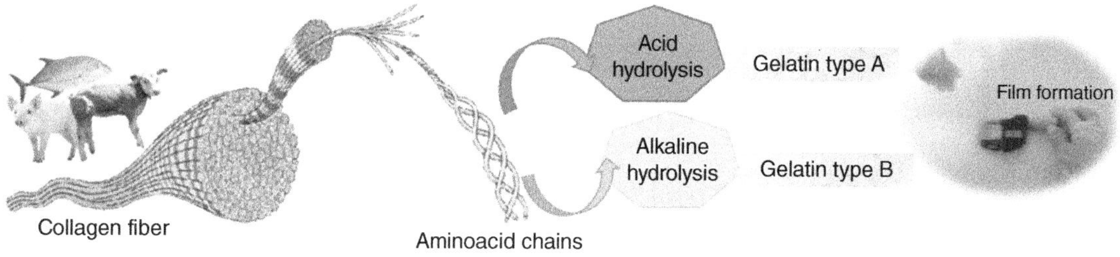

Figure 1.8 Scheme of the gelatin manufacturing from denaturation of collagen for film formation.

with other plant proteins [127, 128]. WG has been used for the production of sustainable food packaging because it is a renewable raw material, available in large quantities, fully biodegradable, and inexpensive. It has properties suitable for applications in edible films and biodegradable packaging. The methods used for processing WG-based bioplastics are casting, extrusion, and compression molding [129, 130]. WG-based films have exceptional oxygen and carbon dioxide barrier properties. However, they have lower water vapor barrier, mechanical strength, and thermal properties in comparison with conventional plastic films [131, 132]. Lignocellulosic reinforcement fillers have been used to improve the properties of WG since they can interact with proteins and lead to the formation of protein–polyphenol complexes, allowing a higher flexural strength, tensile strength, and modulus [133, 134]. Also, WG/montmorillonite (MMT) films were prepared by melt mixing and thermoforming [135]. The introduction of up to 5 wt% of MMT increased the mechanical properties in ways that were not possible by just the variation of the glycerol content and the processing temperature.

1.3.11.3 Soy Protein

Soy protein (SP) is an inexpensive renewable resource, sustainable, abundant, and functional, constituted by different globulins with mainly polar amino acids including acidic and basic amino acids and nonpolar amino acids fractions such as 2S, 7S, 11S, and 15S. The major components of SP are β-conglycinin (7S, nearly 35%) and glycinin (11S, nearly 52%) [136]. The protein with higher 11S fractions produces stronger films with lower water uptake than those richer in 7S, which is attributed to the presence of different sets of amino acids in 7S and 11S fractions [137]. Likewise, different chemical treatments and plasticizers have been explored to improve the intrinsic brittleness and low water resistance of SP-based films. Among them, glycerol, ethylene glycol, and propylene glycol have been found to outperform 1,3-propanediol. Glycerol and water can significantly increase the flexibility of films made of SP, but greatly decrease the tensile strength [138].

1.3.11.4 Corn Zein

Zein is a group of alcohol-soluble prolamin proteins found in the endosperm of corn. It is constituted by hydrophobic and neutral amino acids as well as some sulfur-containing amino acid [139]. Corn zein has a M_w ranging from 18 to 45 kDa and it is soluble in ethanol solutions in water at 60–70 wt% [140, 141]. Indeed, corn zein is produced commercially by extraction with aqueous alcohol and dried to a granular powder. The high proportion of nonpolar amino acid residues confers highly hydrophobic properties and solubility characteristics to zein, allowing its use in food packaging materials [142]. Zein films are formed by dissolving the protein into aqueous ethanol or isopropanol, heating to 65–85 °C, cooling down to 40–50 °C, and casting them by solvent evaporation. The resultant films are, however, translucid and present an intense yellow color (see Figure 1.9). Glycerol is often used to reduce the film brittleness though it tends to easily migrate through the film matrix because of the weak interaction between protein and plasticizer molecules. Migration of glycerol results in loss of flexibility in the film. Zein films have good oxygen barrier properties and

Figure 1.9 A zein film obtained from corn.

are greaseproof, which have been attributed to the helical conformation of the protein, but their mechanical properties and water resistance are low, similar to those of other protein films [143]. In order to overcome these deficiencies, blended films of zein with other biodegradable biopolymers have been widely studied [144–146].

1.3.11.5 Milk Proteins

Biodegradable films can also be formed from milk proteins. The two most important types in the packaging field are casein and whey protein [147]. Casein, comprising 80% of total milk protein, consists of three main components, α, β, and γ with *Mws* in the 19–25 kDa range. It forms colloidal micelles in milk and is stabilized by calcium phosphate bridging. Casein precipitates when milk is acidified to the isoelectric point (pH = 4.6). Acidified casein is converted to functional soluble caseinates, that is, sodium and calcium caseinates, by neutralization through addition of alkali [148]. Biodegradable films based on caseinates can be obtained by solubilization in water followed by casting and drying. Film formation in water is feasible due to its emulsification capability [149]. Films from caseinates are transparent, with good mechanical and oxygen barrier properties but poor water vapor permeability in the range of WG- and SP-based films [150].

Whey protein is the milk protein that remains soluble in milk serum after casein is coagulated during cheese or casein production. It comprises around 20% of the total milk proteins, being based on a mixture of proteins β-lactoglobulin (approximately 57%, M_w of 18 kDa), α-lactalbumin (approximately 20%, M_w of 14 kDa), bovine serum albumin, and immunoglobulins, among others. Formation of whey protein films involves heat denaturation in aqueous solutions, breaks existing disulfide bonds, and forms new intermolecular disulfide and hydrophobic bonds [151]. Films based on whey protein isolate (WPI) have shown promising mechanical properties as well as moderate water vapor permeability and good oxygen barrier properties [152, 153]. Nevertheless, the properties of WPI films are highly affected by relative humidity (RH) and the type and concentration of plasticizer [154, 155].

1.4 Concluding Remarks

Due to their low cost and high-performance characteristics, fossil derived polymers have remained as the most preferable materials for packaging applications.

However, as mentioned throughout this chapter, serious environmental issues along with the limited natural resources have led to an increase in the development and use of biopolymers as potential strategies to generate novel packaging items. Currently, there are still challenges to overcome, since the material performance of biopolymers in terms of mechanical, thermal, and vapor and gas barrier properties still needs to be further improved to be able to compete with the current petrochemical plastics commonly found in the food packaging industry. Furthermore, ethical issues frequently arise when attempts are made to use biomass for industrial purposes.

According to the European Bioplastics, the global production capacities of bio-based plastics, biodegradable or not, amounted to 1.7 million tons in 2014. This translates to approximately 680 000 ha of land. Consequently, the surface required to grow sufficient feedstock for current bioplastics production is only about 0.01% of the global agricultural area of a total of 5 billion hectares. In any case, the bioplastics industry is also researching the use of nonfood crops and biomass derived from algae, that is, the "second and third generation feedstock", respectively, with a view to its further use. Innovative technologies are currently being focused on nonedible by-products or waste materials as the source for bioplastics. In this regard, the trend for the development of the next generation of bioplastics is currently led by the emergence of conventional polymers made from renewable and nonfood sources. This generation feedstock is based on the production of plastics from cellulosic materials derived from food crop by-products such as straw, corn stover, or bagasse, which are usually left on the field where they biodegrade at a quantity far higher than necessary to restore the soil carbon pool. At best they are used to produce the energy for the conversion process to feedstock. This leaves significant potential for using biotechnological processes to create a platform to generate chemicals for industrial purposes, the so-called biorefinery concept, which includes the production of bioplastics. Indeed, the word "sustainable" means to maintain or keep going continuously and the word has been used in connection with forest management for over a century.

Although the renewable source may not be necessarily pursued in packaging applications, it can be still regarded as a remarkable plus, especially in times when oil prices are being increased markedly. However, it is also worthy to note that currently less than 6% of petroleum is addressed to produce polymers in the plastic industry. As a result, the real key property associated with biopolymers from an environmental point of view would be to develop polymers being readily biodegradable. Some plastic parts made of biodegradable polymers can additionally meet the requirements of harmonized standards for compostable materials. Today, a lot of different European and International Standards exist for biodegradable polymer-based materials (e.g. EN 13432 or ASTM D6400). They are mainly dealing with natural degradation under specific environmental conditions, particularly for packaging waste.

Looking at the end-of-life options for biopolymers, different recovery and disposal paths can coexist. In most countries, no separate collection system for biopolymers is established. In USA and Western Europe, highly developed separate collection systems for packaging waste exist and littering of plastic waste present a minor problem. However, waste disposal represents a serious problem for other countries where the recycling rates are still low and waste management is not properly working. Beside residual waste, the mixed plastic fraction from waste sorting can be supplied to waste incineration. During quaternary recycling, the resulting steam can be used in energy production, heating, and solid recovered fuel production. Secondary (mechanical) and, even, tertiary (chemical) recycling of biopolymers, particularly relevant if they are not biodegradable, is also possible. However, these processes would first require appropriate separate collection systems, which are difficult as biopolymers are still present in minor amounts and a wide range of bioplastics and their blends coexist on the market. At this moment, the mechanical recycling of biopolymers is at its beginning and the market penetration is not high enough to make recycling of biopolymers profitable. Furthermore, modern recycling processes, which are state of the art for commercial polymers, will have to be adapted to handle bioplastics properly. For instance, owners of recycling plants should make an effort to ensure that no mixed PLA- and PET-fractions are getting into bottle-to-bottle recycling.

Biodegradable packaging is expected to solve the solid waste problem as well as the litter problem. Nevertheless, the current situation indicates that only a few cities are able to collect and compost green waste. Converting a solid material to a gas via biodegradation or composting could not be sustainable if the collecting system is not prepared. In such cases, it is much better to recycle or recover the embodied energy through incineration. Therefore, there is a need to close the resource loop and make the most out of the material rather than simply use it once. Composting, both industrial or home types, is the expected option for biodegradable polymer-based packaging in a near future. This should be considered after material recycling and before thermal utilization. The production of humus-rich compost from different organic waste is a kind of material use. However, home composting rarely leads to success in terms of complete decomposition in an adequate period of time. Most of the biodegradable polymers are only usable for industrial composting. Nonbiodegradable biopolymers in addition to conventional fossil derived polymers can have a negative impact on the composting process. Additionally, owners of composting plants are not always pleased with the development and introduction of compostable articles either. In recent years, some labeling for compostable packaging materials has been developed, though they do not necessarily give information about the renewable material content.

Overall, although promising steps have been done in the area of biopolymers and new law regulations are pushing toward the use of such materials, more advances need to be carry not only to enhance material performance but also in terms of market possibilities. Current bioplastics should be in a near future more cost-effective, easier to process, and with enhanced performance so that food packaging users will fully benefit from their use and also, in terms of policy regulations, the use of such materials can be standardized in the food packaging industry.

References

1 FAO (2019). The Food and Agriculture Organization of the United Nations.
2 Ahvenainen, R. (2003). *Novel Food Packaging Techniques*. Woodhead Publishing.
3 Torres-Giner, S., Gil, L., Pascual-Ramírez, L., and Garde-Belza, J.A. (2018). Packaging: food waste reduction. *Encyclopedia of Polymer Applications* 3: 1990–2009.
4 Majid, I., Ahmad Nayik, G., Mohammad Dar, S., and Nanda, V. (2018). Novel food packaging technologies: innovations and future prospective. *Journal of the Saudi Society of Agricultural Sciences* 17 (4): 454–462.
5 Jabeen, N., Majid, I., and Nayik, G.A. (2015). Bioplastics and food packaging: a review. *Cogent Food & Agriculture* 1 (1): 1117749.
6 Williams, C.K. and Hillmyer, M.A. (2008). Polymers from renewable resources: a perspective for a special issue of polymer reviews. *Polymer Reviews* 48 (1): 1–10.
7 Xie, Y., Kohls, D., Noda, I. et al. (2009). Poly(3-hydroxybutyrate-co-3-hydroxyhexanoate) nanocomposites with optimal mechanical properties. *Polymer* 50 (19): 4656–4670.
8 Payne, J., McKeown, P., and Jones, M.D. (2019). A circular economy approach to plastic waste. *Polymer Degradation and Stability* 165: 170–181.
9 Tharanathan, R.N. (2003). Biodegradable films and composite coatings: past, present and future. *Trends in Food Science & Technology* 14 (3): 71–78.
10 Ashter, S.A. (ed.) (2016). *In Plastics Design Library, Introduction to Bioplastics Engineering*, 81–151. William Andrew Publishing.
11 Laycock, B., Nikolić, M., Colwell, J.M. et al. (2017). Lifetime prediction of biodegradable polymers. *Progress in Polymer Science* 71: 144–189.
12 Thakur, S., Chaudhary, J., Sharma, B. et al. (2018). Sustainability of bioplastics: opportunities and challenges. *Current Opinion in Green and Sustainable Chemistry* 13: 68–75.
13 Tsang, Y.F., Kumar, V., Samadar, P. et al. (2019). Production of bioplastic through food waste valorization. *Environment International* 127: 625–644.
14 Balan, V. (2014). Current challenges in commercially producing biofuels from lignocellulosic biomass. *ISRN Biotechnology* 2014: 463074–463074.
15 Braunegg, G., Lefebvre, G., and Genser, K.F. (1998). Polyhydroxyalkanoates, biopolyesters from renewable resources: physiological and engineering aspects. *Journal of Biotechnology* 65 (2): 127–161.
16 Cui, S., Borgemenke, J., Liu, Z., and Li, Y. (2019). Recent advances of "soft" bio-polycarbonate plastics from carbon dioxide and renewable bio-feedstocks via straightforward and innovative routes. *Journal of CO2 Utilization* 34: 40–52.
17 Höfer, R. and Selig, M. (2012). Green chemistry and green. *Polymer Chemistry* 10: 5–13.
18 Hatti-Kaul, R., Nilsson, L.J., Zhang, B. et al. (2019). Designing biobased recyclable polymers for plastics. *Trends in Biotechnology* 38: 50–67.

19 Sorrentino, A., Gorrasi, G., and Vittoria, V. (2007). Potential perspectives of bio-nanocomposites for food packaging applications. *Trends in Food Science and Technology* 18 (2): 84–95.

20 Rhim, J.-W., Park, H.-M., and Ha, C.-S. (2013). Bio-nanocomposites for food packaging applications. *Progress in Polymer Science* 38 (10): 1629–1652.

21 Vink, E.T.H., Glassner, D.A., Kolstad, J.J. et al. (2007). The eco-profiles for current and near-future NatureWorks® polylactide (PLA) production. *Industrial Biotechnology* 3 (1): 58–81.

22 Vroman, I. and Tighzert, L. (2009). Biodegradable polymers. *Materials* 2 (2): 307–344.

23 Schué, F. (2000). Biopolymers from renewable resources. Edited by D.L. Kaplan Springer-Verlag, Heidelberg, 1998. Pp 417, Price DM278.00 ISBN 3-540-63567-X. *Polymer International* 49 (5): 472–473.

24 Auras, R., Harte, B., and Selke, S. (2004). An overview of polylactides as packaging materials. *Macromolecular Bioscience* 4 (9): 835–864.

25 Jamshidian, M., Tehrany, E.A., Imran, M. et al. (2010). Poly-lactic acid: production, applications, nanocomposites, and release studies. *Comprehensive Reviews in Food Science and Food Safety* 9 (5): 552–571.

26 Garlotta, D. (2001). A literature review of poly(lactic acid). *Journal of Polymers and the Environment* 9 (2): 63–84.

27 Conn, R.E., Kolstad, J.J., Borzelleca, J.F. et al. (1995). Safety assessment of polylactide (PLA) for use as a food-contact polymer. *Food and Chemical Toxicology* 33 (4): 273–283.

28 Harada, M., Ohya, T., Iida, K. et al. (2007). Increased impact strength of biodegradable poly(lactic acid)/poly(butylene succinate) blend composites by using isocyanate as a reactive processing agent. *Journal of Applied Polymer Science* 106 (3): 1813–1820.

29 Babu R.P., O'Connor, K., and Seeram, R. (2013). Current progress on bio-based polymers and their future trends. *Progress in Biomaterials* 2 (1): 8.

30 Auras, R., Lim, L.T., Selke, S.E.M., and Tsuji, H. (2010). *Poly(Lactic Acid): Synthesis, Structures, Properties, Processing, and Applications.* Wiley, ISBN: 978-0-470-29366-9.

31 Relinque, J.J., de León, A.S., Hernández-Saz, J. et al. (2019). Development of surface-coated polylactic acid/polyhydroxyalkanoate (PLA/PHA) nanocomposites. *Polymers* 11 (3).

32 Rocca-Smith, J.R., Pasquarelli, R., Lagorce-Tachon, A. et al. (2019). Toward sustainable PLA-based multilayer complexes with improved barrier properties. *ACS Sustainable Chemistry & Engineering* 7 (4): 3759–3771.

33 Torres-Giner, S., Montanes, N., Fombuena, V. et al. (2018). Preparation and characterization of compression-molded green composite sheets made of poly(3-hydroxybutyrate) reinforced with long pita fibers. *Advances in Polymer Technology* 37 (5): 1305–1315.

34 Mutlu, G., Calamak, S., Ulubayram, K., and Guven, E. (2018). Curcumin-loaded electrospun PHBV nanofibers as potential wound-dressing material. *Journal of Drug Delivery Science and Technology* 43: 185–193.

35 Reddy, C.S.K., Ghai, R., Rashmi, and Kalia, V.C. (2003). Polyhydroxyalkanoates: an overview. *Bioresource Technology* 87 (2): 137–146.
36 Choi, J.-I. and Lee, S.Y. (1997). Process analysis and economic evaluation for poly(3-hydroxybutyrate) production by fermentation. *Bioprocess Engineering* 17 (6): 335–342.
37 Yeo, J.C.C., Muiruri, J.K., Thitsartarn, W. et al. (2018). Recent advances in the development of biodegradable PHB-based toughening materials: approaches, advantages and applications. *Materials Science and Engineering: C* 92: 1092–1116.
38 McChalicher, C.W.J. and Srienc, F. (2007). Investigating the structure–property relationship of bacterial PHA block copolymers. *Journal of Biotechnology* 132 (3): 296–302.
39 Keshavarz, T. and Roy, I. (2010). Polyhydroxyalkanoates: bioplastics with a green agenda. *Current Opinion in Microbiology* 13 (3): 321–326.
40 Khosravi-Darani, K. and Bucci, D.Z. (2015). Application of poly(hydroxyalkanoate) in food packaging: improvements by nanotechnology. *Chemical and Biochemical Engineering Quarterly* 29 (2): 275–285.
41 Requena, R., Vargas, M., and Chiralt, A. (2017). Release kinetics of carvacrol and eugenol from poly(hydroxybutyrate-co-hydroxyvalerate) (PHBV) films for food packaging applications. *European Polymer Journal* 92: 185–193.
42 Torres-Giner, S., Hilliou, L., Melendez-Rodriguez, B. et al. (2018). Melt processability, characterization, and antibacterial activity of compression-molded green composite sheets made of poly(3-hydroxybutyrate-co-3-hydroxyvalerate) reinforced with coconut fibers impregnated with oregano essential oil. *Food Packaging and Shelf Life* 17: 39–49.
43 Herrera, R., Franco, L., Rodríguez-Galán, A., and Puiggalí, J. (2002). Characterization and degradation behavior of poly(butylene adipate-co-terephthalate)s. *Journal of Polymer Science Part A: Polymer Chemistry* 40 (23): 4141–4157.
44 Li, G., Shankar, S., Rhim, J.-W., and Oh, B.-Y. (2015). Effects of preparation method on properties of poly(butylene adipate-co-terephthalate) films. *Food Science and Biotechnology* 24 (5): 1679–1685.
45 Witt, U., Einig, T., Yamamoto, M. et al. (2001). Biodegradation of aliphatic–aromatic copolyesters: evaluation of the final biodegradability and ecotoxicological impact of degradation intermediates. *Chemosphere* 44 (2): 289–299.
46 Fukushima, K., Wu, M.-H., Bocchini, S. et al. (2012). PBAT based nanocomposites for medical and industrial applications. *Materials Science and Engineering: C* 32 (6): 1331–1351.
47 Xing, Q., Ruch, D., Dubois, P. et al. (2017). Biodegradable and high-performance poly(butylene adipate-co-terephthalate)–lignin UV-blocking films. *ACS Sustainable Chemistry & Engineering* 5 (11): 10342–10351.
48 Wang, X., Peng, S., Chen, H. et al. (2019). Mechanical properties, rheological behaviors, and phase morphologies of high-toughness PLA/PBAT blends

by in-situ reactive compatibilization. *Composites Part B: Engineering* 173: 107028.
49 Someya, Y., Sugahara, Y., and Shibata, M. (2005). Nanocomposites based on poly(butylene adipate-co-terephthalate) and montmorillonite. *Journal of Applied Polymer Science* 95 (2): 386–392.
50 Chivrac, F., Kadlecova, Z., Pollet, E., and Avérous, L. (2006). Aromatic copolyester-based nano-biocomposites: elaboration, structural characterization and properties. *Journal of Polymers and the Environment* 14: 393–401.
51 Mondal, D., Bhowmick, B., Mollick, M.M.R. et al. (2014). Antimicrobial activity and biodegradation behavior of poly(butylene adipate-co-terephthalate)/clay nanocomposites. *Journal of Applied Polymer Science* 131 (7): 40079.
52 Al-Itry, R., Lamnawar, K., and Maazouz, A. (2014). Reactive extrusion of PLA, PBAT with a multi-functional epoxide: physico-chemical and rheological properties. *European Polymer Journal* 58: 90–102.
53 Zehetmeyer, G., Meira, S.M.M., Scheibel, J.M. et al. (2016). Influence of melt processing on biodegradable nisin-PBAT films intended for active food packaging applications. *Journal of Applied Polymer Science* 133 (13).
54 Sousa, G.M., Soares Júnior, M.S., and Yamashita, F. (2013). Active biodegradable films produced with blends of rice flour and poly(butylene adipate co-terephthalate): effect of potassium sorbate on film characteristics. *Materials Science and Engineering: C* 33 (6): 3153–3159.
55 Succinity. 2016. Biobased Polybutylene Succinate (PBS) - an attractive polymer for biopolymer compounds.
56 Xu, J. and Guo, B.-H. (2010). Poly(butylene succinate) and its copolymers: research, development and industrialization. *Biotechnology Journal* 5 (11): 1149–1163.
57 Doug, S. (2010). Bioplastics: Technologies and Global Markets. *BCC research reports PLS050A.*
58 Ravenstijn, J. (2010). *The State-of-the-Art on Bioplastics: Products, Markets, Trends and Technologies.* Polymedia.
59 Bajpai, P. (2019). *Biobased Polymers: Properties and Applications in Packaging.* Elsevier Science.
60 Vytejčková, S., Vápenka, L., Hradecký, J. et al. (2017). Testing of polybutylene succinate based films for poultry meat packaging. *Polymer Testing* 60: 357–364.
61 Jacquel, N., Freyermouth, F., Fenouillot, F. et al. (2011). Synthesis and properties of poly(butylene succinate): efficiency of different transesterification catalysts. *Journal of Polymer Science Part A: Polymer Chemistry* 49 (24): 5301–5312.
62 Eslami, H. and Kamal, M. (2013). Elongational rheology of biodegradable poly(lactic acid)/poly[(butylene succinate)-co-adipate] binary blends and poly(lactic acid)/poly[(butylene succinate)-co-adipate]/clay ternary nanocomposites. *Journal of Applied Polymer Science* 127: 2290–2306.
63 Liu, L., Yu, J., Cheng, L., and Qu, W. (2009). Mechanical properties of poly(butylene succinate) (PBS) biocomposites reinforced with surface

modified jute fibre. *Composites Part A: Applied Science and Manufacturing* 40 (5): 669–674.
64 Liu, L., Yu, J., Cheng, L., and Yang, X. (2009). Biodegradability of poly(butylene succinate) (PBS) composite reinforced with jute fibre. *Polymer Degradation and Stability* 94 (1): 90–94.
65 Zhao, P., Liu, W., Wu, Q., and Ren, J. (2010). Preparation, mechanical, and thermal properties of biodegradable polyesters/poly(lactic acid) blends. *Journal of Nanomaterials* 2010: 8.
66 Xu, J. and Guo, B. (2009). Microbial succinic acid, its polymer poly(butylene succinate), and applications. In: *Plastics from Bacteria. Microbiology Monographs*, vol. 14 (ed. G.Q. Chen), 347–388. Berlin, Heidelberg: Springer.
67 Ayu, R.S., Khalina, A., Harmaen, A.S. et al. (2018). Effect of modified tapioca starch on mechanical, thermal, and morphological properties of PBS blends for food packaging. *Polymers* 10 (11): 1187.
68 Kim, H.-S., Kim, H.-J., Lee, J.-W., and Choi, I.-G. (2006). Biodegradability of bio-flour filled biodegradable poly(butylene succinate) bio-composites in natural and compost soil. *Polymer Degradation and Stability* 91 (5): 1117–1127.
69 Chen, G., Li, S., Jiao, F., and Yuan, Q. (2007). Catalytic dehydration of bioethanol to ethylene over $TiO_2/\gamma\text{-}Al_2O_3$ catalysts in microchannel reactors. *Catalysis Today* 125 (1): 111–119.
70 Torres-Giner, S., Torres, A., Ferrándiz, M. et al. (2017). Antimicrobial activity of metal cation-exchanged zeolites and their evaluation on injection-molded pieces of bio-based high-density polyethylene. *Journal of Food Safety* 37 (4): e12348.
71 Braskem. (2014) I'm green polyethylene. Innovation and differentiation for your product.
72 De Castro Morschbacker, A.L. (2010). A method for the production of one or more olefins, an olefin, and a polymer. US 2010/0069691A1, 18 March 2010.
73 Koopmans, R.J. (2013). Polyolefin-based plastics from biomass-derived monomers. In: *Bio-Based Plastics* (ed. S. Kabasci), 295–310. Chichester: Wiley.
74 Huang, Y.M.L.H., Huang, X.L., Hu, Y.C., and Hu, Y. (2008). Advances of bio-ethylene. *Chinese Journal of Bioprocess Engineering* 6: 1–6.
75 LyondellBasell. (2019) Circulen and Circulen Plus.
76 Robertson, G.L. (2015). *Trends in Food Packaging*. The Journal of the Instutite of Food Science & Technology.
77 Smith, P.B. (2015). Bio-based sources for terephthalic acid. In: *Green Polymer Chemistry: Biobased Materials and Biocatalysis*, vol. 1192 (eds. H.N. Cheng, R.A. Gross and P.B. Smith), 453–469. American Chemical Society.
78 Tsusho, T. (2013). Toyota Tsusho Expanding its New Plant-Derived Plastic Brand Globio. https://www.toyota-tsusho.com/english/press/detail/130326_001840.html#:~:text=Mineral%20Water%20bottles%2D-,Toyota%20Tsusho%20Expanding%20its%20New%20Plant%2DDerived%20Plastic%20Brand%20GLOBIO,Suntory%20Natural%20Mineral&

percnt;20Water%20bottles%2D&text=Since%20Bio%2DPET%20is%20made,the%20atmosphere%20even%20when%20burned. (accessed 09 September 2019).

79 SCG Chemicals. (n.d.) The green plastic "Bio-PET". https://www.scgchemicals.com/en/news-media/feature-story/detail/9 (accessed 09 September 2019).

80 Feldman, R.M.R.G.U., Urano, J., Meinhold, P. et al. (2011). Yeast organism producing isobutanol at a high yield. US Patent 8455239, issued 13 September, 2011.

81 Peters, M.T.J.D., Jenni, M., Manzer, L.E., and Hendon, D.E. (2010). Integrated process to selectively convert renewable isobutanol to *p*-xylene. US 12/899285, filed October 6, 2010.

82 The Coca-Cola Company (2015). Great things come in innovative packaging: an introduction to PlantBottle™ packaging.

83 Siracusa, V. and Rosa, M.D. (2018). Sustainable packaging. In: *Sustainable Food Systems from Agriculture to Industry*, Chapter 8 (ed. C.M. Galanakis), 275–307. Academic Press.

84 Pellis, A., Haernvall, K., Pichler, C.M. et al. (2016). Enzymatic hydrolysis of poly(ethylene furanoate). *Journal of Biotechnology* 235: 47–53.

85 Weinberger, S., Canadell, J., Quartinello, F. et al. (2017). Enzymatic degradation of poly(ethylene 2,5-furanoate) powders and amorphous films. *Catalysts* 7 (11): 318.

86 Rosenboom, J.-G., Hohl, D.K., Fleckenstein, P. et al. (2018). Bottle-grade polyethylene furanoate from ring-opening polymerisation of cyclic oligomers. *Nature Communications* 9 (1): 2701.

87 Kasmi, N., Papageorgiou, G.Z., Achilias, D.S., and Bikiaris, D.N. (2018). Solid-state polymerization of poly(ethylene Furanoate) biobased polyester, II: an efficient and facile method to synthesize high molecular weight polyester appropriate for food packaging applications. *Polymers* 10 (5): 471.

88 Avantium. (n.d.) FDCA-From plant based materials to FDCA and PEF. https://www.avantium.com/yxy/products-applications/ (accessed 09 September 2019).

89 Cruz-Izquierdo, Á., van den Broek, L.A.M., Serra, J.L. et al. (2015). Lipase-catalyzed synthesis of oligoesters of 2,5-furandicarboxylic acid with aliphatic diols. *Pure and Applied Chemistry* 87 (1): 59–69.

90 Jiang, Y., Woortman, A.J.J., Alberda van Ekenstein, G.O.R. et al. (2014). Enzymatic synthesis of biobased polyesters using 2,5-bis(hydroxymethyl)furan as the building block. *Biomacromolecules* 15 (7): 2482–2493.

91 Rudnik, E. (2013). Compostable polymer properties and packaging applications. In: *Plastic Films in Food Packaging*, Chapter 13 (ed. S. Ebnesajjad), 217–248. Oxford: William Andrew Publishing.

92 Pawar, P.A. and Purwar, A.H. (2013). Biodegradable polymers in food packaging. *American Jorunal of Engineering Research* 2 (5): 151–164.

93 García Ibarra, V., Sendón, R., and Rodríguez-Bernaldo de Quirós, A. (2016). Antimicrobial food packaging based on biodegradable materials.

In: *Antimicrobial Food Packaging*, Chapter 29 (ed. J. Barros-Velázquez), 363–384. San Diego, CA: Academic Press.
94 Khalid, S., Yu, L., Feng, M. et al. (2018). Development and characterization of biodegradable antimicrobial packaging films based on polycaprolactone, starch and pomegranate rind hybrids. *Food Packaging and Shelf Life* 18: 71–79.
95 Chan, C.M., Vandi, L.-J., Pratt, S. et al. (2018). Composites of wood and biodegradable thermoplastics: a review. *Polymer Reviews* 58 (3): 444–494.
96 Khan, B.M., Niazi, B.K., Samin, G., and Jahan, Z. (2017). Thermoplastic starch: a possible biodegradable food packaging material—a review. *Journal of Food Process Engineering* 40 (3): e12447.
97 Li, H., Qi, Y., Zhao, Y. et al. (2019). Starch and its derivatives for paper coatings: a review. *Progress in Organic Coatings* 135: 213–227.
98 Mohanty, A.K., Misra, M., and Hinrichsen, G. (2000). Biofibres, biodegradable polymers and biocomposites: an overview. *Macromolecular Materials and Engineering* 276–277 (1): 1–24.
99 Zhu, F. (2015). Composition, structure, physicochemical properties, and modifications of cassava starch. *Carbohydrate Polymers* 122: 456–480.
100 Forssell, P.M., Mikkilä, J.M., Moates, G.K., and Parker, R. (1997). Phase and glass transition behaviour of concentrated barley starch-glycerol-water mixtures, a model for thermoplastic starch. *Carbohydrate Polymers* 34 (4): 275–282.
101 Gaudin, S., Lourdin, D., Le Botlan, D. et al. (1999). Plasticisation and mobility in starch-sorbitol films. *Journal of Cereal Science* 29 (3): 273–284.
102 Ma, X. and Yu, J. (2004). The plastcizers containing amide groups for thermoplastic starch. *Carbohydrate Polymers* 57 (2): 197–203.
103 Nakamura, S. and Tobolsky, A.V. (1967). Viscoelastic properties of plasticized amylose films. *Journal of Applied Polymer Science* 11 (8): 1371–1386.
104 Kalichevsky, M.T., Blanshard, J.M.V., and Tokargzuk, P.F. (1993). Effect of water content and sugars on the glass transition of casein and sodium caseinate. *International Journal of Food Science & Technology* 28 (2): 139–151.
105 Kaseem, M., Hamad, K., and Deri, F. (2012). Thermoplastic starch blends: a review of recent works. *Polymer Science, Series A* 54 (2): 165–176.
106 Wang, Z.-F., Peng, Z., Li, S.-D. et al. (2009). The impact of esterification on the properties of starch/natural rubber composite. *Composites Science and Technology* 69 (11): 1797–1803.
107 Liu, H., Xie, F., Yu, L. et al. (2009). Thermal processing of starch-based polymers. *Progress in Polymer Science* 34 (12): 1348–1368.
108 Dong, Y., Novo, D.C., Mosquera-Giraldo, L.I. et al. (2019). Conjugation of bile esters to cellulose by olefin cross-metathesis: a strategy for accessing complex polysaccharide structures. *Carbohydrate Polymers* 221: 37–47.
109 Chavan, R.B., Rathi, S., Jyothi, V.G.S.S., and Shastri, N.R. (2019). Cellulose based polymers in development of amorphous solid dispersions. *Asian Journal of Pharmaceutical Sciences* 14 (3): 248–264.

110 Petersen, K., Væggemose Nielsen, P., Bertelsen, G. et al. (1999). Potential of biobased materials for food packaging. *Trends in Food Science & Technology* 10 (2): 52–68.

111 Vom Stein, T., Grande, P., Sibilla, F. et al. (2010). Salt-assisted organic-acid-catalyzed depolymerization of cellulose. *Green Chemistry* 12 (10): 1844–1849.

112 Andrade, R., Skurtys, O., and Osorio, F. (2015). Drop impact of gelatin coating formulated with cellulose nanofibers on banana and eggplant epicarps. *LWT Food Science and Technology* 61 (2): 422–429.

113 Azevedo, V.M., Silva, E.K., Gonçalves Pereira, C.F. et al. (2015). Whey protein isolate biodegradable films: influence of the citric acid and montmorillonite clay nanoparticles on the physical properties. *Food Hydrocolloids* 43: 252–258.

114 Ahmad, M., Benjakul, S., Sumpavapol, P., and Nirmal, N.P. (2012). Quality changes of sea bass slices wrapped with gelatin film incorporated with lemongrass essential oil. *International Journal of Food Microbiology* 155 (3): 171–178.

115 Martucci, J.F. and Ruseckaite, R.A. (2010). Biodegradable three-layer film derived from bovine gelatin. *Journal of Food Engineering* 99 (3): 377–383.

116 Tongnuanchan, P., Benjakul, S., and Prodpran, T. (2014). Structural, morphological and thermal behaviour characterisations of fish gelatin film incorporated with basil and citronella essential oils as affected by surfactants. *Food Hydrocolloids* 41: 33–43.

117 Martucci, J.F., Gende, L.B., Neira, L.M., and Ruseckaite, R.A. (2015). Oregano and lavender essential oils as antioxidant and antimicrobial additives of biogenic gelatin films. *Industrial Crops and Products* 71: 205–213.

118 Fakhouri, F.M., Costa, D., Yamashita, F. et al. (2013). Comparative study of processing methods for starch/gelatin films. *Carbohydrate Polymers* 95 (2): 681–689.

119 Gómez-Estaca, J., Gómez-Guillén, M.C., Fernández-Martín, F., and Montero, P. (2011). Effects of gelatin origin, bovine-hide and tuna-skin, on the properties of compound gelatin–chitosan films. *Food Hydrocolloids* 25 (6): 1461–1469.

120 Yilmaz, M.T., Kesmen, Z., Baykal, B. et al. (2013). A novel method to differentiate bovine and porcine gelatins in food products: nanoUPLC-ESI-Q-TOF-MSE based data independent acquisition technique to detect marker peptides in gelatin. *Food Chemistry* 141 (3): 2450–2458.

121 Nur Azira, T., Man, Y.B.C., Raja Mohd Hafidz, R.N. et al. (2014). Use of principal component analysis for differentiation of gelatine sources based on polypeptide molecular weights. *Food Chemistry* 151: 286–292.

122 Andreuccetti, C., Carvalho, R.A., Galicia-García, T. et al. (2011). Effect of surfactants on the functional properties of gelatin-based edible films. *Journal of Food Engineering* 103 (2): 129–136.

123 Díaz, P., Arratia, C., Vásquez, C. et al. (2011). Effect of glycerol on water sorption of bovine gelatin films in the glassy state. *Procedia Food Science* 1: 267–274.

124 Sinha Ray, S. and Bousmina, M. (2005). Biodegradable polymers and their layered silicate nanocomposites: in greening the 21st century materials world. *Progress in Materials Science* 50 (8): 962–1079.

125 Liao, L., Zhang, F.-L., Lin, W.-J. et al. (2019). Gluten–starch interactions in wheat gluten during carboxylic acid deamidation upon hydrothermal treatment. *Food Chemistry* 283: 111–122.

126 Pallos, F.M., Robertson, G.H., Pavlath, A.E., and Orts, W.J. (2006). Thermoformed wheat gluten biopolymers. *Journal of Agricultural and Food Chemistry* 54 (2): 349–352.

127 Nataraj, D., Sakkara, S., Hn, M., and Reddy, N. (2018). Properties and applications of citric acid crosslinked banana fibre-wheat gluten films. *Industrial Crops and Products* 124: 265–272.

128 Zubeldía, F., Ansorena, M.R., and Marcovich, N.E. (2015). Wheat gluten films obtained by compression molding. *Polymer Testing* 43: 68–77.

129 Bootklad, M., Chantarak, S., and Kaewtatip, K. (2016). Novel biocomposites based on wheat gluten and rubber wood sawdust. *Journal of Applied Polymer Science* 133 (30).

130 Tuntachon, S., Sukolrat, A., Numnuam, A., and Kaewtatip, K. (2019). Effect of kaolin content and sonication on the properties of wheat gluten composites. *Powder Technology* 351: 66–70.

131 Chen, L., Reddy, N., Wu, X., and Yang, Y. (2012). Thermoplastic films from wheat proteins. *Industrial Crops and Products* 35 (1): 70–76.

132 Guilbert, S., Gontard, N., Morel, M.H. et al. (2002). Formation and properties of wheat gluten films and coatings. In: *Protein-Based Films and Coatings* (ed. A. Gennadios), 69–122. CRC Press.

133 Cho, S.W., Gällstedt, M., Johansson, E., and Hedenqvist, M.S. (2011). Injection-molded nanocomposites and materials based on wheat gluten. *International Journal of Biological Macromolecules* 48 (1): 146–152.

134 Zárate-Ramírez, L.S., Romero, A., Martínez, I. et al. (2014). Effect of aldehydes on thermomechanical properties of gluten-based bioplastics. *Food and Bioproducts Processing* 92 (1): 20–29.

135 Angellier-Coussy, H., Torres-Giner, S., Morel, M.H. et al. (2008). Functional properties of thermoformed wheat gluten/montmorillonite materials with respect to formulation and processing conditions. *Journal of Applied Polymer Science* 107 (1): 487–496.

136 Koshy, R.R., Mary, S.K., Thomas, S., and Pothan, L.A. (2015). Environment friendly green composites based on soy protein isolate – a review. *Food Hydrocolloids* 50: 174–192.

137 Rani, S. and Kumar, R. (2019). A review on material and antimicrobial properties of soy protein isolate film. *Journal of Polymers and the Environment* 27 (8): 1613–1628.

138 Wihodo, M. and Moraru, C.I. (2013). Physical and chemical methods used to enhance the structure and mechanical properties of protein films: a review. *Journal of Food Engineering* 114 (3): 292–302.

139 Kasaai, M.R. (2018). Zein and zein-based nano-materials for food and nutrition applications: a review. *Trends in Food Science & Technology* 79: 184–197.

140 Shukla, R. and Cheryan, M. (2001). Zein: the industrial protein from corn. *Industrial Crops and Products* 13 (3): 171–192.

141 Sun, H., Shao, X., Jiang, R. et al. (2018). Mechanical and barrier properties of corn distarch phosphate-zein bilayer films by thermocompression. *International Journal of Biological Macromolecules* 118: 2076–2081.

142 Anderson, T.J. and Lamsal, B.P. (2011). REVIEW: zein extraction from corn, corn products, and coproducts and modifications for various applications: a review. *Cereal Chemistry* 88 (2): 159–173.

143 Spasojević, L., Katona, J., Bučko, S. et al. (2019). Edible water barrier films prepared from aqueous dispersions of zein nanoparticles. *LWT* 109: 350–358.

144 Altan, A., Aytac, Z., and Uyar, T. (2018). Carvacrol loaded electrospun fibrous films from zein and poly(lactic acid) for active food packaging. *Food Hydrocolloids* 81: 48–59.

145 Chen, X., Cui, F., Zi, H. et al. (2019). Development and characterization of a hydroxypropyl starch/zein bilayer edible film. *International Journal of Biological Macromolecules* 145: 1175–1182.

146 Qian, J., Ma, J., Su, J. et al. (2016). PHBV-based ternary composite by intermixing of magnesium calcium phosphate nanoparticles and zein: in vitro bioactivity, degradability and cytocompatibility. *European Polymer Journal* 75: 291–302.

147 Kinsella, J.E. and Morr, C.V. (1984). Milk proteins: physicochemical and functional properties. *Critical Reviews in Food Science and Nutrition* 21 (3): 197–262.

148 Rehan, F., Ahemad, N., and Gupta, M. (2019). Casein nanomicelle as an emerging biomaterial—a comprehensive review. *Colloids and Surfaces B: Biointerfaces* 179: 280–292.

149 Picchio, M.L., Linck, Y.G., Monti, G.A. et al. (2018). Casein films crosslinked by tannic acid for food packaging applications. *Food Hydrocolloids* 84: 424–434.

150 Belyamani, I., Prochazka, F., and Assezat, G. (2014). Production and characterization of sodium caseinate edible films made by blown-film extrusion. *Journal of Food Engineering* 121: 39–47.

151 McHugh, T.H., Aujard, J.F., and Krochta, J.M. (1994). Plasticized whey protein edible films: water vapor permeability properties. *Journal of Food Science* 59 (2): 416–419.

152 Kokoszka, S., Debeaufort, F., Lenart, A., and Voilley, A. (2010). Water vapour permeability, thermal and wetting properties of whey protein isolate based edible films. *International Dairy Journal* 20 (1): 53–60.

153 Ramos, Ó.L., Silva, S.I., Soares, J.C. et al. (2012). Features and performance of edible films, obtained from whey protein isolate formulated with antimicrobial compounds. *Food Research International* 45 (1): 351–361.

154 Krochta, J.M., De Mulder-Johnson, C., and Institute of Food Technologists. Expert Panel on Food Safety and Nutrition (1997). Edible and biodegradable polymer films : challenges and opportunities. 51 (2): 61–74.

155 Pérez-Gago, M.B. and Krochta, J.M. (2002). Formation and properties of whey protein films and coatings. In: *Protein-Based Films and Coatings* (ed. A. Gennadios), 159–180. CRC Press.

2

Biopolymers Derived from Marine Sources for Food Packaging Applications

Jone Uranga, Iratxe Zarandona, Mireia Andonegi, Pedro Guerrero, and Koro de la Caba

BIOMAT research group, University of the Basque Country (UPV/EHU), Escuela de Ingeniería de Gipuzkoa, Plaza de Europa 1, Donostia-San Sebastián 20018, Spain

2.1 Introduction

The main function of food packaging is to protect products from chemical, physical, and biological deteriorating effects, thereby extending food shelf life [1], although packaging is also used as a marketing tool to communicate with consumers [2]. Packaging materials include paper, glass, aluminum, steel, and polymers. Among them, polymeric materials show some advantages such as lightness and processability, allowing the production of flexible pouches, bags, or rigid containers of various shapes and sizes [3]. Thus, they have been extensively used to contain and protect fresh food products. Nevertheless, petroleum-based polymeric materials lead to serious environmental issues because they generate extensive volumes of waste after disposal, considering that approximately 40% of plastic products have a service life of less than one month [4]. Moreover, even if there is an increasing awareness of recycling, the most common materials utilized for food packaging have a limited recyclability (e.g. thermoplastics) or they are hardly recyclable (e.g. thermoset plastics) [5]. In that way, 30% of plastic packaging materials may never be eligible for recycling or reuse without a fundamental redesign [6]. In this regard, ongoing research is focused on sustainable alternatives, giving special emphasis to films and coatings derived from renewable sources. Among biopolymers, proteins and polysaccharides, which are long chain hydrophilic polymers known as hydrocolloids, can be found. These hydrocolloids are capable of forming viscous dispersions or gels in water, providing easy handling for film and coating preparation [7]. In general, hydrocolloid films and coatings exhibit good barrier properties to oxygen, carbon dioxide, and lipids [8–10]. In terms of mechanical properties, protein-based films present greater flexibility but lower tensile strength than polysaccharide-based films [11].

Using renewable packaging materials is not a novel strategy, since some ancient human practices of food preservation were based on renewable raw materials. For instance, fruit waxing has been used since the twelfth century in China, meat larding has been employed in England since the sixteenth century, as well as wrapping

Sustainable Food Packaging Technology, First Edition. Edited by Athanassia Athanassiou.

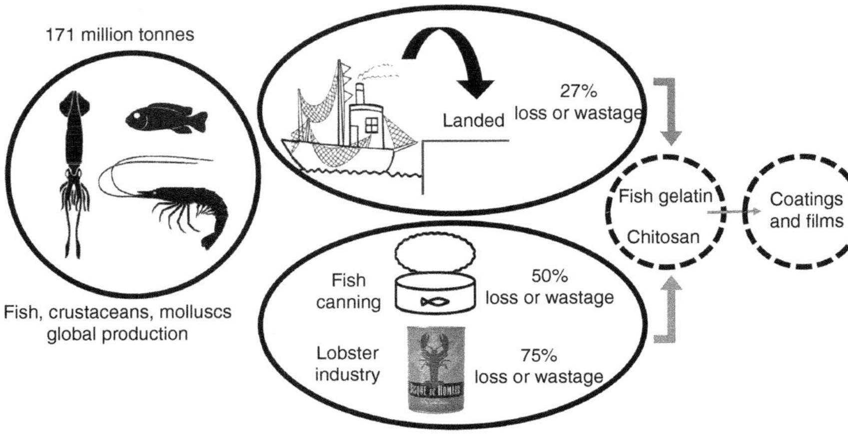

Figure 2.1 Fishery production, wastage, and possible solutions.

ground meats or vegetables with soy films in the Orient [11]. Currently, considerable progress has been made in developing packaging based on renewable materials, among them, films and coatings based on biopolymers derived from marine sources [12]. Thereby, by-products from seafood industry can be used in the development of environmentally friendly packaging solutions to prevent food spoilage and extend shelf life [13].

The global fish production (fish, crustaceans, mollusks, and other aquatic animals, including fisheries and aquaculture) peaked at about 171 million tons in 2016 [14], with an estimated loss or wastage between landing and consumption of 27% of landed fish, crustaceans, and mollusks (Figure 2.1). Additionally, high amounts of solid waste generated in fishery industries must be considered; for instance, in fish canning operations, the amount of waste can be as high as 50% by weight of raw products [15]. In all these fish by-products (skin, bones, scales, tendons), there is a considerable amount of collagen, a fibrous protein with a structural function [16, 17]. Carrying out thermal denaturation or partial hydrolysis of collagen, gelatin is obtained [18], whose properties differ between fish species. In general, cold-water fish gelatins have relatively lower gel strength than warm-water fish gelatins due the lower content of imino acid residues [19]. In this regard, gelatin from salmon skin (cold-water fish) has 166 residues per 1000 total amino acid, whereas gelatin from sea bass skin (warm-water fish) contains around 198–202 residues per 1000 total amino acid [20].

In the case of crustacean and mollusks by-products, shrimp and crab shells can be used to obtain chitin [21]. As an example, lobster processing industry by-products account for around 75% by weight of the starting material (Figure 2.1). Chitin has a similar structural function to that of collagen. Chitin is the second most abundant natural polymer after cellulose, and it is usually transformed into chitosan, a water-soluble polymer, to broaden its application. This transformation process is based on chitin deacetylation under alkaline conditions [22–24]. It must be considered that the functional properties of chitosan depend on structural features, such as average molecular weight and degree of deacetylation [25].

This chapter provides an overview of packaging films and coatings based on biopolymers derived from marine sources, with emphasis on fish gelatin and chitosan materials. The techniques used for extraction of biopolymers, preparation methods of films and coatings, and their characterization, together with diverse novel techniques employed for extending food products shelf life, will be described in this chapter.

2.2 Fish Gelatin Films and Coating

2.2.1 Collagen and Gelatin Extraction

Collagen can be obtained from fish skin, scales, and bones, which are cleaned and reduced in size to facilitate the extraction [26]. The extraction process from fishery wastes consists of two main steps: pretreatment of raw materials and collagen extraction. In general, alkaline pretreatment is carried out to remove impurities such as non-collagenous proteins and lipids, as well as to increase the quality of the final extracted collagen [27]; and it can be carried out using a strong alkali, such as sodium hydroxide [20, 28] or calcium hydroxide [29]. Sometimes, to remove the lipids, alcohols such as ethanol [30, 31], isopropanol [32], or butyl alcohol [33, 34] are also used. Additionally, the demineralization of the raw materials can be carried out using ethylenediaminetetraacetic acid (EDTA); thus, calcium or other inorganic materials are removed [35]. After that, the extraction of collagen is usually carried out.

There are numerous methods reported for collagen extraction, based on three main extraction processes: extraction of salt-soluble collagen (SSC), acid-soluble collagen (ASC), and enzyme-soluble collagen (ESC). These extraction methods directly affect collagen properties and yield [36], and depend on factors such as fish species and age [37]. It is worth noting that all procedures within collagen extraction are performed at low temperature (~4 °C) for 24–48 hours. Although an increasing extraction temperature and time can offer a higher collagen yield, it may not be desirable due to collagen degradation [38].

Regarding extraction in NaCl solution (SSC extraction), lower yields are obtained compared with ASC and ESC extractions [39]; hence, this method is not widely utilized and acid and/or enzymatic extractions are preferred. In terms of acidic extractions, the use of acetic acid medium for extraction and isolation of collagen from by-products such as sabalo skin [26] and sea bass scales [40] has been reported. In addition to acetic acid, collagen can also be extracted using other organic acids, such as citric acid [32], or inorganic acids such as hydrochloric acid [41] and phosphoric acid [42]. However, it must be considered that the solubility of collagen in acidic media is limited owing to many interchain cross-links, covalent bonds formed via the condensation reaction of aldehydes with lysine and hydroxylysine at the telopeptide region [43]. Moreover, habitat and age of animals or collagen by-product source affect directly the content of keto–imine bonds and, hence, the solubility in acidic media [44]. In order to improve solubility and achieve higher yield, extraction parameters can be changed, including acid concentration, ratio of acid solution/raw materials, extraction temperature and time; however, these parameters must be controlled in order to avoid collagen degradation [45]. Depending on the degree of collagen aggregation, the extraction process could also include the use of proteolytic enzymes [46–48], which together with an acidic solution might lead to an increase in the yield, by combining biological and chemical pretreatments. Diverse enzymes have been employed, among them, trypsin and pepsin digestive proteases [49] and bacterial collagenases [27]. Indeed, the most commonly employed enzyme is pepsin and, in this sense, the resultant pepsin-extracted collagen is called pepsin-soluble collagen (PSC) [50]. This digestive protease can remove non-collagenous proteins and increase the purity of collagen. Thus, non-collagenous proteins can be hydrolyzed by pepsin and removed using salt precipitation and dialysis. Additionally, pepsin can also hydrolyze collagen telopeptides, contributing to the treated collagen solubilization in acid media and, thereby increasing the yield of acid-soluble extraction [36]. Several studies have shown an increase in the acidic extraction yield with the aid of pepsin [40, 51, 52]. It is worth noting that commercial pepsin is commonly obtained from porcine gastric mucosa; however, in order to avoid some religious restrictions a huge range of proteolytic enzymes, including pepsin, can be extracted from fish viscera [34]. Furthermore, the choice of suitable enzymes and physiochemical conditions for the enzymatic reaction, such as solution pH, temperature, hydrolysis time, and enzyme concentration, must be optimized for the maximum activity. In recent years, other techniques have been used to improve SSC, ASC, and ESC extraction methods, among them, ultrasonic treatments and homogenization-aided methods. Zou et al. [53] worked with ultrasonic power of 200 W having a single frequency of 24 kHz to obtain collagen extracted from calipash of soft-shelled turtle. On the other hand, Tan and Chang [41] successfully extracted collagen from catfish skin by mixing catfish skins, hydrochloric acid, and pepsins at 7000 rpm for five minutes until homogenization and then the mixture was stirred for one hour at 4 °C. After extraction processes, some final steps must be carried out. Thus, collagen is usually recovered using salt precipitation, centrifugation, dialysis, and freeze-drying. Generally, collagen solution is precipitated using NaCl. The salt concentrations employed for collagen precipitation can be adjusted to maximize the collagen recovery and removal of impurities. Then, centrifugation (around 10 000–20 000 rpm) is used to collect the precipitated collagen. The resultant precipitate is dissolved in acetic acid prior to dialysis against distilled water. The dialysate is finally freeze-dried and the obtained collagen powder is stored [43].

All aforementioned methods are generally time-, energy-, reactant-, and cost-consuming batch type processes. Therefore, alternative methods are emerging, among them, extrusion-hydro-extraction (EHE) processes. Extrusion is widely used in the food industry and it offers many advantages, such as ease of operation, continuous production, high yield, and little waste. Using this method, Huang et al. [54] developed a novel EHE process for collagen extraction from tilapia fish scale, obtaining extruded scale samples with two to three times higher protein extraction yield than that of non-extruded scale samples.

With regard to fish gelatin extraction, this protein is obtained through the hydrolysis of collagen. Many studies in the literature have reported different protocols for the obtainment of fish gelatin. The extraction protocols typically include the use of acid or alkaline chemicals as pretreatments. In this way, two types of gelatins can be differentiated: type A gelatin, derived from collagen by acid pretreatment (most fish gelatins are within this group); and type B gelatin, as a result of an alkaline pretreatment of collagen [55]. After pretreatments, gelatin is extracted using relative high temperature (45–50 °C) [56]. Compared with collagen extraction conditions, gelatin extraction temperatures are higher, since the protein chains need a stronger modification to become gelatin.

2.2.2 Preparation and Characterization of Fish Gelatin Films and Coatings

In order to prepare fish gelatin coatings, solutions can be applied onto food surfaces using different techniques, such as dipping, spraying, brushing, or panning [57], dipping and spraying being the most employed on fish, dairy products, or minimally-processed fruit and vegetables. Abdelhedi et al. [58] evaluated the quality of fresh fish fillet samples immersed into different gelatin solutions. Indeed, samples dipped into black barred halfbeak (*Hemiramphus far*) gelatin solution were more efficient than samples coated with commercial bovine gelatin, and the effect was improved by the addition of its hydrolysate.

Considering gelatin has excellent film-forming capacity, fish gelatin films are often prepared via solution casting for food packaging applications [59]. The process starts with dissolving gelatin along with additives, such as plasticizers or active compounds, in a suitable solvent, usually water or water–alcohol solutions. Heating and/or pH changes alter solution conditions, which affect the final properties of the film. Then, the solution is cast and, finally, solvent evaporation takes place when solutions are subjected to drying processes, leading to film formation [60]. Solution casting is widely used at laboratory scale owing to its simplicity [61, 62]. However, dry methods, such as compression molding [63] and extrusion [64], have also been used [65]. These methods are larger production techniques, faster, and more appropriate for industrial scale production since the pre-existing technology used in the plastic industry can be used for gelatin film production.

In order to assess the suitability of films for food packaging, their properties must be evaluated. The mechanical behavior of films is an important issue, since films intended for food packaging require both resistance and flexibility to facilitate handling and to avoid breaking during packaging, transport, and storage stages [66]. Thus, plasticizers are commonly added to reduce hydrogen bonding and increase the molecular spacing between protein chains [67]. Due to its small size and hydrophilic nature, glycerol is a compatible plasticizer with fish gelatin and it has been widely used for gelatins [68–70]. However, the incorporation of more hydrophobic additives, such as fatty acids, may increase water resistance of fish gelatin films [71]. Physical, chemical, and enzymatic cross-linking have also been explored to reduce water sensitivity of gelatin [65]. In particular, the addition of glucose to promote the chemical cross-linking with amino groups of gelatin, known as Maillard reaction, reduces water solubility and increases water contact angle of fish gelatin films [72]. Furthermore, barrier properties against UV light are improved with Maillard reaction, even if fish gelatin shows high absorption of UV light in the range of 200–300 nm due to the presence of peptide and aromatic amino acid residues [73]. In addition to UV light barrier properties, fish gelatin can provide materials with good oxygen barrier properties, preventing food oxidation and extending food shelf life [74].

In terms of optical properties, film and coating appearance (color, transparency, and gloss) is directly related to consumer acceptability. Films and coatings based on fish gelatin are usually transparent and colorless [75]. Nevertheless, the incorporation of additives into the formulations to improve functional properties can affect the film transparency [76]. For instance, the presence of zinc oxide leads to a whitening effect [77]. Also cross-linking reactions, such as Maillard reaction, produce a browning effect [78]. Regarding gloss, this property is related to the surface morphology [79]. Fish gelatin films showed lower gloss values when cross-linked with citric acid, indicating the formation of a rougher surface as a consequence of the cross-linking reaction [80]. It is worth noting that rougher surfaces provide more convenient properties to print the film for commercial purposes.

In addition to film characterization, other issues, such as contamination of food by migration of additives, are relevant matters due to potential health-related risks [81]. Thus, even if fish gelatin is classified as Generally Recognized As Safe (GRAS) polymer by the Food and Drug Administration (FDA) center for food safety and nutrition, fish gelatin-based materials intended for the packaging of foodstuffs must be approved after assessing the absence of toxic effects to the consumer and organoleptic changes to the food. Furthermore, the assessment of food quality can give detailed information of film and coating suitability. For instance, one of the most widely used means for the evaluation of fish quality, total volatile bases nitrogen (TVBN), is the key index of fish freshness [82, 83].

2.2.3 Food Shelf Life Extension Using Fish Gelatin Films and Coatings

Fish gelatin films and coatings can be employed to improve food quality and extend food shelf life. In recent years, different strategies have been used for this purpose. Cross-linking usually enhances film barrier properties, increasing water, light, and chemical resistance [84]. In this regard, Etxabide et al. [85] employed a commercial cod fish gelatin, using lactose as a cross-linker, and Taghizadeh et al. [86] incorporated riboflavin into fish gelatin film forming solutions to carry out a photosensitizer-induced cross-linking. Both studies showed improved light barrier properties, which highlight the potential of these films to prevent food oxidation caused by light, contributing to extend food shelf life.

Active packaging, antioxidant and/or antimicrobial packaging, is a strategy to extend food shelf life. In the last few years, gelatin of diverse aquatic fish species has been employed on active films and coatings, in which different additives were used as antioxidant and antimicrobial agents. Some of these works are outlined in Table 2.1. Adilah et al. [88] incorporated mango peel extract into fish gelatin

Table 2.1 Fish gelatin-based active films and coatings.

Fish gelatin	Packaging	Bioactives	Bioactivity	References
Silver carp skin	C	Curcumin/β-cyclodextrin	AO + AM	[87]
Tilapia fish skin	F	Mango peel extracts	AO	[88]
Rousselot 200FG	F	Tyrosol, ferulic and caffeic acids + chitosan	AO	[89]
Hagfish skin	F	Cinnamon-bark essential oil	AO + AM	[90]
Tilapia skin	F	Epigallocatechin gallate	AO	[91]
Silver carp skin	F	Curcumin/β-cyclodextrin	AO	[92]
Commercial tilapia (*Oreochromis niloticus*)	C	Tea polyphenol	AO + AM	[93]
Sturgeon skin	F	Esculine	AO	[94]
Atlantic salmon (*Salmo salar*) skin	F	Boldine	AO + AM	[95]
Rousselot 200FG	F	Coumarin + chitosan	AO	[96]
Commercial tilapia	C	Chitosan	AM	[97]

C, coating; F, film; AO, antioxidant; AM, antimicrobial.

film forming solutions, obtaining a material with excellent free radical scavenging activity due to the high presence of polyphenols, carotenoids, phytochemicals, enzymes, vitamin C, and vitamin E on mango peels. It is worth mentioning that mango skins contribute about 7–24% from the whole fruit weight, and so, using this by-product will help in reducing waste. *Aloe vera* was also used as an antioxidant agent with fish gelatin, obtaining films that exhibited antioxidant properties dependent on *A. vera* concentration [98]. Thereby, control and 9% *A. vera* films showed 65.78% and 74.76% 1,1-diphenyl-2-picrylhydrazyl (DPPH) radical as well as 32.59% and 65.24% 2,2′-azino-bis-3-ethylbenzthiazoline-6-sulphonic acid (ABTS) radical scavenging activity values, respectively. Similarly, epigallocatechin gallate, one of the major flavanols obtained from tea extract, was added into tilapia skin gelatin film forming solutions in order to prepare active films [91]. Apart from providing antioxidant activity, epigallocatechin gallate enhanced the UV light barrier properties of films, contributing significantly to the food shelf life extension. Liang et al. [94] employed a commonly used traditional Chinese medicine for the extraction of esculine, a natural antioxidant agent, which was then incorporated into sturgeon skin gelatin film forming solutions to prepare films intended to be used as food packaging materials for long-term preservation. Moreover, films containing this antioxidant formed non-covalent cross-linkages between the hydroxyl group of esculine and amino acid residues of gelatin that enhanced chemical, physical, and mechanical properties.

Some antioxidant additives show limitations, such as the poor solubility of curcumin, the major bioactive compound of curcuma rhizome (*Curcuma zedoaria*), which has been regarded as a spice, food preservative, and flavoring and coloring agent, approved for use in food industry. Wu et al. [92] and Sun et al. [87] encapsulated curcumin in β-cyclodextrin to overcome this problem, and combined the complex with silver carp skin to prepare active films and coatings, respectively. Furthermore, it was observed that curcumin could inhibit microbial propagation. This combination of antioxidant and antimicrobial activities can be observed in several additives used with fish gelatin, for instance, in cinnamon-bark essential oil [90] and in tea polyphenol [93]. On the other hand, some fish gelatin species have inherent antimicrobial or antioxidant activity, as in the case of Atlantic salmon [95] or cuttlefish [99], respectively.

Besides incorporation of cross-linkers and antimicrobial and antioxidant agents into fish gelatin films and coatings, blending with other polymers such as polysaccharides could be another approach to extend food shelf life. For instance, fish gelatin and chitosan, a polymer that has intrinsic antimicrobial properties, have been shown to be compatible. When chitosan is positively charged and gelatin is negatively charged, under proper pH conditions, both biopolymers can be associated through electrostatic and hydrogen bonding. Thus, tilapia fish gelatin--chitosan coatings were analyzed and it was found that these coatings significantly prevented deterioration of golden pomfret fillet at 4 °C, inhibiting the degradation of myosin light chain and myoglobin [97], as well as coumarin [96]. Furthermore, the addition of natural antioxidants (ferulic acid, caffeic acid) showed that films containing caffeic acid or a caffeic–ferulic acid mixture exhibited a high radical scavenging activity [89].

2.3 Chitosan Films and Coatings

2.3.1 Chitin and Chitosan Extraction

Marine wastes, such as crustaceans' shells and squid pens, can be employed for chitin and chitosan extraction [100]. For this purpose, chemical and/or biological treatments have been employed [101]. Generally, chitin is extracted by a two-step process, including deproteinization and demineralization [102]. In the case of shell wastes, another step, depigmentation treatment, is required in order to remove color [101]. After extracting chitin, chitosan is obtained by a deacetylation process. Depending on the aggressiveness of the treatment, the deacetylation degree of chitosan may differ, affecting the final properties of the material [103].

In terms of chemical treatments, the deproteinization step is carried out using an alkali solution (usually NaOH), the demineralization process is performed with an acidic solution (typically HCl), and the depigmentation process is accomplished by soaking the chitin in an alkali solution. Subsequently, an alkali solution (commonly NaOH) is used for the chitosan extraction [104, 105]. In the case of biological treatments, instead of chemical substances, bacteria strains are employed for deproteinization and demineralization. For instance, *Bacillus subtilis*, *Pseudomonas aeruginosa*, and *Serratia marcescens*, bacterial strains that produce protease, have been tested for the deproteinization treatment and *Lactobacillus plantarum* has been analyzed for the demineralization process to remove calcium from chitin [106]. Moreover, for the chitin transformation into chitosan, *Rhizopus japonicus*, a bacteria strain that processes a high intracellular and extracellular deacetylation mechanism, has been used as a biological alternative [107]. It is worth mentioning that no depigmentation process using bacterial strains has been found.

Although chemical treatments are cheap procedures, appropriate for mass production [108], high concentrations and volumes of acid and alkali solutions are used at high temperatures during chemical treatments and, thus, biological treatments could be more suitable processes from the environmental perspective. Nevertheless, they have longer production times, are more expensive, and have not been implemented for mass production; only pilot scale processes have been reported [107].

A combination of both chemical and biological methods can also be employed. For instance, Pachapur et al. [109] proposed a combined enzymatic deproteinization (with *Bacillus licheniformis*) followed by a chemical demineralization process, using seawater during all steps of chitin extraction. Thereby, using large amounts of freshwater and chemicals can be avoided. Furthermore, Lopes et al. [110] compared the chitosan extraction by chemical and biological processes at pilot scale and their environmental impact. The use of a soft alkaline treatment, and the possibility of recovery NaOH, water, and fish protein hydrolysates are some of the reported improvements to decrease the environmental impact associated with the chemical treatments. Additionally, some improvements were proposed for the biological processes in order to mitigate the environmental impacts and the costs associated with the enzyme production; in particular, employing crustacean biomass as peptones for bacterial growth could be a strategy to reduce environmental loads.

Recently, a novel method for chitosan extraction has been proposed [111], using microwave technique for all chitosan extraction steps (demineralization, deproteinization and deacetylation). Although the method uses the same preparation conditions as chemical treatments, it needs less time than the conventional one. Indeed, this novel method reduced the deacetylation time from more than 6 hours to less than 30 minutes, reaching the same deacetylation degree. Furthermore, chitosan structure, morphology, and chemical composition in both methods were similar.

2.3.2 Preparation and Characterization of Chitosan Films and Coatings

Chitosan-based materials can be prepared employing two different processes known as wet or dry techniques. The wet method is the most most employed process when using this polysaccharide because it is a versatile technique [112]. During this process, chitosan is dissolved in acidified water since it is insoluble in basic media [113, 114]. After heating and stirring the solutions, films can be obtained by solution casting and drying, or food coatings can be prepared by spraying or dipping [115, 116]. On the other hand, few works related to dry process to prepare chitosan films have been reported in the literature. Galvis-Sánchez et al. [117] prepared chitosan films by thermal compression. First, chitosan powder was mixed with natural deep eutectic solvents, such as citric acid or lactic acid. The mixture was introduced in an oven at 80 °C for 30 minutes and then an acetic acid solution was added during manual mixing. The resultant mixture was introduced in a hydraulic press and thermo-compressed. In another work, Guerrero et al. [118] thermo-compressed citric acid–chitosan films at 125 °C for two minutes, using glycerol as plasticizer. Chitosan has also been compressed after blending with other polysaccharides such as starch. Valencia-Sullca et al. [119] dispersed chitosan, starch, glycerol, and polyethylene glycol in water and the mixture was melt blended at 160 °C for 30 minutes until a homogeneous paste was obtained before processing by thermal compression.

After preparing chitosan films and coatings, material characterization must be carried out in order to know their suitability for the specific application of food packaging. Mechanical properties are one of the most important parameters of the characterization. Mechanical behavior determines whether the film or coating is suitable for the packaging purpose since it has to support mechanical loads during the logistic process to keep the product intact and avoid food deterioration [120]. Unmodified chitosan films are brittle due to the electrostatic interactions and hydrogen bonding between chitosan chains. Hence, plasticizers are used to lend chitosan better mechanical properties, to provide chitosan chains with mobility, and, thus, increase flexibility. Among plasticizers, glycerol is the most employed in chitosan films [102].

In relation to water-related properties, these are of interest to define which kind of product can be packaged. Since chitosan films and coatings present moderate values of water resistance due to the presence of hydrophilic groups (–OH and $-NH_2$) [113], chitosan properties are modified employing different additives. Wang et al. [121] studied chitosan films enriched with anthocyanins at different concentrations (5%, 10% and 15% by weight based on the chitosan dry basis). Regarding moisture content, this value decreased with the increase of anthocyanin concentration due to the intermolecular interactions by hydrogen bonding between anthocyanins and the hydrophilic groups of chitosan. Concerning water vapor permeability, anthocyanin-added chitosan films exhibited lower values due to the more compact structure formed as a consequence of the interactions between anthocyanin and chitosan.

2.3.3 Food Shelf Life Extension Using Chitosan Films and Coatings

Chitosan films and coatings have been extensively assessed for non-processed food, such as fruits, vegetables, refrigerated fish, and meat as well as for processed food such as sausages or bread. Depending on the product, different coating techniques are used: dipping, spraying, or wrapping. Leceta et al. [122] compared spraying and dipping on ready-to-eat baby carrots and found that both chitosan coatings were effective in maintaining the product safe against microbiological spoilage during a storage period of 15 days. A slightly better antimicrobial activity was shown for dipped samples, whereas moderately better results for weight loss and texture were presented in sprayed samples. Pea pods have also been coated by chitosan solutions, reducing the vegetable weight loss, titratable acidity, and chilling injury compared to the untreated product [123].

In order to provide the packaging with active properties that prolong the shelf life of the packed product, additives are incorporated into chitosan solutions. Some of these studies are summarized in Table 2.2. Lekjing [130] employed clove

Table 2.2 Chitosan-based active films and coatings.

Food packaged	Packaging	Bioactives	Bioactivity	References
Cherry tomatoes and grapes	F	Tannic acid	AO + AM	[124]
White shrimp (*Litopenaeus vannamei*)	C	ε-Polylysine	AM	[125]
Bread slices	F	Apricot kernel essential oil	AO + AM	[113]
Chicken	F	*Schinus terebinthifolius* Raddi (pink pepper) extract/peanut skin extract	AO	[126]
Nile tilapia fillets	C	Peels of pomegranate extracts	AM	[127]
Grass carp (*Ctenopharyngodon idellus*)	C	Glycerol monolaurate	AM	[128]
Yellow croakers	C	Nisin	AM	[129]
Pork sausages	C	Clove oil	AO + AM	[130]
Tomatoes	C	Iodide	AM	[116]
Pork fillets	C	*Origanum vulgare* essential oil	AM	[131]
Pacific mackerel (*Pneumatophorus japonicus*) fillets	C	Gallic acid	AO + AM	[132]
Chicken breast	C	*Zataria multiflora* essential oil/pomegranate juice	AO + AM	[133]

C, coating; F, film; AO, antioxidant; AM, antimicrobial.

oil, an essential oil with antimicrobial and antioxidant properties, to coat pork sausages by dipping and microbiological, physical, and chemical analyses as well as sensory evaluation were carried out. It was demonstrated that the coating inhibited the microbial growth, retarded the lipid oxidation, and extended the shelf life of the product during refrigerated storage. However, negative impacts were found on odor and taste qualities. Other essential oils, such as that extracted from oregano, have also been incorporated into chitosan solutions to provide coatings with active properties. Paparella et al. [131] added oregano essential oil into chitosan solutions to coat fresh pork fillets and found that oregano had antimicrobial properties against meat spoilage bacteria, such as *Brochothrix thermosphacta* and *Pseudomonas* spp. In another study, Alsaggaf et al. [127] analyzed the effect of edible chitosan coatings applied by immersion of Nile tilapia fillets into chitosan solutions with pomegranate peel extract, an extract that has several bioactive phytochemicals for antimicrobial or food preservative applications. Composition, microbiological, antioxidant, total nitrogen volatile base value, and sensory analyses demonstrated that chitosan/pomegranate peel extract coatings were effective at extending the shelf life of fish fillets.

In addition to coatings, chitosan films have been used to wrap food products with the aim of prolonging food shelf life. Serrano-León et al. [126] wrapped chicken products using chitosan films incorporated with peanut peel extract or pink pepper by-products. Peanut peel extract contains proanthocyanidins and procyanidins, phenolic compounds with antioxidant activity, while pink pepper by-products are composed of vitamin C, phenolic compounds, flavonoids, and carotenoids. The characterization of the chicken product was carried out analyzing chicken pH and color, peroxide value, thiobarbituric acid reacting substances (TBARS), and microbiological parameters. None of the additives altered chicken pH or color, but pink pepper by-products showed greater antimicrobial and antioxidant activities than peanut peel extract.

2.4 Future Perspectives and Concluding Remarks

The amount of biowaste generated by food processing industries is huge. This biowaste can be valorized to obtain raw materials for the production of food packaging, reducing the use of nonrenewable and nonbiodegradable materials in line with the sustainability principles. In particular, proteins such as collagen and gelatin can be obtained from fishery waste and by-products, but also active compounds can be extracted from other food processing wastes or by-products in order to develop novel active packaging to extend food shelf life and reduce food losses. A review of the most employed methods to extract biopolymers from fishery waste, comparing benefits and drawbacks of the use of chemical and enzymatic processes, has been provided. In relation to film and coating production, wet (solution casting, spraying, dipping) and dry processes (compression molding, extrusion) have been mentioned in this chapter, highlighting the potential of dry processes for scaling-up production. Also, the characterization of the most relevant properties of the films and coatings developed using fish gelatin and chitosan has been reported in order to give information on an appropriate selection of the most suitable material for specific food products. In particular, properties related to food quality and food life-extension have been highlighted by means of the use of natural antioxidants and antimicrobials. Although further research is still needed, mainly to scale production at competitive costs, the valorization of food processing biowaste to produce active packaging could become a suitable approach toward more sustainable food packaging.

References

1 Szczepańska, N., Kudłak, B., and Namieśnik, J. (2018). Recent advances in assessing xenobiotics migrating from packaging material: a review. *Analytica Chimica Acta* 1023: 1–21.

2 Krishna, A., Cian, L., and Aydınoğlu, N.Z. (2017). Sensory aspects of package design. *Journal of Retailing* 93 (1): 43–54.

3 Kim, D. and Seo, J. (2018). A review: breathable films for packaging applications. *Trends in Food Science & Technology* 76: 15–27.

4 Hahladakis, J.N., Velis, C.A., Weber, R. et al. (2018). An overview of chemical additives present in plastics: migration, release, fate and environmental impact during their use, disposal and recycling. *Journal of Hazardous Materials* 344: 179–199.
5 Geueke, B., Groh, K., and Muncke, J. (2018). Food packaging in the circular economy: overview of chemical safety aspects for commonly used materials. *Journal of Cleaner Production* 193: 491–505.
6 Dilkes-Hoffman, L.S., Lane, J.L., Grant, T. et al. (2018). Environmental impact of biodegradable food packaging when considering food waste. *Journal of Cleaner Production* 180: 325–334.
7 Romani, V.P., Machado, A.V., Olsen, B.D., and Martins, V.G. (2018). Effects of pH modification in proteins from fish (*Whitemouth croaker*) and their application in food packaging films. *Food Hydrocolloids* 74: 307–314.
8 Kurek, M., Ščetar, M., and Galić, K. (2017). Edible coatings minimize fat uptake in deep fat fried products: a review. *Food Hydrocolloids* 71: 225–235.
9 Otoni, C.G., Avena-Bustillos, R.J., Olsen, C.W. et al. (2016). Mechanical and water barrier properties of isolated soy protein composite edible films as affected by carvacrol and cinnamaldehyde micro and nanoemulsions. *Food Hydrocolloids* 57: 72–79.
10 Tavassoli-Kafrani, E., Shekarchizadeh, H., and Masoudpour-Behabadi, M. (2016). Development of edible films and coatings from alginates and carrageenans. *Carbohydrate Polymers* 137: 360–374.
11 Bertuzzi, M.A. and Slavutsky, A.M. (2016). Standard and new processing techniques used in the preparation of films and coatings at the lab level and scale-up. In: *Edible Films and Coatings* (eds. M.P.M. García, M.C. Gómez-Guillén, M.E. López-Caballero and G.V. Barbosa-Cánovas), 21–42. Boca Raton, Florida: CRC Press.
12 de la Caba, K., Guerrero, P., Trung, T.S. et al. (2019). From seafood waste to active seafood packaging: an emerging opportunity of the circular economy. *Journal of Cleaner Production* 208: 86–98.
13 Suresh, P.V., Kudre, T.G., and Johny, L.C. (2018). Sustainable valorization of seafood processing by-product/discard. In: *Waste to Wealth* (eds. R.R. Singhania, R.A. Agarwal, R.P. Kumar and R.K. Sukumaran), 111–139. Singapore: Springer.
14 FAO (2018). *The State of World Fisheries and Aquaculture*. Food and Agriculture Organization of the United States.
15 Kafle, G.K., Kim, S.H., and Sung, K.I. (2013). Ensiling of fish industry waste for biogas production: a lab scale evaluation of biochemical methane potential (BMP) and kinetics. *Bioresource Technology* 127: 326–336.
16 Ferraro, V., Gaillard-Martinie, B., Sayd, T. et al. (2017). Collagen type I from bovine bone. Effect of animal age, bone anatomy and drying methodology on extraction yield, self-assembly, thermal behaviour and electrokinetic potential. *International Journal of Biological Macromolecules* 97: 55–66.
17 Venkatesan, J., Anil, S., Kim, S.K., and Shim, M.S. (2017). Marine fish proteins and peptides for cosmeceuticals: a review. *Marine Drugs* 15 (5): 143.

18 Huang, T., Tu, Z.C., Wang, H. et al. (2017). Pectin and enzyme complex modified fish scales gelatin: rheological behavior, gel properties and nanostructure. *Carbohydrate Polymers* 156: 294–302.
19 Lin, L., Regenstein, J.M., Lv, S. et al. (2017). An overview of gelatin derived from aquatic animals: properties and modification. *Trends in Food Science & Technology* 68: 102–112.
20 Sinthusamran, S., Benjakul, S., and Kishimura, H. (2014). Characteristics and gel properties of gelatin from skin of seabass (*Lates calcarifer*) as influenced by extraction conditions. *Food Chemistry* 152: 276–284.
21 Arya, S.K., Manohar, M., Singh, G., and Siddiqui, W.A. (2017). Chitin and chitosan-complexes and their applications. In: *Chitosan: Derivatives, Composites and Applications* (eds. S. Ahmed and S. Ikram), 151–165. Hoboken, New Jersey: Wiley.
22 Hamed, I., Özogul, F., and Regenstein, J.M. (2016). Industrial applications of crustacean by-products (chitin, chitosan, and chitooligosaccharides): a review. *Trends in Food Science & Technology* 48: 40–50.
23 Sayari, N., Sila, A., Abdelmalek, B.E. et al. (2016). Chitin and chitosan from the Norway lobster by-products: antimicrobial and anti-proliferative activities. *International Journal of Biological Macromolecules* 87: 163–171.
24 Soares, N.M., Fernandes, T.A., and Vicente, A.A. (2016). Effect of variables on the thickness of an edible coating applied on frozen fish-establishment of the concept of safe dipping time. *Journal of Food Engineering* 171: 111–118.
25 Rocha, M.A.M., Coimbra, M.A., and Nunes, C. (2017). Applications of chitosan and their derivatives in beverages: a critical review. *Current Opinion in Food Science* 15: 61–69.
26 Chen, J., Li, L., Yi, R. et al. (2016a). Extraction and characterization of acid-soluble collagen from scales and skin of tilapia (*Oreachromis niloticus*). *LWT Food Science and Technology* 66: 453–459.
27 Ahmed, R., Getachew, A.T., Cho, Y., and Chun, B. (2018). Application of bacterial collagenolytic proteases for the extraction of type I collagen from the skin of bigeye tuna (*Thunnus obesus*). *LWT Food Science and Technology* 89: 44–51.
28 Kittiphattanabawon, P., Benjakul, S., Sinthusamran, S., and Kishimura, H. (2016). Gelatin from clown featherback skin: extraction conditions. *LWT Food Science and Technology* 66: 186–192.
29 See, S.F., Ghassem, M., Mamot, S., and Babji, A.S. (2015). Effect of different pretreatments on functional properties of African catfish (*Clarias gariepinus)* skin gelatin. *Journal of Food Science and Technology* 52 (2): 753–762.
30 Chen, S., Tang, L., Hao, G. et al. (2016c). Effects of $\alpha1/\alpha2$ ratios and drying temperatures on the properties of gelatin films prepared from tilapia (*Tilapia zillii*) skins. *Food Hydrocolloids* 52: 573–580.
31 Tkaczewska, J., Morawska, M., Kulawik, P., and Zając, M. (2018). Characterization of carp (*Cyprinus carpio*) skin gelatin extracted using different pretreatments method. *Food Hydrocolloids* 81: 169–179.

32 Sae-leaw, T., Benjakul, S., and O'Brien, N.M. (2016). Effects of defatting and tannic acid incorporation during extraction on properties and fishy odour of gelatin from seabass skin. *LWT Food Science and Technology* 65: 661–667.

33 Ali, A.M.M., Kishimura, H., and Benjakul, S. (2018). Physicochemical and molecular properties of gelatin from skin of Golden carp (*Probarbus Jullieni*) as influenced by acid pretreatment and prior-ultrasonication. *Food Hydrocolloids* 82: 164–172.

34 Gomez, A.V.A., Gomez, G., Chamorro, E. et al. (2018). Digestive aspartic protease from sábalo (*Prochilodus lineatus*): characterization and application for collagen extraction. *Food Chemistry* 269: 610–617.

35 Chuaychan, S., Benjakul, S., and Nuthong, P. (2016). Element distribution and morphology of spotted golden goatfish fish scales as affected by demineralisation. *Food Chemistry* 197: 814–820.

36 Wang, L., Liang, Q., Chen, T. et al. (2014). Characterization of collagen from the skin of Amur sturgeon (*Acipenser schrenckii*). *Food Hydrocolloids* 38: 104–109.

37 Liao, W., Guanghua, X., Li, Y. et al. (2018). Comparison of characteristics and fibril-forming ability of skin collagen from barramundi (*Lates calcarifer*) and tilapia (*Oreochromis niloticus*). *International Journal of Biological Macromolecules* 107: 549–559.

38 Liu, D., Wei, G., Li, T. et al. (2015). Effects of alkaline pretreatments and acid extraction conditions on the acid-soluble collagen from grass carp (*Ctenopharyngodon idella*) skin. *Food Chemistry* 172: 836–843.

39 Liang, Q., Wang, L., Sun, W. et al. (2014a). Isolation and characterization of collagen from the cartilage of Amur sturgeon (*Acipenser schrenckii*). *Process Biochemistry* 49 (2): 318–323.

40 Chuaychan, S., Benjakul, S., and Kishimura, H. (2015). Characteristics of acid- and pepsin-soluble collagens from scale of seabass (*Lates calcarifer*). *LWT Food Science and Technology* 63 (1): 71–76.

41 Tan, Y. and Chang, S.K.C. (2018). Isolation and characterization of collagen extracted from cannel catfish (*Ictalurus punctatus*) skin. *Food Chemistry* 242: 147–155.

42 Hanjabam, M.D., Kannaiyan, S.K., Kamei, G. et al. (2015). Optimisation of gelatin extraction from unicorn leatherjacket (*Aluterus Monoceros*) skin waste: response surface approach. *Journal of Food Science and Technology* 52 (2): 976–983.

43 Pal, G.K., Nidheesh, T., and Suresh, P.V. (2015). Comparative study on characteristics and *in vitro* fibril formation ability of acid and pepsin soluble collagen from the skin of catla (*Catla catla*) and rohu (*Labeo rohita*). *Food Research International* 76: 804–812.

44 Thuy L.T.M., Okazaki, E., and Osako, K. (2014). Isolation and characterization of acid-soluble collagen from the scales of marine fishes from Japan and Vietnam. *Food Chemistry* 149: 264–270.

45 Pal, G.K. and Suresh, P.V. (2016). Sustainable valorisation of seafood by-products: recovery of collagen and development of collagen-based novel functional food ingredients. *Innovative Food Science & Emerging Technologies* 37: 201–215.

46 Ahmad, T., Ismail, A., Ahmad, S.A. et al. (2017). Recent advances on the role of process variables affecting gelatin yield and characteristics with special reference to enzymatic extraction: a review. *Food Hydrocolloids* 63: 85–96.

47 Jridi, M., Nasri, R., Salem, R.B.S.B. et al. (2015). Chemical and biophysical properties of gelatins extracted from the skin of octopus (*Octopus vulgaris*). *LWT Food Science and Technology* 60 (2): 881–889.

48 Sila, A., Martinez-Alvarez, O., Haddar, A. et al. (2015). Recovery, viscoelastic and functional properties of Barbel skin gelatine: investigation of anti-DPP-IV and anti-prolyl endopeptidase activities of generated gelatine polypeptides. *Food Chemistry* 168: 478–486.

49 Abdollahi, M., Rezaei, M., Jafarpour, A., and Undeland, I. (2018). Sequential extraction of gel-forming proteins, collagen and collagen hydrolysate from gutted silver carp (*Hypophthalmichthys molitrix*), a biorefinery approach. *Food Chemistry* 242: 568–578.

50 Jeevithan, E., Wu, W., Nanping, W. et al. (2014). Isolation, purification and characterization of pepsin soluble collagen isolated from silvertip shark (*Carcharhinus albimarginatus*) skeletal and head bone. *Process Biochemistry* 49 (10): 1767–1777.

51 Chen, S., Chen, H., Xie, Q. et al. (2016b). Rapid isolation of high purity pepsin-soluble type I collagen from scales of red drum fish (*Scianops ocellatus*). *Food Hydrocolloids* 52: 468–477.

52 Liang, Q., Wang, L., He, Y. et al. (2014b). Hydrolysis kinetics and antioxidant activity of collagen under simulated gastrointestinal digestion. *Journal of Functional Foods* 11: 493–499.

53 Zou, Y., Wang, L., Cai, P. et al. (2017). Effect of ultrasound assisted extraction on the physicochemical and functional properties of collagen from soft-shelled turtle calipash. *International Journal of Biological Macromolecules* 105: 1602–1610.

54 Huang, C.Y., Kuo, J.M., Wu, S.J., and Tsai, H.T. (2016). Isolation and characterization of fish scale collagen from tilapia (*Oreochromis sp.*) by a novel extrusion-hydroextraction process. *Food Chemistry* 190: 997–1006.

55 Hattrem, M.N., Molnes, S., Haug, I.J., and Draget, K.I. (2015). Interfacial and rheological properties of gelatin based solid emulsions prepared with acid or alkali pretreated gelatins. *Food Hydrocolloids* 43: 700–707.

56 Liu, Y., Xia, L., Jia, H. et al. (2017). Physiochemical and functional properties of chum salmon (*Oncorhynchus keta*) skin gelatin extracted at different temperatures. *Journal of the Science of Food and Agriculture* 97 (15): 5406–5413.

57 Garrido, T., Uranga, J., Guerrero, P., and de la Caba, K. (2018). The potential of vegetal and animal proteins to develop more sustainable food packaging. In: *Polymers for Food Applications* (ed. T. Gutiérrez), 25–59. Singapore: Springer.

58 Abdelhedi, O., Jridi, M., Nasri, R. et al. (2019). Rheological and structural properties of *Hemiramphus far* skin gelatin: potential use as an active fish coating agent. *Food Hydrocolloids* 87: 331–341.

59 Abdelmalek, B.E., Gómez-Estaca, J., Sila, A. et al. (2016). Characteristics and functional properties of gelatin extracted from squid (*Loligo vulgaris*) skin. *LWT Food Science and Technology* 65: 924–931.
60 Etxabide, A., Garrido, T., Uranga, J. et al. (2018). Extraction and incorporation of bioactives into protein formulations for food and biomedical applications. *International Journal of Biological Macromolecules* 120: 2094–2105.
61 Ananey-Obiri, D., Matthews, L., Azahrani, M.H. et al. (2018). Application of protein-based edible coatings for fat uptake reduction in deep-fat fried foods with an emphasis on muscle food proteins. *Trends in Food Science & Technology* 80: 167–174.
62 Khalil, H.P.S.A., Banerjee, A., Saurabh, C.K. et al. (2018). Biodegradable films for fruits and vegetables packaging application: preparation and properties. *Food Engineering Reviews* 10 (3): 139–153.
63 Uranga, J., Etxabide, A., Guerrero, P., and de la Caba, K. (2018). Development of active fish gelatin films with anthocyanins by compression molding. *Food Hydrocolloids* 84: 313–320.
64 Etxabide, A., de la Caba, K., and Guerrero, P. (2016a). A novel approach to manufacture porous biocomposites using extrusion and injection moulding. *European Polymer Journal* 82: 324–333.
65 Hosseini, S.F. and Gómez-Guillén, M.C. (2018). A state-of-the-art review on the elaboration of fish gelatin as bioactive packaging: special emphasis on nanotechnology-based approaches. *Trends in Food Science & Technology* 79: 125–135.
66 Armentano, I., Fortunati, E., Burgos, N. et al. (2015). Bio-based PLA_PHB plasticized blend films: processing and structural characterization. *LWT Food Science and Technology* 64 (2): 980–988.
67 Mkandawire, M. and Aryee, A.N.A. (2018). Resurfacing and modernization of edible packaging material technology. *Current Opinion in Food Science* 19: 104–112.
68 Bermúdez-Oria, A., Rodríguez-Gutiérrez, G., Vioque, B. et al. (2017). Physical and functional properties of pectin-fish gelatin films containing the olive phenols hydroxytyrosol and 3,4-dihydroxyphenylglycol. *Carbohydrate Polymers* 178: 368–377.
69 Drakos, A., Pelava, E., and Evageliou, V. (2018). Properties of flour films as affected by the flour's source and particle size. *Food Research International* 107: 551–558.
70 Tongnuanchan, P., Benjakul, S., Prodpran, T. et al. (2016). Mechanical, thermal and heat sealing properties of fish skin gelatin film containing palm oil and basil essential oil with different surfactants. *Food Hydrocolloids* 56: 93–107.
71 Amalini, A.N., Norziah, M.H., Khan, I., and Haafiz, M.K.M. (2018). Exploring the properties of modified fish gelatin films incorporated with different fatty acid sucrose esters. *Food Packaging and Shelf Life* 15: 105–112.
72 Kchaou, H., Benbettaïeb, N., Jridi, M. et al. (2018). Enhancement of structural, functional and antioxidant properties of fish gelatin films using Maillard reactions. *Food Hydrocolloids* 83: 326–339.

73 Ahmad, M., Hani, N.M., Nirmal, N.P. et al. (2015). Optical and thermo-mechanical properties of composite films based on fish gelatin/rice flour fabricated by casting technique. *Progress in Organic Coatings* 84: 115–127.
74 Hanani, Z.A.N., Roos, Y.H., and Kerry, J.P. (2014). Use and application of gelatin as potential biodegradable packaging materials for food products. *International Journal of Biological Macromolecules* 71: 94–102.
75 Mohajer, S., Rezaei, M., and Hosseini, S.F. (2017). Physico-chemical and microstructural properties of fish gelatin/agar bio-based blend films. *Carbohydrate Polymers* 157: 784–793.
76 Nilsuwan, K., Benjakul, S., and Prodpran, T. (2018a). Physical/thermal properties and heat seal ability of bilayer films based on fish gelatin and poly(lactic acid). *Food Hydrocolloids* 77: 248–256.
77 Arfat, Y.A., Benjakul, S., Prodpran, T. et al. (2016). Physico-mechanical characterization and antimicrobial properties of fish protein isolate/fish skin gelatin-zinc oxide (ZnO) nanocomposite films. *Food and Bioprocess Technology* 9 (1): 101–112.
78 Riquelme, N., Díaz-Calderón, P., Enrione, J., and Matiacevich, S. (2015). Effect of physical state of gelatin-plasticizer based films on to the occurrence of Maillard reactions. *Food Chemistry* 175: 478–484.
79 Acosta, S., Jiménez, A., Cháfer, M. et al. (2015). Physical properties and stability of starch-gelatin based films as affected by the addition of esters of fatty acids. *Food Hydrocolloids* 49: 135–143.
80 Uranga, J., Leceta, I., Etxabide, A. et al. (2016). Cross-linking of fish gelatins to develop sustainable films with enhanced properties. *European Polymer Journal* 78: 82–90.
81 Guerreiro, T.M., de Oliveira, D.N., Melo, C.F.O.R. et al. (2018). Migration from plastic packaging into meat. *Food Research International* 109: 320–324.
82 Dehghani, S., Hosseini, S.V., and Regenstein, J.M. (2018). Edible films and coatings in seafood preservation: a review. *Food Chemistry* 240: 505–513.
83 Gómez-Estaca, J., López-Caballero, M.E., Martínez-Bartolomé, M.Á. et al. (2018). The effect of the combined use of high pressure treatment and antimicrobial edible film on the quality of salmon carpaccio. *International Journal of Food Microbiology* 283: 28–36.
84 Azeredo, H.M.C. and Waldron, K.W. (2016). Crosslinking in polysaccharide and protein films and coatings for food contact: a review. *Trends in Food Science & Technology* 52: 109–122.
85 Etxabide, A., Urdanpilleta, M., de la Caba, K., and Guerrero, P. (2016b). Control of cross-linking reaction to tailor the properties of thin films based on gelatin. *Materials Letters* 185: 366–369.
86 Taghizadeh, M., Mohammadifar, M.A., Sadeghi, E. et al. (2018). Photosensitizer-induced cross-linking: a novel approach for improvement of physicochemical and structural properties of gelatin edible films. *Food Research International* 112: 90–97.
87 Sun, X., Guo, X., Ji, M. et al. (2019). Preservative effects of fish gelatin coating enriched with CUR/βCD emulsion on grass carp (*Ctenopharyngodon idellus*) fillets during storage at 4 °C. *Food Chemistry* 272: 643–652.

88 Adilah, A.N., Jamilah, B., Noranizan, M.A., and Hanani, Z.A.N. (2018). Utilization of mango peel extracts on the biodegradable films for active packaging. *Food Packaging and Shelf Life* 16: 1–7.

89 Benbettaïeb, N., Tanner, C., Cayot, P. et al. (2018). Impact of functional properties and release kinetics on antioxidant activity of biopolymer active films and coatings. *Food Chemistry* 242: 369–377.

90 Kim, H., Beak, S.E., and Song, K.B. (2018). Development of a hagfish skin gelatin film containing cinnamon bark essential oil. *LWT Food Science and Technology* 96: 583–588.

91 Nilsuwan, K., Benjakul, S., and Prodpran, T. (2018b). Properties and antioxidative activity of fish gelatin-based film incorporated with epigallocatechin gallate. *Food Hydrocolloids* 80: 212–221.

92 Wu, J., Sun, X., Guo, X. et al. (2018). Physicochemical, antioxidant, *in vitro* release, and heat sealing properties of fish gelatin films incorporated with β-cyclodextrin/curcumin complexes for apple juice preservation. *Food and Bioprocess Technology* 11 (2): 447–461.

93 Feng, X., Ng, V.K., Mikš-Krajnik, M., and Yang, H. (2017). Effects of fish gelatin and tea polyphenol coating on the spoilage and degradation of myofibril in fish fillet during cold storage. *Food and Bioprocess Technology* 10 (1): 89–102.

94 Liang, C., Jia, M., Tian, D. et al. (2017). Edible sturgeon skin gelatine films: tensile strength and UV light-barrier as enhanced by blending with esculine. *Journal of Functional Foods* 37: 219–228.

95 López, D., Márquez, A., Gutiérrez-Cutiño, M. et al. (2017). Edible film with antioxidant capacity based on salmon gelatin and boldine. *LWT Food Science and Technology* 77: 160–169.

96 Benbettaïeb, N., Chambin, O., Assifaoui, A. et al. (2016). Release of coumarin incorporated into chitosan-gelatin irradiated films. *Food Hydrocolloids* 56: 266–276.

97 Feng, X., Bansal, N., and Yang, H. (2016). Fish gelatin combined with chitosan coating inhibits myofibril degradation of golden pomfret (*Trachinotus blochii*) fillet during cold storage. *Food Chemistry* 200: 283–292.

98 Chin, S.S., Lyn, F.H., and Hanani, Z.A.N. (2017). Effect of *Aloe vera* (*Aloe barbadensis* Miller) gel on the physical and functional properties of fish gelatin films as active packaging. *Food Packaging and Shelf Life* 12: 128–134.

99 Kchaou, H., Jridi, M., Abdelhedi, O. et al. (2017). Development and characterization of cuttlefish (*Sepia officinalis*) skin gelatin-protein isolate blend films. *International Journal of Biological Macromolecules* 105: 1491–1500.

100 Garrido, T., Etxabide, A., de la Caba, K., and Guerrero, P. (2017). Versatile soy protein films and hydrogels by the incorporation of β-chitin from squid pens (*Loligo* sp.). *Green Chemistry* 19 (24): 5923–5931.

101 Hou, Y., Shavandi, A., Carne, A. et al. (2016). Marine shells: potential opportunities for extraction of functional and health-promoting materials. *Environmental Science and Technology* 46 (11–12): 1047–1116.

102 Muxika, A., Etxabide, A., Uranga, J. et al. (2017). Chitosan as a bioactive polymer: processing, properties and applications. *International Journal of Biological Macromolecules* 105: 1358–1368.

103 Castillo, L.A., Farenzena, S., Pintos, E. et al. (2017). Active films based on thermoplastic corn starch and chitosan oligomer for food packaging applications. *Food Packaging and Shelf Life* 14: 128–136.

104 Baron, R.D., Pérez, L.L., Salcedo, J.M. et al. (2017). Production and characterization of films based on blends of chitosan from blue crab (*Callinectes sapidus*) waste and pectin from orange (*Citrus sinensis Osbeck*) peel. *International Journal of Biological Macromolecules* 98: 676–683.

105 Huang, C.Y., Kuo, C.H., Wu, C.H. et al. (2018). Extraction of crude chitosans from squid (*Illex argentinus*) pen by a compressional puffing-pretreatment process and evaluation of their antibacterial activity. *Food Chemistry* 254: 217–223.

106 Sedaghat, F., Yousefzadi, M., Toiserkani, H., and Najafipour, S. (2017). Bioconversion of shrimp waste *Penaeus merguiensis* using lactic acid fermentation: an alternative procedure for chemical extraction of chitin and chitosan. *International Journal of Biological Macromolecules* 104: 883–888.

107 Zhang, H., Yun, S., Song, L. et al. (2017). The preparation and characterization of chitin and chitosan under large-scale submerged fermentation level using shrimp by-products as substrate. *International Journal of Biological Macromolecules* 96: 334–339.

108 Shavandi, A., Hu, Z., Teh, S.S. et al. (2017). Antioxidant and functional properties of protein hydrolysates obtained from squid pen chitosan extraction effluent. *Food Chemistry* 227: 194–201.

109 Pachapur, V.L., Guemiza, K., Rouissi, T. et al. (2015). Novel biological and chemical methods of chitin extraction from crustacean waste using saline water. *Journal of Chemical Technology and Biotechnology* 91 (8): 2331–2339.

110 Lopes, C., Antelo, L.T., Franco-Uría, A. et al. (2018). Chitin production from crustacean biomass: sustainability assessment of chemical and enzymatic processes. *Journal of Cleaner Production* 172: 4140–4151.

111 El Knidri, H., El Khalfaouy, R., Laajeb, A. et al. (2016). Eco-friendly extraction and characterization of chitin and chitosan from the shrimp shell waste via microwave irradiation. *Process Safety and Environmental Protection* 104: 395–405.

112 Chen, G., Ali, F., Dong, S. et al. (2018). Preparation, characterization and functional evaluation of chitosan-based films with zein coatings produced by cold plasma. *Carbohydrate Polymers* 202: 39–46.

113 Priyadarshi, R., Sauraj, R., Kumar, B. et al. (2018). Chitosan films incorporated with Apricot (*Prunus armeniaca*) kernel essential oil as active food packaging material. *Food Hydrocolloids* 85: 158–166.

114 Zhang, C., Wang, Z., Li, Y. et al. (2019a). The preparation and physiochemical characterization of rapeseed protein hydrolysate-chitosan composite films. *Food Chemistry* 272: 694–701.

115 Bonilla, F., Chouljenko, A., Reyes, V. et al. (2018). Impact of chitosan application technique on refrigerated catfish fillet quality. *LWT Food Science and Technology* 90: 277–282.

116 Limchoowong, N., Sricharoen, P., Techawongstien, S., and Chanthai, S. (2016). An iodine supplementation of tomato fruits coated with an edible film of the iodide-doped chitosan. *Food Chemistry* 200: 223–229.

117 Galvis-Sánchez, A.C., Castro, M.C.R., Biernacki, K. et al. (2018). Natural deep eutectic solvents as green plasticizers for chitosan thermoplastic production with controlled/desired mechanical and barrier properties. *Food Hydrocolloids* 82: 478–489.

118 Guerrero, P., Muxika, A., Zarandona, I., and de la Caba, K. (2019). Crosslinking of chitosan films processed by compression molding. *Carbohydrate Polymers* 206: 820–826.

119 Valencia-Sullca, C., Atarés, L., Vargas, M., and Chiralt, A. (2018). Physical and antimicrobial properties of compression-molded cassava starch-chitosan films for meat preservation. *Food and Bioprocess Technology* 11 (7): 1339–1349.

120 Zhang, J., Zou, X., Zhai, X. et al. (2019b). Preparation of an intelligent pH film based on biodegradable polymers and roselle anthocyanins for monitoring pork freshness. *Food Chemistry* 272: 306–312.

121 Wang, X., Yong, H., Gao, L. et al. (2019). Preparation and characterization of antioxidant and pH-sensitive films based on chitosan and black soybean seed coat extract. *Food Hydrocolloids* 89: 56–66.

122 Leceta, I., Molinaro, S., Guerrero, P. et al. (2015). Quality attributes of map packaged ready-to-eat baby carrots by using chitosan-based coatings. *Postharvest Biology and Technology* 100: 142–150.

123 El-hamahmy, M.A.M., ElSayed, A.I., and Odero, D.C. (2017). Physiological effects of hot water dipping, chitosan coating and gibberellic acid on shelf-life and quality assurance of sugar snap peas (*Pisum sativum* L. var. *macrocarpon*). *Food Packaging and Shelf Life* 11: 58–66.

124 Halim, A.L.A., Kamari, A., and Phillip, E. (2018). Chitosan, gelatin and methylcellulose films incorporated with tannic acid for food packaging. *International Journal of Biological Macromolecules* 120: 1119–1126.

125 Na, S., Kim, J.H., Jang, H.J. et al. (2018). Shelf life extension of Pacific white shrimp (*Litopenaeus vannamei*) using chitosan and ε-polylysine during cold storage. *International Journal of Biological Macromolecules* 115: 1103–1108.

126 Serrano-León, J.S., Bergamaschi, K.B., Yoshida, C.M.P. et al. (2018). Chitosan active films containing agro-industrial residue extracts for shelf life extension of chicken restructured product. *Food Research International* 108: 93–100.

127 Alsaggaf, M.S., Moussa, S.H., and Tayel, A.A. (2017). Application of fungal chitosan incorporated with pomegranate peel extract as edible coating for microbiological, chemical and sensorial quality enhancement of Nile tilapia fillets. *International Journal of Biological Macromolecules* 99: 499–505.

128 Yu, D., Jiang, Q., Xu, Y., and Xia, W. (2017). The shelf life extension of refrigerated grass carp (*Ctenopharyngodon idellus*) fillets by chitosan coating combined with glycerol monolaurate. *International Journal of Biological Macromolecules* 101: 448–454.

129 Hu, G., Liu, W., Feng, H. et al. (2016). Effects of chitosan combined with nisin treatment on storage quality of large yellow croaker (*Pseudosciaena crocea*). *Food Chemistry* 203: 276–282.

130 Lekjing, S. (2016). A chitosan-based coating with or without clove oil extends the shelf life of cooked pork sausages in refrigerated storage. *Meat Science* 111: 192–197.

131 Paparella, A., Mazzarrino, G., Chaves-López, C. et al. (2016). Chitosan boosts the antimicrobial activity of *Origanum vulgare* essential oil in modified atmosphere packaged pork. *Food Microbiology* 59: 23–31.

132 Wu, C., Li, Y., Wang, L. et al. (2016). Efficacy of chitosan-gallic acid coating on shelf life extension of refrigerated pacific mackerel fillets. *Food and Bioprocess Technology* 9 (4): 675–685.

133 Bazargani-Gilani, B., Aliakbarlu, J., and Tajik, H. (2015). Effect of pomegranate juice dipping and chitosan coating enriched with *Zataria multiflora* Boiss essential oil on the shelf-life of chicken meat during refrigerated storage. *Innovative Food Science and Emerging Technologies* 29: 280–287.

3

Edible Biopolymers for Food Preservation

Elisabetta Ruggeri[1,2], *Silvia Farè*[2,3], *Luigi De Nardo*[2,3], *and Benedetto Marelli*[1]

[1] *Massachusetts Institute of Technology, Department of Civil and Environmental Engineering, 77 Massachusetts Avenue, Cambridge, MA 02139, USA*
[2] *Politecnico di Milano, Department of Chemistry, Materials and Chemical Engineering "G. Natta", Piazza Leonardo da Vinci 32, Milan 20133, Italy*
[3] *Consorzio Interuniversitario Nazionale per la Scienza e Tecnologia dei Materiali (INSTM), Local Unit Politecnico di Milano, Via Giusti 9, Firenze 50121, Italy*

3.1 Introduction

Food waste represents a major loss of economic value and resources, and it is an obstacle to economic development. The Food and Agriculture Organization (FAO) of the United Nations estimated that one-third of food produced for human consumption is lost or wasted globally, which amounts to roughly 1.3 billion tons per year [1]. At the global scale, waste generation ranges between 194 and 389 kg per capita per year [2] and European countries are reported to generate 179 kg per person of food waste every year exclusively of agricultural waste [1, 3, 4]. This implies that a great amount of the resources devoted to food production are wasted with food that is never eaten, and that the greenhouse gas emissions due to the production of wasted food could also be deemed avoidable [1]. Food is lost or wasted along the whole supply chain, from the initial agricultural production down to final household consumption. For fruits and vegetables, the FAO estimated a 50% loss of crops during the food supply chain [1], predominantly concentrated in the postharvest, distribution, and end user consumption phases and mainly caused by the premature decay of perishable crops [5].

The source of food loss and waste in medium- to high-income countries mainly relates to a poor coordination between different roles in the supply chain as well as to consumer behavior. Sales agreements may have a role in the quantities of farm crops that are wasted, in fact, retailers want to offer products in the way they believe consumers want them to look like, having the right appearance, size, and shape; consequently, farmers waste large part of their crops instead of supplying it to the retailer [6, 7]. Furthermore, food can be wasted due to quality standards defined by the European Commission, which rejects food that presents defects that affect its general appearance or that does not fulfil weight and shape requirements [8–10]. On the consumer side, bad planning in combination with

Sustainable Food Packaging Technology, First Edition. Edited by Athanassia Athanassiou.

expiring "best-before-dates" and the negligence of people who can afford to waste food also cause sizeable amounts of waste [1].

On the other hand, consumers are becoming increasingly aware of the importance of eating habits and the role of food in improving health, which has led to an increase in the demand for healthy ready-to-eat, minimally processed, and preservative-free food products [11, 12]. For this reason, the production of fresh-cut fruits and vegetables is constantly growing [13, 14]. The term "minimally processed food" refers to food that has been physically modified, by peeling, slicing, or chopping, from its original state, but remains in a fresh and unprocessed state [14, 15]. The production of fresh-cut fruits is a big challenge since the shelf life of fresh-cut fruit is usually one to three days shorter than that of the whole product [15]. In fact, these minimal processing operations accelerate the physiological, physical, and microbial deterioration and raise metabolic activity, which can lead to enzymatic browning, loss of texture, water loss, increased susceptibility to microbial spoilage, and production of undesirable odors and flavors [13–15]. This happens because when fruits are in any way damaged, their tissue responds with a high rise in respiration rate, and consequently, the shelf life and quality of fruits is vastly reduced. As an example, the increase in the respiration rate of fresh-cut products compared with the intact products can range from 15.6%, 23.5%, 25.6% for bananas, peaches, and cucumbers, respectively, to 152.2% and 193.1% for kiwifruit and lettuce [16]. Fresh food products can be damaged also because of their manipulation [12, 17]. During transportation, storage, and handling, products start to deteriorate, dehydrate, and lose appearance, nutritional value, and flavor. If no special protection is provided, food can be damaged within a few hours or days, even if it is not immediately visible [17].

So far, several treatments have been explored to extend the shelf life of food, such as cryopreservation, modified atmosphere packaging, osmotic treatments, synthetic chemical fungicides, and hypobaric and heat treatments [5]. In this scenario, edible packaging has been widely studied as an alternative method to preserve crop freshness by combining extended storage periods with ease of handling [18].

Two types of edible packaging have been developed: coatings and films. The use of edible coatings on minimally processed food consists in the application of a solution of any film-forming edible material directly on the surface of the food [11]. Once dried, the solution forms a thin membrane that coats the food surface and provides a modified atmosphere, reducing moisture and aroma loss, delaying gas transfer, color changes, and improving the general appearance of the product through its enhanced shelf life [14, 19–21]. An edible film consists in a film-forming solution that, once dried, is applied on the food product as a self-standing material in order to perform the desired function [11]. The main difference between these two technologies is that films are first molded as solid sheets and then used as a wrapping for food products, while coatings are applied in liquid form directly on the food as depicted in Figure 3.1 [19, 22]. The edible coating solution is obtained by dispersing and dissolving the film-forming materials in water, alcohol, or other solvents. In this stage, additives such as plasticizers, antimicrobial agents, colors, or flavors can be incorporated. Film solutions can

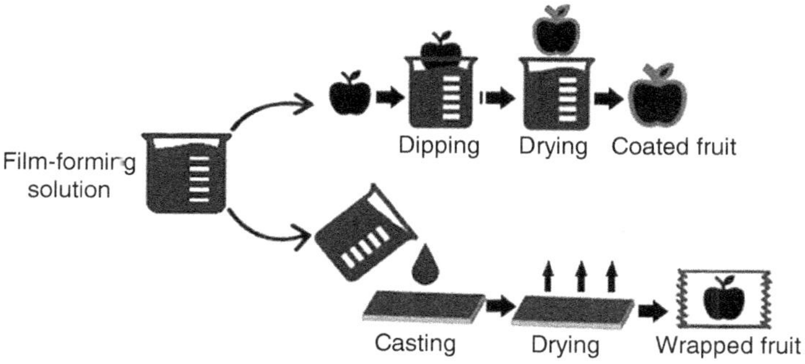

Figure 3.1 Edible film and coating manufacturing.

be applied onto food mainly by dipping or spraying processes followed by drying [23]. Edible films can be obtained through wet processes or dry processes. Wet processes involve the solubilization of all ingredients and subsequent drying to form the film, similarly to the method mentioned earlier for coatings. Dry processes rely upon the thermoplastic properties of film-forming materials, whereby heat is applied to temperatures above the melting point of the selected material in order to use conventional processing techniques, such as extrusion, compression, and injection molding [24, 25].

The main requirements for edible coatings and films are:

1) Relatively low oxygen and carbon dioxide permeability, in order to reduce respiration rate and metabolic activity without creating anaerobic conditions;
2) Low water vapor permeability, in order to reduce weight and firmness loss;
3) Good mechanical properties;
4) Tasteless and odorless in nature, in order to not alter the flavor of the product;
5) Transparency.

Components used for the preparation of edible coatings and films can be classified into three categories: hydrocolloids (proteins and polysaccharides), lipids, and their mixtures [23]. Polysaccharides are generally hydrophilic, which results in poor water vapor barrier properties. Additionally, these substances can act as effective gas barriers because of their well-ordered hydrogen bonded network [24, 26]. Protein-derived coatings are excellent barriers to carbon dioxide and oxygen, but, like polysaccharides, they are weak water vapor barriers and have poor mechanical properties [24]. The primary function of lipids in edible coatings and films is to limit moisture transfer due to their relative low polarity. However, the hydrophobicity of lipids causes the formation of thicker and more brittle films. Consequently, it is common practice to mix lipids with toughening agents (usually a protein or a polysaccharide) to provide mechanical support to the final material [27]. Edible coatings and films composed by a mixture of polysaccharides, protein, and/or lipids may also be designed. This type of materials has been commonly referred to as composite materials – despite the lack of structural organization distinct to plastic composite – and has gained interest due to the possibility of exploiting the distinct functional characteristics of each class of materials thus also minimizing drawbacks [26]. Composite edible materials are applied either in the form of a solution in a common solvent, or of an emulsion of nonmiscible components, or in multilayer structures [24].

The mechanical, nutritional, organoleptic, and functional properties of edible coatings and films can be improved with the incorporation of additives such as plasticizers, antimicrobials, antioxidant agents, preservatives, nutraceuticals, flavors, and colors, giving the possibility of widening the applications of edible coatings for minimally processed food [27]. In particular, the most frequently used additives in edible coatings and films are plasticizers (e.g. glycerol), which are incorporated to improve mechanical properties such as toughness and flexibility [19, 23].

Edible coatings and films have comparable functions as those of conventional packaging: they are water vapor and gases barriers, they improve mechanical-handling properties of foods, and prevent aroma compounds losses. However, edible materials are not supposed to totally replace conventional packaging (e.g. trays, bags, boxes), since external protection is still required for handling and hygiene purposes [28]. Enhanced food preservation can be achieved by combining primary edible materials and secondary nonedible packaging. In this scenario, edible coatings allow the reduction of conventional packaging waste, and since the secondary packaging just provides protection during handling, it must fulfil less requirements, and can also be made of biodegradable materials, which mitigates the use of nonrenewable sources [28, 29]. Edible coatings and films have several advantages with respect to the use of conventional petrochemical-based packaging [30]:

1) Edible coatings can be eaten together with the coated food without leaving any package waste. In the case of edible films, even if not consumed, they are biodegradable and do not pollute.
2) They can be used to package small portions of food, such as nut, peas, and beans, that currently are not individually packaged because of practical reasons.
3) They can improve the nutritional value and the organoleptic properties of food when loaded with nutraceuticals, flavorings, colorings, and antioxidants.

Individual country legislations regulate edible packaging to guarantee consumers' safety, regardless of a products country of origin. Since edible coatings are consumed together with the food products, they should observe all regulations for food ingredients, and be processed according to individual countries' standards for hygiene [21]. In Europe, materials that can be used in the edible coating formulations are regarded as food additives and are included in the Regulation (EC) No. 1331/2008 and Regulation (EU) No. 234/2011 [31, 32] for food additives, enzymes, and flavorings. In the United States, the Food and Drug Administration (FDA) indicates that any ingredient included in edible coatings and films formulations should be generally recognized as safe (GRAS) or regulated as food additive [11, 17]. In Europe, the incorporation of food additives must always be labeled under their category (antioxidant, colorant, etc.) by writing on the secondary packaging the name or E-number of the additive. In the United States, the presence of a coating should be reported on the label to inform the consumers [33]. It is important to acknowledge that edible coatings could be made of substances that can cause allergic reactions, and consequently the presence of any known allergen must be clearly labeled [21].

3.2 Polysaccharides

Polysaccharides are naturally occurring polymers, widely used to prepare edible coatings and films since they are colorless, have an oil-free appearance, have low caloric content, and can be applied to prolong the shelf life of fruits, vegetables, shellfish, or meat products, by significantly reducing dehydration, darkening of the surface, and oxidative rancidity [24]. Polysaccharides are widely available and usually cost-effective [27]. The moisture barrier effect of polysaccharide coatings is negligible (Table 3.1), since they are generally hydrophilic [23], but they have selective permeability for O_2 and CO_2, and resist lipid migration [27]. Examples of polysaccharides used to fabricate edible coatings and films are alginate, carrageenan, cellulose, chitosan, pectin, pullulan, and starch [12, 41], and these are briefly described in Sections 3.2.1–3.2.7.

Table 3.1 Water vapor and oxygen permeability values.

Material	WVP × 10^{11} (g/(m s Pa))	Oxygen permeability (OP) (ml μm/(m^2 d Pa))	References
Polysaccharides			
Chitosan	0.46–15.27	0.0229–0.9	[17, 26, 34]
Hydroxypropyl methyl cellulose	19	0.272	[17, 35]
Methyl cellulose	7.55–12.1	0.097–1.12	[17, 26, 35]
Corn starch	36.8	0.105	[17]
Pullulan	21.79	–	[36]
Alginate	9.5	–	[37]
Carrageenans	12	–	[37]
Proteins			
Wheat gluten	2.81–70	0.00214–1.88	[17, 26, 38]
Gelatin	16–50	0.0024–0.0047	[17, 39]
Zein	8.9–53	0.039–0.25	[17, 26, 35]
Whey protein	54–71	0.053–0.12	[35, 40]
Silk fibroin	6.46–793	0.162–5.8	[5]
Lipids			
Beeswax	0.004–0.06	0.046	[17, 40]
Candelilla wax	0.0017–0.02	0.083	[17, 26]
Shellac	0.36–1.03	0.83	[17, 26]

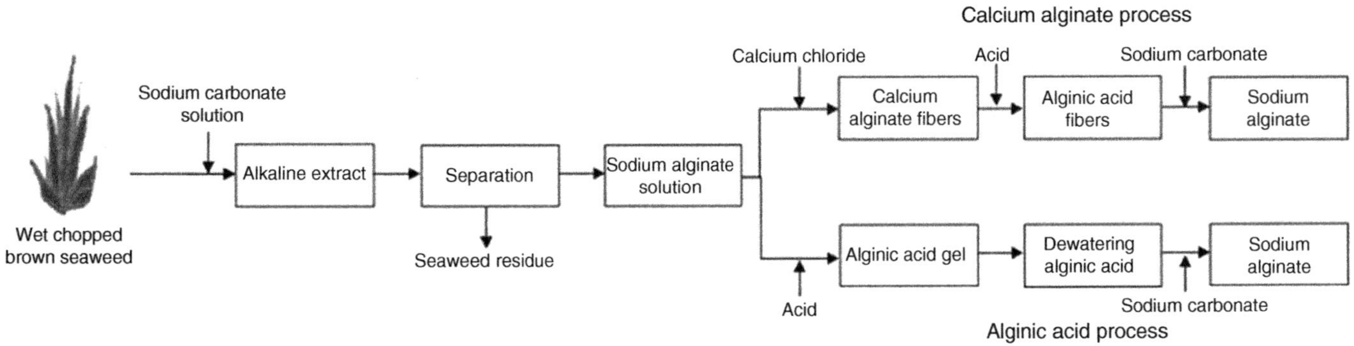

Figure 3.2 Sodium alginate manufacturing process.

3.2.1 Alginate

Alginate is a natural polysaccharide present in brown seaweed (*Phaeophyceae*) as a salt of sodium, calcium, magnesium, strontium, and barium in gelled form. Its extraction consists of an acid treatment to convert alginate into alginic acid and an alkali treatment to produce water-soluble sodium alginate (Figure 3.2) [17]. Alginate is an effective biopolymer in edible coating applications because of its colloidal characteristics such as thickening, suspension forming, gel forming, and emulsion stabilizing properties [42]. It is a linear polymer composed of two uronic acids: β-(1→4)-linked D-mannuronic acid (M) and α-(1→4)-linked L-guluronic acid (G) units [17, 42]. Alginates are block copolymers of both similar and alternating M or G units and the proportion of guluronic (G) to mannuronic (M) acid, the order of these units, the length of each block, and the overall molecular weight of the polymer influence the physical characteristics of alginate gels [24]. One of the most interesting properties of sodium alginate is its capacity to react with polyvalent cations, such as Ca^{2+} ions, and to cross-link forming an egg-box structure when guluronic residues bind to ions [11, 43, 44].

Alginate-based edible coatings have been effective in maintaining postharvest quality of different types of fruits. Four plum cultivars were treated with 1 or 3 wt.%/v alginate edible coating before being stored at 2 °C. Both alginate concentrations were able to inhibit ethylene production and delay weight and acidity losses, softening, and color changes. The treatment increased the plum storage period, depending on the cultivar, by two or three weeks more than untreated controls [45]. Sodium alginate edible coating was found to be effective on sweet cherries. Uncoated cherries stored at 2 °C presented a shelf life of 8 days, while the coating extended the shelf life up to 16 days reducing color changes, respiration rate, and acidity and firmness losses [46]. Maftoonazad, Ramaswamy, and Marcotte [47] studied the effect of sodium alginate coating on peaches stored at 15 °C and 40% RH. The shelf life at 15 °C for control samples was 15 days, while the coated samples maintained an acceptable quality for 21 days. Alginate coatings can also preserve the quality of apple slices stored at 5 °C in 85% RH. They were effective water vapor barriers and decreased the water loss of apple slices, prevented texture and firmness loss and browning, and led to the extension of fruit shelf life without provoking anaerobic respiration [48]. Furthermore, alginate-based edible coatings were applied to fresh-cut pineapples [49], garlic bulbs [50], mozzarella cheese [51], and pork patties [52] showing the effects reported in Table 3.2.

3.2.2 Carrageenans

Carrageenans are sulfated polysaccharides extracted from the cell wall of certain species of red seaweed of the family *Rhodophycae* through a process presented in Figure 3.3 [89]. They are hydrophilic linear sulfated galactans that consist of alternating (1→4)-linked α-D-galactopyranose, (1→3)-linked β-D-galactopyranose, or (1→4)-linked 3,6-anhydro-α-D-galactopyranose [90]. Different seaweeds produce different types of carrageenans; commercially, the three most important types of carrageenan extracts differ in the degree of

Table 3.2 Polysaccharide-based edible coatings.

Coating material	Food	Effect	References
Alginate	Plums	Inhibited ethylene production, delay weight and acidity losses, softening and color changes	[45]
	Sweet cherries	Reduced color changes, respiration rate, and acidity and firmness losses	[46]
	Peaches	Increased the shelf life by 7 days	[47]
	Apple slices	Decreased the water loss of apple slices, texture and firmness loss, and browning	[48]
	Fresh-cut pineapple	Reduced weight loss and respiration rate	[49]
	Garlic bulb	Reduced moisture loss	[50]
	Mozzarella cheese	Preserved physicochemical properties	[51]
	Pork patty	Reduced lipid oxidation	[52]
Carrageenans	Papaya	Preserved firmness and color	[53]
	Strawberries	Preserved firmness and color	[54]
	Pears	Reduced weight and firmness loss	[37]
	Cheese	Reduced weight loss and color changes	[55]
Carboxymethyl cellulose	Strawberries	Improved microbial, physicochemical, and organoleptic properties	[56, 57]
	Carrots	Reduced water and firmness loss	[58]
Hydroxypropyl methyl cellulose	Tomatoes	Delayed softening and color changes	[59]
	Mashed potato balls	61.4% reduction of fat uptake during frying	[60]
Methyl cellulose	Mashed potato balls	83.6% reduction of fat uptake during frying	[60]
Chitosan	Sliced mango	Delayed water loss and drop in sensory quality and inhibited the growth of microorganisms	[61, 62]
	Carrots	Better color retention, sensory and microbial quality	[63]
	Shredded radish	Decreased weight loss, respiration rate, browning, and microbial load	[64]
	Fresh-cut cucumbers	Preserved quality	[65]
	Ricotta	Reduced undesirable acidity and maintained texture	[34]
	Cheddar	Showed antibacterial activity	[66]

(Continued)

Table 3.2 (Continued)

Coating material	Food	Effect	References
	Goat's milk cheese	Prevented weight loss and improved microbial safety	[67]
	Mozzarella cheese	Inhibited the growth of spoilage microorganism	[68]
	Roasted sliced turkey	Showed antibacterial activity	[69, 70]
	Roast beef	Showed antibacterial activity	[71]
	Chicken breast fillets	Reduced lipid oxidation and rancidity	[72]
	Salmon	Delayed lipid oxidation	[73, 74]
Pectin	Lime	Maintained greener color, decreased respiration rate and weight loss, increased vitamin C retention	[75]
	Fresh-cut melon	Increased the water vapor resistance, prevented desiccation, and maintained fruit firmness	[76]
	Cucumbers	Reduced respiration rate and loss of weight, firmness, and color of cucumbers	[77]
	Dried papaya slices	Enhanced vitamin C retention, prevented color degradation and oxidation	[78]
Pullulan	Apple slices	Decreased the weight loss, microbial growth, enzymatic browning, and respiration rate	[79]
	Blueberry	Delayed deterioration, and reduced respiration rate and microbial growth	[80]
	Strawberries	Delayed softening and color changes	[81]
	Kiwi	Preserved color and firmness	[82]
	White asparagus	Reduced weight loss and changes in color	[83]
Cassava starch	Strawberries	Decreased the respiration rate and increased water vapor resistance	[43]
	Pummelo	Maintained physical appearance, external color, ascorbic acid content	[84]
	Avocados	Reduced respiration rate and color changes	[85]
	Tomatoes	Delayed maturation, firmness, and weight loss	[86]
Corn starch	Dried pumpkin slices	Minimized carotenoid degradation during pumpkin drying, maintained better color	[87]
	Dried carrots	Maintained redness and retention of α- and β-carotene	[88]

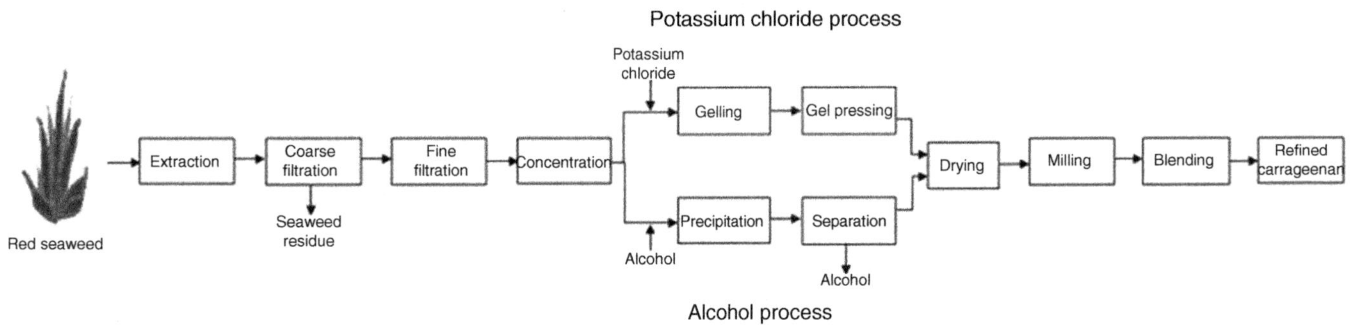

Figure 3.3 Carrageenan manufacturing process.

sulfation: kappa (κ-), iota (ι-), and lambda (λ-) types, going from kappa being the least sulfated (25% sulfate ester content) to lambda being the most sulfated (35% sulfate ester content) [17]. The sulfation level influences the solubility of carrageenans in water, since the sulfate groups are hydrophilic and higher content of ester sulfate leads to lower solubility temperature. Consequently, lambda carrageenan is soluble in cold water, while kappa carrageenan requires heating to solubilize completely [17, 44, 90].

Edible coatings and films made of carrageenans possess compelling characteristics such as good mechanical properties, stabilization of emulsions, aroma reduction, and reduced oxygen transfer [91, 92]. Carrageenan edible coatings and films proved to be effective in reducing firmness and weight loss for papaya stored for 5 days [53], pears stored for 15 days [37], strawberries stored for 6 days [54], and semihard and dry white brined cheeses stored for 46 days [55].

3.2.3 Cellulose

Cellulose is the most abundant biomass resource on earth. It is a water permeable polysaccharide composed of D-glucose units [41] found in plant cell walls of all fruits and vegetables, and it is present in the leaves, trunks, and barks of higher plants as a structural component. It is possible to increase water solubility by immersing cellulose in alkali to swell the structure, followed by reaction with methyl chloride, chloroacetic acid, or propylene oxide to, respectively, yield methylcellulose (MC), carboxymethylcellulose (CMC), hydroxypropyl cellulose (HPC), or hydroxypropyl methylcellulose (HPMC) (Table 3.3) [93]. Chemical modification of cellulose introduces various chemical groups onto the hydroxyl groups of glucose units. Methyl, carboxymethyl, and HPC ethers provide modified derivatives that possess unique functional properties, such as thermal gelation and superior clarity [17]. Cellulose derivatives are made from polymer chain of two repeating units of anhydrous-glucose or β-glucopyranose residue attached through 1,4-glucosidic linkage. Generally, the modified cellulose derivative edible coatings and films are odorless, tasteless, and colorless [89]. The films obtained from aqueous solutions of MC, HPMC, HPC, and CMC are resistant to oils and fats, water-soluble, and provide discrete barriers to oxygen. Despite that, cellulose derivative films are bad water vapor barriers because of the hydrophilic nature of the polysaccharides [94].

CMC is a promising polysaccharide polymer which consists of β-D-glucose and β-D-gluco-pyranosyl-2-*O*-(carboxymethyl)-monosodium salt connected via β-(1,4-glycosidic) bond [95]; it is linear, water-soluble, and anionic [96]. An aqueous solution of CMC 1 wt.%/v was applied with ascorbic acid in order to control surface browning of fresh-cut apples during storage. When applied on apple slices, the coating delayed tissue softening during storage and maintained the fruit's antioxidant properties when compared to the uncoated control [96]. CMC 1%, 1.5%, and 2 wt.%/v aqueous solutions were also found to be effective in extending the shelf life of strawberries [56, 57], citrus fruit [97], and carrots [42, 58].

Table 3.3 Cellulose derivatives solubility properties.

Material	Structure	Properties
Cellulose	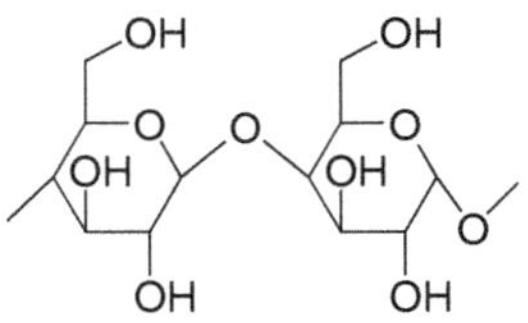	• Not soluble in water
CMC	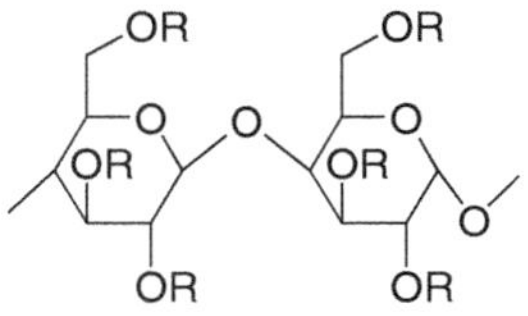 R = H or CH_3CO_2H	• Soluble in neutral water. Solubility increases with the degree of substitution
MC	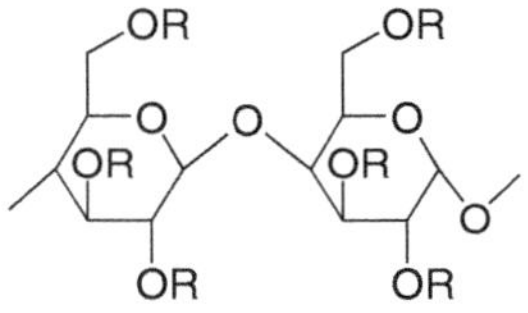 R = H or CH_3	• Soluble in cold water up to 55 °C • Insoluble in alcohol and ether
HPMC	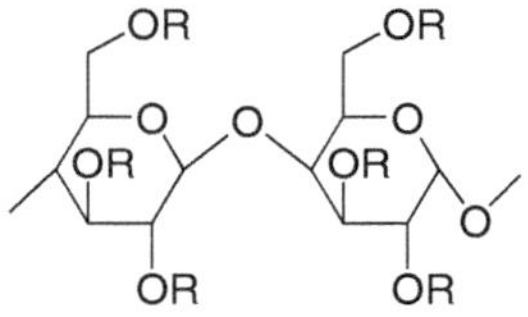 R = H, CH_3, or $CH_3CH(OH)CH_2$	• Soluble in cold water up to 82 °C • Soluble in methanol and ethanol solutions

MC is a chemically-derived cellulose ether with a methyl degree of substitution of 1.6–1.9 on O-6, O-2, or O-3 positions of glucose units. This chemical substitution is responsible for its solubility in cold water, by opening it up for hydration and hydrogen bonding with water. It is linear and nonionic [17].

HPMC is a water-soluble cellulose ether hydrocolloid with good film-forming properties [98]. It possesses a structure similar to methylcellulose with the addition of a methyl substitution and the substitution of some of free hydroxyl groups of glucose with hydroxypropyl ones [17]. Zhuang and Huang [59] studied the effect of application of 3 wt.%/v solution of HPMC in 63% ethanol to mature-green tomatoes. The coating delayed softening of tomatoes for up to

18 days of storage. Firmness of tomatoes coated with HPMC was significantly greater compared to uncoated controls. The study also confirmed that HPMC coatings could significantly delay color changes in tomatoes stored at 20 °C.

HPMC and MC have the ability to form thermally induced gel-like coating [23]; this means that a solution of these derivatives gels above a critical temperature (i.e. gel-like state), and reverses when cooled (i.e. liquid-like state) [99]. This is an interesting feature for their application in thermally treated foods, particularly in fried products, to reduce oil absorption during the frying of various foods (e.g. meat, poultry, starchy foods, doughs) [100]. Mallikarjunan et al. [60] determined the effectiveness in moisture retention and reduction of fat uptake of HPMC and MC edible coatings on a fat-free starchy product during frying. Mashed potato balls (diameter = 47 mm) were coated with film-forming solutions of HPMC and MC 225 wt.%/v in 50% ethanol, showing a reduction in fat uptake during frying of 61.4% and 83.6%, respectively, when compared to the uncoated control.

3.2.4 Chitosan

Chitosan, one of the most abundant polysaccharides found in nature, is a linear polysaccharide consisting of (1,4)-linked 2-amino-deoxy-β-D-glucan and is the deacetylated derivative of chitin [27]. Chitosan is nontoxic, biodegradable [27, 101], and is a very attractive material for edible coatings and films due to its antimicrobial properties against bacteria, molds, and yeasts. The antimicrobial activity is influenced by intrinsic factors such as the form of chitosan (e.g. α- or β-chitosan), molecular weight, degrees of deacetylation and polymerization [102]. Given the coating and film-forming capacity of chitosan-based materials, the polysaccharide was tested on several types of food (e.g. strawberries, mango, carrots, etc.) with the aim of decreasing microbiological growth and thus extending shelf life [103].

Chitosan coating remarkably reduces moisture loss and delays color changes of strawberries, increases the activity of antioxidant enzymes, reduces membrane damage, and prevents flesh browning [104]. Chien et al. [61] manually treated mango slices with aqueous solutions of 0.5%, 1%, or 2% chitosan, using water as control. Chitosan coatings delayed water loss and the drop in sensory quality while also inhibiting biotic spoilage. These data were corroborated by Elsabee and Abdou [62], who also showed that applying a chitosan coating effectively prolongs the desirable attributes and extends the shelf life of sliced mango fruit. Pushkala et al. [63] evaluated the efficacy of a powder coating technique based on chitosan for quality maintenance of shredded carrots. Chitosan-treated samples were characterized by better color retention and superior sensory and microbial quality after 10 days of storage compared to the control, which remained edible only until day 5 of storage at 10 °C. An edible coating based on chitosan powder was also successfully applied to shredded radish by Pushkala et al. [64]. Chitosan-treated samples exhibited lower weight loss, decreased respiration rate, better sensory acceptability, lesser browning, and lower microbial load compared to uncoated control. Olawuyi et al. [65] discovered that the combination of chitosan coating with argon-based modified atmosphere packaging best preserved the quality and prolonged the shelf life of fresh-cut cucumber throughout 12 days of storage.

Chitosan coatings are effective also for the preservation of different cheeses such as ricotta [34], cheddar [66], goat's milk cheese [67], and emmental [103, 105]. Altieri et al. [68] demonstrated that chitosan applied on mozzarella cheese inhibited the growth of some spoilage microorganisms, such as coliforms, whereas it did not influence the growth of other microorganisms, such as *Micrococcaceae*, and lightly stimulated lactic acid bacteria. Chitosan was also applied on meat and fish, for example it showed antioxidant activity when applied on roasted and sliced turkey [69, 70] and roast beef [71], and it delayed lipid oxidation on chicken breast fillets [72], and salmon [73, 74].

3.2.5 Pectin

Pectin is produced from citrus peel or from apple pomace, which are by-products of the extraction of lime, lemon, and orange juices and the extraction of apple juice, respectively [17]. Pectin is a heteropolysaccharide that consists of a homopolymeric linear block of α-(1→4)-D-galacturonic acid units, called "smooth" region, interrupted by branched α-(1→2)-linked α-L-rhamnopyranose residues, called "hairy" region [24, 106]. A portion of the galacturonic acid group may be either free or methylesterified; based on the degree of methyl esterification (DM), pectin can be either high methoxy pectin (HMP) when the degree of methyl esterification is higher than 50%, or low methoxy pectin (LMP) when it is lower than 50% [107]. The DE changes with the source of plant used, time and area in which the plant is harvested, and with the processing conditions (e.g. storage, extraction, isolation, and purification) [108]. Their degree of esterification influences gelling properties: LMP gels in presence of multivalent ions, while HMP forms gel in acidic media in presence of sugars such as sucrose or glucose [109, 110].

Pectin is an ingredient used in the food industry since it is GRAS by the FDA and it can be used in food as gelling, stabilizing, encapsulating, or thickening agent in products such as jams, marmalades, yogurt, and soy drinks [111]. Pectin edible films and coatings have been extensively studied focusing on the improvement of the shelf life of fruit thanks to its good gas barrier properties and ability to delay lipid migration and moisture [112, 113]. Production of edible coatings and films from pectin can be achieved by various methods such as extrusion, solvent casting, and spraying [109].

Maftoonazad and Ramaswamy [75] evaluated the effect of pectin coatings on the quality change and shelf life of lime fruit. The study revealed that the coating was able to extend lime's shelf life by maintaining a greener color, decreasing the respiration rate and the weight loss while increasing the retention of vitamin C. Gas exchange, antioxidant properties, color, firmness, sensory quality, and microbial growth of fresh-cut melon were studied during 15 days at 4 °C by Oms-Oliu, Soliva-Fortuny, and Martín-Belloso [76], comparing the behavior of pectin-coated and uncoated fruit. Pectin maintained melon quality throughout the storage by increasing the water vapor resistance, preventing desiccation, and maintaining fruit firmness. Furthermore, pectin can increase the storage time of cucumber fruits at 23 °C, 40% RH and 12 °C, 85% RH [77]. The coating was able to reduce respiration rate and loss of weight, firmness, and color of cucumbers at both storage conditions.

Pectin coatings were also found to be effective in maintaining the nutritional quality of food during drying. The effects of pectin coatings on the quality of air-dried papaya slices were investigated by Garcia et al. [78]. The coating applied before drying had a protective effect and enhanced vitamin C retention compared to uncoated papaya. Pectin was effective in preventing color degradation and oxidation of the bioactive compounds of papaya without affecting drying efficiency, and while maintaining the cell tissue arrangement of the fresh papaya.

3.2.6 Pullulan

Pullulan is a nonionic exopolysaccharide obtained from the fermentation medium of the fungus-like black yeast *Aureobasidium pullulans* [114, 115]. Its structure, elaborated by Bender [116], can be represented by α-(1→4)-linked repeating trimer maltotriose connected to each other by α-(1→6) glycosidic linkages [11, 114, 117]. Because of this unique linkage pattern, pullulan polymer exhibits distinctive physical properties, including flexibility and considerable mechanical strength, resulting in the ability to form films and fibers [11, 118]. Furthermore, pullulan has other properties that make it suitable for edible coating applications. It is edible, tasteless, odorless, nontoxic, highly soluble in water, and a good barrier to oxygen [114, 119].

Wua and Chen [79] applied pullulan coatings on sliced Fuji apples during hypothermia storage. It decreased the weight loss, inhibited microbial growth, maintained firmness, delayed enzymatic browning, and reduced respiration rate of the apple slices compared to the control. Pullulan was effective in highbush blueberry preservation by delaying deterioration and reducing respiration rate and microbial growth [80]. Furthermore, pullulan was found to be able to extend the shelf life of strawberries [81], kiwi [82], and white asparagus [83] with the effects reported in Table 3.2.

3.2.7 Starch

Starch is a polymeric carbohydrate composed of anhydroglucose units. This is not a uniform material and most starches contain two types of glucose polymers: a linear chain molecule termed amylose and a branched polymer of glucose termed amylopectin [120]. Amylose is a nearly linear polymer of α-1,4-anhydroglucose units that has excellent film-forming ability, rendering strong, isotropic, odorless, tasteless, and colorless film [25], while amylopectin is a highly branched polymer of short α-1,4 chains linked by α-1,6-glucosidic branching points occurring every 25–30 glucose units. Amylose, contained in any type of starch, is more closely associated with the ability to form films and coatings due to its predominant linear nature [17]. Amylose content of starch varies with the botanical source of the starch. For example, amylose content in potato starch varies from 20% to 31%, in rice starch from 5% to 28%, in corn starch from 22% to 32%, in wheat starch from 18% to 30%, and in cassava starch from 15% to 25% [121, 122].

Starch-based edible coatings have been extensively used to extend the shelf life of fruit. Garcia et al. [43] studied an edible coating of cassava starch on strawberries. The coatings did not cause changes in strawberrys mechanical properties and superficial color, decreased the respiration rate, and increased water vapor resistance of the samples, which showed good sensory acceptance. Kerdchoechuen et al. [84] applied two kinds of edible films prepared from rice and cassava starch on minimally processed pummelo. During storage at 4 °C for 21 days, both coatings prolonged pummelo shelf life maintaining physical appearance, external color, and vitamin C content, compared to uncoated fruit. The use of cassava starch coating solution had a better effect compared to rice starch. Cassava starch coatings were also able to extend the shelf life of avocados [85] and tomatoes [86].

Dipping in a 1 wt.%/v citric acid solution and coating with a 2 wt.%/v cassava starch suspension has an effect on quality preservation of fresh-cut mangoes during storage at 5 °C for 15 days [123]. This treatment delayed the deterioration of fresh-cut mangoes due to the low oxygen permeability of cassava starch coatings that decreased the respiration rate. The resulting inhibited metabolic activity prolonged freshness by preserving firmness and color characteristics during storage. The coating did not have an effect in reducing the microbial proliferation with respect to the control. The same treatment was applied by Bierhals et al. [124] on fresh-cut pineapple, where cassava starch coatings were effective in reducing respiration rate, weight loss, and juice leakage and maintained fruit firmness. Nonetheless, the starch-based coatings did not inhibit microbial growth and consequently failed in increasing the shelf life of pineapple.

Starch-based coatings can also be used to prevent the nutrient loss of dried food due to oxygen and high temperature exposure. The aim of the study conducted by Lago-Vanzela et al. [87] was to evaluate the efficacy of starch-based coatings in maintaining the carotenoid content of pumpkin slices dehydrated by hot air. The application of native maize starch coating minimized carotenoid degradation during pumpkin drying and resulted in dehydrated products with better color and with higher retention of trans-α-carotene and trans-β-carotene than uncoated products. Corn starch was also tested on carrots and was found effective in maintaining their redness and improving the retention of α- and β-carotene during storage [88].

3.3 Proteins

Proteins are heteropolymers composed of up to 20 different amino acids. The structure of the side chain is different in each amino acid. The primary structure of the protein is constituted by the sequence of amino acids that are bound by peptide bonds, which are amide linkages formed by condensation of the carboxyl group of one amino acid with the amino group of the next one [11, 125]. The repetitive sequence of N–CH–CO atoms that makes up a continuous chain is called the protein's backbone. The properties of proteins depend also on the way in which protein chains are oriented in space. The most common structures are α-helices or β-sheets, which represent the secondary structure described by the orientation of segments of the peptide backbone into a regular pattern [125, 126]. An α-helix is a right-handed coil of the protein backbone in which the structure is stabilized by hydrogen bonds formed between the hydrogen in the amide N–H groups and the oxygen in the C=O located four residues later. A β-sheet differs from α-helix since the peptide chain is fully extended and the hydrogen bonds occur between residues in adjacent chains [126]. Such structural elements can be packed in different ways forming a 3D structure called tertiary structure. The final protein can contain different polypeptide chains arranged to yield a large aggregate structure, the quaternary structure [11, 125, 126].

According to their structure, proteins are generally classified as either fibrous proteins or globular proteins, which have different behaviors in aqueous solutions [11]. Fibrous proteins consist of polypeptide chains arranged in long filaments, which are fully extended and associated closely with each other in parallel structures, generally through hydrogen bonding, to form fibers. They are insoluble in water or aqueous solutions and are used as structural material in animal tissue [23]. The globular proteins are usually coiled into compact, complicated spherical structures held together by a combination of hydrogen, ionic, hydrophobic, and covalent bonds. They are generally soluble in water. The chemical and physical properties of globular proteins depend on the relative amounts of the component amino acid residues, and their position along the protein polymer chain [23, 126].

Protein films show interesting characteristics for food coating applications, such as transparency and excellent oxygen, carbon dioxide, and lipid barrier properties, particularly at low relative humidity [127]. They are, however, water sensitive and have limited water vapor barrier properties (Table 3.1), and consequently, barrier and mechanical properties are compromised by moisture because of their hydrophilic nature [26, 128]. To improve the mechanical and barrier properties of protein films, it is possible to modify the secondary, tertiary, and quaternary structures using treatments such as chemical and enzymatic methods, mechanical treatments, and irradiation [129, 130]. The most investigated proteins used to fabricate edible coatings and films are casein, collagen, gelatin, silk fibroin, wheat gluten, whey protein, and zein.

3.3.1 Casein

Casein represents 80% of bovine milk proteins. It is an heterogeneous mixture of four principal components α_{s1}- (38%), α_{s2}- (10%), β- (36%), κ-casein (13%) and a minor component γ-casein (3%), each of which has different properties [131–133]. The α-fractions contain more charged residues, less proline and hydrophobic residues content, and as such they can form flexible edible films of high water vapor permeability and low gas permeability [99]. On the other hand, β-fraction is hydrophobic and forms films with low water vapor permeability and low permeability to gases. The κ-fraction produces films with intermediate characteristics due to its intermediate hydrophobicity [134]. When skim milk is brought at pH 4.6 and 20 °C, casein precipitates. Precipitated caseins can be solubilized by neutralization through the addition of alkali. Sodium and calcium caseinates are the most common, but magnesium and potassium caseinates are also commercially available [135, 136].

Casein and caseinate films can be used as edible coatings since they possess low oxygen permeability and good mechanical strength, but it must be taken into consideration that they have low flexibility and are more sensitivity to moisture [137]. Calcium caseinate applied on celery sticks and apples increased the shelf life of the samples by decreasing moisture loss and respiration rate [138]. Le Tien et al. [139] studied the effect of calcium caseinate edible coating as antioxidant and anti-browning agent on apple and potato slices; the coating effectively delayed browning of the samples and acted as an oxygen barrier. Calcium caseinate was also found to have a beneficial effect on the preservation of zucchini, mainly due to the increase of water vapor resistance provided by the coating [140].

3.3.2 Collagen

Collagen is a constituent of skin, bones, tendon, and connective tissues. It is a fibrous protein and represents about 30% of the total mass of the dry body [127]. It is composed of three parallel α-chains, which combine to make a triple-stranded helical structure. The amino acid sequence of collagen is dominated by a repeating trio of Gly-X–Y residues (with X and Y often proline and hydroxyproline, respectively) [17, 141].

Collagen has several characteristics that make it suitable for edible coatings and film applications; it is nontoxic, and can be processed into different forms [142]. Furthermore, collagen films have good mechanical properties and excellent oxygen barrier capabilities at 0% RH, but oxygen permeability increases rapidly with increasing RH [127] given the protein hydrophilicity. Collagen has been extensively used for the production of edible sausage casings replacing their natural gut counterparts [143]. Collagen casing can also be co-extruded around the sausage meat with a continuous process better controlled than the conventional one where meat is stuffed into preformed casing [144, 145]. Collagen has also been applied to other meat products. Farouk and Price [146] wrapped beef round steak with an edible collagen film, which exhibited less fluid exudate than unwrapped controls both in refrigerated and frozen storage conditions.

3.3.3 Gelatin

Gelatin is a protein with high content of amino acids such as proline, hydroxyproline, and glycine and is obtained by controlled hydrolysis of collagen [147]. Gelatin has been widely used for applications within the food industry due to the ease of use, wide availability, biodegradability, compostability, low cost, and film-forming ability [148]. The functionality of gelatin films can be improved by adding plasticizers that give sufficient mechanical stability, good strength, and flexibility, which allow for their use as packaging material [149]. Gelatin films and coatings have clarity and oxygen barrier properties but, being sensitive to moisture, they are bad barrier against water vapor [150, 151].

Gelatin edible coatings have been applied in the meat industry. Antoniewski et al. [152] studied the effect of 20 wt.%/wt. bovine gelatin solution onto beef tenderloins, pork loins, salmon fillets, and chicken breasts stored at 4 °C for two weeks. All the gelatin-coated meat products exhibited a reduction in water loss. Coated beef and pork showed a reduction in color changes, while flavor was not affected by the application of the coating. Porcine gelatin solutions were found to have a beneficial effect on lipid oxidation and color stability of cooked ham and bacon stored for seven months at −20 °C [153]. The application of a porcine gelatin coating was also effective in extending the shelf life of pork loin stored at 4 °C. The coating preserved the appearance of the product by reducing color change, loss of redness, and metmyoglobin formation. Furthermore, it slowed lipid and protein oxidation, thus reducing the formation of off-flavors [154].

3.3.4 Wheat Gluten

Wheat gluten is a globular protein of wheat flour which is composed of a mixture of polypeptide molecules. It is insoluble in water but can be dissolved in aqueous solutions of high or low pH at low ionic strength [155]. Properties such as cohesiveness and elasticity of gluten give integrity to wheat dough and facilitate film formation [23]. Wheat gluten is composed of two main groups, typically referred to as gliadin and glutenin. Gliadins, low molecular weight proteins, are soluble in 70% ethanol, while glutenins, high molecular weight proteins, are not [24, 156]. Films mainly composed of glutenins are stronger and have better water vapor barrier properties than films from gliadins, which, however, present better optical properties [142].

Wheat gluten is very promising for the development of edible coatings and films with selective gas permeability, providing the ability to control respiratory exchanges and improve the conservation of fresh or minimally processed fruits and vegetables. Gluten films are homogenous, transparent, strong, and provide good water barriers, as reported in Table 3.1 [157]. For example, wheat gluten coatings and films were able to extend the shelf life of strawberries up to 12 days and delay senescence with respect to uncoated strawberries used as control, which only lasted 6 days [158]. The fruit coated with gluten also had higher firmness and maintained better visual quality during shelf life. Finally, the consumers approved the taste and flavor of the gluten-coated fruit. Wheat gluten coatings were also applied by Tanada-Palmu et al. [38] on cherry tomatoes and Sharon fruits, with the effect of reducing weight loss during two weeks of storage.

3.3.5 Whey Protein

Whey protein represents approximately 20% of total milk proteins [131]. It contains five main components: α-lactalbumin, β-lactoglobulin, bovine serum albumin, immunoglobulin, and proteose-peptone; the compositional distribution is 19%, 57%, 7%, 13%, and 4%, respectively [99, 137]. Numerous industrial processes such as ultrafiltration, gel filtration, diafiltration, electrodialysis, reverse osmosis, and ion-exchange chromatography are generally used to recovery whey protein and produce whey protein concentrate (WPC, 25–80% protein) or whey protein isolate (WPI, > 90% protein) [127].

Whey protein coatings and films are transparent, flexible, colorless, odorless, and provide excellent oxygen, aroma, and oil barrier at low-to-intermediate values of relative humidity. Moreover, whey protein films present adequate mechanical properties to be used as coatings on food products [137, 159]. Perez-Gago et al. [160] coated fresh-cut apple pieces with an emulsion of WPI; the coating reduced the enzymatic browning, resulting in higher sensory scores than uncoated apples. Tzoumaki et al. [83] applied a whey protein coating on white asparagus stored at 4 °C, and studied the coating effect on parameters such as weight loss, texture, visual appearance, and color. The edible coatings had a positive impact in keeping asparagus quality by delaying weight loss, purple color development, and hardening in the basal part. The use of WPI to improve the rehydration behavior of freeze-dried strawberry pieces was studied by Huang et al. [161], in order to avoid the collapse of the texture after rehydration, which limits the application of freeze-drying to strawberries. WPI was found to be effective in slowing down the rehydration ratio of freeze-dried strawberries and preventing the texture collapse. Furthermore, WPI coatings, acting as oxygen barriers, can delay the development of rancidity of dry-roasted peanuts stored at both intermediate and low relative humidity [162]. WPI decreased water, hardness loss, and color change of cheese during storage [163]. The coating prevented growth of pathogenic microorganisms, but, at the same time, allowed the regular growth of lactic acid bacteria. Finally, the coatings can be used to preserve the quality of frozen or refrigerated fish. This kind of treatment was helpful in preventing moisture loss and delaying lipid oxidation of frozen salmon [164, 165], refrigerated trout [166], and frozen black sea sprat [167].

3.3.6 Silk Fibroin

Silk fibroin is extracted from silk cocoons produced by the *Bombyx mori* silkworm. Raw silk, which consists of two parallel fibroin filaments held together with a layer of sericin, undergoes degumming in order to remove the sericin and obtain fibroin fibers [168]. Silk has high molecular weight (more than 390 kDa) and is composed of a heavy and a light chain linked by a disulfide bond [169, 170]. The heavy chain is rich in glycine and consists of hydrophobic repetitive domains, specifically (Gly–Ala–Gly–Ala–Gly–Ser) units, and hydrophilic non-repetitive domains [171–173]. The repetitive domains of the heavy chains can form β-sheet crystallites, giving the protein excellent mechanical properties and flexibility. The light chain is hydrophilic, amorphous, and relatively elastic [168, 174].

Silk fibroin is a remarkable biomaterial due to the combination of outstanding mechanical properties, biological compatibility, biodegradability, edibility, transparency, and versatility [175, 176]. The polymorphism of the protein can be controlled through water annealing process, which enables the modulation of the mechanical properties and the diffusion of gases through silk fibroin films [177]. This means that by increasing the β-sheet content, it is possible to reduce film oxygen and water vapor permeability, which is a key factor in the preservation of

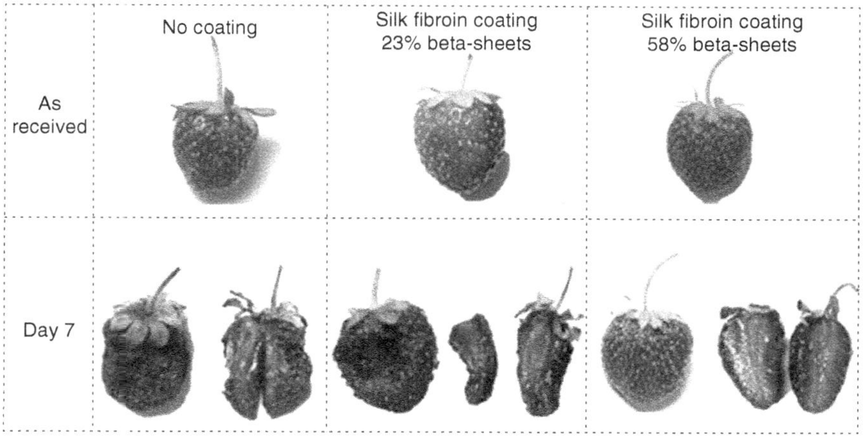

Figure 3.4 Time-lapse of the effect of silk fibroin edible coating on strawberries ripening. Source: Marelli et al. [5]. © 2016, Springer Nature.

food freshness. Marelli et al. [5] applied different silk fibroin edible coatings, having different β-sheet content, on strawberries and bananas stored at 22 °C and 38% RH. All the coatings were able to preserve both kinds of fruit, decreasing the weight loss, respiration rate, water vapor and oxygen diffusion, maintaining firmness and color, and delaying ripening of bananas compared to uncoated control during 14 and 9 days of storage for strawberries and bananas respectively. Increasing the β-sheet content from 36% up to 58%, the coating became more effective (Figures 3.4 and 3.5) [5].

3.3.7 Zein

Zein is the most important protein in corn. It is a mixture of different peptides having diverse molecular weight, solubility, and charge. They are classified according to their relative mass and solubility as α, β, γ, and δ-zein, being α-zein the most abundant fraction which is soluble in 50–95% isopropanol [147]. Zein is a relatively hydrophobic and thermoplastic material. It is composed of a high percentage of nonpolar amino acids and small proportions of basic and acidic amino acids, which determines its insolubility in water [17].

Park et al. [178] applied a corn zein coating on tomatoes during storage at 21 °C. Corn zein coatings delayed weight loss and color change and maintained tomatoes' firmness. The shelf life of the fruits was also extended by six days compared to uncoated samples. Other kinds of fruit, like apricots, were coated with a solution of natural corn zein by dip-coating treatments. Baysal et al. [179] investigated the antimicrobial and antioxidant properties of edible zein film and the ability to maintain the quality of apricots during 10 months at 5 and 20 °C. Zein coating effectively reduced color change and the total viable count of bacteria when compared to the uncoated control. Bai et al. [180] applied zein coatings on apples providing a desirable modified atmosphere, shiny apple surface, and minimal whitening under simulated commercial conditions. Zein coating extended

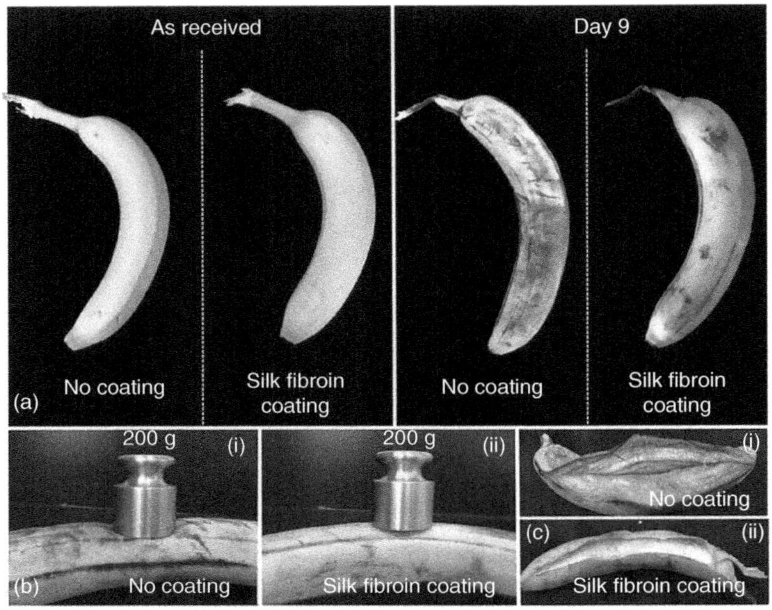

Figure 3.5 (a) Time-lapse photography of silk fibroin coating effect on banana ripening. (b) Investigation of silk fibroin-coated banana turgidity. Source: Marelli et al. [5]. © 2016, Springer Nature.

apple shelf life and produced an effect on fruit that was comparable to the one of commercial wax formulations. A 5 wt.%/v zein coating was used to extend the shelf life of mangoes, as reported by Gol and Ramana Rao [181]. It successfully delayed the ripening of mango fruit by suppressing the activity of softening enzymes. It also had a beneficial impact on soluble solids, titratable acidity, pH, sugar content, and total carotenoids, and resulted in higher retention of ascorbic acid and phenolic content compared with the control as reported in Table 3.4.

3.4 Lipids

Lipids are naturally occurring molecules, slightly soluble in water, than can be extracted from organisms using nonpolar organic solvents [126]. Lipids are indeed hydrophobic and efficient barriers against moisture migration (Table 3.1) and have therefore been used to limit moisture transmission in foods thanks to their low polarity [27]. Depending on their properties and fabrication, lipid-based edible coatings and films can have different behaviors with respect to moisture transfer; in fact, distribution of chemical groups, degrees of unsaturation, and length of the aliphatic chains can affect the polarity and hydrophobicity [23, 24, 27]. Lipids are defined considering solubility, rather than the structure. There are two general types of lipids: those which possess ester linkages and can be hydrolyzed, like waxes, and those which do not have ester linkages and

Table 3.4 Protein-based edible coatings.

Coating material	Food	Effect	References
Calcium caseinate	Celery sticks	Decreased moisture loss and respiration rate	[138]
	Apples	Decreased moisture loss and respiration rate	[138]
	Apple slices	Delayed browning	[139]
	Potato slices	Delayed browning	[139]
	Zucchini	Increased water vapor resistance	[140]
Collagen	Sausage	Acted as casing	[145]
	Beef steak	Reduced fluid exudate	[146]
Gelatin	Beef tenderloins	Reduced color changes and water loss	[152]
	Pork loins	Reduced color changes, water loss, and lipid oxidation	[152, 154]
	Salmon fillets	Reduced water loss	[152]
	Chicken breasts	Reduced water loss	[152]
	Frozen ham	Reduced lipid oxidation and color changes	[153]
	Frozen bacon	Reduced lipid oxidation and color changes	[153]
Silk fibroin	Strawberries	Decreased respiration rate and firmness and weight loss	[5]
	Bananas	Delayed ripening, firmness loss, and browning	[5]
Wheat gluten	Strawberries	Extended the shelf life and retarded senescence	[158]
	Cherry	Reduced weight loss	[38]
	Tomatoes	Reduced weight loss	[38]
	Sharon fruit	Reduced weight loss	[38]
Whey protein	Fresh-cut apples	Reduced enzymatic browning	[160]
	White asparagus	Reduced weight loss and color changes	[83]
	Freeze-dried strawberry	Prevented texture collapse after rehydration	[161]
	Dry-roasted peanuts	Delayed rancidity development	[162]
	Cheese	Decreased water and hardness loss and microorganism growth	[163]
	Frozen salmon	Delayed lipid oxidation	[164]
	Trout	Delayed lipid oxidation	[166]
	Black sea sprat	Delayed lipid oxidation	[167]
Zein	Tomatoes	Delayed the weight and firmness loss and color change	[178]
	Apricots	Reduced color changes and the total viable bacterial count	[179]
	Apples	Made surface shiny and extended shelf life	[180]
	Mango	Delayed ripening and retained ascorbic acid and phenolic content	[181]

cannot be hydrolyzed [126]. Lipid coatings are usually wax-, acylglycerol-, or resin-based coatings [26].

Waxes are esters of long-chain aliphatic acids with long-chain aliphatic alcohols [182]. The alcohol chain normally has an even number of carbons from 24 to 36, while the carboxylic acid has an even number of carbons from 16 to 36 [126]. They are hydrophobic, since they can have no polar constituents or have a hydrophilic part very small and they possess high content of long-chain fatty alcohols and alkanes [183]. Consequently, they are more resistant to water diffusion than major edible film materials [28]. Triglycerides are esters of fatty acids and glycerol. Mono-, di-, and triglycerides can be applied as coating materials. They have higher polarity with respect to waxes, but their chemical structure can significantly modify their functional properties, in particular water vapor permeability [182]. Short-chain molecules are partially water-soluble, while long-chain triglycerides are not soluble in water. Unsaturated fatty acids have increased moisture permeability and lower melting points compared to the corresponding saturated ones [28]. The most used lipid-based materials for edible coatings are beeswax, candelilla wax, carnauba wax, and shellac resins.

Lipids do not have the ability to form cohesive edible materials processable in coatings and films, since they normally form nonuniform and brittle structure due to their hydrophobicity [24]. Therefore, they need to be associated with film-forming agents or with other biopolymers to form edible composite materials [23].

3.4.1 Beeswax

Beeswax is an animal wax and is produced by honeybees. Bees secrete wax from four wax glands, located on the underside of their abdomen [184]. Beeswax is a mixture of hydrocarbons, alcohols, fatty acids, wax esters as well as other materials that exhibits viscoelastic behavior [185, 186]. The melting temperature of beeswax is about 60 ° C, [184]. The high presence of free fatty acids contributes to its viscosity. The hydrolysis of the esters yields C_{26} and C_{28} carboxylic acids and C_{30} and C_{32} straight-chain primary alcohols [185]. Beeswax has been coupled with polysaccharides and proteins to develop effective formulations to preserve fruit and vegetables, and the effect of these solutions will be discussed later in this chapter (Table 3.6).

3.4.2 Candelilla Wax

Candelilla wax is a complex material of vegetable origin. It is produced by a reed-like plant named *Euphorbia antisyphilitica* that grows wild in desertic environments, preferably calcium-rich soils, like those of northwestern Mexico and southern Texas [205]. Its common name derives from the shape of its stems that resemble candles. It has a high content of hydrocarbons (around 50%) with some wax acid esters and alcohols [185]. Consequently, candelilla is not soluble in water but highly soluble in acetone, benzene, chloroform, and other organic solvents [206]. It is recognized by FDA as a "safe" substance (GRAS) for application in the foods industry. Lakshminarayana et al. [207] studied the

Table 3.5 Lipid-based edible coatings.

Coating material	Food	Effect	References
Carnauba wax	Tangor	Reduced mass loss and color changes	[212]
	Eggplant	Reduced mass and firmness loss	[193]
Candelilla wax	Limes	Reduced weight loss and microbial spoilage	[207]
	Oranges	Reduced weight loss and microbial spoilage	[207]
	Grapefruit	Reduced weight loss and microbial spoilage	[207]
	Apples	Reduced changes in colors and weight loss	[209]
	Avocados	Reduced changes in appearance and weight loss	[206]
	Bananas	Reduced weight loss	[210]
Shellac	Oranges	Reduced weight loss and respiration rate	[213]

efficacy of candelilla aqueous wax emulsions in extending the storage life of limes, oranges, and grapefruit in fresh state and stored at ambient conditions. The fruit coated with candelilla wax showed significantly reduced weight loss during 60 days of storage when compared to the uncoated controls. Losses due to microbial spoilage and desiccation were minimized. The efficacy of candelilla wax was also reported by Paredes-López et al. [208]. Furthermore, candelilla wax was found to be able to extend the shelf life of apples [209], avocados [206], and bananas [210].

3.4.3 Carnauba Wax

Carnauba wax is an exudate from the Brazilian plant *Copernica cerifera* [197]. It is composed of C_{32} and C_{34} straight-chained primary alcohols and wax acid esters of C_{24} and C_{28} carboxylic acids [185]. Carnauba wax is the hardest natural commercial wax, and has a high melting temperature [186]. It is considered a GRAS material, and can be used as a coating for different food products, which include fruits and vegetables [211], as also reported in Table 3.5. For example, carnauba wax emulsions applied as coatings were able to reduce mass loss and maintain a greener color in *Ortanique* tangor [212] and reduce mass and firmness loss in eggplants [193].

3.4.4 Shellac

Shellac resin is secreted by the insect *Laccifer lacca* in the forests of India and Thailand [214, 215]. It is composed of several aliphatic polyhydroxy acids present in the form of lactides and inter esters [216]. It is a versatile substance and dissolves in alcohols and alkaline solutions. Shellac can be incorporated into wax formulations, since it is compatible with most waxes, in order to give higher gloss appearance to the coated products [211]. Shellac coatings were applied to oranges by Baldwin et al. [213]. Coated oranges showed less weight loss, lower internal O_2 and higher CO_2 and ethylene concentrations than uncoated samples.

3.5 Edible Composite Materials

Edible composite materials are heterogeneous in nature and consist of a combination of different classes of material [26]. They are normally developed to exploit the different advantages of each component, while minimizing their disadvantages, by combining the distinct functional characteristics of each category of coating or film former [28]. The main goal of engineering edible composite materials (e.g. by mixing carbohydrates and proteins, carbohydrates and lipids, or proteins and lipids [217]) is to improve the permeability or mechanical properties of a material to fulfil the needs of a specific application. In fact, edible coatings and films composed of a single material can perform well in exhibiting a specific property, but may not be able to provide several properties simultaneously. For example, proteins and polysaccharides are good film-forming materials with excellent structural and mechanical properties but are poor moisture barriers. On the other hand, lipids form brittle films but, being hydrophobic, can act as an excellent water vapor barrier [26]. These heterogeneous films can be developed either in the form of successive layers, in bilayer or multilayer structure, or in a suspension or emulsion of the nonmiscible constituents, where the lipid is entrapped in the supporting biopolymer matrix, or in the form of blend from a solution in a common solvent [23, 24].

Many studies related to the combination of polysaccharides and lipids have been conducted (Table 3.6). Velickova et al. [18] coated strawberries with a three-layer coating consisting of a multilayer composed of beeswax-chitosan-beeswax. The coating significantly decreased the respiration rate and weight loss of the fruits and improved color and texture retention. Avocados exposed to *Lasiodiplodia theobromae* infection were coated with an emulsion of pectin and beeswax in order to study its effect on the infection spread [218]. Coated avocados underwent a lower rate of disease spread, respiration rate, and texture and color changes when compared to the controls. An edible composite coating made of HPMC and beeswax extended marketability on plums by delaying softening, internal breakdown, and water loss with respect to uncoated or HPMC-coated controls [187]. This result indicates the efficacy of hydrophobic compounds in enhancing moisture barrier properties. Emulsions of HPMC and beeswax were also effective in preserving the quality of cherry tomatoes during cold storage [188, 219], apples [160], mandarins [189, 190], and oranges [191]. Carnauba wax was used with cassava starch to form emulsion coatings applied on apples reducing respiration rate and weight loss [192], or with sodium alginate to coat eggplants reducing oxidation and firmness loss [193]. Candelilla wax was combined with pectin to form an emulsion, and was able to reduce weight loss and delay deterioration of green peppers [194].

Proteins have been also used in many formulations with lipids to develop edible coatings. Edible emulsion coatings from WPI and beeswax had an anti-browning effect on fresh-cut apples [195], and the same results were obtained with an emulsion of WPI and carnauba wax [160] or by coupling WPC and beeswax and forming an emulsion [196]. Mehyar et al. [197] applied an emulsion of WPI and carnauba wax on pine nuts and walnuts during storage at 25 °C which was effective

Table 3.6 Composite edible coatings.

Coating materials	Structure	Food	Effect	References
Polysaccharides and lipids				
Beeswax (BW) and HPMC	Emulsion	Plums	Delayed softening, internal breakdown, and water loss	[187]
		Cherry tomatoes	Reduced weight and firmness loss	[188]
		Apples	Decreased enzymatic browning	[160]
		Mandarins	Reduced weight and firmness loss	[189, 190]
		Oranges	Showed antimicrobial effect	[191]
Carnauba wax + cassava starch	Emulsion	Apples	Reduced respiration rate and weight loss	[192]
Carnauba wax + sodium alginate	Emulsion	Eggplants	Decreased firmness loss and oxidation	[193]
Carnauba wax + pectin	Emulsion	Green peppers	Reduced weight loss and delayed deterioration	[194]
Proteins and lipids				
BW + WPI	Emulsion	Fresh-cut apples	Reduced browning	[195]
BW + WPC	Emulsion	Fresh-cut apples	Reduced browning	[196]
Carnauba wax + WPI	Emulsion	Fresh-cut apples	Reduced browning	[160]
		Pine nuts Walnuts	Reduced oxidative and hydrolytic rancidity	[197]
BW + WPI + flaxseed oil	Emulsion	Plums	Reduced mass and firmness loss	[40]
Candelilla wax + pea protein	Emulsion	Brussel sprouts	Reduced weight loss and softening	[198]
		Grapes	Reduced weight loss and retained ascorbic acid	[199]
		Broccolis	Improved vitamin C and firmness retention	[200]

(Continued)

Table 3.6 (Continued)

Coating materials	Structure	Food	Effect	References
Polysaccharides and proteins				
Chitosan–CMC–chitosan	Multilayer	Strawberries	Preserved firmness and aroma	[201]
Chitosan–CMC	Bilayer	Mandarins	Improved firmness and gloss appearance	[97]
		Oranges	Improved firmness and gloss appearance	[97]
		Grapefruit	Improved firmness and gloss appearance	
Chitosan–pectin	Multilayer	Mangos	Reduced microbial spoilage, browning and mass loss	[202]
Chitosan–alginate	Bilayer	Fresh-cut melon	Prevented bacteria growth and firmness loss	[203]
Chitosan–gelatin	Bilayer	Trout fillets	Reduced lipid oxidation and rancidity development	[204]

in reducing oxidative and hydrolytic rancidity. Reinoso et al. [40] evaluated the quality of plums coated with WPI composite coatings containing flaxseed oil blended with beeswax. The coating reduced mass and firmness loss and improved sensory acceptability with respect to uncoated samples. Candelilla wax and pea protein coatings were found to be effective in several applications: reduced weight loss and softening and improved overall appearance of Brussels sprouts during storage [198], extended the postharvest life of grapes by improving retention of ascorbic acid and reducing weight loss [199], improved vitamin C, color and firmness retention of broccolis [200].

Several multilayer coating structures based on different hydrocolloids have been studied. A coating of chitosan-CMC-chitosan was found to have an effect in preserving the firmness and aroma volatiles of strawberries [201]. A bilayer coating based on chitosan and CMC was applied on different citrus fruits, such as mandarins, oranges, and grapefruit, improving their firmness and gloss appearance [97, 220]. Medeiros et al. [202] evaluated the effects of a multilayer coating made of chitosan and pectin on the shelf life of mangoes. The coating reduced microbial spoilage, browning, and mass loss. Chitosan was also coupled with alginate in a bilayer coating able to prevent bacteria growth and firmness loss of fresh-cut melon [203]. A bilayer of chitosan and gelatin was found to be helpful in reducing lipid oxidation and rancidity development in rainbow trout fillets during 16 days at refrigerated storage conditions (4 °C) [204].

3.6 Active Coatings

Apart from their traditional function as barrier, active edible coatings and films perform an additional role in preserving the quality and safety of food by interacting with the coated product and the surrounding environment [28]. Active edible coatings and films are systems in which active ingredients have been intentionally included in the protective material to improve the performance of the coating or film [26]. These active ingredients, such as antimicrobials and antioxidant agents, mitigate the agents that cause food deterioration and decay [28].

3.6.1 Antimicrobial Agents

The principal cause of decay for many food products is surface microbial growth. Edible coatings, such as chitosan, cellulose and zein, can act as antimicrobial carriers since they are effective in retaining additives on food surfaces [17, 30]. In fact, the application of antimicrobial substances directly on the product surface has few benefits because the active compounds can diffuse into the food product, reducing the effect of the antimicrobial agent on the surface [21]. The incorporation of antimicrobial agents in edible coatings provides increased inhibitory effects against spoilage because it maintains the concentration of the active ingredients on the food surfaces [17]. There are many categories of antimicrobials that can be added into edible coatings: plant extracts and essential oils, organic acids, fatty acid esters, and polypeptides [21]. Vu et al. [221] extended the shelf life of strawberries by developing bioactive coatings based on chitosan and different essential oils. Red thyme, peppermint, and limonene essential oils showed the highest efficiency. Perdones et al. [222] reported the beneficial effect of chitosan coatings containing lemon essential oil on strawberry preservation. HPMC coatings decorated with oregano and bergamot essential oils were found to be effective in reducing total cell count, respiration rate, ethylene production, and weight loss of plums [223]. Yun et al. [224] investigated the effectiveness of zein-based coatings containing cinnamon and mustard essential oil in reducing populations of *Salmonella enterica serovar Typhimurium* inoculated in cherry tomatoes. Populations of *S. Typhimurium* decreased while increasing essential oil concentration and storage duration. The highest antimicrobial effect was showed by zein coating with 20% w/w of cinnamon. The coating also prevented loss of weight, firmness, and vitamin C (Table 3.7) [224].

3.6.2 Antioxidant Agents

Antioxidant agents are added to edible films in order to increase the stability of lipids, maintain color and nutritional value, avoiding oxidative rancidity and discoloration by slowing down the rate of oxidation reactions as summarized in Table 3.7 [17]. Primary antioxidants are free radical acceptors, such as butylated

Table 3.7 Active edible coatings.

Coating materials	Active ingredient	Food	References
Antimicrobials			
Chitosan	Red thyme	Strawberries	[221]
	Peppermint	Strawberries	[221]
	Limonene	Strawberries	[221]
	Lemongrass	Strawberries	[222]
HPMC	Oregano and bergamot	Plums	[223]
Zein	Cinnamon	Cherry tomatoes	[224]
Antioxidants			
Apple puree	Ascorbic acid	Fresh-cut apples	[225]
Carrageenan	Ascorbic, citric, and oxalic acid	Fresh-cut apples	[226]
Whey protein concentrate	Ascorbic, citric, and oxalic acid	Fresh-cut apples	[226]
WPI and beeswax	Ascorbic acid	Fresh-cut apples	[196]
CMC and soy protein	Ascorbic acid	Fresh-cut apples	[227]
		Sliced potatoes	[227]
MC	Ascorbic and steric acid	Pears	[228]

hydroxytoluene, butylated hydroxyanisole, tertiary butylated hydroxyquinone, and propyl gallate, that delay the initiation or propagation steps of oxidation reactions [17, 30]. Secondary antioxidants, such as citric acid, ascorbic acid, and tartaric acid, delay oxidation by exerting different actions like scavenging O_2, deactivating singlet O_2, absorbing UV rays, or promoting the activity of primary antioxidants [30]. These substances, if incorporated into edible coatings and films, can be maintained on the food–air interface, where the oxidation takes place [17]. Mchugh and Senesi [225] developed a new coating to extend fresh-cut apple shelf life based on apple puree, which acted as oxygen barrier, with the addition of citric and ascorbic acid, which acted as antioxidant agents. Apple puree supplemented with citric acid was able to preserve the color for 12 days at 5 °C. The apple color was also preserved using carrageenan or WPC with the addition of ascorbic, citric, and oxalic acid [226], and an emulsion of WPI, beeswax, and ascorbic acid [196]. The same effect can be achieved with a blend of CMC and soy protein isolate containing ascorbic acid applied both on apple and potatoes [227]. Olivas et al. [228] prolonged the shelf life of pears by retarding browning with a coating of MC and steric acid with the incorporation of ascorbic acid.

3.7 Materials Selection and Application

The selection of the optimal edible coatings and films must take into account the food product specifications. As a general rule, products with high water content need hydrophobic coatings (e.g. lipids) to avoid moisture loss, while products with high lipid content, such as meat, need oxygen barriers (e.g. hydrocolloids) to prevent oxidation [17]. For products that experience growth of molds and bacteria, such as cheese, the incorporation of antimicrobials can be crucial in increasing shelf life [103]. In the case of fruit products, it is important to consider whether or not the fruit is climacteric. The ripening stage of climacteric fruits, such as apples, bananas, and avocados, is characterized by ethylene production and the increase of respiration rates, which make the application of coatings able to reduce gas exchange necessary in order to achieve maturation control [229]. Polysaccharides and proteins are good candidates to fulfil those requirements and their properties can be enhanced through the incorporation of antioxidant agents. In non-climacteric fruits, such as strawberries and citrus fruit, the perishability is mostly related to moisture loss that can be reduced by applying lipid coatings or combinations between lipids and hydrocolloids which perform well both in reducing weight loss and respiration rate.

The efficacy of coatings is also related to the method of application selected between dipping, brushing, and spraying. The procedure must ensure the complete covering of the food and can be chosen based on the size of the product and on the surface characteristics [103]. For products with irregular surfaces, dipping is the most effective and simple choice, while, if the main purpose is to obtain a homogeneous and thin coating, spraying is the better option since it also reduces wastage of coating caused by the big tanks needed by dipping process. Brushing can be effectively implemented in small industries since it does not require big and complex machinery. Edible films can be applied by wrapping the food product; in this case the films are produced before the application as free-standing materials. Edible films can be created by casting the solution on a continuous steel belt that enters a drying chamber to remove water or by melt extrusion, if thermoplastic materials are used [17].

3.8 Conclusions

Edible coatings and films are efficient solutions to increase food shelf life and food security, consequently, they are beneficial in the reduction of food waste and they can also be good candidates against malnutrition in the geographical areas that have limited access to nutritious and fresh food. In order to achieve this goal, an interesting field which recent research has been focusing on is the enhancement of the nutritional value of food by incorporating active compounds and nutrients in edible coatings. In this context, nanotechnological solutions (e.g. nano-encapsulation and nano-multilayered systems) are of increasing interest since they allow to control the release of bioactive compounds while protecting them from heat and moisture that could cause their degradation. In fact, being encapsulated, the active substances can be released slowly in order to keep the functionality of the compound and provide protection to the food during a longer period. Because of these advantages, nano-systems constitute an interesting area of food research and have a high potential in the development of more effective formulations to preserve food while increasing its nutritional value. Due to consumers' increasing interest in eating healthy food, the incorporation of nutrients in coating formulation could be successful on the market.

Edible coatings and films are usually consumed with the coated products, and, therefore, their consumer acceptance should be considered very important, as much as physical or chemical characteristics since poor consumer interest implies poor selling power, even if an edible material has good barrier properties. Under this point of view, hydrocolloids usually generate transparent films, while edible coatings made of lipids generally create an opaque film that could make food unappealing. Coating thickness also has to be considered as barrier properties are enhanced by thicker coatings, which however could induce changes in food texture and taste. The addition of active compounds such as antimicrobials or antioxidants, especially essential oils (EOs), could affect the sensory acceptability of the coated food by imparting unpleasant odor and flavor that can lead to rejection of the product by consumers. Nano-encapsulation could be a solution to this problem, since it can mask the flavor of the active compound and, controlling release during time, allows the use of a smaller amount of it.

Many edible coatings contain ingredients that could cause allergic reactions, such as milk, soy, and wheat derivatives; for this reason their consumption is limited to part of the population. A similar problem arises with coatings and films of animal origin (e.g. chitosan, gelatin, silk fibroin, etc.), which cannot be eaten by people who abstain from the use of specific animal products in their diet due to personal or religious beliefs.

Another drawback of the application of edible coatings and films is the general high cost of the materials they are made of, compared to the cost of classical synthetic packaging. As a consequence, coated foods are more expensive than uncoated ones, and for this reason, close attention should be paid to contain the costs of this technology. The coating should not cause a great rise in food price, thereby turning simple and cheap products into exclusive products.

References

1 FAO (2011). *Global Food Losses and Food Waste*. Food and Agriculture Organization of the United Nations (FAO).

2 Corrado, S. and Sala, S. (2018). Food waste accounting along global and European food supply chains: state of the art and outlook. *Waste Management* 79: 120–131. doi: 10.1016/J.WASMAN.2018.07.032.

3 Garrone, P., Melacini, M., and Perego, A. (2014). Opening the black box of food waste reduction. *Food Policy* 46: 129–139. doi: 10.2139/ssrn.2109594.

4 Garrone, P., Melacini, M., Perego, A., and Sert, S. (2016). Reducing food waste in food manufacturing companies. *Journal of Cleaner Production* 137: 1076–1085. doi: 10.1016/J.JCLEPRO.2016.07.145.

5 Marelli, B., Brenckle, M.A., Kaplan, D.L., and Omenetto, F.G. (2016). Silk fibroin as edible coating for perishable food preservation. *Scientific Reports* 6: 1–11. doi: 10.1038/srep25263.

6 Göbel, C., Langen, N., Blumenthal, A. et al. (2015). Cutting food waste through cooperation along the food supply chain. *Sustainability* 7 (2): 1429–1445. doi: 10.3390/su7021429.
7 Stenmarck, A., Hanssen, O.J., Silvennoinen, K. et al. (2011). Initiatives on Prevention of Food Waste in the Retail and Wholesale Trades Financed by Nordic Council of Ministers. Stockholm, Sweden.
8 Regulation (EC) No. 1182/2007 (2007). European Commission.
9 Regulation (EC) No. 1221/2008 (2008). European Commission. *Official Journal of the European Union.*
10 Regulation (EC) No. 1580/2007 (2007). European Commission.
11 Ribeiro Cerqueira, M.A.P., Correia Pereira, R.N., Leandro da Silva Ramos, O. et al. (eds.) (2016). *Edible Food Packaging*. Boca Raton, FL: CRC Press.
12 Guti, T.J. (ed.) (2018). *Polymers for Food Applications*. Springer Nature.
13 Allende, A., Tomás-Barberán, F.A., and Gil, M.I. (2006). Minimal processing for healthy traditional foods. *Trends in Food Science & Technology* 17: 513–519. doi: 10.1016/j.tifs.2006.04.005.
14 Olivas, G.I. and Barbosa-Cánovas, G.V. (2010). Edible coatings for fresh-cut fruits. *Fruit and Vegetable Processing*. doi: 10.1201/9781439823187.ch16.
15 Mehyar, G.F. and Han, J.H. (2011). Active packaging for fresh-cut fruits and vegetables. In: *Modified Atmosphere Packaging for Fresh-Cut Fruits and Vegetables* (eds. A.L. Brody, H. Zhuang and J.H. Han), 267–284. Wiley-Blackwell.
16 Watada, A.E., Ko, N.P., and Minott, D.A. (1996). Factors affecting quality of fresh-cut horticultural products. *Postharvest Biology and Technology* 9 (2): 115–125. doi: 10.1016/S0925-5214(96)00041-5.
17 Embuscado, M.E. and Huber, K.C. (2009). *Edible Films and Coatings for Food Applications*. doi: 10.1007/978-0-387-92824-1.
18 Velickova, E., Winkelhausen, E., Kuzmanova, S. et al. (2013). Impact of chitosan-beeswax edible coatings on the quality of fresh strawberries (*Fragaria Ananassa* cv Camarosa) under commercial storage conditions. *LWT - Food Science and Technology* 52 (2): 80–92. doi: 10.1016/j.lwt.2013.02.004.
19 Falguera, V., Quintero, J.P., Jiménez, A. et al. (2011). Edible films and coatings: structures, active functions and trends in their use. *Trends in Food Science and Technology* 22 (6): 292–303. doi: 10.1016/j.tifs.2011.02.004.
20 Loizzo, M.R. (2019). New packaging strategies to guarantee the quality of minimally processed fruits: the application of edible coatings. *EC Nutrition* 14 (1): 1–6.
21 Rojas-Graü, M.A., Soliva-Fortuny, R., and Martín-Belloso, O. (2009). Edible coatings to incorporate active ingredients to fresh-cut fruits: a review. *Trends in Food Science and Technology* 20 (10): 438–447. doi: 10.1016/j.tifs.2009.05.002.
22 Zhang, D., Zhang, M., and Xiaoxiao, G. (2018). Seaweed-derived hydrocolloids as food coating and encapsulation agents. In: *Bioactive Seaweeds for Food Applications* (ed. Y. Qin), 153–175. doi: 10.1016/B978-0-12-813312-5.00008-X.

23 Bourtoom, T. (2008b). Review article edible films and coatings: characteristics and properties. *International Food Research Journal* 15: 237–248.

24 Hassan, B., Chatha, S.A.S., Hussain, A.I. et al. (2018). Recent advances on polysaccharides, lipids and protein based edible films and coatings: a review. *International Journal of Biological Macromolecules* 109: 1095–1107. doi: 10.1016/j.ijbiomac.2017.11.097.

25 Jiménez, A., Fabra, M.J., Talens, P., and Chiralt, A. (2012). Edible and biodegradable starch films: a review. *Food and Bioprocess Technology*. doi: 10.1007/s11947-012-0835-4.

26 Yousuf, B., Qadri, O.S., and Srivastava, A.K. (2018). Recent developments in shelf-life extension of fresh-cut fruits and vegetables by application of different edible coatings: a review. *LWT* 89: 198–209. doi: 10.1016/j.lwt.2017.10.051.

27 Dehghani, S., Hosseini, S.V., and Regenstein, J.M. (2018). Edible films and coatings in seafood preservation: a review. *Food Chemistry* 240: 505–513. doi: 10.1016/J.FOODCHEM.2017.07.034.

28 de Azeredo, H.M.C. (2012). Edible coatings. In: *Advances in Fruit Processing Technologies* (eds. S. Rodrigues and F.A.N. Fernandes), 345–356. CRC Press.

29 Debeaufort, F., Quezada-Gallo, J.-A., and Voilley, A. (1998). Edible films and coatings: tomorrow's packagings: a review. *Critical Reviews in Food Science and Nutrition* 38 (4): 299–313. doi: 10.1080/10408699891274219.

30 Robertson, G.L. (2013). *Food Packaging Principles and Practice*. Boca Raton, FL: CRC Press Taylor & Francis Group.

31 Regulation (EC) No. 1331/2008 (2008). European Commission. *Official Journal of the European Union*.

32 Regulation (EU) No. 234/2011 (2011). European Commission. *Official Journal of the European Union*.

33 FDA (2005). *CPG Sec. 562.550 Safety and Labeling of Waxed (Coated) Fruits and Vegetables*. Center for Food Safety and Applied Nutrition Office of Regulatory Affairs of the Food and Drug Administration (FDA).

34 Di Pierro, P., Sorrentino, A., Mariniello, L. et al. (2011). Chitosan/whey protein film as active coating to extend ricotta cheese shelf-life. *LWT - Food Science and Technology* 44 (10): 2324–2327. doi: 10.1016/j.lwt.2010.11.031.

35 Bonilla, J., Atarés, L., Vargas, M., and Chiralt, A. (2012). Edible films and coatings to prevent the detrimental effect of oxygen on food quality: possibilities and limitations. *Journal of Food Engineering* 110 (2): 208–213. doi: 10.1016/j.jfoodeng.2011.05.034.

36 Shih, F.F. (1996). Edible films from rice protein concentrate and pullulan. *Cereal Chemistry* 73 (3): 406–409.

37 De Moraes, K.S., Fagundes, C., Melo, M.C. et al. (2012). Conservation of Williams pear using edible coating with alginate and carrageenan. *Ciência e Tecnologia de Alimentos* 32 (4): 679–684. doi: 10.1590/S0101-20612012005000106.

38 Tanaka-Palmu, P., Helén, H., and Hyvönen, L. (2000). Preparation, properties and applications of wheat gluten edible films. *Agricultural and Food Science in Finland* 9: 23–35.
39 Chiou, B.-S., Avena-Bustillos, R.J., Bechtel, P.J. et al. (2008). Cold water fish gelatin films: effects of cross-linking on thermal, mechanical, barrier, and biodegradation properties. *European Polymer Journal* 44 (11): 3748–3753. doi: 10.1016/J.EURPOLYMJ.2008.08.011.
40 Reinoso, E., Mittal, G.S., and Lim, L.-T. (2008). Influence of whey protein composite coatings on plum (*Prunus domestica* L.) fruit quality. *Food Bioprocess Technology*. doi: 10.1007/s11947-007-0014-1.
41 Sharma, P., Shehin, V.P., Kaur, N., and Vyas, P. (2018). Application of edible coatings on fresh and minimally processed vegetables: a review. *International Journal of Vegetable Science*. doi: 10.1080/19315260.2018.1510863.
42 Villafañe, F. (2017). Edible coatings for carrots. *Food Reviews International* 33 (1): 84–103. doi: 10.1080/87559129.2016.1150291.
43 Garcia, L.C., Pereira, L.M., and Hubinger, M.D. (2010). Selection of an edible starch coating for minimally processed strawberry. doi: 10.1007/s11947-009-0313-9.
44 Tavassoli-Kafrani, E. (2016). Development of edible films and coatings from alginates and carrageenans. *Carbohydrate Polymers* 137: 360–374. doi: 10.1016/J.CARBPOL.2015.10.074.
45 Valero, D., Díaz-Mula, H.M., Zapata, P.J. et al. (2013). Effects of alginate edible coating on preserving fruit quality in four plum cultivars during postharvest storage. *Postharvest Biology and Technology* 77: 1–6. doi: 10.1016/J.POSTHARVBIO.2012.10.011.
46 Díaz-Mula, H.M., Serrano, M., and Valero, D. (2012). Alginate coatings preserve fruit quality and bioactive compounds during storage of sweet cherry fruit. *Food and Bioprocess Technology* 5 (8): 2990–2997. doi: 10.1007/s11947-011-0599-2.
47 Maftoonazad, N., Ramaswamy, H.S., and Marcotte, M. (2008). Shelf-life extension of peaches through sodium alginate and methyl cellulose edible coatings. *International Journal of Food Science & Technology* 43 (6): 951–957. doi: 10.1111/j.1365-2621.2006.01444.x.
48 Olivas, G.I., Mattinson, D.S., and Barbosa-Cánovas, G.V. (2007). Alginate coatings for preservation of minimally processed 'gala' apples. *Postharvest Biology and Technology* 45 (1): 89–96. doi: 10.1016/J.POSTHARVBIO.2006.11.018.
49 Azarakhsh, N., Osman, A., Ghazali, H.M. et al. (2012). Optimization of alginate and gellan-based edible coating formulations for fresh-cut pineapples. *International Food Research Journal* 19: 279–285.
50 Nussinovitch, A. and Hershko, V. (1996). Gellan and alginate vegetable coatings. *Carbohydrate Polymers* 30 (2–3): 185–192. doi: 10.1016/S0144-8617(96)00095-1.
51 Zhong, Y., Cavender, G., and Zhao, Y. (2014). Investigation of different coating application methods on the performance of edible coatings on

mozzarella cheese. *LWT - Food Science and Technology* 56 (1): 1–8. doi: 10.1016/J.LWT.2013.11.006.
52 Wanstedt, K.G. and Seideman, S.C. (1981). Sensory attributes of precooked, calcium alginate-coated pork patties. *Journal of Food Protection* 44 (10): 732–735.
53 Hamzah, H.M., Osman, A., Tan, C.P., and Ghazali, F.M. (2013). Carrageenan as an alternative coating for papaya (*Carica papaya* L. cv. Eksotika). *Postharvest Biology and Technology* 75: 142–146. doi: 10.1016/J.POSTHARVBIO.2012.08.012.
54 Ribeiro, C., Vicente, A.A., Teixeira, J.A., and Miranda, C. (2007). Optimization of edible coating composition to retard strawberry fruit senescence. *Postharvest Biology and Technology* 44 (1): 63–70. doi: 10.1016/j.postharvbio.2006.11.015.
55 Kampf, N. and Nussinovitch, A. (2000). Hydrocolloid coating of cheeses. *Food Hydrocolloids* 14 (6): 531–537. doi: 10.1016/S0268-005X(00)00033-3.
56 Gol, N.B., Patel, P.R., and Ramana Rao, T.V. (2013). Improvement of quality and shelf-life of strawberries with edible coatings enriched with chitosan. *Postharvest Biology and Technology* 85: 185–195. doi: 10.1016/J.POSTHARVBIO.2013.06.008.
57 Shahbazi, Y. (2018). Application of carboxymethyl cellulose and chitosan coatings containing mentha spicata essential oil in fresh strawberries. *International Journal of Biological Macromolecules* 112: 264–272. doi: 10.1016/J.IJBIOMAC.2018.01.186.
58 Hosseini, M. and Rashidi, M. (2009). Effects of coating methods and storage periods on some qualitative characteristics of carrot during ambient storage. *International Journal of Agriculture and Biology* 11: 443–447.
59 Zhuang, R.-Y. and Huang, Y.-W. (2002). Influence of hydroxypropyl methylcellulose edible coating on fresh-keeping and storability of tomato. *Journal of Zhejiang University SCIENCE*: 109–113.
60 Mallikarjunan, P., Chinnan, M.S., Balasubramaniam, V.M. et al. (1997). Edible coatings for deep-fat frying of starchy products. *LWT - Food Science and Technology* 30 (7): 709–714.
61 Chien, P.J., Sheu, F., and Yang, F.H. (2007). Effects of edible chitosan coating on quality and shelf life of sliced mango fruit. *Journal of Food Engineering* 78 (1): 225–229. doi: 10.1016/j.jfoodeng.2005.09.022.
62 Elsabee, M.Z. and Abdou, E.S. (2013). Chitosan based edible films and coatings. *Materials Science and Engineering C* 33: 1819–1841.
63 Pushkala, R., Parvathy, K.R., and Srividya, N. (2012). Chitosan powder coating, a novel simple technique for enhancement of shelf life quality of carrot shreds stored in macro perforated LDPE packs. *Innovative Food Science and Emerging Technologies* 16: 11–20. doi: 10.1016/j.ifset.2012.03.003.
64 Pushkala, R., Raghuram, P.K., and Srividya, N. (2013). Chitosan based powder coating technique to enhance phytochemicals and shelf life quality of radish shreds. *Postharvest Biology and Technology* 86: 402–408. doi: 10.1016/j.postharvbio.2013.07.025.

65 Olawuyi, I.F., Park, J.J., Lee, J.J., and Lee, W.Y. (2019). Combined effect of chitosan coating and modified atmosphere packaging on fresh-cut cucumber. *Food Science & Nutrition* 7 (3): 1043–1052. doi: 10.1002/fsn3.937.
66 Cui, H.Y., Wu, J., Li, C.Z., and Lin, L. (2016). Anti-listeria effects of chitosan-coated Nisin-silica liposome on Cheddar cheese. *Journal of Dairy Science* 99 (11): 8598–8606. doi: 10.3168/jds.2016-11658.
67 Cano, E., Isabel, A., Nácher, M.C. et al. (2017). Quality of goat's milk cheese as affected by coating with edible chitosan-essential oil films. *International Journal of Dairy Technology* 70 (1): 68–76. doi: 10.1111/1471-0307.12306.
68 Altieri, C., Scrocco, C., Sinigaglia, M., and Del Nobile, M.A. (2005). Use of chitosan to prolong mozzarella cheese shelf Life. *Journal of Dairy Science* 88. doi: 10.3168/jds.S0022-0302(05)72946-5.
69 Guo, M., Jin, T.Z., Wang, L. et al. (2014). Antimicrobial films and coatings for inactivation of listeria innocua and salmonella typhimurium on ready-to-eat meat. *Food Control* 40: 215–233.
70 Jiang, Z., Neetoo, H., and Chen, H. (2011). Efficacy of freezing, frozen storage and edible antimicrobial coatings used in combination for control of listeria monocytogenes on roasted Turkey stored at chiller temperatures. *Food Microbiology* 28 (7): 1394–1401. doi: 10.1016/j.fm.2011.06.015.
71 Beverlya, R.L., Janes, M.E., Prinyawiwatkula, W., and No, H.K. (2008). Edible chitosan films on ready-to-eat roast beef for the control of listeria monocytogenes. *Food Microbiology* 25 (3): 534–537. doi: 10.1016/j.fm.2007.11.002.
72 Petrou, S., Tsiraki, M., Giatrakou, V., and Savvaidis, I.N. (2012). Chitosan dipping or oregano oil treatments, singly or combined on modified atmosphere packaged chicken breast meat. *International Journal of Food Microbiology* 156 (3): 264–271. doi: 10.1016/j.ijfoodmicro.2012.04.002.
73 Sánchez-Ortega, I., García-Almendárez, B.E., Santos-López, E.M. et al. (2014). Antimicrobial edible films and coatings for meat and meat products preservation. *The Scientific World Journal* 2014: 248935. doi: 10.1155/2014/248935.
74 Sathivel, S., Liu, Q., Huang, J., and Prinyawiwatkul, W. (2007). The influence of chitosan glazing on the quality of skinless pink salmon (*Oncorhynchus gorbuscha*) fillets during frozen storage. *Journal of Food Engineering* 83 (3): 366–373. doi: 10.1016/j.jfoodeng.2007.03.009.
75 Maftoonazad, N. and Ramaswamy, H.S. (2019). Application and evaluation of a pectin-based edible coating process for quality change kinetics and shelf-life extension of lime fruit (*Citrus aurantifolium*). *Coatings* 9 (5): 285. doi: 10.3390/coatings9050285.
76 Oms-Oliu, G., Soliva-Fortuny, R., and Martín-Belloso, O. (2008). Using polysaccharide-based edible coatings to enhance quality and antioxidant properties of fresh-cut melon. *LWT - Food Science and Technology* 41 (10): 1862–1870. doi: 10.1016/J.LWT.2008.01.007.
77 Moalemiyan, M. and Ramaswamy, H.S. (2012). Quality retention and shelf-life extension in mediterranean cucumbers coated with a pectin-based film. *Journal of Food Research* 1 (3): 159. doi: 10.5539/jfr.v1n3p159.

78 Garcia, C.C., Caetano, L.C., de Souza Silva, K., and Mauro, M.A. (2014). Influence of edible coating on the drying and quality of papaya (*Carica papaya*). *Food and Bioprocess Technology* 7 (10): 2828–2839. doi: 10.1007/s11947-014-1350-6.

79 Wua, S. and Chen, J. (2013). Using pullulan-based edible coatings to extend shelf-Life of fresh-cut 'Fuji' apples. *International Journal of Biological Macromolecules* 55: 254–257.

80 Kraśniewska, K., Ścibisz, I., Gniewosz, M. et al. (2017). Effect of pullulan coating on postharvest quality and shelf-life of highbush blueberry (*Vaccinium corymbosum* L.). *Materials* 10 (8): 965. doi: 10.3390/ma10080965.

81 Treviño-Garza, M.Z., García, S., del Socorro Flores-González, Ma., and Arévalo-Niño, K. (2015). Edible active coatings based on pectin, pullulan, and chitosan increase quality and shelf life of strawberries (*Fragaria ananassa*). *Journal of Food Science* 80 (8): M1823–M1830. doi: 10.1111/1750-3841.12938.

82 Diab, T., Biliaderis, C.G., Gerasopoulos, D., and Sfakiotakis, E. (2001). Physicochemical properties and application of pullulan edible films and coatings in fruit preservation. *Journal of the Science of Food and Agriculture* 81 (10): 988–1000. doi: 10.1002/jsfa.883.

83 Tzoumaki, M.V., Biliaderis, C.G., and Vasilakakis, M. (2009). Impact of edible coatings and packaging on quality of white asparagus (*Asparagus officinalis*, L.) during cold storage. *Food Chemistry* 117 (1): 55–63. doi: 10.1016/J.FOODCHEM.2009.03.076.

84 Kerdchoechuen, O., Laohakunjit, N., Tussavil, P. et al. (2011). Effect of starch-based edible coatings on quality of minimally processed pummelo. *International Journal of Fruit Science* 11 (4): 410–423. doi: 10.1080/15538362.2011.630588.

85 Jimenez, S., Alejandra, D., Salazar, L.N.P. et al. (2016). Coating effect of modified cassava starch in hass avocado. *Producción + Limpia* 10 (2): 31–37. doi: 10.22507/pml.v10n2a3.

86 Barco, P., Burbano, A., Mosquera, S. et al. (2011). Efecto de Recubrimiento a Base de Almidón Sobre La Maduración Del Tomate. *Revista Lasallista de Investigación* 8 (2): 96–103.

87 Lago-Vanzela, E.S., Do Nascimento, P., Fontes, E.A.F. et al. (2013). Edible coatings from native and modified starches retain carotenoids in pumpkin during drying. *LWT - Food Science and Technology* 50: 420–425. doi: 10.1016/j.lwt.2012.09.003.

88 Zhao, Y.P. and Chang, K.C. (1995). Sulfite and starch affect color and carotenoids of dehydrated carrots (*Caucus carota*) during Storage. *Journal of Food Science* 60: 324326.

89 Raghav, P.K., Agarwal, N., and Saini, M. (2016). Edible coatings for fruit and vegetables: a review. *International Journal of Scientific Research and Modern Education (IJSRME)* 1: 188–204.

90 Campo, V.L., Kawano, D.F., da Silva, D.B., and Carvalho, I. (2009). Carrageenans: biological properties, chemical modifications and structural analysis – a review. *Carbohydrate Polymers* 77 (2): 167–180. doi: 10.1016/J.CARBPOL.2009.01.020.

91 Hambleton, A. and Voilley, A. (2011). Transport parameters for aroma compounds through I-carrageenan and sodium alginate-based edible films. *Food Hydrocolloids* 25 (5): 1128–1133. doi: 10.1016/J.FOODHYD.2010.10.010.

92 Hambleton, A. and Fabra, M.-J. (2009). Interface and aroma barrier properties of iota-carrageenan emulsion–based films used for encapsulation of active food compounds. *Journal of Food Engineering* 93 (1): 80–88. doi: 10.1016/J.JFOODENG.2009.01.001.

93 Cazón, P., Velazquez, G., Ramírez, J.A., and Vázquez, M. (2017). Polysaccharide-basedfilms and coatings for food packaging: a review. *Food Hydrocolloids* 68: 136–148. doi: 10.1016/j.foodhyd.2016.09.009.

94 Turhan, N.K. and Sahbaz, F. (2004). Water vapor permeability, tensile properties and solubility of methylcellulose-based edible films. *Journal of Food Engineering* 61. doi: 10.1016/S0260-8774(03)00155-9.

95 Gregorova, A., Saha, N., Kitano, T., and Saha, P. (2015). Hydrothermal effect and mechanical stress properties of carboxymethylcellulose based hydrogel food packaging. *Carbohydrate Polymers* 117: 559–568. doi: 10.1016/j.carbpol.2014.10.009.

96 Saba M.K. and Sogvar, O.B. (2016). Combination of carboxymethyl cellulose-based coatings with calciumand ascorbic acid impacts in browning and quality of fresh-cut apples. *LWT - Food Science and Technology* 8 (5): 55.

97 Arnon, H., Zaitsev, Y., Porat, R., and Poverenov, E. (2014). Effects of carboxymethyl cellulose and chitosan bilayer edible coating on postharvest quality of citrus fruit. *Postharvest Biology and Technology* 87: 21–26. doi: 10.1016/j.postharvbio.2013.08.007.

98 Osorio, F.A. and Molina, P. (2011). Characteristics of Hydroxy propyl methyl cellulose (HPMC) based edible film developed for blueberry coatings. *Procedia Food Science* 1: 287–293.

99 Krochta, J.M., Baldwin, E.A., and Nisperos-Carriedo, M.O. (1994). *Edible Coatings and Films to Improve Food Quality*. Technomic Publ. Co.

100 Varela, P. and Fiszman, S.M. (2011). Hydrocolloids in fried foods. A review. *Food Hydrocolloids* 25: 1801–1812.

101 Li, T. (2013). Quality enhancement in refrigerated red drum (*Sciaenops ocellatus*) fillets using chitosan coatings containing natural preservatives. *Food Chemistry* 138: 821–826. doi: 10.1016/j.foodchem.2012.11.092.

102 Rabea, E.I., Badawy, M.E.-T., Stevens, C.V. et al. (2003). Reviews chitosan as antimicrobial agent: applications and mode of action. *Biomacromolecules* 4 (6): 1457–1465. doi: 10.1021/bm034130m.

103 Costa, M.J., Maciel, L.C., Teixeira, J.A. et al. (2018). Use of edible films and coatings in cheese preservation: opportunities and challenges. *Food Research International* 107: 84–92. doi: 10.1016/J.FOODRES.2018.02.013.

104 Petriccione, M., Mastrobuoni, F., Pasquariello, M. et al. (2015). Effect of chitosan coating on the postharvest quality and antioxidant enzyme system response of strawberry fruit during cold storage. *Food* 4 (4): 501–523. doi: 10.3390/foods4040501.

105 Coma, V., Martial-Gros, A., Garreau, S. et al. (2002). Edible antimicrobial films based on chitosan matrix. *Journal of Food Science* 67 (3): 1162–1169. doi: 10.1111/j.1365-2621.2002.tb09470.x.

106 Pérez, S., Mazeau, K., and du Penhoat, C.H. (2000). The three-dimensional structures of the pectic polysaccharides. *Plant Physiology and Biochemistry* 38 (1–2): 37–55. doi: 10.1016/S0981-9428(00)00169-8.

107 Dhanapal, A. and Sasikals, P. (2013). Edible films from polysaccharides. *Food Science and Quality Management* 3: 9–17.

108 Liu, L.S., Fishman, M.L., and Hicks, K.B. (2007). Pectin in controlled drug delivery - a Review. *Cellulose* 14 (1): 15–24. doi: 10.1007/s10570-006-9095-7.

109 Espitia, P.J.P., Du, W.-X., de Jesús Avena-Bustillos, R. et al. (2014). Edible films from pectin: physical-mechanical and antimicrobial properties - a review. *Food Hydrocolloids* 35: 287–296. doi: 10.1016/J.FOODHYD.2013.06.005.

110 Mishra, R.K., Banthia, A.K., and Majeed, A.B.A. (2012). Pectin based formulations for biomedical applications: a review. *Asian Journal of Pharmaceutical and Clinical Research* 5 (4): 1–7.

111 Laurent, M.A. and Boulenguer, P. (2003). Stabilization mechanism of acid dairy drinks (ADD) induced by pectin. *Food Hydrocolloids* 7. doi: 10.1016/S0268-005X(03)00028-6.

112 Valdés, A., Burgos, N., Jiménez, A., and Garrigós, M. (2015). Natural pectin polysaccharides as edible coatings. *Coatings* 5 (4): 865–886. doi: 10.3390/coatings5040865.

113 Zhang, Y., Rempel, C., and Mclaren, D. (2013). Edible coating and film materials: carbohydrates. In: *Innovations in Food Packaging*, 2nde (ed. J.H. Han), 305–323. Waltham, MA: Academic Press - Elsevier Ltd. doi: 10.1016/B978-0-12-394601-0.00012-6.

114 Farris, S., Unalan, I.U., Introzzi, L. et al. (2014). Pullulan-based films and coatings for food packaging: present applications, emerging opportunities, and future challenges. *Journal of Applied Polymer Science* 131 (13): 1–12. doi: 10.1002/app.40539.

115 Prajapati, V.D., Jani, G.K., and Khanda, S.M. (2013). Pullulan: an exopolysaccharide and its various applications. *Carbohydrate Polymers* 95: 540–549.

116 Bender, H., Lehmann, J., and Kurt, W. (1959). Pullulan, ein extracellulares Glucan von Pullularia pullulans. *Biochimica et Biophysica Acta* 36: 309–316.

117 Leathers, T.D. (2003). Biotechnological production and applications of pullulan. *Applied Microbiology and Biotechnology* 62: 468–473. doi: 10.1007/s00253-003-1386-4.

118 Singh, R.S., Saini, G.K., and Kennedy, J.F. (2008). Pullulan: microbial sources, production and applications. *Carbohydrate Polymers* 73 (4): 515–531. doi: 10.1016/j.carbpol.2008.01.003.

119 Kimoto, T., Shibuya, T., and Shiobara, S. (1997). Safety studies of a novel starch, pullulan: chronic toxicity in rats and bacterial mutagenicity*. *Food and Chemical Toxicology* 35: 323–329.

120 Rodríguez, M., Osés, J., Ziani, K., and Maté, J.I. (2006). Combined effect of plasticizers and surfactants on the physical properties of starch based edible films. *Food Research International* 39 (8): 840–846. doi: 10.1016/j.foodres.2006.04.002.

121 Rolland-Sabaté, A., Sánchez, T., Buléon, A. et al. (2012). Structural characterization of novel cassava starches with low and high-amylose contents in comparison with other commercial sources. *Food Hydrocolloids* 27 (1): 161–174. doi: 10.1016/J.FOODHYD.2011.07.008.

122 Singh, N., Singh, J., Kaur, L. et al. (2003). Morphological, thermal and rheological properties of starches from different botanical sources. *Food Chemistry* 81 (2): 219–231. doi: 10.1016/S0308-8146(02)00416-8.

123 Chiumarelli, M., Pereira, L.M., Ferrari, C.C. et al. (2010). Cassava starch coating and citric acid to preserve quality parameters of fresh-cut 'Tommy Atkins' mango. *Journal of Food Science* 75 (5). doi: 10.1111/j.1750-3841.2010.01636.x.

124 Bierhals, V.S., Chiumarelli, M., and Hubinger, M.D. (2011). Effect of cassava starch coating on quality and shelf life of fresh-cut pineapple (*Ananas Comosus* L. Merril cv "Pérola"). *Journal of Food Science* 76 (1): 62–72. doi: 10.1111/j.1750-3841.2010.01951.x.

125 Brändén, C.-I. and Tooze, J. (1999). Introduction to protein structure. In: *Garland Science*, 2e, 410. New York, NY: Taylor & Francis Groups.

126 McMurry, J. (2011). *Fundamentals of organic chemistry*, 7e, 672. Belmont, CA: Brooks/Cole, Cengage Learning.

127 Lacroix, M. and Vu, K.D. (2014). Edible coating and film materials: proteins. In: *Innovations in Food Packaging*, 2e (ed. J.H. Han), 277–304. Waltham, MA: Academic Press - Elsevier Ltd.

128 Chiralt, A., González-Martínez, C., Vargas, M., and Atarés, L. (2018). Edible films and coatings from proteins. *Proteins in Food Processing*: 477–500. doi: 10.1016/B978-0-08-100722-8.00019-X.

129 Bourtoom, T. (2009). Review article edible protein films: properties enhancement. *International Food Research Journal* 16: 1–9.

130 Han, J.H. (2014). Edible films and coatings: a review. In: *Innovations in Food Packaging*, 2e (ed. J.H. Han), 213–255. Waltham, MA: Academic Press - Elsevier Ltd.

131 Brunner, J.R. (1981). Cow milk proteins: twenty-five years of progress. *Journal of Dairy Science* 64 (6): 1038–1054. doi: 10.3168/jds.S0022-0302(81)82682-3.

132 Dalgleish, D.G. (1982). Milk proteins - chemistry and physics. In: *Food Proteins* (eds. P.F. Fox and J.J. Condon), 155–178. London: Applied Science Publishers Ltd.

133 Southward, C.R. (1994). Utilisation of milk components: casein. In: *Modern Diary Technology - Advances in Milk Processing*, 2e (ed. R.K. Robinson), 375–432. Springer Science+Business Media. doi: 10.1007/978-1-4615-2057-3.

134 Chen, H. (1995). Functional properties and applications of edible films made of milk proteins. *Journal of Dairy Science* 78 (11): 2563–2583. doi: 10.3168/JDS.S0022-0302(95)76885-0.

135 Khwaldia, K., Perez, C., Banon, S. et al. (2004). Milk proteins for edible films and coatings. *Critical Reviews in Food Science and Nutrition* 44: 239–251.

136 Kinsella, J.E. and Morr, C.V. (1984). Milk proteins: physicochemical and functional properties. *CRC Critical Reviews in Food Science and Nutrition* 21 (3): 197–262. doi: 10.1080/10408398409527401.

137 Shendurse, A.M. (2018). Milk protein based edible films and coatings–preparation, properties and food applications. *Journal of Nutritional Health & Food Engineering* 8 (2). doi: 10.15406/jnhfe.2018.08.00273.

138 de Jesus Avena-Bustillos, R., Krochta, J.M., and Saltveit, M.E. (1997). Water vapor resistance of red delicious apples and celery sticks coated with edible caseinate-acetylated monoglyceride films. *Journal of Food Science* 62: 351–354.

139 Le Tien, C., Vachon, C., Mateescu, M.A., and Lacroix, M. (2001). Milk protein coatings prevent oxidative browning of apples and potatoes. *Journal of Food Science: Food Chemistry and Toxicology* 66: 512–512.

140 de Jesús Avena-Bustillos, R., Krochta, J.M., Saltveit, M.E. et al. (1994). Optimization of edible coating formulations on zucchini to reduce water loss. *Journal of Food Engineering* 21 (2): 197–214. doi: 10.1016/0260-8774(94)90186-4.

141 Haug, I.J., Draget, K.I., and Smidsrød, O. (2004). Physical and rheological properties of fish gelatin compared to mammalian gelatin. *Food Hydrocolloids* 18 (2): 203–213. doi: 10.1016/S0268-005X(03)00065-1.

142 Wittaya, T. (2012). Protein-based edible films: characteristics and improvement of properties. *Structure and Function of Food Engineering*. doi: 10.5772/48167.

143 Gennadios, A., Hanna, M.A., and Kurth, L.B. (1997). Application of edible coatings on meats, poultry and seafoods: a review. *LWT - Food Science and Technology* 30 (4): 337–350. doi: 10.1006/fstl.1996.0202.

144 Hashim, P., Ridzwan, M., and Hashim, M. (2015). Collagen in food and beverage industries. *International Food Research Journal* 22 (1): 1–8.

145 Kobussen, J. (1997). Method and means for making coextruded sausage strands. US5951390A, issued October 31, 1997.

146 Farouk, M.M. and Price, J.F. (1990). Effect of an edible collagen film overwrap on exudation and lipid oxidation in beef round steak. *Journal of Food Science* 55 (6): 1510–1512. doi: 10.1111/j.1365-2621.1990.tb03556.x.

147 Serrano, M., Martínez-Romero, D., and Zapata, P.J. (2015). Advances in edible coatings. In: *Advances in Postharvest Fruit and Vegetable Technology* (eds. R.B.H. Wills and J. Golding), 147–166. CRC Press.

148 Molinaro, S., Cruz-Romero, M., Sensidoni, A. et al. (2015). Combination of high-pressure treatment, mild heating and holding time effects as a means of improving the barrier properties of gelatin-based packaging films Using response surface modeling. *Innovative Food Science & Emerging Technologies* 30: 15–23. doi: 10.1016/J.IFSET.2015.05.005.

149 Hanani, Z.A.N., Roos, Y.H., and Kerry, J.P. (2012). Use of beef, pork and fish gelatin sources in the manufacture of films and assessment of their composition and mechanical properties. *Food Hydrocolloids* 29 (1): 144–151. doi: 10.1016/J.FOODHYD.2012.01.015.
150 Farris, S., Schaich, K.M., Liu, L.S. et al. (2009). Development of polyion-complex hydrogels as an alternative approach for the production of bio-based polymers for food packaging applications: a review. *Trends in Food Science & Technology* 20 (8): 316–332. doi: 10.1016/J.TIFS.2009.04.003.
151 Gómez-Guillén, M.C., Pérez-Mateos, M., Gómez-Estaca, J. et al. (2009). Fish gelatin: a renewable material for developing active biodegradable films. *Trends in Food Science & Technology* 20 (1): 3–16. doi: 10.1016/J.TIFS.2008.10.002.
152 Antoniewski, M.N., Barringer, S.A., Knipe, C.L., and Zerby, H.N. (2007). Effect of a gelatin coating on the shelf Life of fresh meat. *Journal of Food Science*. doi: 10.1111/j.1750-3841.2007.00430.x.
153 Villegas, R., O'Connor, T.P., Kerry, J.P., and Buckley, D.J. (1999). Effect of gelatin dip on the oxidative and colour stability of cooked ham and bacon pieces during frozen storage. *International Journal of Food Science and Technology* 34 (4): 385–389. doi: 10.1046/j.1365-2621.1999.00284.x.
154 Herring, J.L., Jonnalongadda, S.C., Narayanan, V.C., and Coleman, S.M. (2010). Oxidative stability of gelatin coated pork at refrigerated Storage. *Meat Science* 85 (4): 651–656. doi: 10.1016/J.MEATSCI.2010.03.019.
155 Krull, L. and Inglett, G. (1971). Industrial uses of gluten. *Cereal Science Today* 16: 232–261.
156 Wieser, H. (2007). Chemistry of gluten proteins. *Food Microbiology* 24 (2): 115–119. doi: 10.1016/J.FM.2006.07.004.
157 Suput, D., Lazic, V., Popovic, S., and Hromis, N. (2015). Edible films and coatings: sources, properties and application. *Food and Feed Research* 42 (1): 11–22. doi: 10.5937/FFR1501011S.
158 Tanada-Palmu, P.S. and Grosso, C.R.F. (2005). Effect of edible wheat gluten-based films and coatings on refrigerated strawberry (*Fragaria ananassa*) quality. *Postharvest Biology and Technology* 36 (2): 199–208. doi: 10.1016/J.POSTHARVBIO.2004.12.003.
159 Cuq, B., Gontard, N., and Guilbert, S. (1998). Proteins as agricultural polymers for packaging production. *Cereal Chemistry Journal* 75 (1): 1–9. doi: 10.1094/CCHEM.1998.75.1.1.
160 Perez-Gago, M.B., Serra, M., Alonso, M. et al. (2005). Effect of whey protein- and hydroxypropyl methylcellulose-based edible composite coatings on color change of fresh-cut apples. *Postharvest Biology and Technology* 36 (1): 77–85. doi: 10.1016/J.POSTHARVBIO.2004.10.009.
161 Huang, L.-L., Zhang, M., Yan, W.-Q. et al. (2011). Rehydration characteristics of freeze-dried strawberry pieces as affected by whey protein edible coatings. *International Journal of Food Science and Technology* 46: 671–677. doi: 10.1111/j.1365-2621.2010.02495.x.
162 Maté, J.I., Frankel, E.N., and Krochta, J.M. (1996). Whey protein isolate edible coatings: effect on the rancidity process of dry roasted

peanuts. *Journal of Agricultural and Food Chemistry* 44: 1736–1740. doi: 10.1021/jf9506646.

163 Ramos, Ó.L., Pereira, J.O., Silva, S.I. et al. (2012). Evaluation of antimicrobial edible coatings from a whey protein isolate base to improve the shelf life of cheese. *Journal of Dairy Science.* doi: 10.3168/jds.2012-5478.

164 Rodriguez-Turienzo, L., Cobos, A., Moreno, V. et al. (2011). Whey protein-based coatings on frozen Atlantic salmon (*Salmo salar*): influence of the plasticiser and the moment of coating on quality preservation. *Food Chemistry* 128 (1): 187–194. doi: 10.1016/J.FOODCHEM.2011.03.026.

165 Stuchell, Y.M. and Krochta, J.M. (1995). Edible coatings on frozen king salmon: effect of whey protein isolate and acetylated monoglycerides on moisture loss and lipid oxidation. *Journal of Food Science* 60 (1): 28–31. doi: 10.1111/j.1365-2621.1995.tb05599.x.

166 Shokri, S., Ehsani, A., and Jasour, M.S. (2015). Efficacy of lactoperoxidase system-whey protein coating on shelf-life extension of rainbow trout fillets during cold storage (4 °C). *Food Bioprocess Technology* 8: 54–62. doi: 10.1007/s11947-014-1378-7.

167 Motalebi, A.A. and Seyfzadeh, M. (2011). Effects of whey protein edible coating on bacterial, chemical and sensory characteristics of frozen common Kilka (*Clupeonellia delitula*). *Iranian Journal of Fisheries Sciences* 11: 132–144.

168 Koh, L.-D., Yu, H.-D., Han, M.-Y. et al. (2015). Structures, mechanical properties and applications of silk fibroin materials. *Progress in Polymer Science* 46: 86–110. doi: 10.1016/j.progpolymsci.2015.02.001.

169 Kundu, B., Rajkhowa, R., Kundu, S.C., and Wang, X. (2013). Silk fibroin biomaterials for tissue regenerations. *Advanced Drug Delivery Reviews.*. doi: 10.1016/j.addr.2012.09.043.

170 Zafar, M.S., Belton, D.J., Hanby, B. et al. (2015). Functional material features of *Bombyx mori* silk light versus heavy chain proteins. *Biomacromolecules* 16 (2): 606–614. doi: 10.1021/bm501667j.

171 DeBari, M.K. and Abbott, R.D. (2019). Microscopic considerations for optimizing silk biomaterials. *Wiley Interdisciplinary Reviews: Nanomedicine and Nanobiotechnology* 11 (2): e1534. doi: 10.1002/wnan.1534.

172 Lotz, B. and Cesari, F.C. (1979). The chemical structure and the crystalline structures of *Bombyx mori* silk fibroin. *Biochimie.* doi: 10.1016/S0300-9084(79)80067-X.

173 Marsh, R.E., Corey, R.B., and Pauling, L. (1955). An investigation of the structure of silk fibroin. *BBA - Biochimica et Biophysica Acta.* doi: 10.1016/0006-3002(55)90178-5.

174 Zhou, C.-Z., Confalonieri, F., Jacquet, M. et al. (2001). Silk fibroin: structural implications of a remarkable amino acid sequence. *Proteins: Structure, Function, and Genetics* 44 (2): 119–122. doi: 10.1002/prot.1078.

175 Kim, D.-H., Viventi, J., Amsden, J.J. et al. (2010). Dissolvable films of silk fibroin for ultrathin, conformal bio-integrated electronics. *Nature Materials* 9 (6): 511–517. doi: 10.1038/nmat2745.

176 Lawrence, B.D., Omenetto, F., Chui, K., and Kaplan, D.L. (2008). Processing methods to control silk fibroin film biomaterial features. *Journal of Materials Science* 43: 6967–6985. doi: 10.1007/s10853-008-2961-y.
177 Hu, X., Shmelev, K., Lin, S. et al. (2011). Regulation of silk material structure by temperature-controlled water vapor annealing. *Biomacromolecules* 12: 1686–1696. doi: 10.1021/bm200062a.
178 Park, H.J., Chinnan, M.S., and Shewfelt, R.L. (1994). Edible coating effects on storage life and quality of tomatoes. *Journal of Food Science* 59 (3): 568–570.
179 Baysal, T., Bilek, S.E., and Apaydın, E. (2010). The effect of corn zein edible film coating on intermediate moisture apricot (*Prunus Armenica* L.) quality. *Gida* 35 (4): 245–249.
180 Bai, J., Alleyne, V., Hagenmaier, R.D. et al. (2003). Formulation of zein coatings for apples (*Malus domestica* Borkh). *Postharvest Biology and Technology* 28 (2): 259–268. doi: 10.1016/S0925-5214(02)00182-5.
181 Gol, N.B. and Ramana Rao, T.V. (2014). Influence of Zein and gelatin coatings on the postharvest quality and shelf life extension of mango. *Fruits* 69 (2): 115. doi: 10.1051/fruits/2014002.
182 Rhim, J.W. and Shellhammer, H. (2005). Lipid-based edible films and coatings. In: *Innovations in Food Packaging* (ed. J.H. Han), 362–383. Waltham, MA: Academic Press - Elsevier Ltd.
183 Morillon, V., Debeaufort, F., Blond, G. et al. (2002). Factors affecting the moisture permeability of lipid-based edible films: a review. *Critical Reviews in Food Science and Nutrition* 42: 67–89.
184 Kameda, T. (2004). Molecular structure of crude beeswax studied by solid-state ^{13}C NMR. *Journal of Insect Science* 4 (1). doi: 10.1093/jis/4.1.29.
185 Shellhammer, T.H., Rumsey, T.R., and Krochta, J.M. (1997). Viscoelastic properties of edible lipids. *Journal of Food Engineering* 33: 305–320.
186 Talens, P. and Krochta, J.M. (2005). Plasticizing effects of beeswax and carnauba wax on tensile and water vapor permeability properties of whey protein films. *Journal of Food Science: Food Egineering and Physical Properties* 70: E239–E243.
187 Navarro-Tarazaga, M.L., Massa, A., and Pérez-Gago, M.B. (2011). Effect of beeswax content on hydroxypropyl methylcellulose-based edible film properties and postharvest quality of coated plums (cv. *Angeleno*). *LWT - Food Science and Technology* 44 (10): 2328–2334. doi: 10.1016/J.LWT.2011.03.011.
188 Fagundes, C., Palou, L., Monteiro, A.R., and Pérez-Gago, M.B. (2015). Hydroxypropyl methylcellulose-beeswax edible coatings formulated with antifungal food additives to reduce alternaria black spot and maintain postharvest quality of cold-stored cherry tomatoes. *Scientia Horticulturae* 193: 249–257. doi: 10.1016/J.SCIENTA.2015.07.027.
189 Navarro-Tarazaga, M.L.L., Del Río, M.A., Krochta, J.M., and Pérez-Gago, M.B. (2008). Fatty acid effect on hydroxypropyl methylcellulose–beeswax edible film properties and postharvest quality of coated 'Ortanique'

mandarins. *Journal of Agricultural and Food Chemistry* 56 (22): 10689–10696. doi: 10.1021/jf801967q.

190 Perez-Gago, M.B., Rojas, C., and DelRio, M.A. (2002). Effect of lipid type and amount of edible hydroxypropyl methylcellulose-lipid composite coatings used to protect postharvest quality of mandarins cv. fortune. *Journal of Food Science* 67 (8): 2903–2910. doi: 10.1111/j.1365-2621.2002.tb08836.x.

191 Valencia-Chamorro, S.A., Pérez-Gago, M.B., del Río, M.Á., and Palou, L. (2009). Effect of antifungal hydroxypropyl methylcellulose (HPMC)–lipid edible composite coatings on postharvest decay development and quality attributes of cold-stored 'Valencia' oranges. *Postharvest Biology and Technology* 54 (2): 72–79. doi: 10.1016/J.POSTHARVBIO.2009.06.001.

192 Chiumarelli, M. and Hubinger, M.D. (2012). Stability, solubility, mechanical and barrier properties of cassava starch – carnauba wax edible coatings to preserve fresh-cut apples. *Food Hydrocolloids* 28 (1): 59–67. doi: 10.1016/J.FOODHYD.2011.12.006.

193 Singh, S., Khemariya, P., Rai, A. et al. (2016). Carnauba wax-based edible coating enhances shelf-life and retain quality of eggplant (*Solanum melongena*) fruits. *LWT* 74: 420–426. doi: 10.1016/J.LWT.2016.08.004.

194 Reyes, E.O., Martínez-Vazquez, G., Saucedo-Pompa, S. et al. (2013). Improvement of shelf life quality of green bell peppers Using edible coating formulation. *Journal of Microbiology, Biotechnology and Food Sciences* 2: 2448–2451.

195 Perez-Gago, M.B., Serra, M., Alonso, M. et al. (2003). Effect of solid content and lipid content of whey protein isolate-beeswax edible coatings on color change of fresh-cut apples. *Food Chemistry and Toxicology* 68: 2186–2191.

196 Perez-Gago, M.B., Serra, M., and del Río, M.A. (2006). Color change of fresh-cut apples coated with whey protein concentrate-based edible coatings. *Postharvest Biology and Technology* 39 (1): 84–92. doi: 10.1016/j.postharvbio.2005.08.002.

197 Mehyar, G.F., Al-Ismail, K., Han, J.H., and Chee, G.W. (2012). Characterization of edible coatings consisting of pea starch, whey protein isolate, and carnauba wax and their effects on oil rancidity and sensory properties of walnuts and pine nuts. *Journal of Food Science* 77 (2): E52–E59. doi: 10.1111/j.1750-3841.2011.02559.x.

198 Kowalczyk, D. (2011). Effect of edible protein-wax coating on post-harvest stability of Brussels sprouts stored under simulated commercial storage conditions. *Food Science Technology Quality* 18 (6): 177–191. doi: 10.4315/0362-028X-67.4.833.

199 Kowalczyk, D. and Pikula, E. (2010). Effect of edible protein-wax coating on storage quality of table grapes (*Vitis vinifera* L.). *Food Science Technology Quality* 17 (5): 67–76. doi: 10.1016/j.foodchem.2003.12.003.

200 Kowalczyk, D., Pikula, E., and Baraniak, B. (2010). Effect of edible protein-wax coating on quality of cold-stored broccoli. *Food. Science Technology Quality* 17 (6): 120–133. doi: 10.1016/S0308-8146(02)00485-5.

201 Yan, L., Luo, Z., Ban, Z. et al. (2019). The effect of the layer-by-layer (LBL) edible coating on strawberry quality and metabolites during Storage. *Postharvest Biology and Technology* 147: 29–38. doi: 10.1016/J.POSTHARVBIO.2018.09.002.

202 Medeiros, B.G.d.S., Pinheiro, A.C., Carneiro-da-Cunha, M.G., and Vicente, A.A. (2012). Development and characterization of a nanomultilayer coating of pectin and chitosan – evaluation of its gas barrier properties and application on 'Tommy Atkins' mangoes. *Journal of Food Engineering* 110 (3): 457–464. doi: 10.1016/J.JFOODENG.2011.12.021.

203 Poverenov, E., Danino, S., Horev, B. et al. (2014). Layer-by-layer electrostatic deposition of edible coating on fresh cut melon model: anticipated and unexpected effects of alginate–chitosan combination. *Food and Bioprocess Technology* 7 (5): 1424–1432. doi: 10.1007/s11947-013-1134-4.

204 Nowzari, F., Shábanpour, B., and Ojagh, S.M. (2013). Comparison of chitosan–gelatin composite and bilayer coating and film effect on the quality of refrigerated rainbow trout. *Food Chemistry* 141 (3): 1667–1672. doi: 10.1016/J.FOODCHEM.2013.03.022.

205 Saucedo-Pompa, S., Jasso-Cantu, D., Ventura-Sobrevilla, J. et al. (2007). Effect of candelilla wax with natural antioxidants on the shelf life quality of fresh-cut fruits. *Journal of Food Quality* 30 (5): 823–836. doi: 10.1111/j.1745-4557.2007.00165.x.

206 Saucedo-Pompa, S., Rojas-Molina, R., Aguilera-Carbó, A.F. et al. (2009). Edible film based on candelilla wax to improve the shelf life and quality of avocado. *Food Research International* 42: 511–515. doi: 10.1016/j.foodres.2009.02.017.

207 Lakshminarayana, S., Luis, S., Isabel Ortiz, J., and Gabriel Siade, R. (1974). Extension of storage life of citrus fruits by application of candelilla wax emulsion and comparison of its efficiency with tag and flavorseal. *Proceedings of the Florida State Horticultural Society* 87: 325–330.

208 Paredes-López, O., Camargo-Rubio, E., and Gallardo-Navarro, Y. (1974). Use of coatings of candelilla wax for the preservation of limes. *Journal of the Science of Food and Agriculture* 25 (10): 1207–1210. doi: 10.1002/jsfa.2740251003.

209 Aguilera-Carbo, A., Augur, C., Prado-Barragan, L.A. et al. (2008). Microbial production of ellagic acid and biodegradation of ellagitannins. *Applied Microbiology and Biotechnology* 78 (2): 189–199. doi: 10.1007/s00253-007-1276-2.

210 Siade, G. and Pedraza, E. (1977). Extension of storage life of banana (*Giant cavendish*) using natural wax candelilla. *Acta Horticulturae* 62: 327–334. doi: 10.17660/ActaHortic.1977.62.35.

211 Alleyne, V. and Hagenmaier, R.D. (2000). Candelilla-shellac: an alternative formulation for coating apples. *HortScience* 35 (4).

212 Ligia, F., Machado, C., Costa, M.C., and Batista, N. (2012). Application of carnauba-based wax maintains postharvest quality of 'Ortanique' Tangor Aplicação de Cera de Carnaúba Mantém a Qualidade Pós-Colheita Do

Tangor 'Ortanique. *Ciência e Tecnologia de Alimentos* 32 (2): 261–266. doi: 10.1590/S0101-20612012005000048.

213 Baldwin, E., Nisperos-Carriedo, M., Shaw, P.E., and Bums, J.K. (1995). Effect of coatings and prolonged storage conditions on fresh orange flavor volatiles, degrees brix, and ascorbic acid levels. *Journal of Agricultural and Food Chemistry* 43: 1321–1331. doi: 10.1021/jf00053a037.

214 Chauhan, O.P., Nanjappa, C., Ashok, N. et al. (2013). Shellac and aloe Vera gel based surface coating for shelf life extension of tomatoes. *Journal of Food Science and Technology*. doi: 10.1007/s13197-013-1035-6.

215 Srivastava, S. and Thombare, N. (2017). Safety assessment of shellac as food additive through long term toxicity study. *Trends in Biosciences* 10 (2): 733–740.

216 Weinberger, H. and Gardner, H. (1938). Chemical composition of shellac. *Industrial and Engineering Chemistry*. doi: 10.1021/ie50340a022.

217 Bourtoom, T. (2008a). Edible films and coatings: characteristics and properties. *International Food Research Journal* 15: 237–248.

218 Maftoonazad, N. and Ramaswamy, H.S. (2007). Effect of pectin-based edible emulsion coating on changes in quality of avocado exposed to *Lasiodiplodia theobromae* infection. *Carbohydrate Polymers* 68: 341–349.

219 Fagundes, C., Palou, L., Monteiro, A.R., and Pérez-Gago, M.B. (2014). Effect of antifungal hydroxypropyl methylcellulose-beeswax edible coatings on gray mold development and quality attributes of cold-stored cherry tomato fruit. *Postharvest Biology and Technology* 92: 1–8. doi: 10.1016/J.POSTHARVBIO.2014.01.006.

220 Arnon, H., Granit, R., Porat, R., and Poverenov, E. (2015). Development of polysaccharides-based edible coatings for citrus fruits: a layer-by-layer approach. *Food Chemistry* 166: 465–472. doi: 10.1016/J.FOODCHEM.2014.06.061.

221 Vu, K.D., Hollingsworth, R.G., Leroux, E. et al. (2011). Development of edible bioactive coating based on modified chitosan for increasing the shelf life of strawberries. *Food Research International* 44 (1): 198–203. doi: 10.1016/J.FOODRES.2010.10.037.

222 Perdones, Á., Escriche, I., Chiralt, A., and Vargas, M. (2016). Effect of chitosan–lemon essential oil coatings on volatile profile of strawberries during storage. *Food Chemistry* 197: 979–986. doi: 10.1016/J.FOODCHEM.2015.11.054.

223 Choi, W.S., Singh, S., and Lee, Y.S. (2016). Characterization of edible film containing essential oils in hydroxypropyl methylcellulose and its effect on quality attributes of 'Formosa' plum (*Prunus Salicina* L.). *LWT - Food Science and Technology* 70: 213–222. doi: 10.1016/J.LWT.2016.02.036.

224 Yun, J., Fan, X., Li, X. et al. (2015). Natural surface coating to inactivate salmonella enterica serovar typhimurium and maintain quality of cherry tomatoes. *International Journal of Food Microbiology* 193: 59–67. doi: 10.1016/J.IJFOODMICRO.2014.10.013.

225 Mchugh, T.H. and Senesi, E. (2000). Apple wraps: a novel method to improve the quality and extend the shelf life of fresh-cut apples. *Journal of Food Science: Food Egineering and Physical Properties* 65: 480–485.

226 Lee, J.Y., Park, H.J., Lee, C.Y., and Choi, W.Y. (2003). Extending shelf-life of minimally processed apples with edible coatings and antibrowning agents. *LWT - Food Science and Technology* 36 (3): 323–329. doi: 10.1016/S0023-6438(03)00014-8.
227 Baldwin, E.A., Nisperos, M.O., Chen, X., and Hagenmaier, R.D. (1996). Improving storage life of cut apple and potato with edible coating. *Postharvest Biology and Technology* 9 (2): 151–163. doi: 10.1016/S0925-5214(96)00044-0.
228 Olivas, G.I., Rodriguez, J.J., and Barbosa-Cánovas, G.V. (2003). Edible coatings composed of methylcellulose, stearic acid, and additives to preserve quality of pear wedges. *Journal of Food Processing and Preservation* 27 (4): 299–320. doi: 10.1111/j.1745-4549.2003.tb00519.x.
229 Flores-López, M.L., Cerqueira, M.A., De Rodríguez, D.J., and Vicente, A.A. (2016). Perspectives on utilization of edible coatings and nano-laminate coatings for extension of postharvest storage of fruits and vegetables. *Food Engineering Reviews* 8. doi: 10.1007/s12393-015-9135-x.

Part II

Food Packaging Based on Individual Biopolymers and their Composites

4

Polylactic Acid (PLA) and Its Composites: An Eco-friendly Solution for Packaging

Swati Sharma

Indian Institute of Technology Delhi, Department of Textile and Fiber Engineering, Hauz Khas, New Delhi 110016, India

4.1 Introduction

Liberalization of foreign trade policy, technological advancement in manufacturing and shipping, and increasing international demands have boosted centralized production. The increased transportation capabilities have given corporates an option to ship products far and to a broad range of consumers [1–3]. However, to make distant shipping a success, the storage space and product quality preservation play a vital role, making quality packaging the heart and soul of world trade [1, 4–6]. The introduction of polymeric materials has revolutionized the storage and packing industry by not just offering a cost-effective method of goods transportation but also by addressing the problem of quality preservation. Therefore, polymers have become a quintessential part of the packaging industry [7].

The successful deliveries of chemically unstable goods such as food and pharmaceuticals are much dependent on the nature of packaging. The material does not just have to protect the goods from external environment but should also be nonreactive to the packaged goods. For this purpose, many variants of polymers, namely polyethylene terephthalate (PET), polypropylene (PP), and polyethylene (PE), have been developed, which demonstrate excellent performance as packaging materials [8]. The biggest drawback of using these polymers is their nonbiodegradable nature and limited recyclability. So, this increasing dependence on polymeric materials has raised the issue of waste dumping and has many adverse environmental impacts. Thus, during the recent decades, lot of research interest has developed for replacing unsustainable polymeric materials with suitable biodegradable alternatives [9].

Polylactic acid (PLA), a biodegradable bio-based aliphatic polyester, is produced from natural renewable sources such as corn sugar, sugarcane, potato, etc. Due to the high cost and low production, the earlier research on PLA was mainly limited to its use in medical applications, but recent advancements in polymer science and manufacturing techniques have made mass production of PLA possible. NatureWorks LLC, a leading PLA producer, has a production capacity of around 150 000 metric tons per year in its Nebraska facility [10]. More recently,

Sustainable Food Packaging Technology, First Edition. Edited by Athanassia Athanassiou.

Total Corbion, a 50/50 joint venture between Total and Corbion, has started a PLA production plant in Rayong, Thailand, with a production capacity of 75 000 tons per year (https://www.total-corbion.com). Due to increasing attention toward environmental issues and decrease in the cost of production, research on PLA has diversified and increased drastically over the last 25 years. Also, the introduction of high M_w PLA and the development of various processing techniques have enabled PLA derivatives to replace fossil-based polymers in various applications [11]. The biocompatibility of PLA with the human body has dramatically affected medical research, and it is widely used in various medical applications such as medical devices, implants, and sutures. PLA also has comparable mechanical, thermal, and optical properties to many commodity polymers such as polyethylene PET, PP, etc. that are typically used in food packaging [12].

4.2 Synthesis of PLA and Its Properties

PLA, a thermoplastic high-strength and high-modulus polymer, is manufactured from annually replenishable resources and finds vast applications in the packaging industry. Among the biopolymers, PLA has been possibly the most successful in developing alternatives and replacing the materials based on nonbiodegradable fossil polymers [13]. PLA is classified as a member of the aliphatic polyester family extracted from α-hydroxy acids. The basic building block of PLA is lactic acid (2-hydroxy propionic acid), which is derived from corn starch by fermentation or using chemical synthesis. Lactic acid is then polymerized to PLA either by stepwise polycondensation or via ring-opening polymerization of di-lactide intermediate (Figure 4.2). Lactic acid can take the form of optically active L (+) and D (−) isomers. Thus, by varying the enantiomers, it is possible to obtain PLAs with different properties. Such flexibility in material selection helps to tune and obtain performance requirements of PLA-based materials, depending on the application. Finally, PLA inherently has comparable optical, mechanical, thermal, and barrier properties to fossil-based alternatives [14]. Figure 4.1 shows the chemical structure of PLA. In terms of tensile strength, tensile modulus, and gas (O_2 and CO_2) barrier properties, PLA is superior to PP; however, toughness and impact strength are lower than those of many commercial polymers. The tensile strength of PLA ranges between 48 and 53 MPa, and modulus is around 3500 MPa [15].

For the production of lactic acid, the fermentation process of carbohydrates offers advantages over chemical synthesis in terms of reaction yield and lower by-products [11]. Lactic acid can be produced using various synthesis techniques, including azeotropic dehydrative condensation, direct condensation polymerization, and/or polymerization through lactide formation. But all these techniques

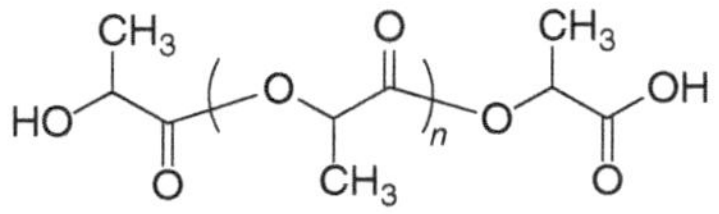

Figure 4.1 Structural formula of PLA.

are not economical and require big amounts of solvents during the reaction process. Currently, commercially available PLA is being produced by fermentation, which results in a mixture of D and L-lactide followed by a ring-opening polymerization. PLA is a thermoplastic semicrystalline polymer and exhibits both glass transition (T_g) and melting point (T_m). Above its T_g, PLA shows rubbery characteristics and it is capable of creeping well below its T_g until its β-transition temperature (−45 °C) [16] (Figure 4.2).

This chapter reviews and describes new strategies for the development of improved PLA-based composite materials with a primary focus on potential packaging applications. First, the review discusses the most significant researches and development in natural fiber-based PLA composites for packaging applications. The second part of this chapter then focuses on PLA composites with nanoparticles, nanofibers, microfibers, and whiskers.

4.3 Properties Required for Food Packaging

In the current scenario of globalization, products produced in specific countries are being consumed by distant and various types of customers. So, packaging is required to be designed in such a way that it can provide shelf life and maintain a good quality of packaged product during the storage, transportation, and sale, as the packaged products are prone to be affected by different factors, such as attack by microbes, oxygen, humidity, light, etc. Thus, the packaging material is not just required to have essential properties, like optimum thermal and mechanical strength, but is also expected to inhibit the growth of microbes and hinder the exposure to water, vapors, and gases. Among biodegradable polymers, there are many alternatives available for the packaging application, but the material selection requires judicious consideration of product quality preservation. Other critical factors influencing the selection are the rate of degradation of the polymer and the ability to manufacture the polymer at an industrial scale. PLA has proven to be one of the most suitable polymers for packaging and thus widely accepted as a biodegradable packaging material. Also, the recent advancement in PLA manufacturing process has helped to produce PLA at a much higher capacity with economical price. Currently, much research is being carried out in improving thermal, barrier, and tensile properties of PLA. In the following subtopics, the properties required for packaging materials especially in the food sector are discussed in detail (Figure 4.3).

4.3.1 Barrier Properties

Barrier properties of materials used for food packaging are mainly associated with their ability to hinder the interaction with water vapor, light, and gases, elongating the shelf life, and preserving the quality of the product stored. The selection of the material for packaging depends on the end use of the packaging and the nature of the product stored, as each product requires a different degree of preservation [17]. PLA has shown to have better barrier properties when compared

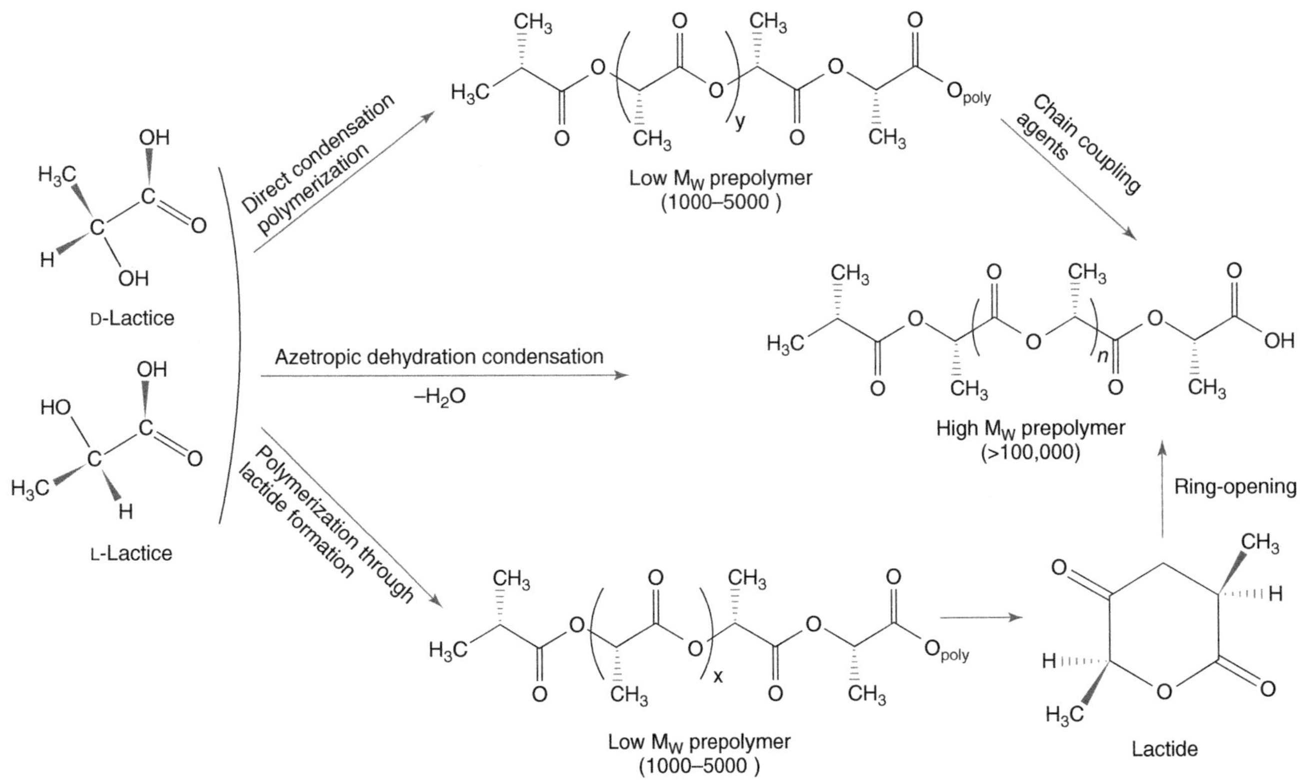

Figure 4.2 Synthesis of PLA from L- and D-lactic acids. Source: From Auras et al. [11]. © 2004, John Wiley & Sons.

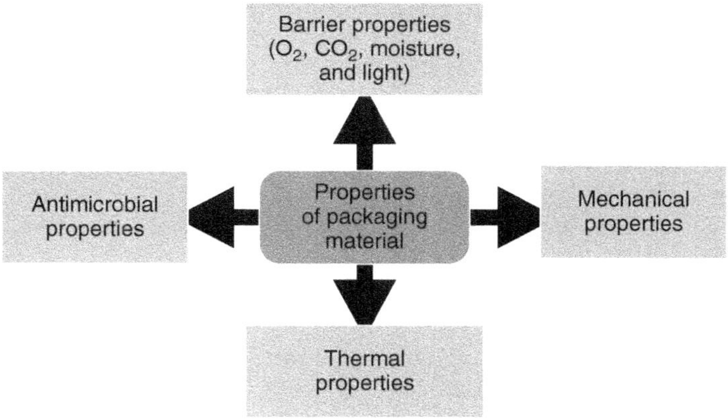

Figure 4.3 Properties required for packaging material.

to polyethylene, polystyrene, and many other dominating petro-based polymers [18]. This is especially true for oxygen that greatly influences the shelf life of fresh products. In fact, lowering the oxygen permeability coefficient of the packaging can significantly reduce the oxidation process, resulting in preserving the product quality and increasing its shelf life [19].

Auras et al. [20, 21] compared the oxygen and water vapor permeability of PLA, PET, and polystyrene (PS) and found that water vapor barrier properties of PLA are poorer than both PET and PS, but in case of O_2 and CO_2 barrier properties, PLA showed lower permeability coefficient than PS and comparable to that of PET. Although PLA is a polar polymeric material, its water vapor permeability coefficient does not vary significantly with the change in humidity. To improve the barrier properties, different additives such as clays, natural fibers as fillers, metal and metal oxides, nano cellulose, or other polymers have been added in PLA [22].

4.3.2 Optical Properties

The average transparency of food packaging films in the current market is around 95% [23]. Transparent materials are those that allow visible light of wavelength 400–700 nm to pass through [24]. PLA is a semicrystalline polymer and is transparent in its parent form; also, it has high luminosity and transparency value. Transparent films are preferred as packaging material for products as they help in product appraisal to customers. On the other hand, the addition of a compatibilizer agent, such as maleic anhydride, and other additives like plasticizers, clays, or metal oxides can change the transparency of PLA depending upon the amount of additive used [25]. Limonene, a natural compound found in many citrus fruits, was blended with PLA by melt extrusion and it was observed that the blend maintained its transparency even after the addition of the limonene with improved flexibility and it presented sufficient properties to be used in packaging application [26].

4.3.3 Mechanical Properties

The mechanical properties of polymeric material, such as the durability and strength, are vital for the packaging application, and thus, play a crucial role in the selection of the material. The variation of PLA crystallinity can significantly affect mechanical properties, such as softness and elasticity. Typical PLA resin shows a tensile strength of 50–70 MPa, flexural modulus around 5 GPa, flexural strength of 100 MPa, and elongation at break of 4%. Despite having excellent mechanical strength, the brittleness of PLA hinders its use in the packaging application. Therefore, different additives have been used by researchers to improve its brittleness. Oyama [27] blended PLA with ethylene copolymer by reactive blending, making it super tough without any adverse effect on other mechanical and thermal properties. Sharma et al. [12] and Arrieta et al. [28] used poly(ethylene glycol) (PEG) as a plasticizer for PLA and found that the tensile elongation improved significantly, but the tensile strength and thermal properties showed downward trends. Apart from PEG, bio-based plasticizers such as citrate ester, starch, oligomeric lactic acid [29, 30], palm oil, soybean oil, and limonene oil have been also explored for packaging application [31]. A disadvantage of improving the flexibility of PLA with plasticizers is associated with drastic fall in tensile strength and modulus [31–34]. To balance the ductility and strength properties of PLA, a combination of plasticizers and reinforcement agents has been proposed by many researchers [35, 36]. Nanoclays and nanocellulose were found to compensate the strength loss due to the addition of plasticizers [37, 38].

Various researchers have mixed PLA with other biodegradable polymers, such as chitosan, poly(hydroxybutyrate) which improved the ductility of PLA [28, 39]. Thermoplastic starch (TPS) plasticized with glycerol has been used by Ayana et al. [40] to improve the toughness of PLA along with nanoclays, and they found that the PLA composites with clays and TPS showed improved mechanical properties. The application of natural fibers to improve the mechanical properties of PLA is also practiced, and cellulosic fibers that are obtained from natural renewable sources are predominately used. Mathew et al. [41] and Li et al. [42] mixed nanocellulose with PLA through melt extrusion and found that the tensile modulus, elongation and impact strength of PLA composites were improved compared to that of neat PLA. In general, compared to other commodity polymers, PLA matrix may have lower mechanical properties, but these properties can be further enhanced by adding different fibers, fillers, or additives to PLA polymer.

4.3.4 Thermal Properties

PLA applications in packaging are usually restricted due not only to mechanical but also to thermal properties. PLA shows reduced heat stability that can be attributed to the presence of ester bonds that can decompose by heating. Many technologies, such as the addition of nucleating agents, annealing, introduction of fibers, and addition of nanofillers, are used for enhancing the thermal properties of PLA material [43]. Yang et al. [44] developed a crosslinking PLA by using a small amount of crosslinking agent triallyl isocyanurate (TAIC) with an initiator dicumyl peroxide (DCP). The thermal stability of the crosslinked PLA was improved, which was indicated by thermo gravimetric analysis (TGA) and dynamic mechanical analysis (DMA) results, along with improved mechanical properties. Quynh et al. [45] also used supercritical carbon dioxide (sc-CO_2) treatment to crosslink TAIC with a blend of poly (L-lactide) (PLLA) and poly (D-lactide) (PDLA). It was found that crosslinking happened in both the amorphous and crystalline regions, hence increasing the thermal stability of crosslinked PLA as compared to neat PLA.

Different types of nanofillers such as nanocellulose, chitosan, clay, and metal oxides have been also used to improve the thermal stability of PLA. Urayama et al. [46] used calcium silicate, potassium titanate, and aluminum borate in particle and whisker form to study the thermal properties of PLA. Whisker-reinforced PLA samples showed superior thermal properties than the particle-reinforced PLA due to high aspect ratio and better interfacial properties between whiskers and PLA. Jonoobi et al. [47] used cellulose nanofibers (CNFs) extracted from kenaf to reinforce PLA through melt blending at three different concentrations, i.e. 1, 3, and 5 wt%. They reported that the thermal stability of PLA increased significantly with 5 wt% of CNFs.

The recent developments in PLA manufacturing techniques make it an excellent economic and eco-friendly alternative to conventional plastics packaging. The work on modified PLA has increased its potentiality. The PLA fillers and their concentration can be selected based on the quality requirements. Thus, the application of PLA as food packaging material can be significantly influenced by the selection of the right fillers.

4.3.5 Antibacterial Properties

Koide and Shi [48] compared the effect of three different packagings, i.e. PLA films, low density polyethylene (LDPE) films, and perforated LDPE films on the microbial and physicochemical quality of freshly harvested green pepper. The gas barrier properties of these films were tested, and antimicrobial activity was recorded after storing the pepper for seven days at 10 °C. The results showed that the bacterial colony count was lower in the case of green pepper, which was stored in the PLA film.

Incorporation of active compounds, e.g. components with antibacterial properties, into the packaging material has further increased the product quality preservation, thereby having a positive impact on the shelf life of stored goods. The antimicrobial packaging is especially useful in the storage of food, which is susceptible to bacterial action and in which contamination can induce detrimental effects. Many studies have suggested that the addition of active agents not only improves the antibacterial activity but also enhance other vital properties like mechanical, thermal and barrier. Thus, such addition can significantly improve the product shelf life and quality [49]. Different types of nanometals, nanoclay, natural oils, and nanocellulose have been used by different researchers to create packaging with antimicrobial properties. Shankar et al. [50] used zinc oxide nanoparticles in PLA and checked the antimicrobial activity of PLA/ZnO nanocomposite films against two types of foodborne pathogenic bacteria, *Listeria monocytogenes* and *Escherichia coli*. The PLA/ZnO composite films exhibited potent antibacterial activity against foodborne pathogenic bacteria, *L. monocytogenes* and *E. coli*. They decreased the bacterial count on minced fish paste and increased the shelf life. Marra et al. [1] also investigated the effect of ZnO on the morphology, thermal degradation stability, mechanical and barrier properties, and antibacterial properties (*E. coli*) of PLA, with 1, 3, and 5% of ZnO, respectively. This study concluded that 5 wt% of ZnO in PLA is suitable for packaging applications with excellent antibacterial properties and good mechanical strength. Li et al. [51] used titanium dioxide (TiO_2) and silver nanoparticles to prepare different PLA nanocomposites for packaging applications. They claimed that these nanofillers are suitable for packaging applications with excellent antimicrobial properties, and they are risk-free to be used in food packaging applications.

The food packaging industry has widely accepted the use of bioactive materials based on plants as additives. Thymol, garlic oil, chitosan, etc. have been widely used as active antimicrobial agents in packaging. Bonilla et al. [52] prepared PLA/chitosan nanocomposite films by melt extrusion using different sizes of chitosan nanopowder. They reported improved antimicrobial efficiency of PLA/chitosan films when used as packaging for minced meat. Tawakkal et al. [53] used thymol as a bioactive agent in PLA/kenaf composites for active packaging application. Similarly, Radusin et al. [54] used wild garlic extracts as antimicrobial agents in PLA films for sustainable packaging, and this study provided sufficient evidence that the wild garlic extract has potential as antimicrobial compound for effective PLA packaging solution.

4.4 General Reinforcements for PLA

Reinforcement used in PLA composites can be of natural or synthetic origin. If the reinforcement of PLA originates from renewable bioresources, then the composite is classified as green composite or biocomposite [55]. Green PLA composites containing natural fibers such as flax, jute, hemp, etc. are fully biodegradable. Moreover, they have significant advantages such as low density, recyclability, low production cost, and functional matrix–fiber compatibility, with some inherent disadvantages associated with the use of natural fibers, such as high moisture absorption, and average mechanical properties. In contrast, the synthetic fiber variants of PLA composites often show better mechanical properties, namely tensile and impact strength and shear stress, but the high costs of manufacturing limit their applications.

4.4.1 Natural Fibers

Natural fiber–PLA composites offer a green and sustainable solution for many applications such as packaging, where high load-carrying capacity is not required. These composites at the end of their life can be easily disposed without deleterious effects for the environment. Natural fiber composites also offer other advantages such as reduced dependence on nonrenewable energy/material sources, lower pollutant emissions, lower greenhouse gas emissions, enhanced energy recovery, and biodegradability.

Natural fibers usually exhibit hydrophilic surface due to the presence of hydroxyl groups. This results in poor interaction between fiber, and hydrophobic PLA matrix. Chemical and physical modifications are widely used to improve these interactions. Researchers have used different approaches to improve the interfacial strength of the composites. Kim et al. [56] modified jute fiber with glycidoxypropyl-trimethoxysilane and poly(vinyl alcohol) (PVA) at 5 wt% concentration. They have found that both PVA and silane are effective in improving the compatibility between jute and PLA, improving the flexural strength and creep deformation behavior of jute/PLA composites. Zafar et al. [57] also treated jute fibers with only alkali and alkali followed by 3-aminopropyltriethoxysilane (APTS). They have reported that the mechanical and thermomechanical properties of alkali- and silane-treated composites were higher than that of neat and only alkali-treated composites.

Foruzanmehr et al. [58] have used a novel approach to improve the interaction between PLA and flax fiber. They have grafted TiO_2 particles on oxidized/non-oxidized and as-received flax fibers to improve the interaction between fiber and matrix and found that the oxidized flax fibers offer better adhesion with TiO_2 particles than the non-oxidized and as-received flax fibers. Mechanical testing results (viz. short beam shear and tensile testing) showed significant improvement in the properties in the case of oxidized flax fiber due to the improved adhesion between fiber and matrix. Aydın et al. [59] treated flax with alkali (NaOH) to study the effect on mechanical, morphological, and thermal properties of PLA/flax composites. Mechanical and thermal testing revealed that Young's modulus and glass transition temperature of alkali-treated flax/PLA composites decreased due to the presence of superficial alkali, which acts as a plasticizer.

Baghaei et al. [60, 61] prepared hemp/PLA composites from PLA/hemp Co-wrapped hybrid yarn prepregs (Figure 4.4) by laying them in different angles (0°, 45°, and 90°). They also carried out alkali treatment of the hemp fibers to improve the adhesion between PLA and hemp. The testing confirmed that alkali-treated resulted in hemp/PLA composites with higher tensile strength, Young's modulus, and flexural strength compared to the untreated hemp/PLA composites. It was also found that the angle of the laying of fibers in the composites dramatically influences the mechanical properties (Figure 4.5).

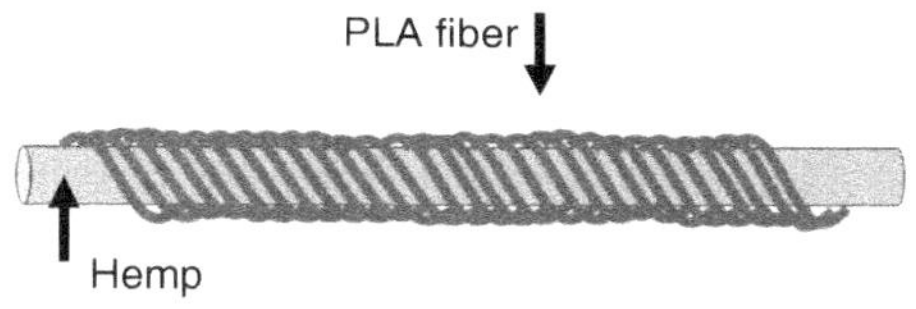

Figure 4.4 Hemp/PLA Co-wrapped hybrid yarn.

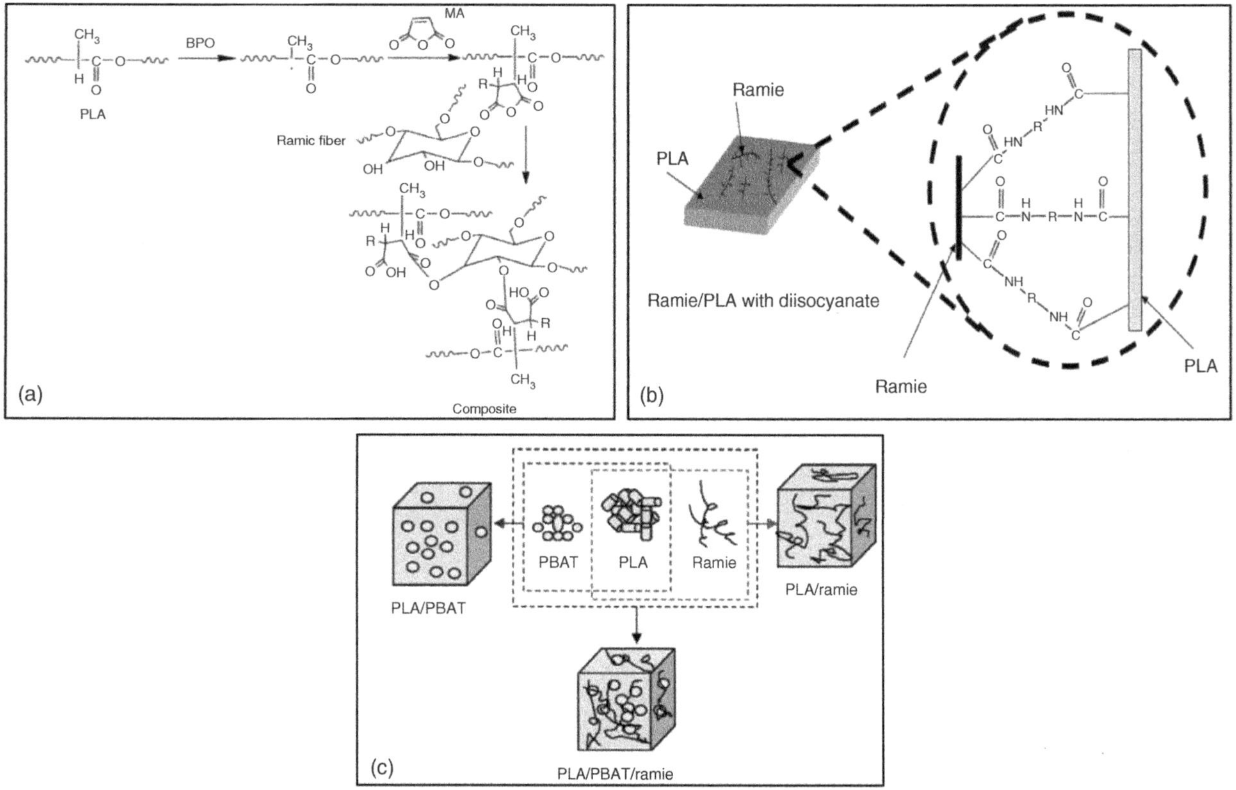

Figure 4.5 Mechanism of enhancement in the interface of ramie and PLA by using (a) maleic anhydride (MA), (b) diisocyanates, and (c) poly(butylenes adipate-*co*-terephthalate). Source: (a) Yu et al. 2014 [65]; (b) Yu et al 2015 [66]; (c) Yu and Li [67]. © 2014, Elsevier.

Hao et al. [62] used Joncryl ADR®-4368, a multi-epoxy-functionalized styrene-acrylic oligomer to improve the interaction between sisal fiber and PLA. The testing results revealed that the tensile and impact strength improved drastically due to improved adhesion between sisal fiber and PLA. Orue et al. [63] studied the effect of alkaline and/or silane treatment of the sisal fibers on the mechanical properties of sisal/PLA composites. The alkaline treatment resulted in inferior tensile strength of the sisal fiber, but the strength of sisal/PLA composite improved significantly as compared to untreated sisal/PLA composites.

Compatibilizers such as maleic anhydride, diisocyanates, and epoxy functionalized agents are widely used to improve the interaction between natural fibers and PLA matrix [62, 64–66]. Yu et al. [65–67] used maleic anhydride, diisocyanates, and poly(butylene adipate-*co*-terephthalate) (PBAT) to improve the interface between ramie and PLA. Maleic anhydride and diisocyanates worked excellently and improved the tensile strength, flexural strength, and impact strength of ramie/PLA composites. On the contrary, the addition of PBAT did not improve the tensile strength due to weak interfacial bonding between ramie and PLA.

Cellulose-based nanofibers, microfibers, and nanowhiskers are widely explored as a reinforcement for PLA-based packaging application. Wang and Drzal [68] prepared PLA and CNF composites by the solvent-free method. By using a small amount of surfactant, PLA was homogeneously mixed with CNFs. Due to the hydrophilic nature of cellulose, it has weak interaction with PLA. This poor interaction results in agglomeration of CNFs. Hence in this study author suggested that the use of surfactant in small amount improved the dispersion of CNF in PLA for composite manufacturing. Chemical modification of CNF offers an alternative to decrease their hydrophilicity and can result in an improved interaction between PLA and CNF. Jonoobi et al. [47], used acetylation to modify the CNF derived from kenaf. The modified and unmodified CNF were further used for manufacturing PLA nanocomposites through a solvent mixture followed by melt extrusion. But these modified CNF did not show any improvement in the properties of the composite as expected. Frone et al. [69] used silane-treated and acid-hydrolyzed and raw CNF as reinforcement for PLA. differential scanning calorimetry (DSC) measurements revealed a higher degree of crystallinity for the composites containing untreated nanofibers and demonstrated the role of CNFs as nucleating agents, whereas improved adhesion between PLA and silane-modified CNF results in lower crystallinity. Coltelli et al. [70] studied the mechanical and thermomechanical properties of PLA and chitin nanofiber composites. The tensile strength did not show any reinforcing effect, but the elongation at break improved drastically. Numerous studies have been reported by different researchers about the PLA composites with different filler and some of them are summarized in Table 4.1. PLA composites with natural fillers are promising candidates for food packaging due to the combination of improved mechanical properties, biocompatibility and biodegradability at the end of life, and they can be explored more systematically to develop active and smart packaging films.

Table 4.1 Summary of different research work done on PLA-composites.

Filler and matrix	Main findings	References
Montmorillonite (MMT)/PLA/starch	Nanoclay act as a compatibilizer between PLA and starch. Mechanical, thermal, and water barrier properties were improved	[40]
Halloysite nanotube/PLA (melt extrusion)	A moderate increase in mechanical properties was observed, and the fire resistance performance of PLA was significantly improved with	[71]
MMT clay/PLA/biopolyethylene	The impact strength of the composite improved, but the tensile strength decreased abruptly	[72]
Poly(butylene adipate-*co*-terephthalate) (PBAT) and carbon nanotubes (CNTs-COOH)/PLA	Tensile strength and elongation of the composites improved. Glass transition temperature also showed an upward shift	[6]
Graphene nanoplatelet (GNP)/PLA	GNPs with larger lateral size, displayed highest mechanical properties with improved electrical conductivity	[73]
Cellulose nanocrystal/PLA	Mechanical properties are adversely affected, and the composites were more brittle	[74]
Halloysite nanotube/PLA	The addition of halloysite nanotube (HNT) slightly increased the glass transition temperature	[75]
Cellulose nanocrystals/silver nanoparticles/PLA	Ag nanoparticles and cellulose nanocrystals have an efficient reinforcing effect on the thermal, mechanical, and water vapor barrier properties, and antibacterial activity of PLA nanocomposite films for fresh food packaging applications	[3]
Microfibrillated cellulose/PLA	The addition of cellulose increased the tensile modulus of crystallized neat PLA by 42% and the strength by 14% at an micro fibrillated cellulose (MFC) content of 20 wt%	[76]
Banana fiber (BF)/PLA	Maleic anhydride (MA) and glycerol triacetate ester (GTA) were used as compatibilizers. Tensile strength, modulus, elongation, and impact strength were improved in the case of PLA/BF/GTA/MA composites	[77]
Banana flour/PLA	PLA-*g*-glycidyl methacrylate (GMA) was synthesized by grafting GMA onto PLA. Composites with PLA-g-GMA improved tensile and thermal properties	[78]
Banana fiber (BF)/PLA	Silane treatment of fiber and mixing with organically modified clay improved the tensile properties of PLA/BF composites	[79]

(Continued)

Table 4.1 (Continued.)

Filler and matrix	Main findings	References
Kenaf fiber (KF)/PLA	Thymol-based KF/PLA composites decreased the tensile properties with excellent antibacterial properties	[80]
KF/PLA	Silane-treated KF fibers extruded with multi-walled carbon nanotubes (MWCNTs) loaded PLA. The study showed positive effect on tensile and thermal properties	[81]
Basalt fiber (BF)/PLA	Two fillers, namely talc and sepiolite, have been used with basalt fiber and extruded with PLA. Tensile, flexural, and impact results showed that talc and basalt additives efficiently enhanced the moduli of the composites	[82]
Basalt fiber (BF)/PLA	Silane treated with BF compression molded with PLA and the tensile strength improved by 26% than that of untreated BF/PLA counterparts	[83]
Chicken feather/PLA	The addition of chicken feather improved modulus and storage modulus of PLA up to 8% fiber content	[84]
Coir/PLA	Alkali treatment of coir fiber enhanced the interfacial shear strength (IFSS) of coir/PLA composite	[85]

4.4.2 Synthetic Fibers

Synthetic fibers have found various applications in high-performance textiles and are being increasingly used as a replacement for traditional reinforcements in composite manufacturing. The inherent advantage of excellent mechanical properties and tailorable manufacturing techniques gives synthetic fibers a competitive edge over natural fibers for their use in composites. The development of carbon and glass fibers (GFs) has increased the use of fiber-based composites in various challenging areas. Due to exceptional mechanical properties, excellent thermal resistance, and low manufacturing cost, glass fiber is among the most widely used reinforcement in modern fiber composites. On the other hand, these fibers have found limited application in food packaging because of the inherent risk of toxicity. Huda et al. [86] compared the properties of PLA composites reinforced with chopped glass fiber and recycled newspaper cellulose fiber (RNCF). The result of the study concluded that the RNCF fiber could be used to replace glass fiber in many low load-bearing applications. Wang et al. [87] used annealing to develop strong and thermally stable biocomposites based on GFs and PLA. Glass fibers significantly improved the crystallization behavior of PLA during annealing which resulted in enhanced mechanical properties of GF/PLA composites.

Carbon fibers have also been explored as reinforcement for PLA-based composites. Many researchers have tried to incorporate short carbon fibers in PLA for 3D printing. Tian et al. [88] have successfully carried out the 3D printing of carbon fiber and PLA composites and found that with optimized parameters, the maximum flexural strength of 335 MPa and flexural modulus of 30 GPa can be achieved in these composites. Similarly, Ferreira et al. [89] have fabricated 3D printing-based carbon fiber/PLA composites and studied their mechanical and flexural properties. The addition of carbon fiber in PLA increased the tensile modulus and shear modulus by 2.00 and 1.16 times, respectively, in the printing direction.

Regenerated cellulose has also been widely explored as an alternative reinforcement for PLA-based composites. Shibata et al. [90] used lyocell fabric as reinforcement for the PLA using the compression molding technique, and the result showed that with the increase in fiber content of the composite both tensile strength and Young's modulus improved. The Izod impact strength of multilayer PLA–lyocell laminate also increased considerably, and the impact strength approximately tripled in comparison to that of neat PLA matrix. Graupner and Müssig [91] also used lyocell fiber with different fineness to reinforce PLA. They used multilayered mats and needle punched nonwoven mats of lyocell fiber as reinforcement. Results proved that needle felts can be compressed better than multilayer webs, hence resulting in excellent mechanical characteristics.

4.4.3 Functional Fillers

The applications and usage of different fillers are increasing in composite manufacturing. The fillers offer additional benefit without hindering the inherent properties of composites and can also be carefully selected to obtain desired properties. Various research works have been conducted on the incorporation of fillers in PLA-based composites. Recent research for packaging industries is focused on the development of biocomposites with functional packaging applications, like active packaging, being at the same time environmental-friendly. The use of active agents in packaging material can revolutionize food storage and transportation. Active packaging films contain antimicrobial agents to reduce the foodborne microorganisms and increase the shelf life of the food material. Currently, the films used for food storage are of synthetic origins, but much progress has been made toward the implementation of bio-based films for packaging applications.

4.4.3.1 Clay/PLA Composites

Among the most used nanofillers, organic-modified clays have been extensively studied, and the resulting PLA/clay nanocomposites are of great interest to the packaging industry because of their superior mechanical, thermal, gas barrier, and optical properties even at low clay content. The application of even 1–5% w/w of nanoclays can improve the gas barrier functionality of the PLA composite as well as the mechanical, thermal, and optical properties [2]. Lai et al. [92] used organic-modified clay with PLA and tested the mechanical and thermal properties of the resulting composite. They have found that the addition of the clays improves the mechanical properties of PLA nanocomposites, and the intercalated clay also served as a nucleation agent that increased the crystallinity of the PLA matrix. Cele et al. [93] studied optical properties of PLA/clay nanocomposites and found that the incorporation of the nanoparticles to PLA results in red-shift and reduction in transmittance. The reduction in the transmittance was due to the absorption of light by the clay particles. Darie et al. [94] assessed the effect of nanoclays on the hydrophilicity of PLA, and also found that the barrier and antimicrobial properties of PLA improved after the addition of clays. Fukushima et al. [95] reported the effect of two-layered silicate clays (montmorillonite and fluoro-hectorite) with different filler content on the thermal, mechanical properties, and degradation process of the composite of PLA. The clay content at 10 wt% improved the thermal and thermomechanical properties of PLA as compared to that of neat PLA. It was also found that the addition of clay particles accelerates the degradation process of PLA nanocomposites.

4.4.3.2 Metal-oxide/PLA Composites

Metals and metal oxides are best known for their antibacterial and UV resistance properties. Among all metal oxides, zinc oxide (ZnO) and magnesium oxide (MgO) are the safest and best solution for food packaging [1]. Patani et al. [96] used silane-coated special ZnO particles to improve interaction and dispersion in PLA. Among several results, it was claimed that PLA with 1, 2, and 3 wt% of ZnO had an effective antibacterial activity after seven days against *E. coli*, whereas only the biocomposite with 3 wt% of ZnO was found functional against *Staphylococcus aureus* at the seventh day. Marra et al. [1] studied the properties of ZnO/PLA films for food packaging application and concluded that 5 wt% ZnO is suitable for food packaging because it shows, compared to neat PLA, improved tensile and gas barrier properties, whereas the antibacterial properties were excellent for all the PLA/ZnO films. Shankar et al. [50] developed ZnO nanoparticles-incorporated PLA films using a solution casting method and tested their antimicrobial, gas barrier, mechanical, and thermal properties. These composite films had excelled antimicrobial and UV protective properties and suit very well for packaging applications [50]. Chu et al. [97] used ZnO and silver (Ag) nanoparticles to develop PLA nanocomposite films for packaging applications. They found that Ag- and ZnO-based PLA composite films, due to their excellent antimicrobial properties, have enormous potential to be used in packaging applications.

4.5 Biodegradability of PLA

Biopolymers such as PLA, poly(hydroxybutyrate) (PHB), starch, etc. are derived from natural resources, and at the end of their life cycle, they tend to decompose to form fossil fuel, compost, water, and carbon dioxide. The time required for the degradation of the polymer depends upon several parameters such as humidity, temperature, light, and type of bacteria responsible for degradation. The degradation of biopolymers at the industrial scale requires 6–12 weeks to convert it into water, CO_2, and biomass (https://www.european-bioplastics.org/bioplastics/). Packaging materials based on PLA are sustainable, eco-friendly, require no treatment at the end of the product life, and are neutral to greenhouse gas emission as the CO_2 generated during the degradation process is neutralized by the amount of CO_2 consumed during the corn growth [11]. Arrieta et al. [98] studied the disintegration of PLA plasticized with PEG, acetyl-tri-*n*-butyl citrate (ATBC) and PHB by soil burial method. The study suggested that the presence of plasticizer in PLA favored the surface hydrolysis of PLA. Lv et al. [99] also studied the decomposition process of starch/PLA/wood composites, by soil burial method followed by thermal and chemical characterization. The study found that the thermal stability of composites changes during the degradation process, and the number of short molecular chains increased. PLA can also be recycled, thus reducing the need for new products and the associated environmental and financial expenses. The recycling process of other polymers such as PET and nylon may result in contamination and hence lead to poor properties but in the case of PLA, it can be readily hydrolyzed to lactic form using water hydrolysis followed by filtration of the monomer and finally polymerization [100]. The recycling of PLA by controlled hydrolysis is preferred over the biodegrading of the polymer.

4.6 Conclusions and Future Prospects

PLA-based composites are proving to be a sustainable solution to various issues of petroleum-based materials. The successful application of PLA in packaging and food storage has boosted its use in this sector. In Germany, Dannon and McDonald's have successfully used PLA in yogurt cups and cutlery. Bio4Pack, Celebration Packaging Ltd, and BioPak are commercially using PLA in packaging materials manufacturing. As the PLA price drops and new facilities produce higher volumes of PLA, new applications will be pursued. Although PLA is a relatively new polymer, it is possible to manipulate its physical, mechanical, and barrier properties by changing its chemical composition and varying its molecular characteristics. It is also possible to blend PLA with other polymers and fillers, making it an excellent biodegradable alternative for use in plastic packaging.

In the future, PLA and its composites are expected to have competitive thermal, mechanical, and functional properties. A lot of research work today focuses on optimizing these properties. The mass production and cost-effectiveness will act as catalysts in boosting the usage of PLA in the packaging industry. Future work should focus on the improvement of different properties of the PLA composites essential for food packaging, such as the flexibility, gas barrier properties, resistance to high temperatures and humidity, and fast biodegradation.

References

1 Marra, A. et al. (2016) 'Polylactic acid/zinc oxide biocomposite films for food packaging application', *International Journal of Biological Macromolecules*, 88, pp. 254–262. doi: https://doi.org/10.1016/j.ijbiomac.2016.03.039.

2 Silvestre, C., Duraccio, D. and Cimmino, S. (2011) 'Food packaging based on polymer nanomaterials', *Progress in Polymer Science*, 36(12), pp. 1766–1782. doi: https://doi.org/10.1016/j.progpolymsci.2011.02.003.

3 Yu, H. Y. et al. (2016) 'Fabrication of multifunctional cellulose nanocrystals/poly(lactic acid) nanocomposites with silver nanoparticles by spraying method', *Carbohydrate Polymers*, 140, pp. 209–219. doi: https://doi.org/10.1016/j.carbpol.2015.12.030.

4 Liu, M., Zhang, Y. and Zhou, C. (2013) 'Nanocomposites of halloysite and polylactide', *Applied Clay Science*, 75–76, pp. 52–59. doi: https://doi.org/10.1016/j.clay.2013.02.019.

5 Yang Z. et al. (2010) 'Crystallization behavior of poly(ε-caprolactone)/layered double hydroxide nanocomposites', *Journal of Applied Polymer Science*, 116(5), pp. 2658–2667. doi: https://doi.org/10.1002/app.

6 Zhou, Y. et al. (2018) 'Preparation and characterization of polylactic acid (PLA) carbon nanotube nanocomposites', *Polymer Testing*, 68, pp. 34–38. doi: https://doi.org/10.1016/j.polymertesting.2018.03.044.

7 Cirillo, G., Kozlowski, M.A., and Spizzirri, U.G. (2017). Composites Materials for Food Packaging, Scrivener Publishing. Hoboken, NJ, USA: John Wiley & Sons, Inc. . doi: https://doi.org/10.1002/9781119160243.

8 Armentano, I. Bitinis, N., Fortunati, E., Mattioli, S., (2013) 'Multifunctional nanostructured PLA materials for packaging and tissue engineering', *Progress in Polymer Science*, 38(10–11), pp. 1720–1747. doi: https://doi.org/10.1016/j.progpolymsci.2013.05.010.

9 Jamshidian, M. et al. (2010) 'Poly-lactic acid: production, applications, nanocomposites, and release studies', *Comprehensive Reviews in Food Science and Food Safety*, 9(5), pp. 552–571. doi: https://doi.org/10.1111/j.1541-4337.2010.00126.x.

10 Raquez, J.-M. et al. (2013) 'Polylactide (PLA)-based nanocomposites', *Progress in Polymer Science*, 38(10–11), pp. 1504–1542. doi: https://doi.org/10.1016/j.progpolymsci.2013.05.014.

11 Auras, R., Harte, B. and Selke, S. (2004) 'An overview of polylactides as packaging materials', *Macromolecular Bioscience*. 4(9), pp. 835–864. doi: https://doi.org/10.1002/mabi.200400043.

12 Sharma, S. et al. (2019) 'Tailoring the mechanical and thermal properties of polylactic acid-based bionanocomposite films using halloysite nanotubes and polyethylene glycol by solvent casting process', *Journal of Materials Science*, 54(12), pp. 8971–8983. doi: https://doi.org/10.1007/s10853-019-03521-9.

13 Lim, L. T., Auras, R. and Rubino, M. (2008) 'Processing technologies for poly(lactic acid)', *Progress in Polymer Science (Oxford)*, pp. 820–852. doi: https://doi.org/10.1016/j.progpolymsci.2008.05.004.

14 Garlotta, D. (2001) 'A literature review of poly(lactic acid)', *Journal of Polymers and the Environment*. 9(2), pp. 63–84. doi: https://doi.org/10.1023/A:1020200822435.

15 Sinclair, R. G. (1996) 'The case for polylactic acid as a commodity packaging plastic', *Journal of Macromolecular Science, Part A: Pure and*

Applied Chemistry, 33(5), pp. 585–597. doi: https://doi.org/10.1080/10601329608010880.

16 Jamshidi, K., Hyon, S. H. and Ikada, Y. (1988) 'Thermal characterization of polylactides', *Polymer*, 29(12), pp. 2229–2234. doi: https://doi.org/10.1016/0032-3861(88)90116-4.

17 Suyatma, N. E. et al. (2004) 'Mechanical and barrier properties of biodegradable films made from chitosan and poly(lactic acid) blends', *Journal of Polymers and the Environment*, 12(1), pp. 1–6. doi: https://doi.org/10.1023/B:JOOE.0000003121.12800.4e.

18 Hamad, K. et al. (2015) 'Properties and medical applications of polylactic acid: a review', *Express Polymer Letters*, 9(5), pp. 435–455. doi: https://doi.org/10.3144/expresspolymlett.2015.42.

19 Chaudhary, M. (2006) Nano-technology: applications in food industry, *Indian Food Industry* doi: https://doi.org/10.1016/B978-0-12-811942-6.00005-4.

20 Auras, R. A., Harte, B., Selke, S., Hernandez, R., (2003) 'Mechanical, physical, and Barrier properties of poly(lactide) films*', *Journal of Plastic Film & Sheeting*, 19, pp. 123–135. doi: https://doi.org/10.1177/8756087903039702.

21 Auras, R. A., Singh, S. P. and Singh, J. J. (2005) 'Evaluation of oriented poly(lactide) polymers vs. existing PET and oriented PS for fresh food service containers', *Packaging Technology and Science*, 18(4), pp. 207–216. doi: https://doi.org/10.1002/pts.692.

22 Rodríguez-Núnez, J. R. et al. (2017). 'Composite materials based on PLA and its applications in food packaging', in *Composites Materials for Food Packaging*, pp. 355–399. doi: https://doi.org/10.1002/9781119160243.ch12.

23 Abbas Ali, N. (2017) 'Study the optical properties of biodegradable polylactic acid/silver nanocomposites', *International Journal of Advanced Biological Research*, 7(2), pp. 391–395.

24 Intawiwat, N. et al. (2011) 'Packaging materials with tailor made light transmission properties for food protection', *Polymer Engineering and Science*, 52(9), pp. 2015–2024. doi: https://doi.org/10.1002/pen.

25 Du, J. et al. (2017) 'Styrene-assisted maleic anhydride grafted poly(lactic acid) as an effective compatibilizer for wood flour/poly(lactic acid) bio-composites', *Polymers*, 9(11). doi: https://doi.org/10.3390/polym9110623.

26 Arrieta, M. P., López, J., Ferrándiz, S., Peltzer, M.A., (2013) 'Characterization of PLA-limonene blends for food packaging applications', *Polymer Testing*, 32(4), pp. 760–768. doi: https://doi.org/10.1016/j.polymertesting.2013.03.016.

27 Oyama, H. T. (2009) 'Super-tough poly(lactic acid) materials: reactive blending with ethylene copolymer', *Polymer*, 50(3), pp. 747–751. doi: https://doi.org/10.1016/j.polymer.2008.12.025.

28 Arrieta, Marina P., Samper, M.D., López, J., Jiménez, A., (2014) 'Combined effect of poly(hydroxybutyrate) and plasticizers on polylactic acid properties for film intended for food packaging', *Journal of Polymers and the Environment*, 22(4), pp. 460–470. doi: https://doi.org/10.1007/s10924-014-0654-y.

29 Martino, V. P., Ruseckaite, R. A. and A. Jiménez (2006) 'Thermal and mechanical characterization of plasticized poly(l-lactide-co-d,l-lactide) films

for food packaging', *Journal of Thermal Analysis and Calorimetry*, 86(3), pp. 707–712.

30 Sharma, S. et al. (2020) 'Harnessing the ductility of polylactic acid/halloysite nanocomposites by synergistic effects of impact modifier and plasticiser', *Composites Part B Engineering*, p. 107845. doi: https://doi.org/10.1016/j.compositesb.2020.107845.

31 Giita Silverajah, V. S. et al. (2012) 'A comparative study on the mechanical, thermal and morphological characterization of poly(lactic acid)/epoxidized palm Oil blend' *International Journal of Molecular Sciences*, 13(5), pp. 5878–98. doi: https://doi.org/10.3390/ijms13055878.

32 Choi, K. M. Choi, M.C., Han, D.H., Park, T.S., Ha, C.S., (2013) 'Plasticization of poly(lactic acid) (PLA) through chemical grafting of poly(ethylene glycol) (PEG) via in situ reactive blending', *European Polymer Journal*, 49(8), pp. 2356–2364. doi: https://doi.org/10.1016/j.eurpolymj.2013.05.027.

33 Hassouna, F. et al. (2011) 'New approach on the development of plasticized polylactide (PLA): grafting of poly(ethylene glycol) (PEG) via reactive extrusion', *European Polymer Journal*, 47(11), pp. 2134–2144. doi: https://doi.org/10.1016/j.eurpolymj.2011.08.001.

34 Ljungberg, N. and Wessle, B. (2005) 'Preparation and properties of plasticized poly(lactic acid) films', *Biomacromolecules*, 6, pp. 1789–1796. doi: https://doi.org/10.1021/bm050098f.

35 Herrera, N., Mathew, A. P. and Oksman, K. (2015) 'Plasticized polylactic acid/cellulose nanocomposites prepared using melt-extrusion and liquid feeding: mechanical, thermal and optical properties', *Composites Science and Technology*, 106, pp. 149–155. doi: https://doi.org/10.1016/j.compscitech.2014.11.012.

36 Ozdemir, E. and Hacaloglu, J. (2017) 'Characterizations of PLA-PEG blends involving organically modified montmorillonite', *Journal of Analytical and Applied Pyrolysis*, 127, pp. 343–349. doi: https://doi.org/10.1016/J.JAAP.2017.07.016.

37 Chen, G.-X., Kim, H.-S., Kim, E.-S., Yoon, J.-S., (2005) 'Compatibilization-like effect of reactive organoclay on the poly(l-lactide)/poly(butylene succinate) blends', *Polymer*, 46(25), pp. 11829–11836. doi: https://doi.org/10.1016/J.POLYMER.2005.10.056.

38 Mohapatra, A. K., Mohanty, S. and Nayak, S. K. (2014) 'Effect of PEG on PLA/PEG blend and its nanocomposites: a study of thermo-mechanical and morphological characterization', *Polymer Composites*, 35(2), pp. 283–293. doi: https://doi.org/10.1002/pc.22660.

39 Râpă, M. et al. (2016) 'Influence of chitosan on mechanical, thermal, barrier and antimicrobial properties of PLA-biocomposites for food packaging', *Composites Part B: Engineering*, 102, pp. 112–121. doi: https://doi.org/10.1016/j.compositesb.2016.07.016.

40 Ayana, B., Suin, S., Khatua, B.B., (2014) 'Highly exfoliated eco-friendly thermoplastic starch (TPS)/poly(lactic acid)(PLA)/clay nanocomposites using unmodified nanoclay', *Carbohydrate Polymers*, 110, pp. 430–439. doi: https://doi.org/10.1016/j.carbpol.2014.04.024.

41 Mathew, A. P., Oksman, K. and Sain, M. (2005) 'Mechanical properties of biodegradable composites from poly lactic acid (PLA) and microcrystalline cellulose (MCC)', *Journal of Applied Polymer Science*, 97(5), pp. 2014–2025. doi: https://doi.org/10.1002/app.21779.

42 Li, J. et al. (2017) 'Excellent rheological performance and impact toughness of cellulose nanofibers/PLA/ionomer composite', *RSC Advances*, 7(46), pp. 28889–28897. doi: https://doi.org/10.1039/c7ra04302c.

43 Ahemd, J., Zhang, J.-X., Song, Z., and Varshney, S.K. (2009). Thermal properties of polylactides Effect of molecular mass and nature of lactide isomer. *Journal of Thermal Analysis and Calorimetry.* 95: 957–964. https://doi.org/https://doi.org/10.1007/s10973-008-9035-x.

44 Yang, S., Wu, Z.-H., Yang, W., and Yang, M.B. (2008). Thermal and mechanical properties of chemical crosslinked polylactide (PLA). *Polymer testing Test.* 27: 957–963. https://doi.org/10.1016/j.polymertesting.2008.08.00.

45 Quynh, T.M., Mitomo, H., Zhao, L., and Asai, S. (2008). The radiation crosslinked films based on PLLA/PDLA stereocomplex after TAIC absorption in supercritical carbon dioxide. *Carbohydrate Polymer* 72: 673–681. https://doi.org/10.1016/j.carbpol.2007.10.010.

46 Urayama, H., Ma, C. and Kimura, Y. (2003) 'Mechanical and thermal properties of poly(l-lactide) incorporating various inorganic fillers with particle and whisker shapes', *Macromolecular Materials and Engineering*, 288(7), pp. 562–568. doi: https://doi.org/10.1002/mame.200350004.

47 Jonoobi, M. et al. (2010) 'Mechanical properties of cellulose nanofiber (CNF) reinforced polylactic acid (PLA) prepared by twin screw extrusion', *Composites Science and Technology*, 70(12), pp. 1742–1747. doi: https://doi.org/10.1016/j.compscitech.2010.07.005.

48 Koide, S. and Shi, J. (2007) 'Microbial and quality evaluation of green peppers stored in biodegradable film packaging', *Food Control*, 18(9), pp. 1121–1125. doi: https://doi.org/10.1016/j.foodcont.2006.07.013.

49 Rhim, J.W., Hong, S.I., and Ha, C.S. (2009). Tensile, water vapor barrier and antimicrobial properties of PLA/nanoclay composite films. *LWT - Food Science and Technology.* 42: 612–617. https://doi.org/10.1016/j.lwt.2008.02.015.

50 Shankar, S., Wang, L. F. and Rhim, J. W. (2018) 'Incorporation of zinc oxide nanoparticles improved the mechanical, water vapor barrier, UV-light barrier, and antibacterial properties of PLA-based nanocomposite films', *Materials Science and Engineering C*, 93, pp. 289–298. doi: https://doi.org/10.1016/j.msec.2018.08.002.

51 Li, W. et al. (2017) 'Development of antimicrobial packaging film made from poly(lactic acid) incorporating titanium dioxide and silver nanoparticles', *Molecules*, 22(7), p. 1170. doi: https://doi.org/10.3390/molecules22071170.

52 Bonilla, J., Fortunati, E., Vargas, M., Chiralt, A., Kenny, J.M.M., (2013) 'Effects of chitosan on the physicochemical and antimicrobial properties of PLA films', *Journal of Food Engineering*, 119(2), pp. 236–243. doi: https://doi.org/10.1016/j.jfoodeng.2013.05.026.

53 Tawakkal, I. S. M. A., Cran, M. J. and Bigger, S. W. (2016) 'Release of thymol from poly(lactic acid)-based antimicrobial films containing kenaf fibres

as natural filler', *LWT - Food Science and Technology*, 66, pp. 629–637. doi: https://doi.org/10.1016/j.lwt.2015.11.011.

54 Radusin, T. et al. (2019) 'Hybrid Pla/wild garlic antimicrobial composite films for food packaging application', *Polymer Composites*, 40(3), pp. 893–900. doi: https://doi.org/10.1002/pc.24755.

55 Zini, E. and Scandola, M. (2011) 'Green composites: an overview', *Polymer Composites*, pp. 1905–1915. doi: https://doi.org/10.1002/pc.21224.

56 Kim, H.J., Miyamoto, S., Takada, Y., and Takemura, K. (2011). Effect of Surface Modification on Flexural Properties of Jute Fiber Green Composites. https://pdfs.semanticscholar.org/395d/c784731da24e708a9337da1a5d5aac14f190.pdf (accessed 2 September 2019).

57 Zafar, M. T., Maiti, S. N. and Ghosh, A. K. (2016) 'Effect of surface treatments of jute fibers on the microstructural and mechanical responses of poly(lactic acid)/jute fiber biocomposites', *RSC Advances*, 6(77), pp. 73373–73382. doi: https://doi.org/10.1039/c6ra17894d.

58 Foruzanmehr, Mr. et al. (2016) 'Physical and mechanical properties of PLA composites reinforced by TiO2 grafted flax fibers', *Materials & Design*, 106, pp. 295–304. doi: https://doi.org/10.1016/j.matdes.2016.05.103.

59 Aydın, M., Tozlu, H., Kemaloglu, S., Aytac, A., Ozkoc, G., (2011) 'Effects of alkali treatment on the properties of short flax fiber-poly(lactic acid) eco-composites', *Journal of Polymers and the Environment*, 19(1), pp. 11–17. doi: https://doi.org/10.1007/s10924-010-0233-9.

60 Baghaei, B., Skrifvars, M., Salehi, M., Bashir, T., Rissanen, M., (2014) 'Novel aligned hemp fibre reinforcement for structural biocomposites: porosity, water absorption, mechanical performances and viscoelastic behaviour', *Composites Part A: Applied Science and Manufacturing*, 61, pp. 1–12. doi: https://doi.org/10.1016/j.compositesa.2014.01.017.

61 Baghaei, B., Skrifvars, M. and Berglin, L. (2013) 'Manufacture and characterisation of thermoplastic composites made from PLA/hemp co-wrapped hybrid yarn prepregs', *Composites Part A: Applied Science and Manufacturing*, 50, pp. 93–101. doi: https://doi.org/10.1016/j.compositesa.2013.03.012.

62 Hao, M. et al. (2018) 'Interface bond improvement of sisal fibre reinforced polylactide composites with added epoxy oligomer', *Materials*, 11(3). doi: https://doi.org/10.3390/ma11030398.

63 Orue, A. et al. (2016) 'The effect of alkaline and silane treatments on mechanical properties and breakage of sisal fibers and poly(lactic acid)/sisal fiber composites', *Composites Part A: Applied Science and Manufacturing*, 84, pp. 186–195. doi: https://doi.org/10.1016/j.compositesa.2016.01.021.

64 Birnin-yauri, A. U., Ibrahim, N.A., Zainuddin, N., Abdan, K., Then, Y.Y., (2017) 'Effect of maleic anhydride-modified poly(lactic acid) on the properties of its hybrid fiber biocomposites', *Polymers*, 9(5), pp. 1–16. doi: https://doi.org/10.3390/polym9050165.

65 Yu, T. et al. (2015) 'Effect of diisocyanates as compatibilizer on the properties of ramie/poly(lactic acid) (PLA) composites', *Composites Part A: Applied Science and Manufacturing*, 76, pp. 20–27. doi: https://doi.org/10.1016/j.compositesa.2015.05.010.

66 Yu, T., Jiang, N. and Li, Y. (2014) 'Study on short ramie fiber/poly(lactic acid) composites compatibilized by maleic anhydride', *Composites Part A: Applied Science and Manufacturing*, 64, pp. 139–146. doi: https://doi.org/10.1016/j.compositesa.2014.05.008.

67 Yu, T. and Li, Y. (2014) 'Influence of poly(butylenes adipate-co-terephthalate) on the properties of the biodegradable composites based on ramie/poly (lactic acid)', *Composites Part A: Applied Science and Manufacturing*, 58, pp. 24–29. doi: https://doi.org/10.1016/j.compositesa.2013.11.013.

68 Wang, T. and Drzal, L.T. (2012). Cellulose-nanofiber-reinforced poly(lactic acid) composites prepared by a water-based approach. *ACS Appl. Mater. Interfaces* 4: 5079–5085. https://doi.org/10.1021/am301438g.

69 Frone, A.N., Berlioz, S., Chailan, J.-F., and Panaitescu, D.M. (2013). Morphology and thermal properties of PLA–cellulose nanofibers composites. *Carbohydrate. Polymer.* 91: 377–384. https://doi.org/10.1016/J.CARBPOL.2012.08.054.

70 Coltelli, M.B., Cinelli, P., Gigante, V. et al. (2019). Chitin nanofibrils in poly(Lactic acid) (PLA) nanocomposites: Dispersion and thermo-mechanical properties. *International journal of molecular Science* 20: 504. https://doi.org/10.3390/ijms20030504.

71 Stoclet, G. et al. (2014) 'Elaboration of poly(lactic acid)/halloysite nanocomposites by means of water assisted extrusion: structure, mechanical properties and fire performance', *RSC Advances*, 4(101), pp. 57553–57563. doi: https://doi.org/10.1039/C4RA06845A.

72 Araújo, A. P. M. Agrawal, P., Cavalcanti, S.N., Alves, A.M., (2014) 'Bio-nanocomposite obtained from poly(lactic acid)/biopolyethylene blend and clay', *Macromolecular Symposia*, 343(1), pp. 59–64. doi: https://doi.org/10.1002/masy.201300190.

73 Gao, Y. et al. (2017) 'Influence of filler size on the properties of poly(lactic acid) (PLA)/graphene nanoplatelet (GNP) nanocomposites', *European Polymer Journal*, 86, pp. 117–131. doi: https://doi.org/10.1016/j.eurpolymj.2016.10.045.

74 Mariano, M. et al. (2017) 'Preparation of cellulose nanocrystal-reinforced poly(lactic acid) nanocomposites through noncovalent modification with PLLA-based surfactants', *ACS Omega*, 2(6), pp. 2678–2688. doi: https://doi.org/10.1021/acsomega.7b00387.

75 Teo, Z. X. and Chow, W. S. (2016) 'Impact, thermal, and morphological properties of poly(lactic acid)/poly(methyl methacrylate)/halloysite nanotube nanocomposites', *Polymer-Plastics Technology and Engineering*, 55(14), pp. 1474–1480. doi: https://doi.org/10.1080/03602559.2015.1132464.

76 Suryanegara, L., Nakagaito, A. N. and Yano, H. (2009) 'The effect of crystallization of PLA on the thermal and mechanical properties of microfibrillated cellulose-reinforced PLA composites', *Composites Science and Technology*, 69(7–8), pp. 1187–1192. doi: https://doi.org/10.1016/j.compscitech.2009.02.022.

77 Majhi, S. K. et al. (2010) 'Mechanical and fracture behavior of banana fiber reinforced polylactic acid biocomposites', *International Journal of Plastics Technology*, 14, pp. 57–75. doi: https://doi.org/10.1007/s12588-010-0010-6.

78 Wang, Y., Weng, Y. and Wang, L. (2014) 'Characterization of interfacial compatibility of polylactic acid and bamboo flour (PLA/BF) in biocomposites', *Polymer Testing*, 36, pp. 119–125. doi: https://doi.org/10.1016/j.polymertesting.2014.04.001.

79 Jandas, P. J., Mohanty, S. and Nayak, S. K. (2013) 'Surface treated banana fiber reinforced poly(lactic acid) nanocomposites for disposable applications', *Journal of Cleaner Production*, 52, pp. 392–401. doi: https://doi.org/10.1016/j.jclepro.2013.03.033.

80 Tawakkal, I. S. M. A., Cran, M. J. and Bigger, S. W. (2014) 'Effect of kenaf fibre loading and thymol concentration on the mechanical and thermal properties of PLA/kenaf/thymol composites', *Industrial Crops and Products*, 61, pp. 74–83. doi: https://doi.org/10.1016/j.indcrop.2014.06.032.

81 Chen, P.-Y. Lian, H.-Y., Shih, Y.-F., Chen-Wei, S.-M., Jeng, R.-J., (2017) 'Preparation, characterization and crystallization kinetics of Kenaf fiber/multi-walled carbon nanotube/polylactic acid (PLA) green composites', *Materials Chemistry and Physics*, 196, pp. 249–255. doi: https://doi.org/10.1016/j.matchemphys.2017.05.006.

82 Zotti, A. et al. (2018) 'Effects of 1D and 2D nanofillers in basalt/poly(lactic acid) composites for additive manufacturing', *Composites Part B: Engineering*, 153, pp. 364–375. doi: https://doi.org/10.1016/j.compositesb.2018.08.128.

83 Kurniawan, D. et al. (2013) 'Effect of silane treatment on mechanical properties of basalt fiber/polylactic acid ecofriendly composites', *Polymer - Plastics Technology and Engineering*, 52(1), pp. 97–100. doi: https://doi.org/10.1080/03602559.2012.722740.

84 Cheng, S. Lau, K., Liu, T., Zhao, Y., Lam, P.-M., Yin, Y., (2009) 'Mechanical and thermal properties of chicken feather fiber/PLA green composites', *Composites Part B: Engineering*, 40(7), pp. 650–654. doi: https://doi.org/10.1016/j.compositesb.2009.04.011.

85 Nam, T. H., Ogihara, S. and Kobayashi, S. (2012) 'Interfacial, mechanical and thermal properties of coir fiber-reinforced poly(lactic acid) biodegradable composites', *Advanced Composite Materials*, 21(1), pp. 103–122. doi: https://doi.org/10.1163/156855112X629559.

86 Huda, M. S. et al. (2006) 'Chopped glass and recycled newspaper as reinforcement fibers in injection molded poly(lactic acid) (PLA) composites: a comparative study', *Composites Science and Technology*, 66(11–12), pp. 1813–1824. doi: https://doi.org/10.1016/j.compscitech.2005.10.015.

87 Wang, G. et al. (2019) 'Strong and thermal-resistance glass fiber-reinforced polylactic acid (PLA) composites enabled by heat treatment', *International Journal of Biological Macromolecules*, 129, pp. 448–459. doi: https://doi.org/10.1016/j.ijbiomac.2019.02.020.

88 Tian, X. et al. (2016) 'Interface and performance of 3D printed continuous carbon fiber reinforced PLA composites', *Composites Part A: Applied Science and Manufacturing*, 88, pp. 198–205. doi: https://doi.org/10.1016/j.compositesa.2016.05.032.

89 Ferreira, R. T. L. et al. (2017) 'Experimental characterization and micrography of 3D printed PLA and PLA reinforced with short carbon fibers',

Composites Part B: Engineering, 124, pp. 88–100. doi: https://doi.org/10.1016/j.compositesb.2017.05.013.

90 Mitsuhiro Shibata, Shingo Oyamada, Shin-ichi Kobayashi, Daisuke Yaginuma (2004) 'Mechanical properties and biodegradability of green composites based on biodegradable polyesters and lyocell fabric', *Journal of Applied Polymer Science*, 92(6), pp. 3857–3865.

91 Graupner, N. and Müssig, J. (2009) 'Man-made cellulose fibres as reinforcement for poly(lactic acid) (PLA) composites', *Journal of Biobased Materials and Bioenergy*, 3(3), pp. 249–261. doi: https://doi.org/10.1166/jbmb.2009.1032.

92 Lai, S.-M. et al. (2014) 'Macromolecular nanotechnology unusual mechanical properties of melt-blended poly(lactic acid) (PLA)/clay nanocomposites', *European Polymer Journal*, 52, pp. 193–206. doi: https://doi.org/10.1016/j.eurpolymj.2013.12.012.

93 Cele, H. M., Ojijo, V., Chen, H., Kumar, S., Land, K., Joubert, T., (2014) 'Material properties effect of nanoclay on optical properties of PLA/clay composite films', *Polymer Testing*, 36, pp. 24–31. doi: https://doi.org/10.1016/j.polymertesting.2014.03.010.

94 Darie, R. N. Pâslaru, E., Sdrobis, A., Pricope, G.M., Hitruc, G.E., (2014) 'Effect of nanoclay hydrophilicity on the poly(lactic acid)/clay nanocomposites properties', *Industrial and Engineering Chemistry Research*, 53(19), pp. 7877–7890. doi: https://doi.org/10.1021/ie500577m.

95 Fukushima, K. et al. (2013) 'Effect of clay type and loading on thermal, mechanical properties and biodegradation of poly(lactic acid) nanocomposites', *Reactive and Functional Polymers*, 73(3), pp. 540–549. doi: https://doi.org/10.1016/j.reactfunctpolym.2013.01.003.

96 Pantani, R. and Turng, L. S. (2015) 'Manufacturing of advanced biodegradable polymeric components', *Journal of Applied Polymer Science*, 132(48). doi: https://doi.org/10.1002/app.42889.

97 Chu, Z. Zhao, T., Li, L., Fan, J., Qin, Y., (2017) 'Characterization of antimicrobial poly(lactic acid)/nano-composite films with silver and zinc oxide nanoparticles', *Materials*, 10(6), p. 659. doi: https://doi.org/10.3390/ma10060659.

98 Arrieta, M. P. López, J., Rayón, E., Jiménez, A., (2014) 'Disintegrability under composting conditions of plasticized PLA–PHB blends', *Polymer Degradation and Stability*. 108, pp. 307–318. doi: https://doi.org/10.1016/j.polymdegradstab. 2014.01.034.

99 Lv, S. et al. (2017) 'Microstructure analysis of polylactic acid-based composites during degradation in soil', *International Biodeterioration and Biodegradation*, 122, pp. 53–60. doi: https://doi.org/10.1016/j.ibiod.2017.04.017.

100 Mohanty, A. K., Misra, M. and Drzal, L. T. (eds) (2005) *Natural Fibers, Biopolymers, and Biocomposites*. 1. Taylor & Francis. doi: https://doi.org/10.1201/9780203508206.ch6.

5

Green and Sustainable Packaging Materials Using Thermoplastic Starch

Anshu A. Singh and Maria E. Genovese

Smart Materials, Istituto Italiano di Tecnologia, Via Morego 30, Genova 16163, Italy

The environmental and health risks associated with the emission of greenhouse gases and plastic pollution, combined with the increasing price and future shortage of petroleum, have been the main driving forces for the development of greener materials from renewable resources. Although such increased emphasis on sustainability has transversally come across all the technological sectors and end-markets of mass production goods, it is especially the packaging industry, traditionally dominated more than others by disposable fossil-based plastics, that has been facing a growing pressure to provide and implement commercially attractive and competitive alternatives, with an optimal balance in terms of functionality, costs, eco-efficiency, product availability, and production capacity. In this regard, public organizations, industrial manufacturers, and enterprises from different industries, ranging from food packaging to agriculture to medical care, are joining efforts aimed at the large-scale production and commercialization of biodegradable packaging solutions that can meet the legislative requirements imposed by the restrictions in the use of disposable plastics and the demand of a more responsible waste management arising from an increased environmental awareness.

Among natural materials, starch, an edible polysaccharide, represents one of the most promising candidates to replace conventional nonbiodegradable polymers used in packaging applications owing to its low cost, abundancy, high biodegradability, non-toxicity, and the possibility to undergo melt processing through standard techniques commonly used for plastic materials.

In the present chapter, an overview of the research literature and commercial solutions based on thermoplastic starch and its composites and their suitability as packaging materials will be presented. In particular, in Section 5.1, some economical and legislative framework related to sustainability will be provided and the current natural and green materials used in sustainable packaging will be introduced. In Section 5.2, the physicochemical properties, processing, applications, and drawbacks of starch will be elucidated with a particular focus on thermoplastic starch. In Section 5.3, the most relevant studies assessing the performance of thermoplastic starch composites in terms of mechanical properties, gas and

Sustainable Food Packaging Technology, First Edition. Edited by Athanassia Athanassiou.

water vapor permeability, and biodegradability for packaging applications will be described along with commercially mature and existing packaging solutions. In Section 5.4, the challenges and future trends of starch and its derivatives in the context of green and sustainable packaging will be discussed.

5.1 Sustainability and Packaging: Toward a Greener Future

5.1.1 The Plastic Threat

Plastics, glass, and metals are the most commonly used materials for packaging applications, with plastic being the most ubiquitous one. In particular, conventional petrochemical-based plastics such as polyethylene (PE), polypropylene (PP), polyethylene terephthalate (PET), and polystyrene (PS) have been used universally as packaging materials, because of well-demonstrated advantages, such as wide availability, high production throughput, versatility, light-weight, durability, low costs, attractive mechanical, thermoforming, optical, and barrier properties, resistance to water and oil, heat sealability, etc. [1].

However, globalization of plastic-based packaging materials is a major concern because of their dependency on nonrenewable petrochemical resources and especially because of the increasing environmental issues due to their resistance to biodegradation. The estimated global production of plastic materials reached 350 million tons in 2017, with packaging accounting for around 40% of the total plastic demand [2, 3]. It has been reported that, only in 2012, 17 million tons of plastic waste were generated in the European Union of which approximately 88% (i.e. 15 million tons) was from packing materials [2]. As a consequence of plastic massive production and waste generation combined with its improper disposal, thousands of tons of waste plastic materials end up in the landfills and in the oceans with a dramatic impact on the environment. For example, out of the 8.3 billion tons of plastic produced globally since the 1950s, only about 9% has been recycled, 12% has been burned, and the remaining 79% has ended up in landfills or the environment [4]. It was estimated that 5.25 trillion macro and microplastic pieces are floating in gyres in the open ocean weighing up to 269 000 tons and with more than 8 million tons of plastic entering the marine environment every year, equivalent to dumping the contents of one garbage truck into the ocean every minute [5–7]. Since conventional plastics are not biodegradable nor completely recyclable their accumulation and high persistency raise serious environmental and health concerns including global warming, pollution, and toxicity to humans and other living organisms. Plastics are in fact a primary cause of death for sealife through starvation, suffocation, infection, drowning, and entanglement, and bioaccumulation throughout the food chain. According to Greenpeace, 700 marine species are affected by ocean plastic [6]. It was estimated that up to 9 out of 10 seabirds and 1 out of 3 sea turtles have ingested plastic, while 1 out 3 fish caught for human consumption now contains plastic [8].

5.1.2 The Call for Sustainability

It is thus of utmost importance to find environmentally friendly, nontoxic, and safe packaging materials alternative to conventional petrochemical-based plastics to drastically reduce the environmental impact caused by the disposal of plastic waste and the greenhouse emissions of its production and incineration. In this context, the use of biopolymers or biodegradable polymers obtained from renewable sources for the development of greener packaging solutions has gained widespread attention due to a growing awareness of the impact of plastic on the environment and the need to reduce the dependency on fossil resources as well as to meet the demand for continuous advancements and innovation. These efforts have been unanimously encouraged by political organizations, industrialists, and consumers. In fact, over the last few years, the legal framework has reflected the demand for an increased environmental awareness and sustainability through the adoption of stricter regulations concerning the use of disposable and packaging plastic-based materials. More than 30 countries have already imposed either regional or country-wide bans on plastic bags and on different single-use and packaging plastic items [6]. In Europe, the European Commission welcomed a ban that will be active from 2021 on selected non-biodegradable plastic products including articles such as cups, food and beverage containers and imposed the incorporation of 25% of recycled plastic in bottles by 2025 [9].

From a market perspective, although many challenges exist and only few commercial breakthroughs have occurred, biodegradable packaging materials have been projected as one of the most lucrative segments of the so-called bioeconomy as the demand keeps rising and more sophisticated biopolymers, applications, and products emerge. In fact, according to the latest forecast, the bioplastic market will reach US$68.5 billion by 2024, registering a compound annual growth rate (CAGR) of 18.8% during 2018–2024, with rigid packaging applications accounting for more than 40% of the revenue [3].

5.1.3 Biomaterials for Sustainable Packaging

The major function of packaging is to maintain the quality as well as safety of the packaged products during storage and transportation and, therefore, packaging materials should exhibit appropriate mechanical, thermal, and optical properties (Figure 5.1) [10]. When it comes to food packaging, in addition to such properties, suitable materials should preserve food freshness, protect the food against microbial or chemical contaminations, impede the gain or loss of moisture, provide barrier against oxygen, carbon dioxide, water vapor, and any aroma, and should not release any toxic substance by prolonged contact with the food item.

The first patent on the biodegradable packaging films to extend the shelf life of foods as well as films for frozen meat, poultry, and seafood using fats, starches, gums, and alginates dates back to the 1950s [11]. Since then, scientists and researchers have exploited and developed several biopolymeric materials that are environmentally friendly and can be used in the packaging sector.

Depending on the origin or source and manufacturing process, biodegradable biopolymers can be divided into different types (Figure 5.2) [10, 13]:

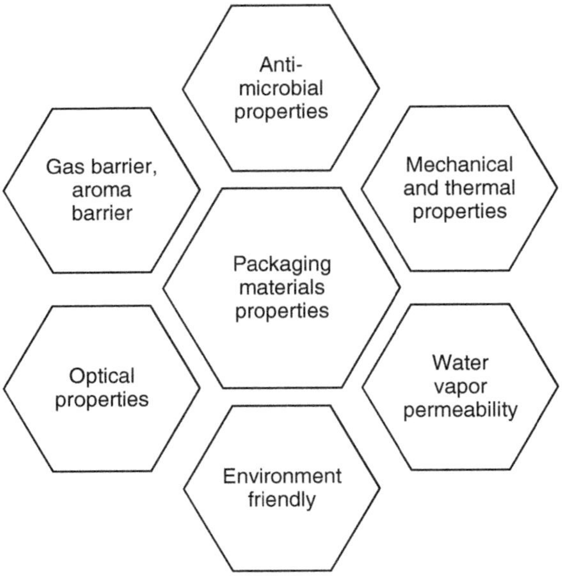

Figure 5.1 General properties required for properties of packaging materials. Source: Adapted from Rhim et al. 2013 [10].

(i) natural biopolymers obtained from biomass upon extraction or other methods, such as cellulose, starch, and proteins;
(ii) biopolymers produced by microorganisms, such as polyhydroxyalkanoates (PHAs);
(iii) synthetic biopolymers from biomass precursors or bio-derived monomers, such as polylactic acid (PLA) which is chemically synthesized from lactic acid obtained by fermentation of carbohydrate feedstock;
(iv) synthetic biodegradable polymers from petrochemicals (Figure 5.2), such as polyesters like polycaprolactone (PCL) and polyvinyl alcohols (PVAs).

The polymers of the first three categories are obtained from renewable resources, while the others are fossil-based [13].

Natural biopolymers are further classified into four different categories [1, 13]:

(i) hydrocolloids – they include cellulose, starch, pectin, algal gum, and proteins (e.g. wheat gluten, soy protein, casein, zein);
(ii) lipids – they include waxes, acylglycerols, and fatty acids;
(iii) resins – they include shellac and wood resin;
(iv) composites which contain both lipid and hydrocolloid components in the form of a bilayer or emulsion.

Among the abovementioned materials, natural biopolymers are especially advantageous as they are abundant, readily available, and generally more cost-effective compared to biopolymers produced by fermentation and chemical synthesis [12–14]. Of all the natural biopolymers available, starch and its derivatives are considered among the most promising materials for packaging applications due to starch abundancy, renewability, cost effectiveness, and thermoplasticity. Moreover, starch is considered eco-friendly and safe because it

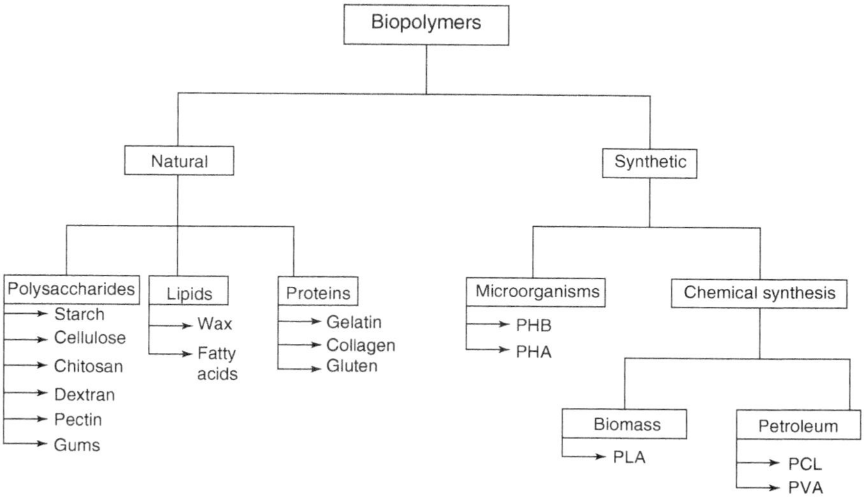

Figure 5.2 Classifications of biopolymers. Source: Rhim et al. 2013 and Averous 2004 [10, 12].

is edible and completely biodegradable [1, 12, 14]. According to the latest market data, starch blends currently represent roughly 18% of the global bioplastic production capacity with a steady 200 000 tons for use in the packaging field [3].

5.2 Thermoplastic Starch

5.2.1 Starch: Physicochemical Properties, Processing, Applications

Starch is one of the most commonly available natural polysaccharides and can be obtained in granular form from potato, corn, rice, wheat, barley, oats, tapioca, and pea. Its chemical composition (i.e. α-glucan, lipid, protein, mineral content), shape (i.e. round, polygonal, lenticular), granule size (1–100 μm), distribution (unimodal, bimodal), and association (individual or cluster) varies depending on its origin [15]. It is economical, edible, and biodegradable in a wide variety of environments.

Starch typically contains 20–25 wt% of amylose and 75–80% of amylopectin (Figure 5.3), although the amount may vary depending on the source of plant, as reported in Table 5.1. In addition, starch also contains approximately 2 wt% of free fatty acid, 0.6% proteins, and 0.4% minerals (calcium, magnesium, phosphorus, potassium, and sodium) [15].

Amylose is a linear polymer chain of α-1,4-anhydroglucose units, having a molecular weight of approximately 1×10^5–1×10^6 g/mol and a degree of polymerization (DP) by number (DP_n) of 324–492, with approximately 9–20 branch points, equivalent to 3–11 chains per molecule, where each chain contains approximately 200–700 glucose residues. On the other hand, amylopectin is a highly branched polymer of α-1,4-anhydroglucose units linked by α-1,6-glycosidic bonds with a molecular weight of 1×10^7–1×10^9 g/mol

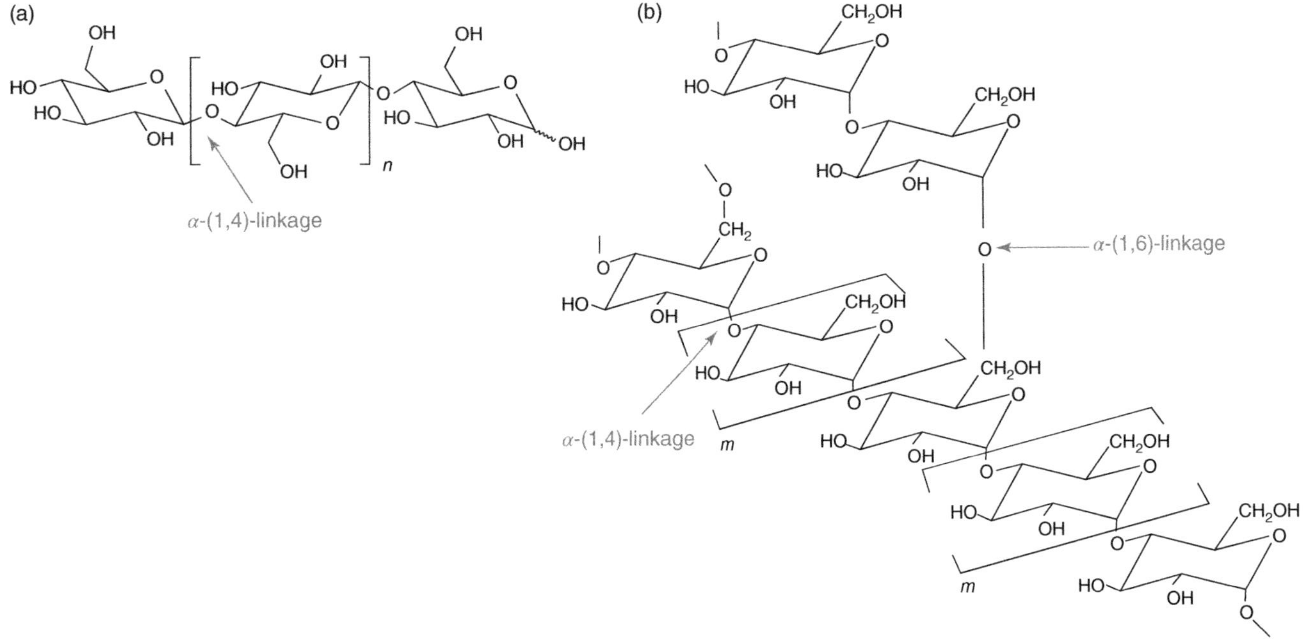

Figure 5.3 Chemical structure of (a) amylose and (b) amylopectin. Source: Petersen et al. 1999 [16] with permission from Elsevier.

Table 5.1 Composition of starch from different sources.

Source	Amylopectin (%)	Amylose (%)
Maize	71–73	26–28
Corn	75–83	17–25
Potato	76–83	17–24
Wheat	75–80	20–25
Rice	65–85	15–35
Chickpeas	60–70	30–40
Tapioca	78–81	19–22
Banana	76–83	17–24

Source: Adapted from [17].

and DP_ns in the range of 9600–15 900 [15, 17, 18]. The chemical structure of amylose and amylopectin are shown in Figure 5.3. Starch is unique among carbohydrates as it occurs naturally as discrete granules and this is due to the presence of short-branched amylopectin chains that form helical structures [19]. The degree of crystallinity in starch ranges from 15% to 45%, depending on the ratio of amylose to amylopectin, and has an important effect on properties such as biodegradation rate, solubility, and mechanical strength. The amylose and the branching points of amylopectin form the amorphous regions of the starch while the short-branched chains of amylopectin form the crystalline regions [20, 21]. From investigations of the crystalline organization of starch it was determined it may occur in four different polymorphs, i.e. A, B, C, and V crystalline forms [22]. Type A is mainly associated with cereals (such as rice, wheat, and corn) and is established in dry and warm conditions, while type B is obtained from tubers, fruit, and corn in which the amylose content is high. Both A and B types consist of parallel-packed left-handed double helices with six glucose units per turn and A has a monoclinic unit cell, whereas B has a hexagonal unit cell. Type C is a mixture of both A and B and occurs naturally in most vegetables, e.g. pea and various legume seeds. Type V starch is found when starch forms complexes with substances such as aliphatic fatty acids, emulsifiers, butanol, and iodine and it consists of single left-handed helices with the guest complexed molecules located inside the helix channel. Such structure is prone to be formed by thermal processing like extrusion [22–24]. Hizukuri studied the distributions of chain lengths in amylopectin of 20 different starch specimens and reported that amylopectin of type A starch has shorter chains than type B, while in type C they are intermediate [25]. In the X-ray diffractograms, the type A crystals show three strong patterns at 5.8, 5.2, and 3.8 Å, while for type B typical patterns are at 15.8–16.0 Å, a broad medium intensity signal at 5.9 Å, a strong line at 5.2 Å, and a medium intensity doublet at 4.0 and 3.7 Å. The X-ray diffraction (XRD) pattern of Type C is similar to type A except for the addition of the medium to strong pattern at 16.0 Å. Hydrated and dehydrated type V exhibit patterns at 2.0, 6.8, and 4.4 Å and at 11.3, 6.5, and 4.3 Å respectively [26].

In terms of chemical reactivity, the abundancy of hydroxyl groups in the pyranose structure of starch renders it highly hydrophilic and allows for the formation of strong intermolecular associations via hydrogen bonding and for a number of chemical modifications including oxidation, reduction, polymer grafting, and the conversion of the alcohol moieties into esters and ethers [27]. Moreover, starch can be easily biodegraded by microorganisms or enzymes upon hydrolysis to obtain glucose molecules, which are further converted into non-harmful substances like carbon dioxide and water.

Like other polysaccharides, native starch tends to absorb environmental moisture and it is thermally sensitive, having a melting point (220 °C) higher than its thermal degradation temperature. At temperatures above 150 °C, in fact, the glycosidic bonds start to disrupt and above 250 °C the granules subside endothermally [13]. These characteristics determine a poor dimensional and thermal stability that restrict its direct utilization as biomaterial in packaging and biomedical applications. On the other hand, starch films show excellent oxygen barrier properties because of the presence of a tightly packed and highly ordered hydrogen bond network structure in which amylose and amylopectin form alternate layers of crystalline and non-crystalline regions [14, 20, 26, 28].

In terms of physicomechanical properties, starch films are usually stronger and more flexible when the amylose content is higher, while the branched structure of amylopectin leads to films with poorer mechanical properties characterized by brittleness and low strength [29].

As starch is highly hydrophilic in nature, its structure, intermolecular interactions, mechanical properties as well as its glass transition temperature (T_g) not only depend on the amylose/amylopectine ratio but they are also strongly influenced by the water content. For example, the T_g of native starch can be as low as 60–80 °C when the weight fraction of water is in the range of 0.12–0.14 [27].

Starch granules are insoluble in cold water but when dispersed in hot water, partial solubilization occurs due to the strong interaction between the water molecules with the hydroxyl group of amylose and amylopectin. Specifically, the granules of starch swell and undergo a transition process known as gelatinization. The gelatinization process involves two steps: (i) hydration and diffusion of the solvent into the starch granules and (ii) melting of starch crystals [20, 28]. Lai and Kokini reported that, during gelatinization, addition of water causes the diffusion of amylose out of the granules and disrupts their crystalline structure [30]. The granules containing mostly amylopectin then collapse in a matrix of amylose forming a gel. The gelatinization temperature, i.e. the temperature at which the starch granules start to gelatinize, depends on the source of starch, ratio of amylose to amylopectin, and content of water. In particular, the gelatinization temperature can be increased to an extent, and thus the water solubility of starch reduced, by increasing the crystallinity or the content of amylopectin and by using a limited amount of water [30].

Owing to its attractive properties, starch is widely used in the food industry and in pharmaceutical and cosmetic applications as a filler, thickening, stiffening, gluing, and binding agent to modify and enhance the physical properties of food and cosmetic products, like texture and viscosity, for gel and film formation, and as a precursor to dextrins, glucose, ethanol, and sugars for processed food.

It is also used in the production of biofuel and as adhesive in the papermaking process [13]. Starch and starch composites have also been developed for packaging applications both at the research stage and as commercially mature solutions and will be discussed in Section 5.3.

With regard to processing methods for starch and its composites, generally, two distinct approaches can be used: (i) solvent casting or wet method and (ii) thermoplastic/thermal processing or dry method. In the former, starches are solubilized upon heating at 65–100 °C in an excess of water or other solvents, e.g. liquid ammonia, chloroacetic acid, dimethyl sulfoxide, formic acid, and formamide to achieve irreversible gelatinization. This process is highly affected by the type of solvent used and the ratio of starch to solvent [28]. Films can be obtained with this method upon evaporation of the solvent. In the latter approach, the starch is plasticized by heating with a low water content. The dry method is greener as it requires only low amounts of water, is more feasible industrially, and is more versatile as it can be used for film formation as well as for the fabrication of items of different shape and geometry with conventional thermoforming technologies used for plastic materials [20, 28]. Such method is the most suitable for the large-scale production of packaging products based on starch and its blends and will be further described in Section 5.2.2.

5.2.2 From Starch to Thermoplastic Starch

To summarize Section 5.2.1, the brittleness, hydrophilicity, and low thermal decomposition temperature of native starch limit its applications in the packaging sector [31, 32]. Moreover, due to the strong inter- and intramolecular hydrogen bonding between amylose and amylopectin, it does not melt before degradation occurs and for this reason it cannot be considered as a thermoplastic polymer.

In order to overcome such drawbacks, starch granules can be rendered thermoplastic with the combined effect of plasticizers, temperature, and shear, to yield thermoplastic starch (TPS) [32, 33]. TPS exhibits viscoelastic behavior and physicochemical properties similar to synthetic polymers and is, thus, considered a promising material for application as a biodegradable alternative to plastic. TPS is very attractive as it can be either used alone or blended with other polymers such as PVA and PCL through melt compounding. In particular, due to their inherent non-toxicity, TPS based films and coatings offer tremendous potential as packaging materials for food, such as fruits, vegetables, grains, candy, meat, seafood, and poultry [34].

In order to produce TPS, granular starch is first mixed with plasticizers, such as water, glycerol, urea, etc., and then processed using kneading, extrusion, injection molding, blow molding, or heating and casting in excess of water or other solvents together with plasticizers [1]. The final properties of TPS are dependent on its molecular mass and amylose/amylopectin ratio as well as on the processing conditions (i.e. temperature profile range, shear stress) and method (i.e. extrusion, kneading, injection molding) and on the chemical nature and concentration of the chosen plasticizers [14, 28, 35].

5.2.3 Plasticizers of Starch

Plasticization refers to the changes in the mechanical and thermal properties of a particular polymer that occur in the presence of plasticizer molecules. Plasticizers are known to reduce the glass transition temperature, tensile strength, and modulus but improve the polymer chain flexibility, toughness (impact strength), fracture resistance, and dielectric constant. Plasticizers are generally low molecular weight molecules that can occupy the intermolecular spaces between polymer chains and improve their processability by lowering their processing temperature. Several theories have been proposed to explain the mechanism of plasticization. For example, according to the "lubricity theory," a plasticizer acts as a lubricant facilitating the movement of macromolecules over each other. "Gel theory" discusses the disruption of polymer–polymer interactions via hydrogen bonds, van der Waals or ionic forces. The "free volume theory" states that plasticizers increase the free volume between the polymers and in this way, they are able to lower the glass transition temperature [36].

In starch, plasticizers are used to overcome its brittleness, improve flexibility, and make it thermoplastic by breaking the hydrogen bonds in its structure, thus increasing the mobility of macromolecular chains [37]. Suitable plasticizers should be rather polar and hydrophilic in nature to be chemically compatible with starch, have a boiling temperature and a degradation temperature higher than the material processing temperature, in order to prevent evaporation or decomposition, and should not be sticky or build up deposits in the die region during extrusion. In addition, especially for food packaging applications, plasticizers should possess a good stability over time and low migration.

Although water can be regarded as a process aid since the viscosity of the melt is lowered with increasing water content and is itself a plasticizer for starch, it is volatile and cannot be used to produce utilizable plastic-like materials. Moreover, high amounts of water restrict the upper processing temperature because steam generation in the material leads to bubbles and foaming, which are not desirable [38]. Therefore, water is used in combination with other plasticizers. Common plasticizers for starch include polyols (glycerol, sorbitol, ethylene glycol, maltitol), monosaccharides (glucose, mannose, fructose, galactose), citric acid, urea, amides, and amino acids. Each type has been found to have a strong impact on the physical and functional properties of the resulting TPS [19, 39–41]. Monosaccharides were reported to be more effective than polyols as plasticizers, not only because of their hydroxyl groups, but also due to their similarity in the structure and conformation to anhydroglucose in starch molecules. Zhang and Han reported that such structural compatibility and interactions yield denser materials with a reduced free volume and segmental motions of the starch chains compared to starch plasticized with polyols, resulting in improved moisture barrier properties [39].

To date, glycerol is the most commonly used plasticizer for TPS production due to its high boiling point, low cost, and ready availability [19]. One of the drawbacks when glycerol is used as a plasticizer for starch is the retrogradation of starch. In fact, TPS is generally amorphous but over time it can recrystallize forming strong hydrogen bonds that cause embrittlement and loss of mechanical strength in the material. This effect is further enhanced by the poor stability of plasticizers such as glycerol, which tend to migrate upon storage. Xiaofei et al. reported that the formamide as well as the combination of urea and formamide can effectively reduce the retrogradation of the TPS and make it more flexible [40]. This might be due to the formation of more stable and stronger hydrogen bond between urea, formamide, and hydroxyl groups of starch molecules with respect to glycerol. However, plasticizers containing amide groups are not recommended for use in food packaging applications [42]. Pushpadass et al. [37] reported the use of stearic acid, sucrose, and urea as secondary plasticizers for the corn starch–glycerol (3 : 1 ratio) mix and studied the effect of extrusion temperature and concentrations of the plasticizers on the film-forming characteristics of starch. Of all the three secondary plasticizers used, urea was found to be the most effective, due to its low molecular weight and formation of more stable and stronger hydrogen bonds with the hydroxyl group of starch molecules than those of stearic acid and sucrose. However, when the urea content exceeded 5 wt%, films appeared to be slightly murky with traces of undissolved urea granules on the film surface. In spite of possessing the largest number of hydroxyl groups, sucrose was not a good plasticizer, because of its hygroscopic and recrystallization properties. Consequently, TPS films containing 5 wt% or more of sucrose tended to be much stiffer than those plasticized with stearic acid and urea. Similarly, stearic acid yielded poor extensibility and contents above 1 wt% structurally affected the films. This is because stearic acid is bulkier than sucrose and urea and has less hydroxyl groups; therefore it cannot easily penetrate the polymeric chains in starch and interfere with their packing through hydrogen bond interactions.

Another important consideration in the choice of plasticizers beside their physicochemical properties and chemical affinity with starch is the amount to use, as a too low content might not be effective in plasticization, whereas a high amount would impact the mechanical properties of the material. Furthermore, such amount is also a function of the specific amylose/amylopectin ratio and of the process (i.e. extrusion, injection molding). With glycerol, the best results were typically obtained when the content is around 15–20 wt%. For example, de Graaf et al. reported that different types of starch plasticized using 15 wt% of glycerol exhibited tensile strengths of 10–35 MPa, comparable to low-density polyethylene and polystyrene, moduli ranging from 700 to 900 MPa, comparable to high density polyethylene, and elongation at break from 10% (low amylose starches) up to 100% (high amylose starches) [43]. Conversely, when the glycerol content was 25 wt%, only the elongation at break significantly increased up to 200% whereas the modulus and tensile strength dropped to 7 and 2 MPa, respectively, in high amylose-containing starches [43].

5.2.4 Processing of Thermoplastic Starch

Solvent casting method of starch and water-soluble plasticizers is the most commonly used technique to prepare TPS films. Specifically, starch is heated in an excess of water to induce gelatinization and then the resulting aqueous solution of starch is cast to evaporate the water, until a dry film is obtained. However, the solvent casting process requires a great amount of water, long drying times, and can be relatively energy consuming. Therefore, such method is practical only at the laboratory scale for research purposes and is not suitable for large-scale production. Moreover, several factors may affect the formation of TPS when solvent casting is used. For example, Koch et al. studied the effect of glycerol content and processing conditions, i.e. heating temperature and time of starch films prepared by solvent casting on molecular structure, morphology, and mechanical properties of the resulting materials [44]. The heating temperature and time used for the preparation of high amylose starch films varied from 130 to 150 °C and from 5 to 30 minutes, respectively. They reported the formation of fragmented films as a result of high temperature and longer heating times, which might be due to the degradation of starch.

For industrial applications, plasticization of starch and compounding using extrusion at a temperature ranging from 100 to 150 °C constitutes a more realistic and productive approach to fabricate end products such as strands, films, sheets, bags, and containers. During extrusion, a mixture of starch, water, plasticizers, and additives are fed from the hopper to the screw, and as the starch is transported through the barrel, it is subjected to friction, compression, and heating, ensuring a homogeneous mixing of the molten compound. The extruded TPS can be either molded into desired shapes or blown into films or sheets, after it comes out from the extruder die and before it solidifies [38, 45, 46]. In particular, during the extrusion, the water present in the starch and other plasticizers destroy the crystalline structure of starch, which undergoes fragmentation and is able to flow like thermoplastic polymers. The main parameters that influence the processing of TPS by extrusion are the moisture content in the materials, the composition of ingredients, the amount and type of plasticizers, as well as operational conditions, such as the screw speed, the barrel temperature, the diameter to length ratio of the die, and the pressure. It was found that a high amount of liquid plasticizer (more than 25–35 wt%) like glycerol might cause exudation, while lower amounts resulted in difficulties during processing like pressure build up at the die. Moreover, the ratio of amylose to amylopectin also affects the extrusion process. In fact, it is more difficult to extrude high amylose starch compared to normal starch because of its higher melt viscosity [47]. Thunwall et al. extruded high amylose potato (HAP) starch and normal potato starch (NPS) with glycerol and water in different proportions, at different screw speeds, 20 and 100 rpm [45]. A fluoroelastomer lubricant was used as a processing aid to reduce the tendency of the material to stick on the die and prevent its clogging and dextrin was used to lower the viscosity. It was reported that the thermoplastic HAP was difficult to extrude with respect to thermoplastic NPS, as it exhibited a higher moisture content and melt viscosity and thus a higher extrusion temperature (160 °C) and amount of plasticizer (45 parts per 100 parts of starch) were needed.

In addition to extrusion, other commonly used thermoforming techniques for the fabrication of TPS films are compression molding, blow molding, and injection molding [38, 43, 45, 46]. For example, thin films of TPS having different glycerol and water content were prepared by Thunwall et al. using compression molding in a press heated at 140 °C for five minutes by applying a pressure of

9 MPa [46]. Prior to compression molding, starch, glycerol, and water in 10 : 3 : 4.5 and 10 : 0 : 8 proportions were premixed by blending in a Brabender at 140 °C for three minutes at 50 rpm. The same authors fabricated TPS films using film blowing, a commonly used thermoforming method to fabricate self-standing films, in which a hollow tube was extruded at 95–140 °C and subsequently expanded to form a thin film by increasing the pressure inside the tube [38]. The best results were achieved using a blend comprising about 16 wt% of glycerol, 75 wt% of starch, and 9 wt% of water. The melt exhibited a satisfactory tenacity for the expansion when processed at temperatures below 140 °C. It was also observed that by increasing the amount of water and using higher temperatures, bubbles were formed causing the disruption of the material, whereas an increase in the amount of glycerol resulted in processing difficulties related to the stickiness of the resulting extrudate. TPS from four different types of starch was also successfully processed through injection molding by de Graaf et al., but the results of the work indicate that the choice of the starch in terms of amylose/amylopectin ratio, as well as the amount of glycerol plasticizer, are fundamental to limit the shrinkage effect due both to retrogradation and to the elongation flows during the injection [43]. In particular, the authors correlated the shrinkage to the chain mobility, which in turn depends on both glycerol content and amylose/amylopectin ratio. In samples, with a glycerol amount lower than 15 wt% and high amylopectin, the shrinkage was negligible, while the highest shrinkage (18%) was observed in samples with 25 wt% of glycerol and low amylopectin.

5.3 Thermoplastic Starch-Based Materials in Packaging

5.3.1 Technical and Legal Requirements for Packaging Materials

The basic role of food packaging materials is to preserve the properties and quality of a packaged item from the action of external chemical, biological, and physical agents, satisfying specific industrial and legislative requirements and consumer demands [21].

For this reason, the main technical requirements of packaging materials include mechanical strength, a certain degree of rigidity or flexibility, thermal stability in a range of hot and cold temperatures, chemical and UV light resistance, low water vapor and oxygen permeability. In terms of sustainability, packaging materials should be biodegradable, of low carbon footprint, recyclable, and compostable.

Biodegradation can occur in various conditions such as marine, soil, freshwater, and composting environments by the action of naturally occurring microorganisms to yield water, CO_2, biomass, and inorganic compounds and leave no visible distinguishable or toxic residue [48]. According to EN 13432, composting is defined as a biodegradation process controlled by time, under specific humidity, temperature, and aeration conditions. It can take place in an industrial compost plant or in a domestic device. A classical composting cycle in a plant is spread over 6–12 weeks. The European standard covers four fields of metrics to assess compostability: biodegradability, disintegration, absence of heavy metals and ecotoxicity [49]. Moreover, several factors can influence the rate of degradation of materials under composting conditions such as sample size, sample composition, crystallinity, polarity, etc. [50].

TPS packaging materials are mostly made from corn, potato starch, and maize and they are entirely biodegradable when disposed in soil, sewage, sea, and lakes [48]. As studied by Li et al., the degradation rate of TPS is significantly higher than that of native starch due to its amorphous structure [50]. Moreover, it was reported that the glycerol content in TPS films has an effect on the microbial growth, as it promotes swelling, and thereby, the mass transport of water [48]. Biodegradation studies on cassava starch films containing glycerol, agar, and sorbitan mono-oleate showed that the degradation of the material was complete in 30 days under indoor soil composting method [51].

Although TPS is biodegradable, compostable, and cost-effective, it is generally not used alone for packaging applications due to challenges related to its processability, retrogradation, and plasticizer migration, and to its poor mechanical properties, dimensional and thermal stability, and high water absorbancy [27]. In order to improve TPS performance and suitability in the packaging industry, different additives, chemical modifications, and blends with other biodegradable polymers have been proposed and will be reviewed in Sections 5.3.2–5.3.5.

5.3.2 Composites of TPS with Fillers

Different types of natural, mineral, and synthetic fillers have been incorporated into TPS mainly to induce a mechanical reinforcement, to improve its water vapor properties, and to impart additional functionalities for applications, such as smart packaging.

TPS-based biodegradable films containing cassava starch, glycerol, and green tea or basil extract were prepared by C. Medina-Jaramillo et al. by a casting method [52]. The extract of green tea contains polyphenols, for example, catechins, epicatechins, gallic acid, etc., while the basil, an aromatic herb, contains active antioxidant compounds such as chlorogenic acid, methyl chavicol, and linalool. Addition of green tea and basil extract delayed the oxidation of food products, thus avoiding their faster deterioration. Chlorophyll and carotenoids present in both the extracts showed change in the color when exposed to different pH, and therefore the TPS films with green tea and basil extract could be potentially used as smart packaging materials to monitor and preserve food quality. Moreover, the obtained films had lower water vapor permeability (WVP) and biodegraded in the soil in less than two weeks.

TPS functional nanostructured films with antimicrobial properties and improved mechanical and barrier properties were successfully fabricated by Abreu et al. through a casting method by adding nanoclays and in situ generated silver nanoparticles (NPs) to starch. Moreover, the authors found that the migration of components from the nanostructured starch films, assessed by food contact tests, was minor and within the legal limits. Such properties render the obtained films promising for food packaging applications [53]. Another biodegradable TPS/clay nanocomposite film was prepared by homogeneously dispersing montmorillonite nanoparticles (4 wt%) via polymer melt processing techniques by Avella et al. [54]. Although a good intercalation of starch phase into the clay layers was achieved, the mechanical reinforcement was not too remarkable. The highest Young's modulus and tensile strength for reinforced films were reported to be 1135 and 22 MPa, respectively, whereas the values for the unloaded samples were 979 and 19 MPa, respectively. However, the authors demonstrate the conformity of the films with the European directive EN13432 on biodegradable materials and their suitability in the food packaging sector owing to low overall migration. Other studies showed that the addition of nanoclays can enhance the rate of biodegradation due to the presence of Lewis acid sites which increase the polarity of the matrix, thereby causing more water absorption and promoting hydrolysis phenomena [51].

Concerning water vapor barrier properties, TPS reinforced with halloysites nanofillers has shown a drastic reduction in the WVP from $2.91 \pm 0.53 \times 10^{-11}$ to $0.83 \pm 0.11 \times 10^{-11}$ g/m s Pa with a filler content of 5 wt%. At the same time, the oxygen permeability also decreased from 76.88 ± 2.4 to 38.2 ± 0.81 cm^3 μm/m^2 d. This effect was ascribed to the enhanced tortuosity provided by nanoparticles that hinders the permeation of gaseous molecules [55]. In another study by López et al., the incorporation of 5 wt% of talc particles in TPS films had a similar effect, promoting a 54% and a 26% reduction of the WVP and oxygen permeability, respectively, with respect to pure TPS [56]. The use of zinc oxide nanorods fillers in low content (i.e. below 5 wt%) decreased the WVP of TPS composites to less than one-third, while considerable reductions in WVP, water absorption, and water solubility also occurred for TPS films loaded with SiO_2 nanoparticles [51].

The use of nanofillers, especially clays, for the enhancement of TPS properties could be a suitable approach for the development of novel starch packaging materials but, especially for food contact applications, rigorous toxicology assessments would be required before commercialization. Other aspects to be considered to evaluate the feasibility of such system are the challenges related to the filler dispersion in the matrix and the changes in the mechanical properties and color/transparency of the resulting composites.

5.3.3 Composites of Thermoplastic Starch with Polysaccharides

Cellulose and polymers such as chitosan and chitin are natural, biodegradable, and abundant polysaccharide materials that could be used in composites with starch for packaging applications. Due to the similarity in the chemical structure, they can be mixed with TPS ensuring a good dispersion and interfacial adhesion even at higher amounts compared to other types of fillers and serving as reinforcing and barrier properties enhancement agents [57]. Composites of TPS reinforced with 30 wt% of flax cellulose nanocrystals exhibited a tensile strength of 12 MPa and a Young's modulus of 499 MPa, with an improvement of 205% and of 1450% with respect to pure TPS. However, the elongation at break was observed to decrease drastically from 68.2% for control TPS to 7.2% in the reinforced samples [58]. Liu et al. have produced TPS-reinforced composites containing 1–20 wt% of bamboo cellulose crystals. With 8 wt% of bamboo filler, the tensile strength of the material increased from 2.5 to 12.8 MPa and the Young's modulus went from 20 to 210 MPa [59]. Ghanbarzadeh et al. incorporated citric acid and carboxymethyl cellulose (CMC) into TPS films with a 10% and 20% weight ratio, respectively. The synergistic effect of the two additives increased the ultimate strength of the film up to 16 MPa, while the elongation at break was relatively high with a value of 60% [60].

TPS reinforced with starch nanofillers represents another promising composite material. The tensile strengths of TPS films improved with addition of rice starch nanocrystals up to 20 wt% due to their high specific surface area and good dispersion, allowing an efficient stress transfer from the matrix to the particle [51]. However, beyond such content, the films became brittle and their tensile strength decreased due to the rigidity and the aggregation of the fillers. Acetylated cassava starch NPs were much more effective in improving the mechanical properties of TPS. In particular, with just 0.5 wt% of nanofiller, the tensile strength increased from 47 to 69.5 MPa and the Young's modulus from 316 to 829 MPa, whereas the elongation at break was found almost unchanged [51].

Regarding the barrier properties, upon addition of polysaccharide fillers and nanofillers, a decrease in the WVP and water absorption has been generally observed. This is ascribed to the hydrogen bonding interactions between the hydroxyl groups of starch and those of the filler, which decrease the TPS hydrophilic character and limit its polymer chains mobility, reducing swelling upon contact with water [58]. For example, the water uptake of pure TPS was 70 wt% at 98% relative humidity after 72 hours and this value markedly decreased to 50 wt% upon addition of 30 wt% of flax cellulose nanocrystals [58]. Similarly, Wattanakornsiri et al. reported an improved water resistance for cellulose fibers incorporated in corn starch. The presence of 8% wt cellulose fibers decreased water absorption by 63% when compared with control TPS [51]. The addition of chitosan and chitin into TPS has been reported to have a remarkable effect on the water barrier properties of starch [51]. Specifically, the WVP value of TPS (1.33 ± 0.09 g/s m Pa) decreased to 0.87 ± 0.04 and 0.59 ± 0.02 g/s m Pa when 10%wt chitosan and 10%wt chitin were incorporated, respectively. Chitin was much more effective in improving the water vapor barrier properties of TPS due to its N-acetylglucosamine groups in its structure. The improvement of water barrier properties was achieved also with the incorporation of CMC and citric acid [60]. In fact, the WVP decreased from 4.64×10^{-7} to 2.35×10^{-7} g/Pa h m when 10 wt% and 15 wt% of citric acid and CMC were added to TPS, respectively. The reason for the reduced WVP is both chemical and physical. On one hand, the interactions of TPS with citric acid and the nature of CMC increase the hydrophobicity of the system, and on the other, the fillers provide for a tortuous path that limits water vapor diffusivity. In the same way, other fillers and nanofillers, have been used to create longer diffusive paths to oppose water vapor penetration. For example, TPS reinforced with 20%wt of starch nanocrystals exhibited a WVP decrease 2.56 g mm/m^2 d kPa from its original value of 6.25 g mm/m^2 d kPa [51].

Although, polysacaccharides have been demonstrated to actively improve TPS properties and the obtained composites constitute the most sustainable route for the development of green starch based packaging materials, these technologies are still not viable for industrial production. The major concerns include the excessive rigidity imparted to the materials by such fillers, the thermal degradation at high temperatures that limit their processability and use, and the acid-based hydrolysis procedures to obtain nanofillers from raw materials, which has an impact on the production costs, reproducibility, and production yield.

5.3.4 Composites of Thermoplastic Starch with Polyesters

Blends of TPS with biodegradable polyesters have been widely investigated over the last few decades, especially for packaging applications, leading to the development of innovative biodegradable compostable packaging materials currently available on the market. Blending starch with biodegradable hydrophobic polyesters such as PCL, PLA, and PHAs can be performed through melt processing in mixers or extruders either in one step or by a first production of TPS pellets that are subsequently melt blended with one or a combination of said polymers. The obtained composites combine the advantages of both types of materials. In fact, the incorporation of polyesters improves the thermal stability, processing window, mechanical and water vapor barrier properties of TPS and, at the same time, the presence of starch enhances the degradation rate and reduces the costs of the final material [48, 61].

Due to the lower price and greater availability compared to other biopolymers, PLA, an aliphatic thermoplastic polyester widely used in food packaging, is one of the most suitable polyesters to combine with TPS for the development of biodegradable sustainable rigid packaging materials with good moisture and barrier properties.

Physical blends of TPS with PLA were first studied by Martin and Averous [62]. They reported that the properties of the extruded blends were highly dependent on the amount of glycerol used. The TPS/PLA blends with lower glycerol content appeared rigid and brittle (glycerol/starch ratio of 0.14), while a higher glycerol content (glycerol/starch ratio of 0.26 and 0.54) imparted a better flexibility. However, the lack of affinity between TPS and PLA led to overall poor mechanical properties even at lower PLA contents (i.e. 10 and 25 wt%). In particular, the elongation at break and impact resistance of the blends reached values lower than those of pure components. Sarazin et al. improved the mechanical properties of TPS and PLA through the incorporation of PCL, a biodegradable polyester, characterized by a high elongation at break. The materials were extruded and subsequently injection molded [63]. Owing to a synergistic effect that exceeded the results obtained for any of the binary pairs, the ternary PLA/PCL/TPS 40/10/50 blend exhibited an elongation at break of 55% and a substantial increase by three times in the impact energy compared to pure PLA. Poly(hydroxyester ether) and PVA have also been used as third components to improve the poor interfacial interaction between the hydrophilic starch and the hydrophobic PLA polymer [64, 65]. The addition of PVA to a starch and PLA blend (50 : 50, w/w) was reported to increase the tensile strength of the blends. However, PVA was effective in forming a continuous phase with starch only at proportions greater than 30 wt% and this resulted in an increased water absorption due to PVA hydrophilicity, which would render the blends humidity sensitive and thus not suitable for packaging. In addition, in a study on the anaerobic degradability of a series of TPS/PVA blends, it was shown that PVA can markedly inhibit the degradation of the starch, limiting its biodegradability in either digesters or bioreactor landfills [66].

Averous and Fringant explored different formulations of TPS with various biodegradable low melting polyesters, namely PCL, polyester amide (PEA), polybutylene succinate adipate (PBSA), and polybutylene adipate-*co*-terephtalate (PBAT) [67]. The blends were obtained by extrusion and followed by injection molding at 100–130 °C. It was shown that no compatibilizer was needed in the blends and that the addition of polyesters even in 10% weight proportion to TPS improved its dimensional post-injection stability against shrinkage and drastically increased the hydrophobic character of the material. The different combinations of TPS and polyesters yielded a wide range of mechanical characteristics that could serve for specific applications. In general, blends with PBAT and PEA yielded better results with respect to PCL and PBSA, as they seemed to present a higher interphase compatibility.

Blends of PCL, a petroleum-based biodegradable flexible polyester, and starch have received a widespread attention both at the research level and commercially. Shin et al. fabricated PCL/TPS composites through extrusion and compression molding [68]. From the tensile measurements, they observed that TPS improved the Young's modulus of PCL from 93 up to 180 MPa, with a content of TPS ranging from 0 to 60 wt%. The elongation at break and tensile strength decreased with increasing TPS content, although at 40 wt% of TPS, the elongation was still relatively high (150%). Moreover, all the blends had strengths and moduli comparable to those of low-density polyethylene. The authors also investigated the thermal properties of the obtained composites. It was found that although PCL and TPS might form compatible blends as a result of the hydrogen bonding interaction between the ester carbonyl of PCL and the OH groups on starch, they are thermodynamically immiscible and the melting point of PCL (60 °C) was relatively unchanged by the addition of TPS. Averous et al. produced different compositions of wheat TPS and PCL by extrusion and injection molding, with a PCL concentration up to 40 wt% [69]. They reported that the addition of PCL to TPS matrix allows to overcome the weaknesses of pure TPS such as low resilience, high moisture sensitivity, and high shrinkage even at low PCL concentration (10 wt%). Myllymäki et al. prepared TPS/PCL stretched films in which the elongation at break of TPS increased from 28% to 340% and the tensile strength from 1.3 to 35 MPa upon addition of 30 wt% of PCL [70]. The authors also investigated the barrier properties of the proposed films. The composite with 30 wt% PCL exhibited the lowest WVP (11 g mm/kPa h m^2), with a drastic reduction compared to pure TPS (190 g mm/kPa h m^2). Improved water barrier properties were also measured by Ortega-Toro et al. in TPS/PCL films obtained through compression molding [71]. The WVP was markedly reduced from 18 to 2.7 g mm/kPa h m^2 when PCL was varied from 0 wt% to 80 wt%. On the other hand, the oxygen permeability was observed to increase with the content of PCL, exceeding the acceptable values required for food packaging. Overall, TPS/PCL blends are suitable especially for flexible packaging but their main limitations are the low melting point of PCL, which translates to lower temperature for the final usage, the lack of mechanical strength, and the fossil origin of this polymer.

Recently, PHAs have been emerging as very promising polyester polymers owing to their excellent mechanical properties and biodegradability. Flexible TPS blends with polyhydroxybutyrate-*co*-valerate (PHBV) using PBAT as a compatibilizer were developed by Parulekar and Mohanty [72]. In particular, the formulation with PHBV/TPS/PBAT 50 : 20 : 30 exhibited suitable mechanical properties for a potential use in flexible packaging, as the reported tensile strength and modulus were 17 and 660 MPa, respectively, and the elongation at break reached 284%. Moreover, the thermoplastic blend was also processed through film extrusion, yielding films with further enhanced elongation of 368%. Aging, an inherent problem with starch-based materials as well as PHA-based materials, was effectively reduced by limiting the moisture uptake of the starch component, hindering the leaching of glycerol and inhibiting the secondary crystallization of the PHA component in the films. More modest results were achieved by using polyethylene oxide (PEO) as a compatibilizer in PHBV and starch blends [73]. In fact, the maximum elongation at break was 22%.

In terms of biodegradability, the combination of TPS with polyesters is generally beneficial, as the biodegradation is faster compared to pure polyesters due to several factors including the increased surface area and water absorption. This property renders such materials promising candidates for sustainable biodegradable packaging applications. However, the results can greatly differ depending on the processing method, crystallinity, nature of additives and methods, conditions and microorganisms used to assess the biodegradation. Li et al. studied the biodegradation of extruded blends of TPS and PLA 50 : 50 in powder form in compost at room temperature for 14 weeks [50]. Although the contribution to biodegradation in the blends came exclusively from the TPS component, due to the slow biodegradation of PLA, the co-continuous morphology increased the surface area of TPS, accelerating its degradation rate so that it reached a plateau after 6 weeks compared to the 10 weeks required for the pure TPS. Vikman et al., investigated the enzymatic hydrolysis of 50 : 50 TPS/PCL blends at 37 °C in an aqueous medium [61]. As the enzymatic hydrolysis proceeded from the surface of the material inward, it was more rapid for milled TPS–PCL samples compared to TPS–PCL in pieces, due to the higher surface area of the former. In fact, 49–69% of the original TPS was hydrolyzed after 11 days when samples were in pieces, whereas in milled samples 36–53% of starch was hydrolyzed in only 3 hours, reaching 93–97% after 172 hours. Shin et al. tested PCL/TPS blends with 0–60% weight content of TPS in a compost environment for 45 days [68]. They concluded that the biodegradability of the blends increased with increasing TPS content. Also, blends having TPS contents greater than 50% showed a faster degradation rate in the earlier days as compared to cellulose reference and PCL. This result could be attributed to the fact that during composting, TPS degraded primarily and created holes, which accelerated the degradation of PCL. After 45 days, a biodegradability of 70% was determined for the blends, reaching almost the value of pure TPS (80%) and a much higher biodegradability with respect to pure PCL (30%). Imam et al., demonstrated that starch can accelerate PHBV degradation in a compost environment [74].

5.3.5 Composite of TPS Based on Chemical Modification

Chemical derivatization of starch (i.e. esterification, acylation, hydroxylation, epoxidation, silylation, cross-linking) although costly and not always straightforward has also been investigated as a possibility to improve its chemical compatibility to produce flexible water-resistant biodegradable materials [75–77]. In this regard, grafting of thermoplastic polymers onto starch or starch cross-linking with modified polyesters are the most common approaches to obtain starch–polymer composites through reactive processes that can be performed directly in the molten state using standard melt compounding equipment (i.e. extruders). Compared to physical blends, such composites exhibit an increased chemical and thermal stability, enhanced mechanical properties, water resistance, and processability. In addition, phase separation is unlikely to occur and their structure can be precisely designed to impart desired functionalities for the targeted application [78].

Narayan has developed a process technology involving reactive processing of plasticized starch with modified PCL in a twin-screw extruder [78]. Wang et al. obtained TPS/PLA blends using maleic anhydride (MA) as a compatibilizer through a one-step reactive extrusion process [79]. The formation of branched and cross-linked macromolecules caused by the reaction of the carboxyl groups of PLA-graft-MA with the hydroxyl groups of starch improved the dispersion and compatibility between PLA and TPS. As a result, the tensile strength of compatibilized blends was higher than that of the physically mixed composites. However, the process was relatively slow as the reactive compounding was carried out in a single screw extruder at a rate of 10 rpm, with a yield of only 15 g/min. PHA/TPS blends with improved interfacial adhesion were obtained by grafting PHA in situ onto starch using dycumil peroxide as initiator. The blends exhibited better mechanical properties and thermal stability compared to the physical blends of the same materials [80]. Similar approaches based on the compatibilization with reactive functional groups that are able to enhance the interfacial interaction of polyesters/starch blends include the use of methylene diphenyl diisocyanate, PLA-graft-(acrylic acid), PLA-graft-starch. [81, 82]. The main drawbacks associated with the proposed chemical functionalizations in the context of packaging applications are the reproducibility of the obtained materials, the optimization required for a large-scale production (i.e. reaction times and yields, number of processing steps, energy consumption, and production capacity), and the assessment of the safety of the additives and compatibilizers used. Moreover, due to the strong covalent bonds and the low water absorption, grafted and chemically modified starch blends might be more difficult to biodegrade as reported by Feng Zuo et al. [83].

5.3.6 Commercial Packaging Materials Based on Thermoplastic Starch

A few TPS-based materials have reached commercial availability and are currently used for many applications ranging from agriculture to catering and packaging materials. A biodegradable and compostable TPS–PCL blend was reported by Novamont (Italy) and commercialized under the trade name

Mater-Bi. The company has since further expanded their portfolio of products from renewable resources introducing more diversified blends of Mater-Bi comprising starches, celluloses, vegetable oils, and their combinations. These biomaterials are compliant with the compostability and biodegradability regulations and can be thermally processed with standard techniques industrially used for plastic transformation, such as extrusion, injection molding, blow molding, and foaming, for a wide range of applications.

Other manufacturers of starch/biodegradable polyester blends include Biotec, Earthshell, and Biop [78]. Biotec is a German company that manufactures and commercializes plasticizer-free bioplastics that contain a high portion of biologically sourced raw materials, including starch, under the brand name Bioplast. The biobased carbon in the blends share varies from 23% to 100%. The resins are biodegradable, recyclable, compostable, and can be processed through various melt processing technologies including extrusion blow and injection molding. The applications target both rigid and flexible food and cosmetic packaging such as shopper bags, blisters, cosmetic containers, food trays, etc. Earthshell, based in the United States, developed biodegradable disposable food containers made of starch from potato waste. The main characteristic of these items are: grease, soak and cut resistance, leak proof, thermal stability for both freezing and microwaving, thermal insulation of hot and cold food, compostability, and low carbon footprint. Moreover, according to their life cycle inventory, the greenhouse gas emissions were less than half compared to paper and conventional plastics.

5.4 Conclusions

Legislation that pertains to sustainability, global warming, and growing concerns related to the environmental impact of conventional plastics, both in terms of their persistency when discarded in landfills and high carbon footprint and pollution levels associated with their production and incineration, have stimulated a widespread research interest in biodegradable polymers from renewable resources as alternatives to plastics especially for short term, disposable, and single-use applications, in particular packaging. From a sustainability perspective, natural biopolymers are clearly advantageous compared to conventional plastics owing to their renewability, availability, and biodegradability. Moreover, life cycle analyses demonstrate that they can significantly reduce CO_2 emissions compared to traditional plastics. Starch-based materials, in particular, have received considerable attention as alternatives to synthetic polymers obtained from petroleum due to attributes such as natural occurrence, non-toxicity, and low cost. During the last decade, several developments were made to convert native starch into a plastic-like material. The so-called TPS was obtained in the presence of water and plasticizers, under the effect of heat and thermo-mechanical energy, by the disruption of native starch granules. Further efforts have led to the improvement of TPS inherent hydrophilicity and moisture sensitivity as well as its poor water vapor barrier and mechanical properties to allow its utilization for several end applications. The most common approaches relied, on one hand, on chemical modifications and reactive processes, and, on the other, on physically blending TPS with other biopolymers and nanosized fillers through melt processing or film casting.

In the next few years, due to the reduction of the traditional packaging stream and to a rising demand for more sophisticated green solutions, the market of biodegradable packaging is expected to continuously grow and evolve. The possibility to produce specifically designed starch materials with tailored properties will be fundamental for such materials to compete with new emerging products. However, the design and engineering of high-performance starch packaging materials with the required technical characteristics (thermoforming properties, mechanical strength, transparency, gas, and moisture barrier properties) still present a number of challenges. A primary issue regarding the large-scale production of starch-based materials is the troublesome standardization, as the composition and ratio between linear amylose and branched amylopectin in starch change considerably depending on the botanical origin, giving rise to significant differences in its structural and physical properties and functionalities. Moreover, particular attention has to be devoted to the processing and handling of TPS. This is because its behavior and its stability over time are greatly influenced by several factors, such as the plasticizer nature, the water content, the crystalline structure and total crystallinity, the storage temperature, and relative humidity. In particular, aging and retrogradation can dramatically affect the properties and applications of TPS-based materials. Prior to commercialization of starch packaging products, the influence of physicochemical and biochemical parameters such as temperature, humidity, and microbial and chemical agents on the material stability will thus need to be addressed. Specifically, for food packaging, the materials will need to meet the requirements for each food product and comply with food packaging regulatory aspects, which are often not harmonized globally and change frequently with new limits and restrictions. In this regard, low migration and the use of additives that belong to approved components for food contact are the most important challenges to overcome. Regarding the biodegradability, the degradation rates of starch-based packaging products will need to be assessed in real environments and under different conditions. This is true especially for blends with hydrophobic polymers, since their effective degradation occurs only when the starch concentration exceeds the percolation threshold and significant pathways for moisture penetration and microbial invasion are generated. Furthermore, existing starch blends will need to be further optimized to decrease their production costs by exploring the utilization of agricultural and food waste as a starch source rather than its cultivation and by increasing the content of biobased carbon in the blend.

In order to overcome such challenges, joint efforts will be required by leading industry participants, consumers with strong purchasing power and a high degree of environmental awareness, and institutions, who should provide the opportune legislative and regulatory frameworks to actively promote sustainable development and aligned innovations.

References

1 Zhang, Y., Rempel, C., and Liu, Q. (2014). Thermoplastic starch processing and characteristics – a review. *Critical Reviews in Food Science and Nutrition* 54 (10): 1353–1370.

2 Rodriguez-Perez, S., Serrano, A., Pantión, A.A. et al. (2018). Challenges of scaling-up PHA production from waste streams – A review. *Journal of Environmental Management* 205: 215–230.

3 Dawande, R. (2018). *Bioplastics Market by Type (Biodegradable Plastic and Non-biodegradable Plastic) and Application (Rigid Packaging, Flexible Packaging, Textile, Agriculture & Horticulture, Consumer Good, Automotive, Electronic, Building & Construction, and Others)-Global Opportunity Analysis and Industry Forecast, 2018–2024.* Allied Market Research https://www.alliedmarketresearch.com/bioplastics-market.

4 Geyer, R., Jambeck, J.R., and Law, K.L. (2017). Production, use, and fate of all plastics ever made. *Science Advances* 3 (7): e1700782.

5 Jambeck, J.R., Geyer, R., Wilcox, C. et al. (2015). Plastic waste inputs from land into the ocean. *Science* 347 (6223): 768–771.

6 Bullock, S. (2018). *Key Facts About Plastic Pollution*. Greenpeace https://www.greenpeace.org/usa/key-facts-about-plastic-pollution.

7 (2016). Elllen MacArthur Foundation The New Plastics Economy: rethinking the future of plastics. The World Economic Forum and the Ellen MacArthur Foundation. https://www.newplasticseconomy.org.

8 Lusher, A.L., Mchugh, M., and Thompson, R.C. (2013). Occurrence of microplastics in the gastrointestinal tract of pelagic and demersal fish from the English Channel. *Marine Pollution Bulletin* 67 (1–2): 94–99.

9 European Commission Press release Database. (2019). Circular Economy: Commission welcomes European Parliament adoption of new rules on single–use plastics to reduce marine litter. http://europa.eu/rapid/press-release_STATEMENT-19-1873_en.htm.

10 Rhim, J.W., Park, H.M., and Ha, C.G. (2013). Bio-nanocomposites for food packaging applications. *Progress in Polymer Science* 38 (10–11): 1629–1652.

11 Guilbert, S., Gontard, N., and Gorris, L.G.M. (1996). Prolongation of the shelf-life of perishable food products using biodegradable films and coatings. *LWT - Food Science and Technology* 29 (1–2): 10–17.

12 Averous, L. (2004). Biodegradable multiphase systems based on plasticized starch: a review. *Journal of Macromolecular Science, Part C: Polymer Reviews* 44 (3): 231–274.

13 Janssen, L. and Moscicki, L. ed. (2009). *Thermoplastic Starch: A Green Material for Various Industries*. Wiley.

14 Mali, S., Grossmann, M.V.E., Garcı´ a, M.A. et al. (2005). Mechanical and thermal properties of yam starch films. *Food Hydrocolloids* 19 (1): 157–164.

15 Tester, R.F., Karkalas, J., and Qi, X. (2004). Starch-composition, fine structure and architecture. *Journal of Cereal Science* 39 (2): 151–165.

16 Petersen, K., Nielsen, P.V., Bertelsen, G. et al. (1999). Potential of biobased materials for food packaging. *Trends in Food Science & Technology* 10 (2): 52–68.

17 Abdullah, Z.W. and Dong, Y. (2018). Recent advances and perspectives on starch nanocomposites for packaging applications. *Journal of Material Sciences* 53 (22): 15319–15339.

18 Buleon, A., Colonna, P., Planchot, V. et al. (1998). Starch granules: structure and biosynthesis. *International Journal of Biological Macromolecules* 23 (2): 85–112.

19 Kaseem, M., Hamad, K., and Deri, F. (2012). Thermoplastic starch blends: a review of recent works. *Polymer Science, Series A* 54 (2): 165–176.

20 Cazon, P., Velazquez, G., Ramírez, J.A. et al. (2017). Polysaccharide-based films and coatings for food packaging: a review. *Food Hydrocolloids* 68: 136–148.

21 Khan, B., Niazi, M.B.K., Samin, G. et al. (2017). Thermoplastic starch: a possible biodegradable food packaging material – a review. *Journal of Food Process Engineering* 40 (3): e12447.

22 Tan, I., Flanagan, B.M., Halley, P.J. et al. (2007). A method for estimating the nature and relative proportions of amorphous, single, and double-helical components in starch granules by ^{13}C CP/MAS NMR. *Biomacromolecules* 8 (3): 885–891.

23 Cheetham, N.W.H. and Tao, L. (1998). Variation in crystalline type with amylose content in maize starch granules: an X-ray powder diffraction study. *Carbohydrate Polymers* 36 (4): 277–284.

24 Xie, F., Luckman, P., Milne, J. et al. (2014). Thermoplastic starch: current development and future trends. *Journal of Renewable Materials* 2 (2): 95–106.

25 Hizukuri, S. (1985). Relationship between the distribution of the chain length of amylopectin and the crystalline structure of starch granules. *Carbohydrate Research* 141 (2): 295–306.

26 Ribotta, P.D., Cuffini, S., Leon, A.E. et al. (2004). The staling of bread: an X-ray diffraction study. *European Food Research and Technology* 218 (3): 219–223.

27 Lu, D.R., Xiao, C.M., and Xu, S.J. (2009). Starch-based completely biodegradable polymer materials. *Express Polymer Letters* 3 (6): 366–375.

28 Jiménez, A., Fabra, M.J., Talens, P. et al. (2012). Edible and biodegradable starch films: a review. *Food and Bioprocess Technology* 5 (6): 2058–2076.

29 Tharanathan, R.N. (2003). Biodegradable films and composite coatings: past, present and future trends. *Trends in Food Science and Technology* 14 (3): 71–78.

30 Lai, L.S. and Kokini, J.L. (1991). Physicochemical changes and rheological properties of starch during extrusion (a review). *Biotechnology Progress* 7 (3): 251–266.

31 Shogren, R.L., Lawton, J.W., Tiefenbacher, K.F. et al. (1998). Starch-poly(vinyl alcohol) foamed articles prepared by a baking process. *Journal of Applied Polymer Science* 68 (13): 2129–2140.

32 Curvelo, A.A.S., De Carvalhoa, A.J.F., and Agnelli, J.A.M. (2001). Thermoplastic starch-cellulosic fibers composites: preliminary results. *Carbohydrate Polymers* 45 (2): 183–188.
33 Ma, X.F., Yu, J.G., and Wang, N. (2007). Fly ash reinforced thermoplastic starch composites. *Carbohydrate Polymers* 67 (1): 32–39.
34 Debeaufort, F., Quezada-Gallo, J.A., and Voilley, A. (1998). Edible films and coatings: tomorrow's packagings: a review. *Critical Reviews in Food Science and Nutrition* 38 (4): 299–313.
35 Soest, J.J.G.V., Benes, K., and Wit, D.D. (1995). The influence of acid hydrolysis of potato starch on the stress-strain properties of thermoplastic starch. *Starch* 47 (11): 429–434.
36 Gioia, L.D. and Guilbert, S. (1999). Corn protein-based thermoplastic resins: effect of some polar and amphiphilic plasticizers. *Journal of Agricultural and Food Chemistry* 47 (3): 1254–1261.
37 Pushpadass, H.A., Marx, D.B., and Hanna, M.A. (2008). Effect of extrusion temperature and plasticizers on the physical and functional properties of starch films. *Starch* 60 (10): 527–538.
38 Thunwall, M., Kuthanova, V., Boldizar, A. et al. (2008). Film blowing of thermoplastic starch. *Carbohydrate Polymers* 71 (4): 583–590.
39 Zhang, Y. and Han, J.H. (2006). Plasticization of pea starch films with monosaccharides and polyols. *Journal of Food Science* 71 (6): E253–E261.
40 Xiaofei, M., Jiugao, Y., and Jin, F. (2004). Urea and formamide as a mixed plasticizer for thermoplastic starch. *Polymer International* 53 (11): 1780–1785.
41 Stein, T.M. and Greene, R.V. (1997). Amino acids as plasticizers for starch-based plastics. *Starch* 49 (6): 245–249.
42 Yu, J., Chang, P.R., and Ma, X. (2010). The preparation and properties of dialdehyde starch and thermoplastic dialdehyde starch. *Carbohydrate Polymers* 79 (2): 296–300.
43 de Graaf, R.A., Karman, A.P., and Janssen, L.P. (2003). Material properties and glass transition temperatures of different thermoplastic starches after extrusion processing. *Starch* 55 (2): 80–86.
44 Koch, K., Gillgren, T., Stading, M. et al. (2010). Mechanical and structural properties of solution-cast high-amylose maize starch films. *International Journal of Biological Macromolecules* 46 (1): 13–19.
45 Thuwall, M., Boldizar, A., and Rigdahl, M. (2006). Extrusion processing of high amylose potato starch materials. *Carbohydrate Polymers* 65 (4): 441–446.
46 Thunwall, M., Boldizar, A., and Rigdahl, M. (2006). Compression molding and tensile properties of thermoplastic potato starch materials. *Biomacromolecules* 7 (3): 981–986.
47 Mehyar, G.F. and Han, J.H. (2004). Physical and mechanical properties of high-amylose rice and pea starch films as affected by relative humidity and plasticizer. *Journal of Food Science* 69 (9): E449–E454.
48 Muthuraj, R., Misra, M., and Mohanty, A.K. (2018). Biodegradable compatibilized polymer blends for packaging applications: a literature review. *Journal of Applied Polymer Science* 135 (24): 45726.

49 ASTM D6400-12 (2012). *Standard Specification for Labeling of Plastics Designed to be Aerobically Composted in Municipal and Industrial Facilities*, ASTM, West Conshohocken, PA, USA.

50 Li, G., Sarazin, P., Orts, W.J. et al. (2011). Biodegradation of thermoplastic starch and its blends with poly(lactic acid) and polyethylene: influence of morphology. *Macromolecular Chemistry and Physics* 212 (11): 1147–1154.

51 Niranjana Prabhu, T. and Prashantha, K. (2018). A review on present status and future challenges of starch based polymer films and their composites in food packaging applications. *Polymer Composites* 39 (7): 2499–2522.

52 Medina-Jaramillo, C., Ochoa-Yepes, O., Bernal, C. et al. (2017). Active and smart biodegradable packaging based on starch and natural extracts. *Carbohydrate Polymers* 176: 187–194.

53 Abreu, A.S., Oliveira, M., de Sá, A. et al. (2015). Antimicrobial nanostructured starch based films for packaging. *Carbohydrate Polymers* 129: 127–134.

54 Avella, M., De Vlieger, J.J., Errico, M.E. et al. (2005). Biodegradable starch/clay nanocomposite films for food packaging applications. *Food Chemistry* 93 (3): 467–474.

55 Sadegh-Hassani, F. and Nafchi, A.M. (2014). Preparation and characterization of bionanocomposite films based on potato starch/halloysite nanoclay. *International Journal of Biological Macromolecules* 67: 458–462.

56 López, O.V., Castillo, L.A., Garcia, M.A. et al. (2015). Food packaging bags based on thermoplastic corn starch reinforced with talc nanoparticles. *Food Hydrocolloids* 43: 18–24.

57 Taguet, A., Huneault, M.A., and Favis, B.D. (2009). Interface/morphology relationships in polymer blends with thermoplastic starch. *Polymer* 50 (24): 5733–5743.

58 Cao, X., Chen, Y., Chang, P.R. et al. (2008). Starch-based nanocomposites reinforced with flax cellulose nanocrystals. *Express Polymer Letters* 2 (7): 502–510.

59 Liu, D., Zhong, T., Chang, P.R. et al. (2010). Starch composites reinforced by bamboo cellulosic crystals. *Bioresource Technology* 101 (7): 2529–2536.

60 Ghanbarzadeh, B., Almasi, H., and Entezami, A.A. (2011). Improving the barrier and mechanical properties of corn starch-based edible films: effect of citric acid and carboxymethyl cellulose. *Industrial Crops and Products* 33 (1): 229–235.

61 Vikman, M., Hulleman, S.H.D., Van Der Zee, M. et al. (1999). Morphology and enzymatic degradation of thermoplastic starch–polycaprolactone blends. *Journal of Applied Polymer Science* 74 (11): 2594–2604.

62 Martin, O. and Averous, L. (2001). Poly(lactic acid): plasticization and properties of biodegradable multiphase systems. *Polymer* 42 (14): 6209–6219.

63 Sarazin, P., Li, G., Orts, W.J. et al. (2008). Binary and ternary blends of polylactide, polycaprolactone and thermoplastic starch. *Polymer* 49 (2): 599–609.

64 Ke, T. and Sun, X.S. (2003). Starch, poly(lactic acid), and poly(vinyl alcohol) blends. *Journal of Polymers and the Environment* 11 (1): 7–14.

65 Shogren, R.L., Doane, W.M., Garlotta, D. et al. (2003). Biodegradation of starch/polylactic acid/poly(hydroxyester-ether) composite bars in soil. *Polymer Degradation and Stability* 79 (3): 405–411.

66 Russo, M.A., O'Sullivan, C., Rounsefell, B. et al. (2009). The anaerobic degradability of thermoplastic starch: polyvinyl alcohol blends: potential biodegradable food packaging materials. *Bioresource Technology* 100 (5): 1705–1710.

67 Averous, L.U.C. and Fringant, C. (2001). Association between plasticized starch and polyesters: processing and performances of injected biodegradable systems. *Polymer Engineering & Science* 41 (5): 727–734.

68 Shin, B.Y., Lee, S.I., Shin, Y.S. et al. (2004). Rheological, mechanical and biodegradation studies on blends of thermoplastic starch and polycaprolactone. *Polymer Engineering & Science* 44 (8): 1429–1438.

69 Averous, L., Moro, L., Dole, P. et al. (2000). Properties of thermoplastic blends: starch–polycaprolactone. *Polymer* 41 (11): 4157–4167.

70 Myllymäki, O., Myllärinen, P., Forssell, P. et al. (1998). Mechanical and permeability properties of biodegradable extruded starch/polycaprolactone films. *Packaging Technology and Science: An International Journal* 11 (6): 265–274.

71 Ortega-Toro, R., Contreras, J., Talens, P. et al. (2015). Physical and structural properties and thermal behaviour of starch–poly(ε-caprolactone) blend films for food packaging. *Food Packaging and Shelf Life* 5: 10–20.

72 Parulekar, Y. and Mohanty, A.K. (2007). Extruded biodegradable cast films from polyhydroxyalkanoate and thermoplastic starch blends: fabrication and characterization. *Macromolecular Materials and Engineering* 292 (12): 1218–1228.

73 Shogren, R.L. (1995). Poly(ethylene oxide)-coated granular starch-poly(hydroxybutyrate-*co*-hydroxyvalerate) composite materials. *Journal of Environmental Polymer Degradation* 3 (2): 75–80.

74 Imam, S.H., Chen, L., Gordon, S.H. et al. (1998). Biodegradation of injection molded starch–poly(3-hydroxybutyrate-*co*-3-hydroxyvalerate) blends in a natural compost environment. *Journal of Environmental Polymer Degradation* 6 (2): 91–98.

75 Moad, G. (2011). Chemical modification of starch by reactive extrusion. *Progress in Polymer Science* 36 (2): 218–237.

76 Fang, J.M., Fowler, P.A., Tomkinson, J. et al. (2002). The preparation and characterisation of a series of chemically modified potato starches. *Carbohydrate Polymers* 47 (3): 245–252.

77 Rutot, D., Degée, P., Narayan, R. et al. (2000). Aliphatic polyester-grafted starch composites by in situ ring opening polymerization. *Composite Interfaces* 7 (3): 215–225.

78 Fang, J. and Fowler, P. (2003). The use of starch and its derivatives as biopolymer sources of packaging materials. *Journal of Food, Agriculture and Environment* 1: 82–84.

79 Wang, N., Yu, J., and Ma, X. (2007). Preparation and characterization of thermoplastic starch/PLA blends by one-step reactive extrusion. *Polymer International* 56 (11): 1440–1447.

80 Xu, P., Zeng, Q., and Cao, Y. (2017). Interfacial modification on polyhydroxyalkanoates/starch blend by grafting in-situ. *Carbohydrate Polymers* 174: 716–722.

81 Wu, C.S. (2005). Improving polylactide/starch biocomposites by grafting polylactide with acrylic acid–characterization and biodegradability assessment. *Macromolecular Bioscience* 5 (4): 352–361.

82 Chen, L., Qiu, X., Xie, Z. et al. (2006). Poly(L-lactide)/starch blends compatibilized with poly(L-lactide)-*g*-starch copolymer. *Carbohydrate Polymers* 65 (1): 75–80.

83 Feng Zuo, Y., Gu, J., Qiao, Z. et al. (2015). Effects of dry method esterification of starch on the degradation characteristics of starch/polylactic acid composites. *International Journal of Biological Macromolecules* 72: 391–402.

6

Cutin-Inspired Polymers and Plant Cuticle-like Composites as Sustainable Food Packaging Materials

Susana Guzmán-Puyol[1], Antonio Heredia[3], José A. Heredia-Guerrero[1], and José J. Benítez[2]

[1] *Instituto de Hortofruticultura Subtropical y Mediterránea La Mayora, Universidad de Málaga - Consejo Superior de Investigaciones Científicas, Departamento de Mejora Genética y Biotecnología, Estación Experimental La Mayora, Avenida Dr. Wienberg, s/n, Algarrobo-Costa, Málaga E- 29750, Spain*
[2] *Centro Mixto CSIC-Universidad de Sevilla, Instituto de Ciencia de Materiales de Sevilla (ICMS), Americo Vespucio 49, Isla de la Cartuja, Seville 41092, Spain*
[3] *IHSM-CSIC-UMA, Departamento de Bioquímica y Biología Molecular Facultad de Ciencias Universidad de Málaga Campus de Teatinos, Bulevar Louis Pasteur 31, Málaga 29071, Spain*

You're a piece of plastic! You were made to be thrown away!
Toy Story 3 (2010)

6.1 Introduction

6.1.1 Bioplastics as Realistic Alternatives to Petroleum-Based Plastics

Petroleum-based plastics are considered the materials of the twenty-first century. In fact, these materials can be found everywhere: bags, bottles, packaging, furniture, etc. According to the European association of plastic manufacturers "PlasticsEurope" the world production of plastics reached 348 million tons in 2017, with an increment of around 55% in 13 years, as illustrated in Figure 6.1 [1].

In terms of economy, plastics are driving the development of pioneering innovations to meet society's biggest challenges. Plastics improve quality of life, facilitating resource efficiency and climate protection, for instance. Around 1.45 million people work in more than 62 000 companies that generate an estimated annual turnover of ~350 billion euros in EU-28 [1]. The "big six" plastic types that stand out in terms of their market share are: polyethylene (PE), polypropylene (PP), polyvinyl chloride (PVC), polyurethane (PUR), polyethylene terephthalate (PET), and polystyrene solid (PS), which together account for around 80% of the overall plastics demand in Europe. All of them are petroleum-based plastics with remarkable physical properties, easy processing, attractive cost/performance ratio, high versatility, excellent corrosion resistance, and low weight [2]. These characteristics make them ideal materials for a wide variety of applications including packaging (40%), building and construction (20%), automotive (9%),

Sustainable Food Packaging Technology, First Edition. Edited by Athanassia Athanassiou.

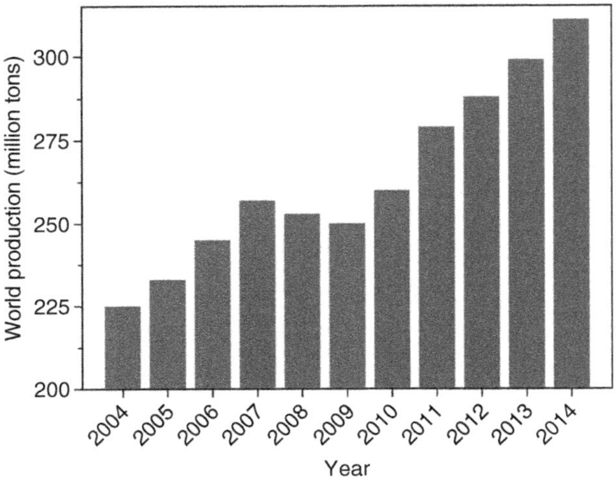

Figure 6.1 World plastic production 2004–2017. Source: Modified from Association of Plastics manifacturors 2018 [1].

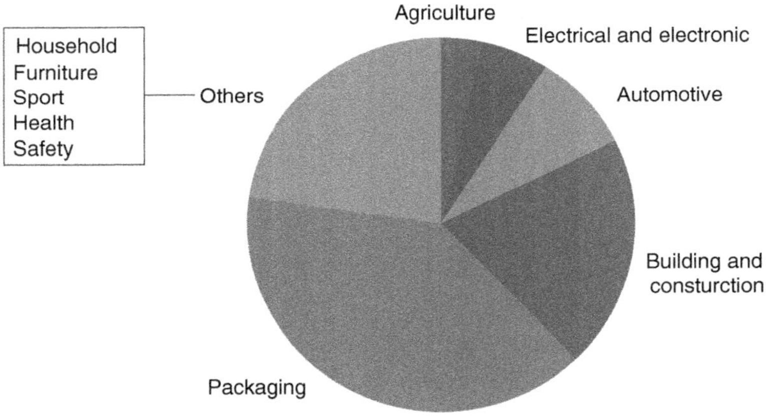

Figure 6.2 Distribution of plastics demand by segments. Source: Modified from Association of Plastics manifacturors 2018 [1].

electrical and electronic (6%), agriculture (3%), and others (23%), as represented in Figure 6.2 [1].

Despite the above described advantages of plastics, the problems related to their massive use are becoming evident. Each step of the plastics life cycle, from extraction of fossil resources to waste disposal, including monomer synthesis and large-scale production, has negative effects on the environment [3]. In addition, it is estimated that only 9% is recycled, while 12% is incinerated, and 79% is accumulated in landfills and oceans [4]. The advantages and drawbacks of plastics are summarized in Table 6.1.

Most likely, the main and most obvious negative impact of plastic use is the huge waste generation. Only in Europe in 2014, 25.8 million tons of

Table 6.1 Main advantages and drawbacks of petroleum-based plastics.

Advantages	Drawbacks
Remarkable physical properties	Environmentally unfriendly
Facile processing	Price fluctuation
Attractive cost/performance ratio	Depletion of fossil resources
High versatility	
Excellent corrosion resistance	
Low weight	

post-consumer plastics ended up in the waste upstream [1]. Plastic wastes can persist for hundreds or even thousands of years in the environment without biodegrading. Moreover, they can also break down over time into small toxic fragments (because of environmental conditions such as UV light or heat) that finally widespread within the marine environment [5, 6]. Among the large list of harmful components of these fragments are substances such as bisphenol A (BPA), phthalates, and polychlorinated biphenyls (PCBs), known for their endocrine disruptor functions and absorption of organic pollutants [7–9]. Furthermore, waste management is another important problem since most of plastics are not properly managed according to official practices, finishing as litter in rivers, lakes, and seas [10]. The plastic debris has significant harmful effects on marine biota [11]. In fact, 44% of all seabird species are known to ingest plastic and 267 species of marine organisms are known to have been affected by plastic debris, a number that will increase as smaller organisms are assessed [12, 13]. Marine plastic debris can be divided into two categories: macro (>5 mm) and micro (<5 mm). While macro-debris may sometimes be traced to its origin by object identification or markings, micro-debris is difficult to trace. In this way, plastics are present in every level of the food chain, ranging from plankton to marine mammals [14–16]. Figure 6.3 shows the degradation time of some common plastic commodities. As observed, a cigarette butt can persist for 1–5 years, while a plastic bag around 10–20 years. Plastics film containers, usually used in food packaging, degrade in 20–30 years. Foamed plastic cups and plastic buoy persist for 50 and 80 years, respectively, whereas disposable diapers and plastic beverage bottles remain for 450 years. Finally, plastics from filament fishing line (usually nylon) do not degrade until around 600 years. All these degradation times are very high in comparison with natural sources such as banana peel (two to five weeks) or the apple core (two months). In consequence, the substitution of these petroleum-based plastics is becoming a need and the search for environmentally friendly alternatives is gaining more and more attention.

The production of bioplastics from renewable feedstocks is an emerging field of research that is receiving considerable attention as an alternative to petroleum-based plastics [17]. It is estimated that bioplastics from biopolymers can replace up to 90% of petroleum-based homologues [18]. Cellulose, lignin,

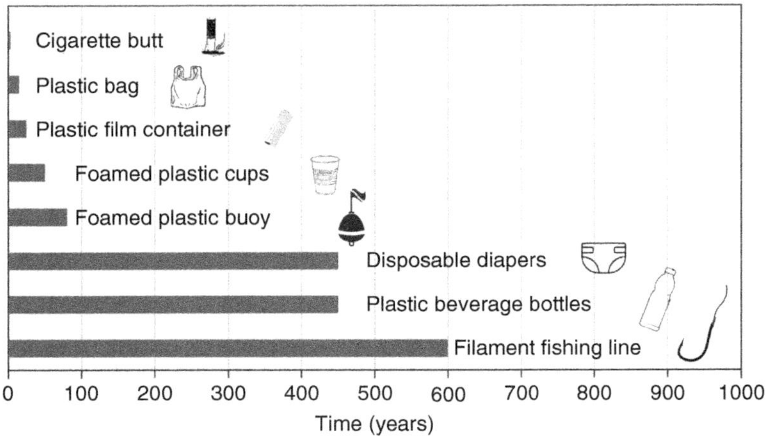

Figure 6.3 Degradation time of selected fuel-based plastics.

Table 6.2 Main advantages and drawbacks of the use of biopolymers.

Advantages	Drawbacks
Renewable resources	Competition with food production
Use of agricultural and food wastes and by-products	Intensified farming and extensive use of fertilizers
No toxicity and health hazards	Threaten the biodiversity
Biodegradation	No biodegradation in absence of water and oxygen

pectin, proteins, and polyesters are some examples of natural sources that have been proposed for the production of polymeric materials [19–24]. The main advantages and problems concerning the use of biopolymers for the production of bioplastics are listed in Table 6.2. Among the advantages is the renewable nature of these resources, which makes them practically inexhaustible, together with the potential use of agricultural and food wastes and by-products. In addition, all these sources present no toxicity, no health hazards, and a complete biodegradation in the short or medium term. Nevertheless, important drawbacks are implicit in the use of natural sources. For example, the substitution of fossil-fuel based bioplastics with renewable ones requires intensified farming, competing in this way with the food production. A direct consequence of intensified farming is the massive use of fertilizers, resulting in an increase of the greenhouse gases. In addition, intensified farming can reduce the biodiversity because of the production of few crops with high associated productions. Finally, the biodegradation of renewable plastics is possible only in the presence of water and oxygen.

Bio-based, biodegradable polymers can be classified based on their origin into three main groups. The first group is related to biomass and includes cellulose, lignin, starch, chitosan, terpenes, cutin, and proteins. The second

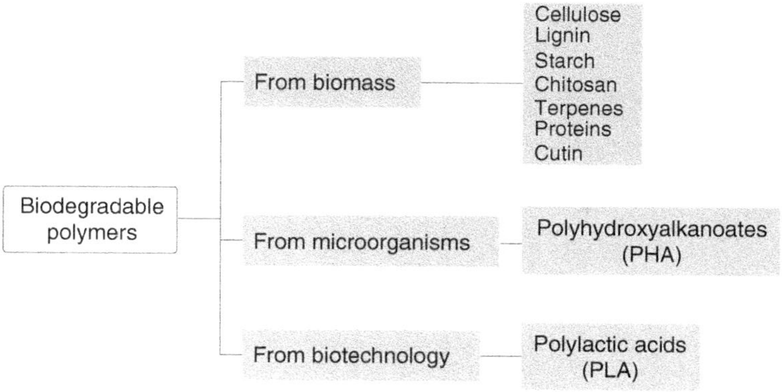

Figure 6.4 Classification of bio-based, biodegradable polymers based on their origin.

one includes the biodegradable polymers from microorganisms, this is, polyhydroxyalkanoates (PHAs). The last one is associated with the biopolymers from biotechnology, which includes the polylactic acids (PLAs). Figure 6.4 presents a general scheme about these bio-based, biodegradable polymers.

Regarding biodegradable polymers from biomass, cellulose is the most abundant biopolymer in the world with an annual biomass production of $\sim 1.5 \times 10^{12}$ tons. It is considered an almost inexhaustible source of raw material [25]. This polysaccharide is found in the plant cell walls, including some bacteria, certain marine algae, tunicates, and fungi to which it provides physical structure and strength [26–28]. Cellulose consists of a linear chain of several β(1→4) linked D-glucose units and has a number of important properties such as complete biodegradability, light weight, and good mechanical strength [25, 29–31]. Cellulose fibers are commonly used in various technological products such as packaging paper, membranes, medical products, textiles, and so on [32]. Another important polysaccharide is starch. It is the major carbohydrate of tuber and root crops [33]. The molecular structure of starch consists of a combination of linear macromolecules known as amylose and branched macromolecular chains known as amylopectin [34]. Chitosan is also a linear polysaccharide composed of randomly distributed β-(1→4)-linked D-glucosamine and *N*-acetyl-D-glucosamine. It is found in the chitin shells of shrimp and other crustaceans [35]. On the other hand, lignin is another of the main components of plant cell walls. It is the most abundant form of aromatic carbon in the biosphere [36]. Lignin is the generic term for a large group of complex phenolic polymers that are generated through radical polymerization of phenolic compounds, especially the monolignols *p*-coumaryl, coniferyl, and sinapyl alcohols [37, 38]. Terpenes are biosynthetically derived from units of isoprene, which has the molecular formula C_5H_8. They are the major components of resins produced by plants, particularly conifers. Terpenes may be classified by the number of isoprene units in the molecule. Thus, polyterpenes consist of long chains of many isoprene units. Natural rubber consists of polyisoprene in which the double bonds are cis. Some plants produce a polyisoprene with trans double bonds, known as gutta-percha. Proteins are macromolecules consisting of one or more long chains of amino acid residues [39]. The main proteins used as raw materials for the production of bioplastics are soy protein, zein, wheat protein, and casein. Finally, cutin (the main topic of this chapter) is the main component of plant cuticles, constituting the framework that supports the rest of cuticle components. This biopolymer consists of esterified bi- and trifunctional C16 and C18 fatty acids. Despite its ubiquity in terrestrial plants, it has been underutilized as a raw material due to its insolubility and lack of melting point [40]. On the other hand, biopolyesters are biodegradable, bio-based polymers from microorganisms and biotechnology. They include PHA, polyhydroxybutyrates (PHBs), and PLAs. In nature, they are produced by bacterial fermentation of sugars and lipids [41].

Table 6.3 summarizes the main bio-based, biodegradable polymers, their monomeric units, and main applications.

6.1.2 The Plant Cuticle and Cutin: The Natural Food Packaging of the Plant Kingdom

The plant cuticle is the outermost membrane that covers and protects the epidermis of non-lignified aerial parts of plants such as leaves, fruits, stems, and flowers [42]. It can be considered as a thin film since its thickness usually ranges between 0.02 and 200 μm [43]. The plant cuticle is a composite material that consists of cutin, a polyfunctional long-chain polyester, that acts as a branched, amorphous, and flexible three-dimensional polymer matrix where other components such as polysaccharides (cellulose, hemicellulose, and pectin) from the cell wall, waxes (a complex mixture of long-chain alkanes, alcohols, esters, aldehydes, and carboxyl groups and some cyclic compounds such as triterpenoids and hydroxycinnamic acids) located on the surface (epicuticular waxes) or spread inside the cuticle (intracuticular waxes), and phenolic compounds (mainly, cinnamic acids and flavonoids) are deposited [42, 44–49]. In some species, a recalcitrant macromolecule denominated cutan (an ether-linked network of methylene chains, double bonds, and carboxyl groups) can partially or completely substitute cutin [50]. The chemical composition and location of the cuticle components can be different among plants, organs, and developmental stages [51] although a heterogeneous spatial distribution as displayed in Figure 6.5 is generally accepted [47].

As the interface between the plant and the environment, plant cuticles prevent massive water loss from internal tissues, regulate gas exchange, defend against pathogens, and protect the plant from abiotic and biotic stresses such as mechanical damage and severe changes in relative humidity, light intensity, and quality (e.g. harmful UV radiation) and temperature [52]. In addition, cutin is a non-toxic, biodegradable, waterproof, UV-blocking, amorphous, insoluble, and infusible biopolyester, being considered a potential substitute for conventional plastics in specific uses, as part of a wider strategy of reducing marine plastic litter and microplastics, as reported by the United Nations [53, 54]. These characteristics make the cuticle and cutin ideal systems to be mimicked by bio-based food packaging materials. In this sense, there is an increasing interest in the production of plant cuticle-like and cutin-based materials with similar properties to standard plastics for food packaging [40, 55].

Table 6.3 Main bio-based, biopolymers, their monomeric units, and main applications.

Biopolymer	Monomeric units	Main applications
Cellulose	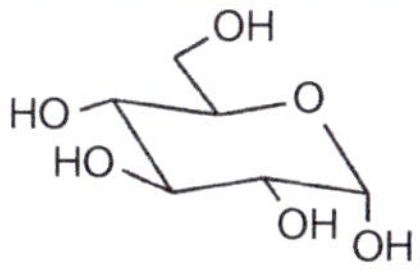 D-Glucose	• Paper manufacturing • Building • Textiles • Additives • Fiber reinforcement
Lignin	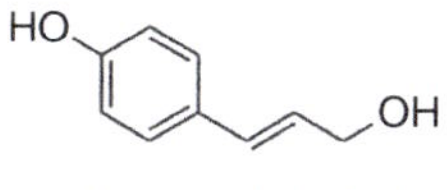 *p*-Coumaryl alcohol 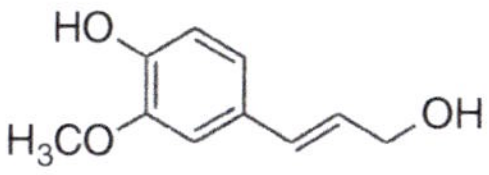 Coniferyl alcohol 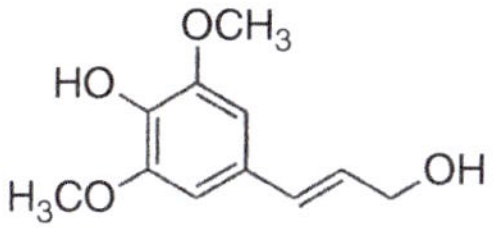 Sinapyl alcohol	• Paper manufacturing • Dispersing agent • Biorefinery • Furniture • Automotive parts
Starch	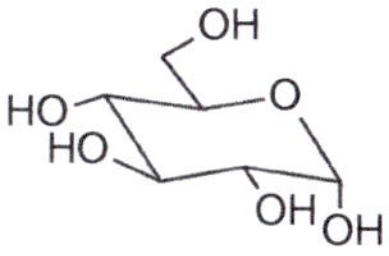 D-Glucose	• Food industry • Cosmetics • Adhesives • Textile • Biotechnology

(*continued overleaf*)

Table 6.3 (Continued)

Biopolymer	**Monomeric units**	**Main applications**
Chitosan/chitin	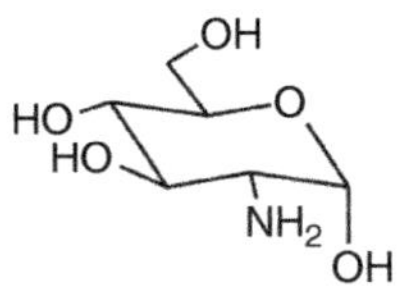 D-Glucosamine 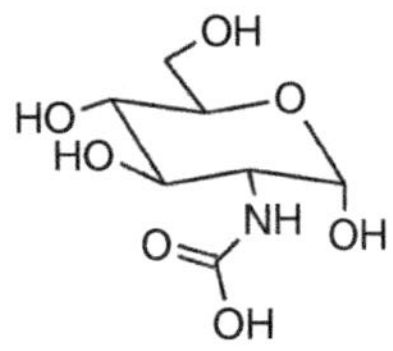 *N*-Acetyl-D-glucosamine	• Drug release • Encapsulation • Wound dressings • Water purification
Terpenes	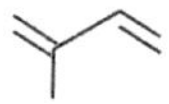 Isopropene	• Chemical industry • Adhesives • Epoxy resins
Cutin	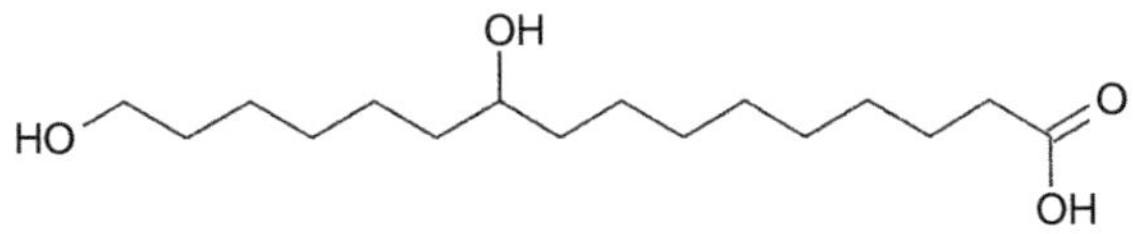 Polyhydroxylated fatty acid	• Food packaging • Cosmetics • Pharmaceutical industry
Proteins	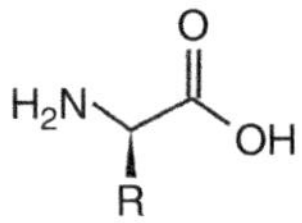 Aminoacids	• Textile industry • Drug encapsulation • Adhesives
Polyesters	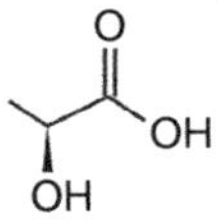 Lactic acid 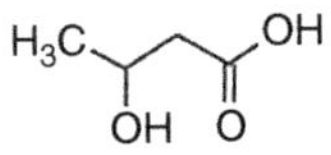 Hydroxy butyrate	• Biomedical applications • Pharmaceutical industry

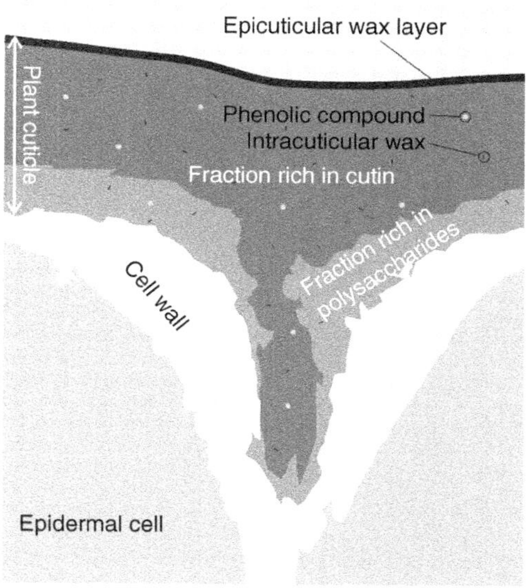

Figure 6.5 Schematic representation of a transversal section of the outer part of a plant epidermis, displaying a plant cuticle between two epidermal cells. Main components and their space distribution are shown. Source: From Heredia-Guerrero et al. [47]. © 2014, Frontiers Media.

As described above, cutin consists of C16, C18, or a mixture of both fatty acids [56]. These monomers are typically mid-chain (epoxy, oxo, hydroxy, vicinal diol, and double bonds) and end-chain (OH groups) functionalized [49]. Main monomers of C16-based cutins are 9(10),16-dihydroxyhexadecanoic and 16-hydroxyhexadecanoic acids, whereas 16-hydroxy-10-oxo-hexadecanoic and 16-oxo-9 or 10-hydroxyhexadecanoic acids are uncommon. In contrast, C18-based cutins are rich in 18-hydroxy-9,10-epoxyoctadecanoic and 9,10,18-trihydroxyoctadecanoic acids and their monounsaturated homologues. In some species, other cutin monomers are glycerol and cinnamic compounds [49, 57]. Table 6.4 reports a generic cutin composition together with the specific chemical description of tomato (*Solanum lycopersicum*) and pepper (*Capsicum annuum*) fruit cutins [49, 58, 59].

As observed, dihydroxylated C16 fatty acids are the main monomers of these cutins. Interestingly, there is a commercially available fatty acid with a similar chemical structure. It is 9,10,16-trihydroxyhexadecanoic or aleuritic acid. This polyhydroxylated fatty acid is derived from a renewable natural lac resin denominated *shellac* with a relatively low annual commercial production of ~150 tons per year [60, 61]. Aleuritic acid is used in many applications as a substitute for α-hydroxy acids in cosmetic formulations, food additive, and coating material for tablets and capsules [40]. In addition, as described in Sections 6.2 and 6.3, aleuritic acid has been often used to fabricate plant cuticle-like and cutin-inspired polymers.

6.1.3 A Comparison of Cutin with Commercial Plastics and Bioplastics

Mechanical, hydrodynamic, and thermal properties are key parameters to estimate the potential applicability and the specific uses of a material. In this sense, Table 6.5 reports a comparison of cutin to PLA, poly-3-hydroxybutyrate (P3HB),

Table 6.4 Main composition of a generic cutin and specific composition of tomato and pepper fruit cutins.

General composition for cutin [49]			Tomato fruit cutin [59]		Pepper fruit cutin [58]	
Monomer type	Common monomers	Abundance (%)	Monomer	Abundance (%)	Monomer	Abundance (%)
Unsubstituted fatty acids	C16:0, C18:0, C18:1, C18:2	1–25	Hexadecanoic acid	0.2		
			Octadecanoic acid	0.2		
ω-Hydroxy fatty acids	C16:0, C18:1, C18:2	1–32	16-Hydroxyhexadecanoic acid	2.9	16-Hydroxyhexadecanoic acid	2.7
			18-Hydroxyoctadecanoic acid	0.1	18-Hydroxyoctadecenoic acid	1.9
			24-Hydroxytetracosanoic acid	0.2		
α,ω-Dicarboxylic acids	C16:0, C18:0, C18:1, C18:2	0–5	1,16-Hexadecanedioic acid	0.5	1,16-Hexadecanedioic acid	1.0
			9-Hydroxy-1,16-hexadecanedioic acid	2.7		
Epoxy-fatty acids	C18:0 (9,10-epoxy) C18:1 (9,10-epoxy)	0–34	9,10-Epoxy-18-hydroxyoctadecenoic acid	0.2	9,10-Epoxyoctadecanoic acid	9.3
			9,10-Epoxy-18-hydroxyoctadecanoic acid	0.1	9,10-Epoxyoctadecenoic acid	1.3
Polyhydroxy-fatty acids	C16:0 (10,16-Dihydroxy), C18:0 (9,10,18-trihydroxy)	16–92	9(10),16-Dihydroxyhexadecanoic acid	81.6	9(10),16-Dihydroxyhexadecanoic acid	63.8
			9(10),18-Dihydroxyoctadecanoic acid	1.4	9(10),18-Dihydroxyoctadecanoic acid	1.3
			9,10,18-Trihydroxyoctadecenoic acid	0.2	9,10,18-Trihydroxyoctadecanoic acid	6.5
			9,10,18-Trihydroxyoctadecanoic acid	0.4		
Fatty alcohols	C16:0, C18:1	0–13				
Glycerol		1–14	Glycerol	0.6		
Phenolic compounds		0–1	*p*- and *m*-Coumaric acids	2.8	*p*- and *m*-Coumaric acids	2.3
			Unidentified compounds	5.8	Unidentified compounds	9.9

Table 6.5 Young's modulus, stress and elongation at break, water contact angle, and glass transition temperature of cutin and other bio-based polyesters, cellulose, and cellulose derivatives.

Parameter[a)]	Cutin	PLA	P3HB	Cellulose	CA	EC	CMC
Young's modulus (MPa)	45 [63]	1925 [118]	2600 [118]	2620 [119]	1650 [64]	1430 [120]	1935 [121]
Stress at break (MPa)	12 [63]	41 [118]	32 [118]	48 [119]	30 [64]	10 [120]	70 [121]
Elongation at break (%)	27 [63]	4 [118]	3 [118]	8 [119]	40 [64]	1 [120]	12 [121]
W-CA (°)	86[b)]	83 [122]	89 [123]	32 [124]	55 [125]	85 [125]	18 [126]
T_g (°C)	23 [65]	32 [118]	7 [118]	—	205 [127]	143 [120]	55 [128]

a) Data provided in this table are averaged. Important differences can be found as a function of molecular weight, crystallinity, degree of substitution, etc.
b) Data from authors' lab.

cellulose, cellulose acetate (CA), ethyl cellulose (EC), and sodium carboxymethyl cellulose (CMC). The mechanical properties of tomato fruit cutin have been well characterized in many conditions, and it has been found that they change with the temperature and relative humidity and depend on the fruit development [62, 63]. As an example, cutin of red ripe tomatoes at 23 °C and 40% relative humidity (RH) shows a Young's modulus, stress at break, and elongation at break values of 45.0 MPa, 12.3 MPa, and 27.0%, respectively. When compared to the other polymers of Table 6.5, Young's modulus and stress at break are low. In fact, cutin shows a rigidity similar to those of common elastomers such as PUR, polydimethylsiloxane, and acrylonitrile-butadiene rubber [64]. On the other hand, the elongation at break of cutin is higher than that of less ductile polymers such as PLA, P3HB, cellulose, EC, and CMC. Interestingly, despite the presence of polar groups (mainly free hydroxyl groups), cutin is relatively hydrophobic with a water contact angle (W-CA) value of 86°. This W-CA is in the range of other biopolyesters such as PLA (83°) P3HB (89°) as well as of the ethyl cellulose (85°). Other more polar polymers (e.g. cellulose, CA, and CMC) show much higher hydrophilic behavior. Finally, the phase transitions of cutin have been extensively analyzed [65]. Cutin does not show any melting point due to its cross-linking polymer network (thermoset behavior, infusibility). However, cutin has a glass transition temperature (the temperature at which segment motion of macromolecules becomes thermally activated) c. 23 °C, which is close to that of PLA (32 °C) and slightly higher than the T_g of P3HB (7 °C). The glass transition of cellulose derivatives is usually much higher, hindering their industrial processability.

In addition to physical properties, biodegradability is another important characteristic to be considered. For instance, petroleum-based polymers typically show very low biodegradation rates. Thus, plastics such as polyethylene, polystyrene, PVC, and urea formaldehyde resin buried over 30 years exhibit insignificant degradation [66]. Cutin is biodegraded (mainly by cutinases secreted by microorganisms and fungi) in soil after three to eight months depending on the conditions (i.e. season, characteristics of the soil, and microbiota) [67]. This weight loss with time is much higher than those of PLA (weight loss of ~0.1% after 24 weeks) and similar to the biodegradation rates of cellulose (weight loss of ~36% after 4 weeks) and P3HB (weight loss of ~50% in 42 days) [68–70].

6.1.4 Tomato Pomace is the Main and Most Sustainable Cutin Renewable Resource

Among the different cutin resources, the most sustainable ones are those that come with a high cutin percentage and can be derived from non-reusable wastes and by-products and are accessible (i.e. they can be found concentrated in few places such as agro-food processing industries) and abundant. Tomato pomace, the by-product after tomato fruit food processing with a cutin content close to 20 wt%, is considered the best choice. In fact, the global tomato production reached 171 million tons in 2014, with China (53 million tons) and Europe (23 million tons, being Spain and Italy the main tomato producers) as leaders, followed by India, United States, and Turkey (data from FAOSTAT, 2017). The global tomato processing market reached a volume of 36 million tons in the same year. Tomato fruits are mainly processed for the preparation of peeled tomatoes, purées, juices, concentrates, sauces, etc. Peeling is performed before the manufacture of these products, with a final residue (tomato peel) rich in cutin [71]. Considering the average amount of cutin and the area and weight of a commercial tomato fruit, the volume of cutin from this agro-waste is ~10^5–10^6 tons per year (comparable to lignin stock, but considerably lower than cellulose annual biomass production). Tomato pomace is composed of peels (~48 wt%), seeds (~38 wt%), and residual fibrous material (~14 wt%) [72, 73]. On a dry basis, it represents about 2–3% of the fresh processed weight and constitutes a potential renewable global feedstock of about 1.2 M tons per year, of which ~0.3 M tons per year are generated in Europe. Nowadays, this inexpensive by-product (whose disposal is valuated to be ~€4 per ton) is wasted or, to a lesser extent, used for animal feeding or to extract lycopene [74, 75].

Cutin is usually depolymerized after cuticle isolation. For this, plant peels or pomace are suspended into enzymatic solutions of pectinases, cellulases, and hemicellulases to remove the polysaccharides from the cell wall. Other harder chemical procedures with ammonium oxalate/oxalic acid or $ZnCl_2$/HCl can be also employed [76–79]. After isolation, plant cuticles are dewaxed by immersion in organic solvents (usually, chloroform, methanol, hexane, or different mixtures of them) and remnant polysaccharides are removed by acid hydrolysis with HCl, thus purifying cutin. The rupture of the ester bonds to obtain cutin monomers is typically achieved by alkaline hydrolysis (NaOH or KOH in water or methanol), transesterification with methanol containing BF_3 or $NaOCH_3$, reductive cleavage by exhaustive treatment with $LiAlH_4$ in tetrahydrofuran, or with trimethylsilyl iodide in organic solvents [80]. However, these conventional, multi-step methods are time consuming and expensive for their industrial implementation. In this sense, other simpler, cheaper, and greener alternatives are preferred. For example, cutin monomers have been obtained from methodologies involving sodium carboxylate or sodium hydroxide/hydrogen peroxide and sequential hydrolysis in KOH and H_2SO_4 [74, 81]. More interestingly, tomato pomace has been directly hydrolyzed by alkaline solutions (NaOH in water) [82]. In this case, in addition to polyhydroxylated monomers from cutin, unsaturated fatty acids (mainly oleic and linoleic acids) from seeds are also present at the end of the process [72]. By following this process, a lipid fraction (36 wt% of tomato pomace) formed by cutin monomers and unsaturated fatty acids in a ratio 4 : 3 (w:w) can be obtained. This methodology has been successfully scaled to pilot plant, which has allowed the production of large quantities of multifunctional fatty acids in short times, as required by the industry.

6.1.5 Toward a Sustainable Industrial Production of Cutin-Inspired Commodities

Cutin-inspired polyesters have been polymerized by using different types of catalysts such as *p*-toluene sulfonic or 4-dodecylbenzoic acids, lipases in organic media, and ionic liquids, or by self-assembly [83–89]. These methodologies are not adequate for large production technologies due to low yields and high costs, toxicity, and unsafety of solvents and catalysts. However, in recent years, more sustainable approaches based on direct melt polycondensation have been investigated [58, 60, 90–95]. This procedure consists in the melt of the polyhydroxylated fatty acids and the reaction of hydroxyl and carboxyl groups to produce water that is quickly removed by evaporation. Such polymerization presents several advantages. For instance, no solvents and catalysts are necessary, thereby reducing costs and making the process greener. In addition, water from the esterification reaction is the single by-product. Moreover, the synthesis is carried out at mild temperatures (~150 °C) in air or inert atmosphere and can be considered finished, depending on the sample thickness, after a few hours. For industrial upscaling, it must be considered that resultant cutin-inspired polymers are insoluble and infusible, and therefore techniques applied to thermosets such as compression molding should be employed. On the other hand, spray of cutin monomers from organic or aqueous solutions (where can be blended with other components), before thermal curing, can be used in the formation of coatings and composites [40, 94]. This last technique allows to cover high areas in short times.

6.2 Synthesis of Cutin-Inspired Polyesters

6.2.1 The Influence of the Monomer Architecture in the Physical and Chemical Properties of Cutin-Inspired Polyhydroxyesters

As described above, monomers that can be extracted by the processing of cuticles of commercial fruits are mainly fatty (C16 and C18) mono and di acids as well as the corresponding mono and poly hydroxylated homologues. The separation and isolation of each component is feasible at the laboratory scale but is a complex and expensive process to be upsized to an industrial level. Thus, the development

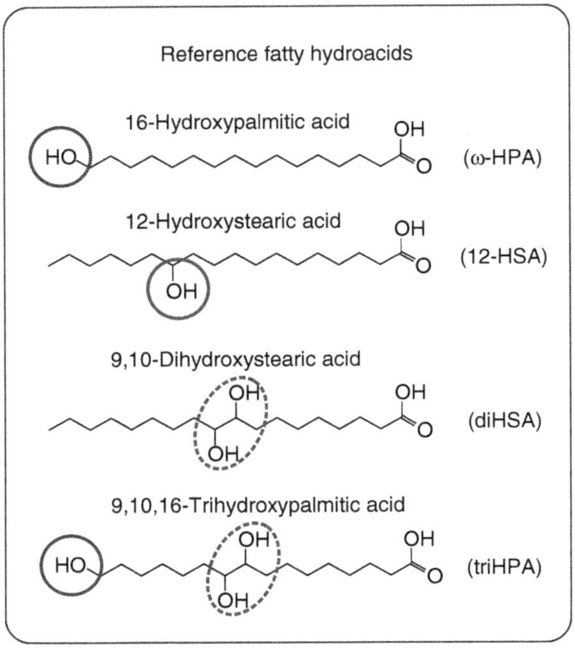

Figure 6.6 Chemical structure of the reference fatty hydroxyesters used.

of viable polymeric materials from such feedstock imposes the use of a mixture of monomers with a varied type and degree of chemical functionalization [53]. In this sense, reference pure hydroxyacids have been used to qualitatively draw specific reactivity patterns in their self-polycondensation reaction. See Figure 6.6.

For instance, primary and secondary hydroxyls show a well-differentiated esterification reactivity [96]. To address this issue, a primary (16-hydroxypalmitic, ω-HPA) acid and a mid-chain secondary (12-hydroxystearic, 12-HSA) monohydroxylated acid have been selected. The development of characteristic ν(C=O) and ν_a(C—O—C) IR bands around 1730 and 1170 cm^{-1}, respectively, reveals the formation, in a few hours, of the corresponding esters at moderate temperature (150 °C) by self-polycondensation in molten state directly in air. See Figure 6.7.

Solid-state nuclear magnetic resonance (NMR) data confirm IR results. See Figure 6.8. The ^{13}C spectrum of poly ω-HPA shows the carbonyl and methylene fingerprints of a primary ester at 174.0 and 65.3 ppm, respectively. In the same reaction conditions, the esterification of the secondary hydroxyl is less effective and a free acid ^{13}C peak is observed at 178.3 ppm [60].

Structurally, poly ω-HPA is solid at room temperature, while poly 12-HSA is an amorphous thick resinous product. The quite low Mw values for poly 12-HSA and the presence of free acid indicate the formation of cyclic oligomeric units (n = 1–5) by intramolecular esterification. On the other hand, poly ω-HPA is a semi-crystalline linear low molecular weight polyester ($M_w \sim 20$ K) with average chain lengths of 50–100 units [90]. Thermally, primary and secondary esterification can also be clearly distinguished. Thermogravimetric analysis (TGA) shows that the maximum thermal decomposition rate of neat secondary esters is

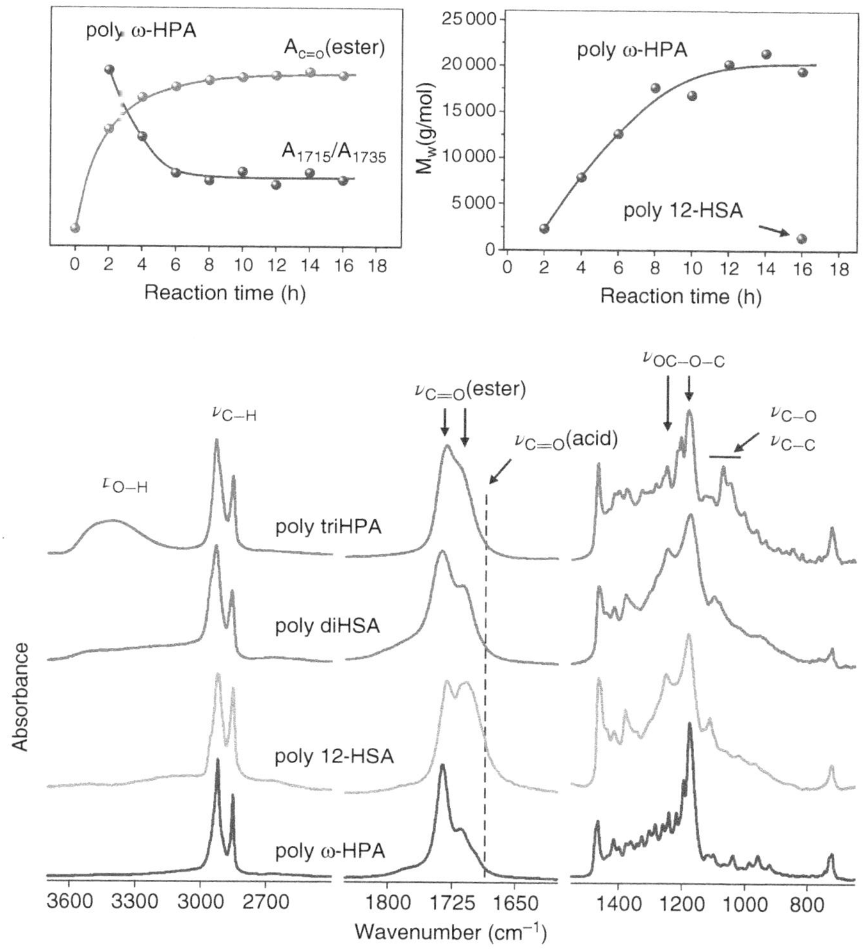

Figure 6.7 (bottom) Transmission IR spectra of reference polyhydroxyesters obtained by self-polycondensation in air at 150 °C for 16 hours. (top-left) IR band and molecular weight evolution vs. reaction time for poly ω-HPA at 150 °C.

around 335 °C while the temperature is 75 °C higher (~410 °C) for primary ones. See Figure 6.9. This result is consistent with the description of the ester bond scission as the main thermal decomposition mechanism and sets a distinctive trait of polymers to be obtained from primary or secondary hydroxylated fatty acids [97].

When primary and secondary hydroxyls coexist in the structure of polyhydroxylated monomers (as in triHPA), solid-state ^{13}C NMR data reveal that primary esterification is favored [92]. Quantification of the ^{13}C signal of the methylene carbon bearing the esterified and the non-esterified primary hydroxyl group (65.3 and 63.1 ppm, respectively) reveals that primary esterification is around 85–90%. Such preferential esterification explains that the thermal stability of poly triHPA is more similar to poly ω-HPA than to poly diHSA and poly 12-HSA.

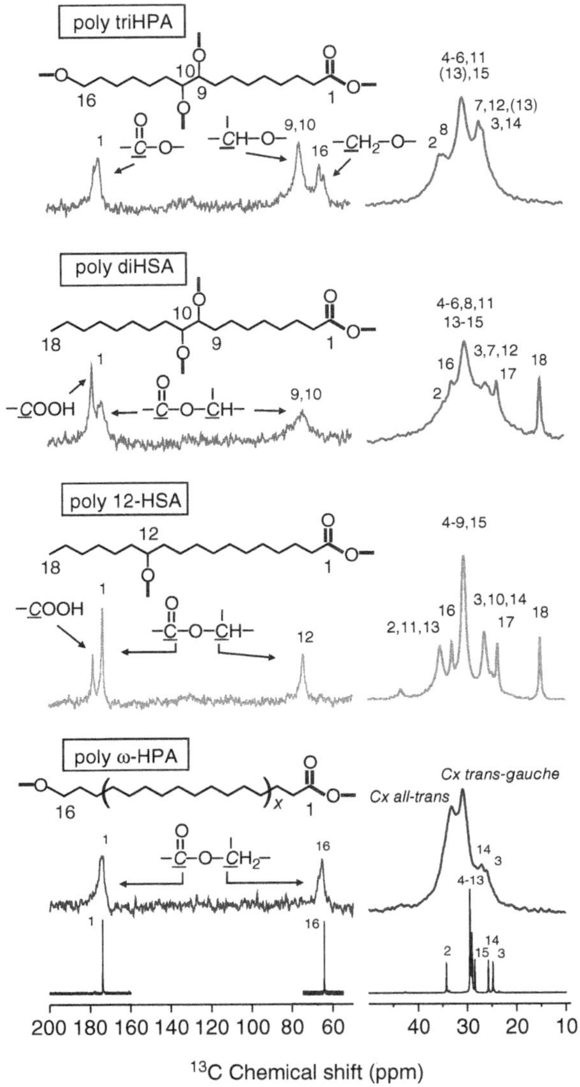

Figure 6.8 Solid-state ^{13}C NMR of reference polyhydroxyesters prepared in air for 16 hours at 150 °C.

However, secondary esterification is a source of branching in the structure of polyhydroxyesters and, despite its low extent in poly triHPA, is enough to condition the packing of chains and to drastically diminish crystallinity. In turn, such structural disruption is responsible for the strong reduction of stiffness of polyhydroxyesters with respect to poly ω-HPA and the development of a ductile behavior. See Figure 6.10. Besides, the Young's modulus and the stress and elongation at break of amorphous polyhydroxyesters have been found to increase with the hydroxylation degree of the monomers and are related to the development of a hydrogen-bonding secondary network from surplus hydroxyls.

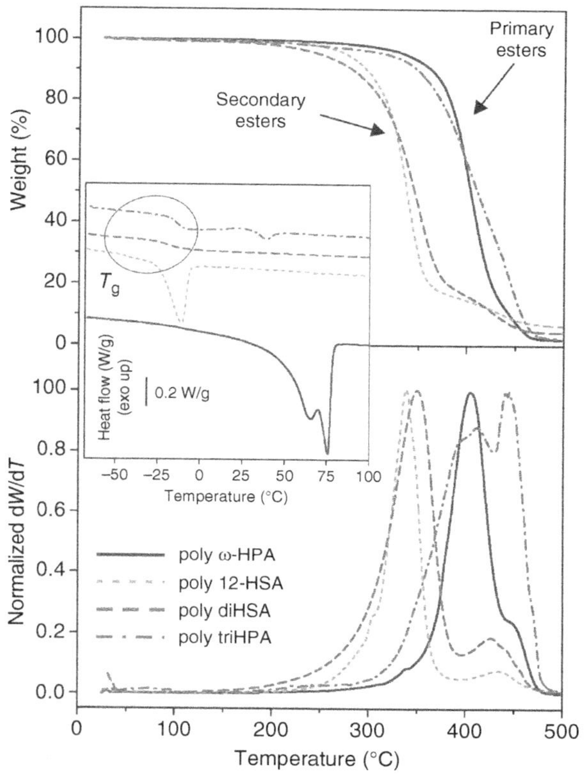

Figure 6.9 (top) TGA and (bottom) TGA-derivative profiles of the reference polyhydroxyesters. The inset shows the first heating cycle of the differential scanning calorimetry (DSC).

6.2.2 The Effect of Oxidation in the Structure and Properties of Cutin-Inspired Fatty Polyhydroxyesters

In the quest for an affordable synthesis route of fatty polyhydroyesters, the polycondensation has been conducted directly in air. The aim of this section is to analyze and state the effect of such oxidative atmosphere on the structure and properties of polyesters obtained from the series of reference hydroxyacids.

IR transmission spectra detects the presence of characteristic oxidation products ($CO_{(ox)}$) such as peracids and peroxyesters (band at $1780\,cm^{-1}$) as well as anhydrides (doublet at 1805 and $1745\,cm^{-1}$) and diacyl peroxides (doublet at 1805 and $1780\,cm^{-1}$). See Figure 6.7. The main oxidation products in monohydroxylated polyesters are peroxy species resulting from the degradation of either acid or already formed ester groups. However, in poly diHSA and poly triHPA, anhydrides and diacyl peroxides are generated by the condensation of new –COOH groups produced by the oxidative cleavage of the vicinal diol moiety, Figure 6.11.

The formation of $CO_{(ox)}$ species is more pronounced in poly diHSA and poly ω-HPA. In the first case, the oxidative diol cleavage is intense because the consumption of secondary hydroxyls to form an ester has been found to be an ineffective process. On the other hand, poly ω-HPA is a primary polyester and the only one fusible within the series. See Figure 6.9. Oxidation is a surface reaction and mass convection within the melt renews the access of oxygen to the

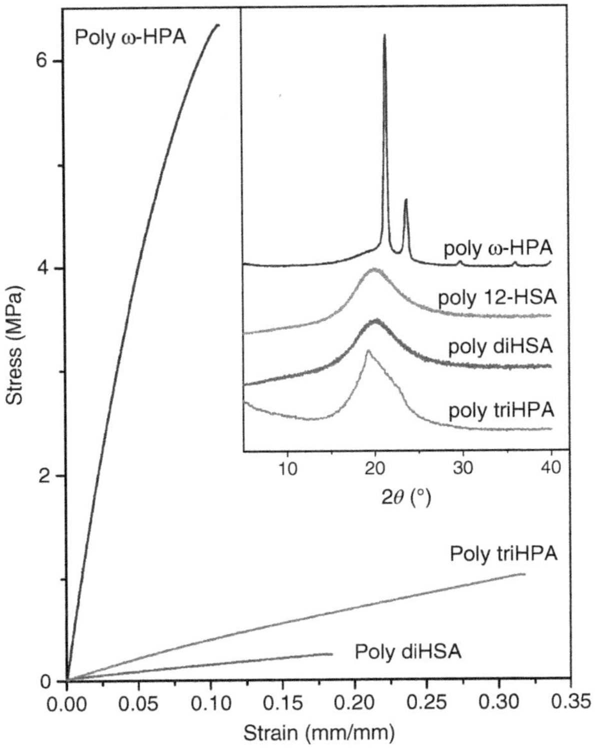

Figure 6.10 Tensile stress-train curves of the reference polyhydroxyesters prepared at 150 °C for 16 hours by self-polycondensation in air. (Inset) X-ray diffraction (XRD) diagrams.

already-formed polyester bulk and enlarges the process. Interestingly, the lower reactivity of secondary hydroxyls toward esterification is also manifested in the oxidation of the secondary ester and the presence of $CO_{(ox)}$ in poly 12-HSA is the lowest of the series. Finally, poly triHPA is an intermediate case. The presence of a primary hydroxyl leads to a fast formation of a viscous ester phase (with also some degree of branching) that hinders the diffusion of oxygen molecules and the attack to the diol moiety. The consequence is a moderated presence of oxidation products, much lower than in poly diHSA.

Though the oxidation process is evident, the relative intensities of ν(C=O) Fourier Transform infra red (FTIR) bands indicate a quite moderate extent. Nevertheless, the main effect on the structure of these polyhydroxyesters is the amorphization of films prepared in air. Compared to those obtained under an inert atmosphere, the degree of crystallinity (X_c) decreased noticeably, particularly when the monomer contains a vicinal diol as in poly diHSA and poly triHPA. See Table 6.6. Most likely, the formation of anhydride and/or peroxide species introduces a lattice mismatch in the ester network that leads to less compact arrangement of linear chains. Such effect accounts for the relatively small reduction of crystallinity in poly ω-HPA. However, in poly diHSA and poly triHPA, the amorphization is stronger. In these cases, the diol cleavage and further esterification causes chain branching and cross-linking that impedes their packing into crystalline domains to the point that fully amorphous solids can be obtained. In parallel to such amorphization, polyhydroxyesters prepared in air become highly

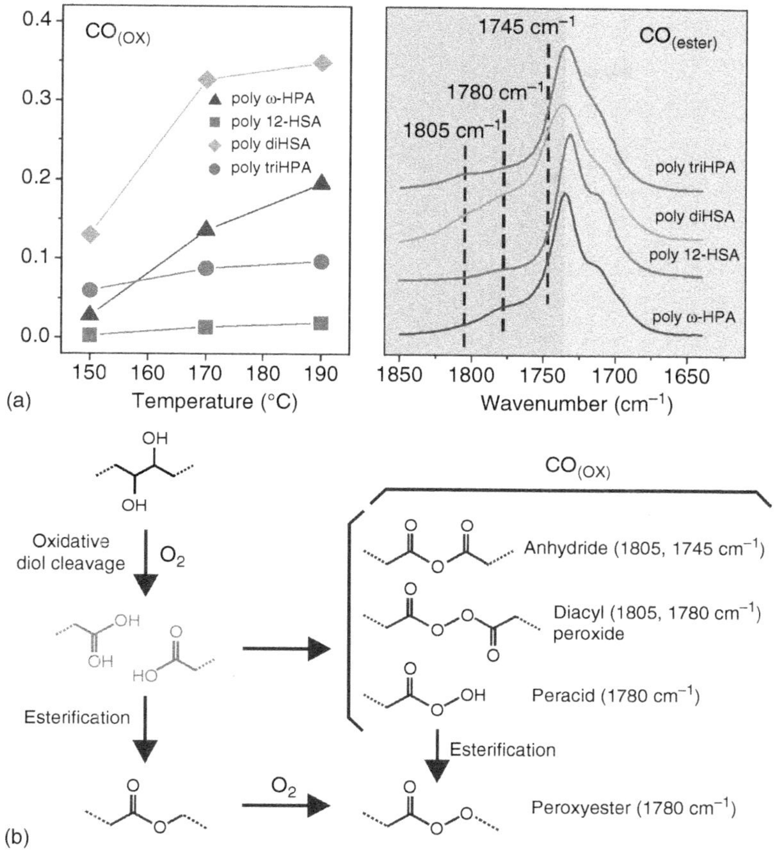

Figure 6.11 (a) ATR-FTIR quantification of the amount of oxidized species at the air-exposed side of polyhydroxyester films obtained. (b) Reaction paths for the oxidative cleavage and $CO_{(ox)}$ formation.

Table 6.6 Thermal parameters and crystallinity of references polyhydroxyesters prepared under oxidative and non-oxidative atmospheres at 170 °C for 16 hours.

	Air				N_2			
	T_g	T_m	ΔH_m	X_c	T_g	T_m	ΔH_m	X_c
poly ω-HPA	−15	50	73	58	—	95	145	71
poly 12-HSA	−44	−16	21	0	−44	−13	42	0
poly diHSA	−15	—	0	0	−15	79	38	25
poly triHPA	−12	47	19	14	−10	49	42	33

T_g (glass transition, °C), T_m (melting temperature, °C), ΔH_m (melting enthalpy, J/g), X_c (crystallinity at room temperature, %).

Table 6.7 Tensile parameters of references polyhydroxyesters films prepared under oxidative and non-oxidative atmospheres at 170 °C for 16 hours.

	Air			N_2		
	E	σ_{max}	ε_{max}	*E*	σ_{max}	ε_{max}
poly ω-HPA	66	4.1	71	312	23	660
poly diHSA	3	0.6	23	104	5.3	7
poly triHPA	21	2.6	31	130	13	390

E (Young's modulus, MPa), σ_{max} (rupture stress, MPa), ε_{max} (rupture strain, %).

insoluble in organic solvents such as chloroform. The insoluble fraction ranges from 75% to 95% depending on the type and preparation conditions.

Mechanical performances of fatty polyhydroxyesters are drastically affected by oxidation. See Table 6.7. Films of primary hydroxyesters such as poly ω-HPA and poly triHPA obtained under N_2, display a typical plastic behavior with strain softening, plastic flow, and strain hardening stages. On the contrary, the secondary polyhydroxyester (poly diHSA) is quite brittle with very little capacity to release mechanical stress through deformation, likely, because of a high concentration of defects. Partial oxidation also eliminates the plastic behavior because of amorphization. Samples become moderately ductile and much softer with lower Young's modulus and breaking stress and strain.

6.2.3 Surface vs. Bulk Properties

Transmission IR spectroscopy provides bulk information. However, to better characterize the oxidation process, a surface IR sensitive technique such as attenuated total reflectance (ATR-FTIR) has been used. Both the air-exposed (top) and the air-preserved (mold facing) sides have been studied. The carbonyl region (1850–1600 cm^{-1}) has been processed to allow a quantitative comparative analysis of both faces of poly ω-HPA, poly diHSA, and poly triHPA (in the case of poly 12-HSA, no self-standing film can be obtained). The comparison of spectra from the air-exposed and air-preserved sides allows the definition of three oxidation patterns in terms of bulk penetration. See Figure 6.12. In poly diHSA, a high penetration oxidation is characterized by the higher intensity of the carbonyl band as well as the similar $C_{(ox)}$ content in both the air-exposed and air-preserved sides. The lack of a primary hydroxyl slows down the development of an ester network capable of blocking the oxygen diffusion into the bulk and, consequently, the air-preserved side shows a similar oxidation degree as the air-exposed one. In poly ω-HPA the oxidation is moderated and the alteration of the air-preserved side is milder than in poly diHSA because of the absence of the vicinal diol. Still, accessibility of oxygen to the air-preserved side is noticeable and it is very likely caused by mass convection within the melt. However, in poly triHPA the oxidation is clearly restricted to the air-exposed side as demonstrated by the lack of oxidation species at the air-preserved region. Furthermore, the alteration is located at the outermost layers because the intensity of the carbonyl

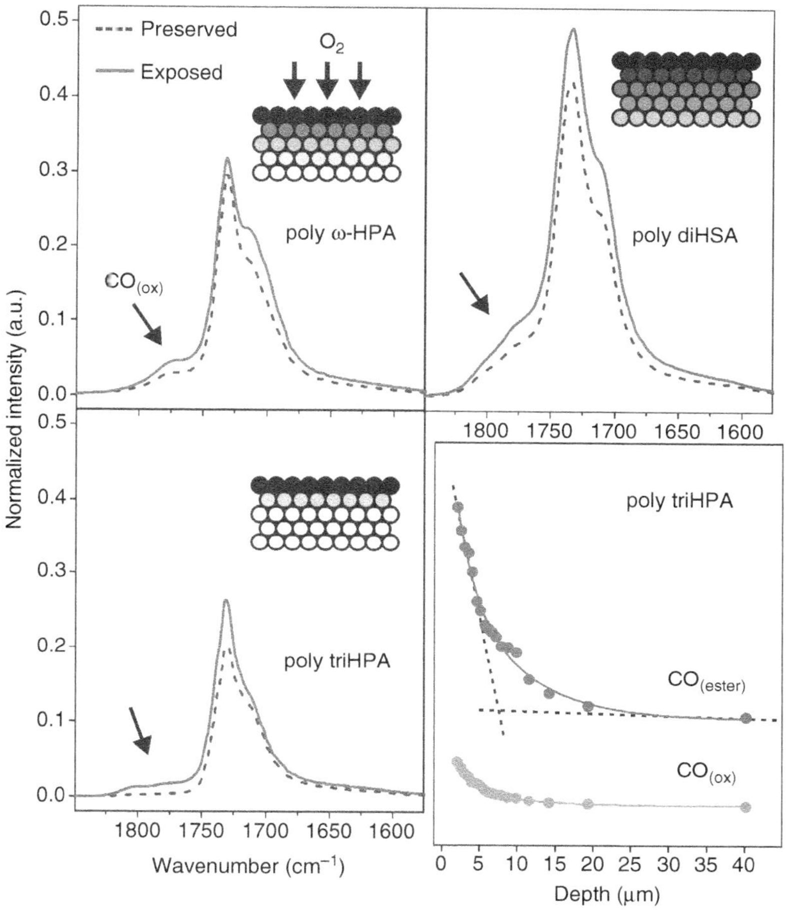

Figure 6.12 ATR-FTIR spectra of the air-exposed and air-preserved sides of polyhydroxyester films prepared in air at 150 °C for 24 hours. (bottom-right) Composition depth profile obtained by variable angle ATR-FTIR of the air-exposed side of poly triHPA synthesized at 190 °C for 22 hours.

signal is quite low compared to those of the rest of the series and very similar to the one corresponding to the unaltered air-preserved side.

To confirm this hypothesis, depth profiling by variable incidence angle ATR-FTIR has been used. Results show that the oxidized layer in poly triHPA extends to about 7–10 μm deep under severe reaction conditions (22 hours at 190 °C) in air. See Figure 6.12. This conclusion is of relevance because the thickness of coatings used, for instance varnish of metal food containers, is typically of several microns and, if prepared in analogous conditions, they may fall within the size of the altered layer. Consequently, their physical and chemical properties will differ from those of the corresponding neat polyhydroxyester.

Mechanical properties of the oxidized layer have been studied by indentation. The Young's modulus of the air-exposed side is an order of magnitude higher that the air-preserved one (385 MPa vs. 38 MPa). Such framework reinforcement is due to the increment of ester bond density derived from the diol cleavage process, as shown by the $CO_{(ester)}$ depth profile. Thanks to this effect, the local mechanical performance of the oxidized layer approaches a suitable level to be used as a protective coating in food containers.

Surface properties are also extremely important because they control the interaction between the solid and the environment. In polymers to be used as food packaging materials, the manipulation of the interface in contact with either the environment or the foodstuff can drastically modify the overall performances of the whole protective layer [98]. Thus, permeability to water vapor of these polyhydroxyesters prepared under N_2 is around 10^{-6} g/h m Pa. When synthesized in air, the value increases by a factor of 4. The lower permeability of the series obtained under non-oxidative conditions is likely due to a higher crystallinity. The compact packing of chains in crystalline domains is an obstacle to water molecule diffusion and transference is mainly through the amorphous inter-regions. The permeability value of natural dewaxed tomato cuticle (fully amorphous) is around 5×10^{-6} g/h m Pa, which supports the fact that water diffusion is enhanced in amorphous polyhydroxyesters. Thus, the control of crystallinity by the selection of the oxidation degree is a tool to tune water vapor permeability to a required value.

Wettability is an important parameter to be considered for the evaluation of the water barrier properties of films. Surface hydrophobicity of fatty polyhydroxyesters is found to depend on the OH/COOH balance in the corresponding monomers as well as on the surface oxidation state. Thus, the W-CA is the highest for poly ω-HPA and decreases as the OH/COOH ratio increases to 2 and 3 in poly diHSA and poly triHPA, respectively. See Table 6.8. The high hydrophilicity of surplus free hydroxyls is proposed as the major cause for the reduction of hydrophobicity within the series. The air-exposed side of films prepared in air shows appreciably lower W-CA values than those prepared under N_2. The presence of oxidized species such as peroxides adds a contribution to the polarity of surfaces and, consequently, hydrophobicity decreases.

Another key property of food preserving materials is the capacity to block light. Visible and, in particular, UV radiation is a well-known food deterioration factor and packaging material should combine a good UV–vis blocking capacity with an acceptable transparency to make the product attractive for consumers. Films

Table 6.8 Static water contact angle values of reference polyhydroxyester films obtained under oxidative and non-oxidative atmospheres.

	W-CA (°)	
	Air	**N_2**
poly ω-HPA	83	104
poly diHSA	80	98
poly triHPA	74	90

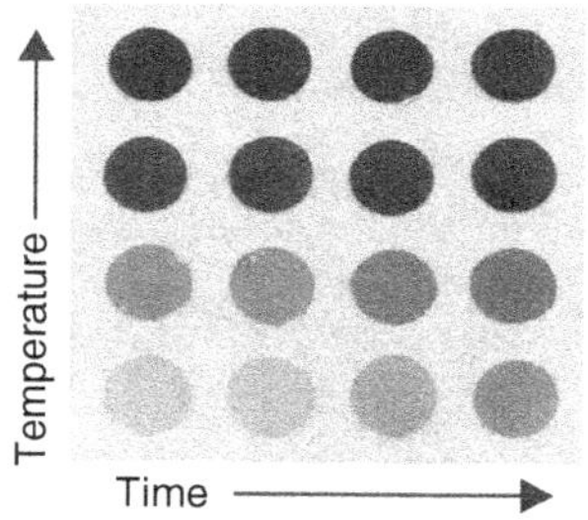

Figure 6.13 Visual aspect of poly triHPA prepared in air at variable time and temperature. Times are 8, 12, 16, and 20 hours. Temperature values are 150, 170, 190, and 200 °C.

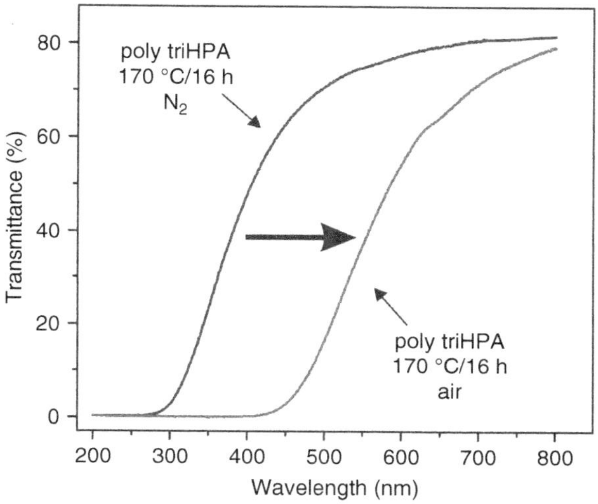

Figure 6.14 Total transmittance curves of poly triHPA prepared under oxidative and non-oxidative atmospheres.

of fatty polyhydroxyesters prepared in air gain gradual coloration as time and reaction temperature are raised, Figure 6.13.

UV–vis transmission curves of films prepared in air show a full blocking capacity for UV radiation, Figure 6.14. The zero-transmittance plateau already extends below 350 nm at the less severe preparation conditions used (150 °C, eight hours). Coloration caused by oxidation products progressively displaces the curve toward higher wavelengths and provides a better protection against the harmful components of the UV–vis light. Such shielding levels cannot be achieved by films prepared under an inert atmosphere. Despite such UV–vis blocking capacity, films 300–400 μm thick can keep a good transparency level (above 50% if obtained below 170 °C).

6.3 Cutin-Based and Cutin-like Coatings and Composites

6.3.1 Cutin-Inspired Coatings on Metal Substrates

Metal packaging plays an important role in the process of food preservation and distribution, safely conserving foodstuffs without microbiological, chemical, and physical deterioration for long periods [99]. Many metal packaging containers such as cans, caps, closures, etc. are usually coated. The inside coating (in direct contact with food) is referred to as an internal lining or lacquer. The main function of these internal coatings is to minimize the food–metal interaction throughout the shelf life of the packaged product [100–102]. In addition, there is an environmentally and cost-driven development of metal packaging into thinner containers (i.e. lower weight of metal) without compromising the protective aspects. It results in more demanding performances regarding chemical inertness, sustainability, flexibility, strength, adherence, viscosity, lubricity, good thermal resistance, and economic viability of the employed coatings. Can coatings are applied to metal surfaces and, after thermal curing, form a dry and thin (thickness between 1 and 10 μm) film [103]. Their major constituents are resins, cross-linking agents, additives, and solvents. Most of them are harmful, toxic, and non-sustainable.

There is a limited number of available chemical functionalities for direct food-contact coatings, resulting in few types of resins that can be used for metal packaging. Among them, epoxy resins, based on BPA and epichlorohydrin, have been largely used in coatings for light metal packaging since the 1950s. In particular, epoxy-phenolic coatings (blends of epoxy and phenolic resins, usually in weight ratios 3 : 1 or 6 : 1) are the most widely employed for metal packaging, achieving yearly market shares close to 95%. However, BPA resins show important issues related to sustainability and human health. On one hand, the industrial synthesis of BPA involves the bulk condensation of acetone and phenol initiated by thiols and catalyzed by strong acids [104]. Obviously, this is a non-sustainable, non-green, environmentally harmful process with high labor risks for operators during production. On the other hand, BPA causes human reproductive anomalies and has been positively associated with cardiovascular disease, cancer, and diabetes, likely because the molecule has a shape similar to estrogens and thus may act as an endocrine disrupter [105, 106]. In the past decade, consumers and health experts have raised concerns about the use of BPA in food packaging, since canned food consumption is the main contributor to BPA concentration in the body [107–110]. In fact, the BPA migration from resin to food has been experimentally quantified with values of 10–70 and 1–23 μg/kg for liquid and solid food, respectively [111]. Currently, the European Food Safety Authority (EFSA) is considering further restrictions to the use of this substance in the fabrication of food packaging materials. In this sense, food companies, with consumers as a priority, are eager to move away from packaging based on BPA.

Synthetic candidates to replace BPA resins are being developed by important chemical companies such as AkzoNobel, Eastman Chemical, BASF, 3 M, and Dow Chemical. They must have minimal cost, avoid health and environmental impacts, and work across different types of food as a drop-in material in high-speed manufacturing lines. These alternatives fall largely into three main chemical categories: acrylics, polyesters, and their respective blends with vinyls, urethanes, and other additives. In addition, plant-derived oleoresins are being also investigated and some of them are currently used with low-acid foods such as beans. However, none of these alternatives achieves high marks in all categories of performance and safety [112]. For instance, acrylics used in food-contact applications are brittle and their monomers (usually ethyl acrylate) have a noticeable odor in low quantities. Common oleoresins do not adhere well to metal substrates and do not resist corrosion when used with chemically reactive foods such as tomatoes. Proposed synthetic polyesters can also fail when attacked by acidic foods. Moreover, consumers wary of BPA are not likely to embrace vinyl-based replacements. And any compound similar to BPA, such as bisphenol S, will also be considered risky. In this scenario, the production of biodegradable, safe, and innocuous materials inspired by nature with similar properties to BPA resins can be considered a potential alternative. In particular, the design of biomimetic systems is giving rise to green and commercially available systems [113]. In this sense, cutin-inspired polymers are ideal systems due to their insolubility (mitigating migration phenomena), infusibility (avoiding any adverse thermal event during canned food processing), non-toxicity, and full biodegradability.

Polyhydroxyester coatings can be prepared on metallic substrates in tens of minutes at temperatures ranging from 140 to 200 °C. They show an excellent adherence to metals, a smooth texture, and a low wettability (W-CA ~ 100°). Specular ATR-FTIR spectroscopy confirms the formation of the corresponding esters by the development of bands at 1738 and 1715 cm^{-1}. See Figure 6.15.

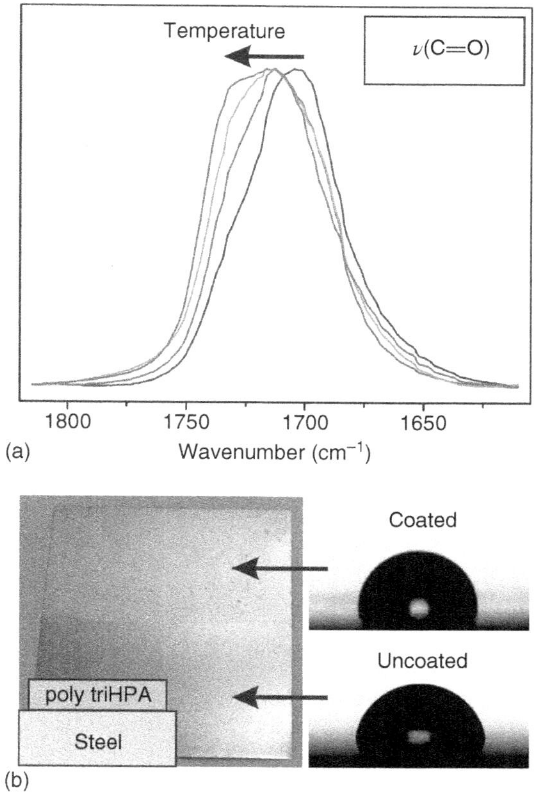

Figure 6.15 (a) Specular ATR-FTIR spectra of polyaleuritate films on stainless steel prepared for 15–20 minutes at variable temperature (140–200 °C). The formation of the ester is confirmed by peaks at 1735 and 1715 cm^{-1}. (b) Left: visual aspect of the coating and right: static water contact angle (WCA) tests on both regions.

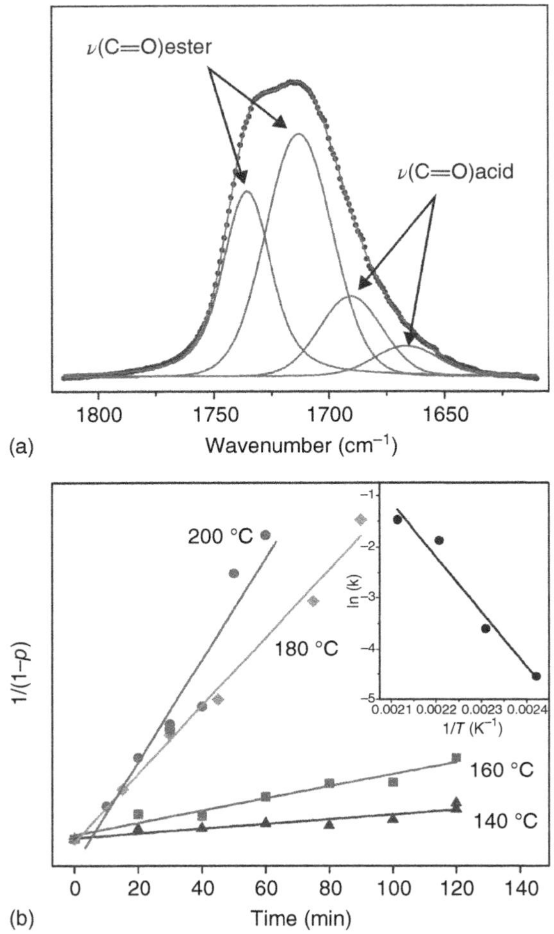

Figure 6.16 (a) Specular ATR–FTIR ν(C=O) band fitting pattern for polyaleuritate coatings on stainless steel. (b) Kinetic data fitting to a second order law and (inset) Arrhenius plot to calculate the activation energy.

The esterification progression (p) has been determined from the quantification of the ester and acid components of the ν(C=O) band. See Figure 6.16. The kinetics can be empirically fitted to a second order reaction law as in homogeneous esterification processes [114]. The activation energy for the esterification was calculated from the Arrhenius plot (ln(k) vs. $1/T$) to be 89 ± 13 kJ/mol, which is about 70% higher than the value reported for the catalyzed esterification in solution [115–117].

6.3.2 Plant Cuticle-like Film Composites

The fabrication of plant cuticle-like films, mainly designed for food packaging, has been carried out through a blending of polyhydroxylated fatty acids, polysaccharides, and waxes. In a first approach, aleuritic acid and carnauba wax were mixed in chloroform:methanol co-solvent system, sprayed onto commercial cellulose substrates (viz, paper), and polymerized by melt polycondensation

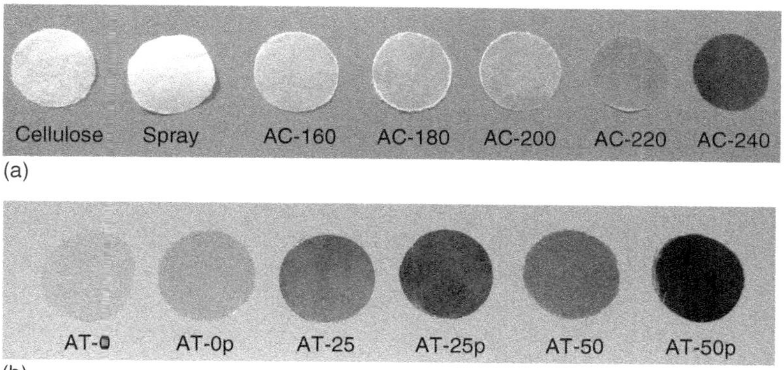

(a)

(b)

Figure 6.17 Photographs of the samples of the plant-cuticle film composites from, (a) mixtures of cellulose, aleuritic acid, and carnauba wax and (b) sodium alginate, tomato pomace monomers, and beeswax.

using a hot-press [40]. Different temperatures and times for the thermal curing were tested, finding that 200 °C for 30 minutes were the optimal conditions. Low degrees of polymerization were achieved at lower temperatures, while for $T > 200\,°C$ cellulose was thermally decomposed. Figure 6.17a shows a photo of the paper used, the sample after spray, and the resultant films (labeled as AC followed by the temperature employed during polymerization) after thermal curing. As observed, AC samples were semitransparent. The colors shifted from white (hot-press at 160 °C) to dark brown (hot-press at 240 °C). The brown color was associated with the thermal degradation of cellulose and the formation of polyaleuritate. The presence of polyaleuritate and its extent of polymerization (ruled by the temperature during the thermal curing) improved mechanical properties, opacity, and wettability. On the other hand, low thermal conductivity was controlled by the cellulose substrate. Both polyaleuritate and cellulose polymers interacted by H-bonds between hydroxyl–hydroxyl and ester–hydroxyl functional groups. The incorporation of carnauba wax significantly increased hydrophobicity and the water and oxygen barrier properties of the films. A variation of this methodology was used for the fabrication of sustainable and flexible electronics [93]. In this case, aleuritic acid was mixed with zein, a corn-derived protein, and graphene, sprayed onto cellulose substrates and hot-pressed. Resultant composites with low electrical resistance and exceptional mechanical and fatigue resilience were fabricated. These plant-based materials were used for different high-performance electronic applications such as terahertz electromagnetic shielding, flexible gigahertz antenna construction, and flexible solar cell electrode.

The second strategy to fabricate plant cuticle-like composites consisted of the blending of sodium alginate, the fatty acid fraction of tomato pomace, and beeswax in green solvents (water and ethanol) and a subsequent thermal treatment (150 °C, eight hours) to polymerize unsaturated and polyhydroxylated fatty acids [95]. Figure 6.17b displays photographs of the composites (labeled as AT followed by the percentage of the lipid fraction of tomato pomace) before and after (labeled as p) polymerization. While pure alginate (AT-0) was completely transparent, AT-25 and AT-50 films exhibited an orange and opaque color due to the presence of tomato pomace monomers in the structure. After thermal curing, colors became darker because of the polymerization. The analysis of chemical structure by solid-state NMR and FTIR spectroscopy showed that the thermal process induced the condensation of the fatty acids and that the resultant polyester interacted with alginate and beeswax by H-bonds and van der Waals forces, respectively. The mechanical properties depended significantly on the tomato pomace monomers content and the water loss produced by the thermal treatment. W-CA, uptake, and barrier properties were improved with the proportion of tomato pomace monomers and the thermal treatment because of the intrinsic hydrophobicity of unsaturated and polyhydroxylated fatty acids as well as the lower presence of polar hydroxyl and carboxyl functional groups after the melt polycondensation. Finally, the addition of beeswax reinforced the structure increasing mechanical properties and improved the wettability.

6.4 Concluding Remarks

Cutin-inspired fatty polyhydroxyesters are sustainable materials from renewable resources (for example, inexpensive and underutilized tomato pomace by-product) that can be processed by standard industrial techniques. They can be synthesized by self-polycondensation in molten state directly in air and with no catalyst in an industrially reasonable range of reaction time and temperature. This esterification is particularly effective when the monomer bears a primary hydroxyl. The presence of oxygen in the reaction atmosphere gives rise to a series of oxidized species that concentrates on the near surface regions of the air-exposed side of films, particularly when the fatty polyhydroxyacid contains a vicinal diol moiety. Such altered layer is few microns thick and its development triggers the amorphization of the polyesters. On the other hand, the local oxidation has shown to be a quite effective barrier to UV radiation. The synthesis in air produces insoluble and infusible polyesters similar to cutin with enhanced UV–vis blocking capacity while maintaining reasonable water barrier properties. Both wettability and permeability can be improved by reducing the excess of free hydroxyl groups by the introduction of reactive molecules. Alternatively, the fabrication of plant cuticle-like composites by the blending of cutin-inspired polymers with waxes and polysaccharides is an interesting strategy to improve mechanical properties and hydrophobicity.

A realistic forecast is that cutin-inspired polymers can replace petroleum-based plastics and other bio-based polymers in some specific applications (mainly related to food packaging) in the coming future. For this reason, in our opinion, it would be reasonable to reach a greater commitment regarding physical performances and productivity levels of cutin monomers. This fits very well with different Sustainable Development Goals of the United Nations in terms of carbon footprint, sustainable food and agriculture, and circular economy. Maybe, the adventurous toys of Toy Story were right respect to petroleum-based plastics had just expressed the current paradigm about linear economy ("take, make, and dispose") with their own words (see initial quote). Until a shift occurs, we hope these cutin-inspired polymers can be produced in a large quantity in green and safe ways for a more sustainable world.

Acknowledgments

Dr. Heredia-Guerrero acknowledges the funding by the Spanish "Ministerio de Ciencia, Innovación y Universidades," project number RTI2018-096896-J-I00.

References

1 Association of Plastics Manifacturors (2018). Plastics -the facts 2018. In: *Plastics Europe*. https://doi.org/10.1016/j.marpolbul.2013.01.015.
2 Andrady, A.L. (2015). *Plastics and Environmental Sustainability*. https://doi.org/10.1002/9781119009405.
3 Perugini, F., Mastellone, M.L., and Arena, U. (2005). A life cycle assessment of mechanical and feedstock recycling options for management of plastic packaging wastes. *Environmental Progress* https://doi.org/10.1002/ep.10078.
4 Geyer, R., Jambeck, J.R., and Law, K.L. (2017). Production, use, and fate of all plastics ever made. *Science Advances* https://doi.org/10.1126/sciadv.1700782.
5 Ruiz-Orejón, L.F., Sardá, R., and Ramis-Pujol, J. (2016). Floating plastic debris in the Central and Western Mediterranean Sea. *Marine Environmental Research* https://doi.org/10.1016/j.marenvres.2016.08.001.
6 Yamamoto, T. and Yasuhara, A. (1999). Quantities of bisphenol A leached from plastic waste samples. *Chemosphere* https://doi.org/10.1016/S0045-6535(98)00464-0.
7 Lithner, D., Larsson, A., and Dave, G. (2011). Environmental and health hazard ranking and assessment of plastic polymers based on chemical composition. *Science of the Total Environment* https://doi.org/10.1016/j.scitotenv.2011.04.038.
8 Rochman, C.M., Browne, M.A., Halpern, B.S. et al. (2013). Policy: classify plastic waste as hazardous. *Nature* https://doi.org/10.1038/494169a.
9 Teuten, E.L., Saquing, J.M., Knappe, D.R.U. et al. (2009). Transport and release of chemicals from plastics to the environment and to wildlife. *Philosophical Transactions of the Royal Society, B: Biological Sciences* https://doi.org/10.1098/rstb.2008.0284.
10 Eriksson, O. and Finnveden, G. (2009). Plastic waste as a fuel – CO_2-neutral or not? *Energy and Environmental Science* https://doi.org/10.1039/b908135f.
11 Wright, S.L., Thompson, R.C., and Galloway, T.S. (2013). The physical impacts of microplastics on marine organisms: a review. *Environmental Pollution* 178: 483–492.

12 Rios, L.M., Moore, C., and Jones, P.R. (2007). Persistent organic pollutants carried by synthetic polymers in the ocean environment. *Marine Pollution Bulletin* https://doi.org/10.1016/j.marpolbul.2007.03.022.

13 Ryan, P.G., Moore, C.J., Van Franeker, J.A., and Moloney, C.L. (2009). Monitoring the abundance of plastic debris in the marine environment. *Philosophical Transactions of the Royal Society, B: Biological Sciences* https://doi.org/10.1098/rstb.2008.0207.

14 Barnes, D.K.A., Galgani, F., Thompson, R.C., and Barlaz, M. (2009). Accumulation and fragmentation of plastic debris in global environments. *Philosophical Transactions of the Royal Society, B: Biological Sciences* https://doi.org/10.1098/rstb.2008.0205.

15 Law, K.L., Morét-Ferguson, S., Maximenko, N.A. et al. (2010). Plastic accumulation in the North Atlantic subtropical gyre. *Science* https://doi.org/10.1126/science.1192321.

16 Morritt, D., Stefanoudis, P.V., Pearce, D. et al. (2014). Plastic in the Thames: a river runs through it. *Marine Pollution Bulletin* https://doi.org/10.1016/j.marpolbul.2013.10.035.

17 Gandini, A. (2008). Polymers from renewable resources: a challenge for the future of macromolecular materials. *Macromolecules* https://doi.org/10.1021/ma801735u.

18 Shen, L., Worrell, E., and Patel, M. (2010). Present and future development in plastics from biomass. *Biofuels, Bioproducts and Biorefining* https://doi.org/10.1002/bbb.189.

19 Cui, S., Yao, B., Sun, X. et al. (2016). Reducing the content of carrier polymer in pectin nanofibers by electrospinning at low loading followed with selective washing. *Materials Science and Engineering C* https://doi.org/10.1016/j.msec.2015.10.086.

20 Hardy, J.G., Römer, L.M., and Scheibel, T.R. (2008). Polymeric materials based on silk proteins. *Polymer* https://doi.org/10.1016/j.polymer.2008.08.006.

21 Shalwan, A. and Yousif, B.F. (2013). In state of art: mechanical and tribological behaviour of polymeric composites based on natural fibres. *Materials and Design* https://doi.org/10.1016/j.matdes.2012.07.014.

22 Thakur, V.K., Thakur, M.K., Raghavan, P., and Kessler, M.R. (2014). Progress in green polymer composites from lignin for multifunctional applications: a review. *ACS Sustainable Chemistry & Engineering* https://doi.org/10.1021/sc500087z.

23 Verlinden, R.A.J., Hill, D.J., Kenward, M.A. et al. (2007). Bacterial synthesis of biodegradable polyhydroxyalkanoates. *Journal of Applied Microbiology* https://doi.org/10.1111/j.1365-2672.2007.03335.x.

24 Vilela, C., Sousa, A.F., Fonseca, A.C. et al. (2014). The quest for sustainable polyesters-insights into the future. *Polymer Chemistry* https://doi.org/10.1039/c3py01213a.

25 Klemm, D., Heublein, B., Fink, H.P., and Bohn, A. (2005). Cellulose: fascinating biopolymer and sustainable raw material. *Angewandte Chemie, International Edition* https://doi.org/10.1002/anie.200460587.

26 Haneef, M., Ceseracciu, L., Canale, C. et al. (2017). Advanced materials from fungal mycelium: fabrication and tuning of physical properties. *Scientific Reports* https://doi.org/10.1038/srep41292.
27 Klemm, D., Schumann, D., Udhardt, U., and Marsch, S. (2001). Bacterial synthesized cellulose - artificial blood vessels for microsurgery. *Progress in Polymer Science (Oxford)* https://doi.org/10.1016/S0079-6700(01)00021-1.
28 Michel, G., Tonon, T., Scornet, D. et al. (2010). The cell wall polysaccharide metabolism of the brown alga *Ectocarpus siliculosus*. Insights into the evolution of extracellular matrix polysaccharides in Eukaryotes. *New Phytologist* https://doi.org/10.1111/j.1469-8137.2010.03374.x.
29 Angles, M.N. and Dufresne, A. (2001). Plasticized starch/tunicin whiskers nanocomposite materials. 2: Mechanical behavior. *Macromolecules* https://doi.org/10.1021/ma001555h.
30 Huber, T., Müssig, J., Curnow, O. et al. (2012). A critical review of all-cellulose composites. *Journal of Materials Science* https://doi.org/10.1007/s10853-011-5774-3.
31 Siró, I. and Plackett, D. (2010). Microfibrillated cellulose and new nanocomposite materials: a review. *Cellulose* https://doi.org/10.1007/s10570-010-9405-y.
32 Bayer, I.S., Fragouli, D., Attanasio, A. et al. (2011). Water-repellent cellulose fiber networks with multifunctional properties. *ACS Applied Materials and Interfaces* https://doi.org/10.1021/am200891f.
33 Zobel, H.F. (1988). Molecules to granules: a comprehensive starch review. *Starch - Stärke* https://doi.org/10.1002/star.19880400203.
34 Hoover, R. (2001). Composition, molecular structure, and physicochemical properties of tuber and root starches: a review. *Carbohydrate Polymers* https://doi.org/10.1016/S0144-8617(00)00260-5.
35 Ravi Kumar, M.N.V. (2000). A review of chitin and chitosan applications. *Reactive and Functional Polymers* https://doi.org/10.1016/S1381-5148(00)00038-9.
36 Rosa, M.F., Medeiros, E.S., Malmonge, J.A. et al. (2010). Cellulose nanowhiskers from coconut husk fibers: effect of preparation conditions on their thermal and morphological behavior. *Carbohydrate Polymers* https://doi.org/10.1016/j.carbpol.2010.01.059.
37 Bunzel, M., Schüßler, A., and Tchetseubu Saha, G. (2011). Chemical characterization of Klason lignin preparations from plant-based foods. *Journal of Agricultural and Food Chemistry* https://doi.org/10.1021/jf2031378.
38 Vanholme, R., Demedts, B., Morreel, K. et al. (2010). Lignin biosynthesis and structure. *Plant Physiology* https://doi.org/10.1104/pp.\ignorespaces110.155119.
39 Albert, L.L. and Lehninger, A.L. (2005). *Lehninger Principles of Biochemistry* (eds. D.L. Nelson and M.M. Cox). Principles of Biochemistry.
40 Heredia-Guerrero, J.A., Benítez, J.J., Cataldi, P. et al. (2017). All-natural sustainable packaging materials inspired by plant cuticles. *Advanced Sustainable Systems* https://doi.org/10.1002/adsu.201600024.

41 Mülhaupt, R. (2013). Green polymer chemistry and bio-based plastics: dreams and reality. *Macromolecular Chemistry and Physics* https://doi.org/10.1002/macp.201200439.

42 Heredia, A. (2003). Biophysical and biochemical characteristics of cutin, a plant barrier biopolymer. *Biochimica et Biophysica Acta, General Subjects* https://doi.org/10.1016/S0304-4165(02)00510-X.

43 Nawrath, C. (2006). Unraveling the complex network of cuticular structure and function. *Current Opinion in Plant Biology* https://doi.org/10.1016/j.pbi.2006.03.001.

44 Chatterjee, S., Matas, A.J., Isaacson, T. et al. (2016). Solid-state ^{13}C NMR delineates the architectural design of biopolymers in native and genetically altered tomato fruit cuticles. *Biomacromolecules* https://doi.org/10.1021/acs.biomac.5b01321.

45 Domínguez, E., Heredia-Guerrero, J.A., and Heredia, A. (2015). Plant cutin genesis: unanswered questions. *Trends in Plant Science* https://doi.org/10.1016/j.tplants.2015.05.009.

46 España, L., Heredia-Guerrero, J.A., Segado, P. et al. (2014). Biomechanical properties of the tomato (*Solanum lycopersicum*) fruit cuticle during development are modulated by changes in the relative amounts of its components. *New Phytologist* https://doi.org/10.1111/nph.12727.

47 Heredia-Guerrero, J.A., Benítez, J.J., Domínguez, E. et al. (2014). Infrared and Raman spectroscopic features of plant cuticles: a review. *Frontiers in Plant Science* https://doi.org/10.3389/fpls.2014.00305.

48 Luque, P., Bruque, S., and Heredia, A. (1995). Water permeability of isolated cuticular membranes: a structural analysis. *Archives of Biochemistry and Biophysics* https://doi.org/10.1006/abbi.1995.1183.

49 Pollard, M., Beisson, F., Li, Y., and Ohlrogge, J.B. (2008). Building lipid barriers: biosynthesis of cutin and suberin. *Trends in Plant Science* https://doi.org/10.1016/j.tplants.2008.03.003.

50 Villena, J.F., Domínguez, E., Stewart, D., and Heredia, A. (1999). Characterization and biosynthesis of non-degradable polymers in plant cuticles. *Planta* https://doi.org/10.1007/s004250050548.

51 Fernández, V., Guzmán-Delgado, P., Graça, J. et al. (2016). Cuticle structure in relation to chemical composition: re-assessing the prevailing model. *Frontiers in Plant Science* https://doi.org/10.3389/fpls.2016.00427.

52 Heredia-Guerrero, J.A., Guzman-Puyol, S., Benítez, J.J. et al. (2018). Plant cuticle under global change: biophysical implications. *Global Change Biology* https://doi.org/10.1111/gcb.14276.

53 Benítez, J.J., Castillo, P.M., del Río, J.C. et al. (2018). Valorization of tomato processing by-products: fatty acid extraction and production of bio-based materials. *Materials* https://doi.org/10.3390/ma11112211.

54 United Nations Environment program (2018). Exploring the potential for adopting alternative materials to reduce marine plastic litter. United Nations Environment Programme (2017). https://doi.org/10.13140/RG.2.2.26075.87849.

55 Zhang, B.X. and Uyama, H. (2016). Biomimic plant cuticle from hyperbranched poly(ricinoleic acid) and cellulose film. *ACS Sustainable Chemistry & Engineering* https://doi.org/10.1021/acssuschemeng.5b01392.
56 Dominguez, E., Heredia-Guerrero, J.A., and Heredia, A. (2011). The biophysical design of plant cuticles: an overview. *New Phytologist* https://doi.org/10.1111/j.1469-8137.2010.03553.x.
57 Graça, J., Schreiber, L., Rodrigues, J., and Pereira, H. (2002). Glycerol and glyceryl esters of ω-hydroxyacids in cutins. *Phytochemistry* https://doi.org/10.1016/S0031-9422(02)00212-1.
58 Parsons, E.P., Popopvsky, S., Lohrey, G.T. et al. (2013). Fruit cuticle lipid composition and water loss in a diverse collection of pepper (*Capsicum*). *Physiologia Plantarum* https://doi.org/10.1111/ppl.12035.
59 Philippe, G., Gaillard, C., Petit, J. et al. (2016). Ester cross-link profiling of the cutin polymer of wild-type and cutin synthase tomato mutants highlights different mechanisms of polymerization. *Plant Physiology* https://doi.org/10.1104/pp.\ignorespaces15.01620.
60 Benítez, J.J., Heredia-Guerrero, J.A., De Vargas-Parody, M.I. et al. (2016). Biodegradable polyester films from renewable aleuritic acid: surface modifications induced by melt-polycondensation in air. *Journal of Physics D: Applied Physics* https://doi.org/10.1088/0022-3727/49/17/175601.
61 Pal, G., Jaiswal, A.K., Bhattacharya, A., and Yogi, R.K. (2011). *Lac Statistics at a Glance*. Indian Institute of Natural.
62 Domínguez, E., Cuartero, J., and Heredia, A. (2011). An overview on plant cuticle biomechanics. *Plant Science* https://doi.org/10.1016/j.plantsci.2011.04.016.
63 López-Casado, G., Matas, A.J., Domínguez, E. et al. (2007). Biomechanics of isolated tomato (*Solanum lycopersicum* L.) fruit cuticles: the role of the cutin matrix and polysaccharides. *Journal of Experimental Botany* https://doi.org/10.1093/jxb/erm233.
64 Mark, J.E. (2007). *Physical Properties of Polymers Handbook*, vol. 1076. Springer.
65 Matas, A.J., Cuartero, J., and Heredia, A. (2004). Phase transitions in the biopolyester cutin isolated from tomato fruit cuticles. *Thermochimica Acta* https://doi.org/10.1016/S0040-6031(03)00357-5.
66 Otake, Y., Kobayashi, T., Asabe, H. et al. (1995). Biodegradation of low-density polyethylene, polystyrene, polyvinyl chloride, and urea formaldehyde resin buried under soil for over 32 years. *Journal of Applied Polymer Science* https://doi.org/10.1002/app.1995.070561309.
67 De Vries, H., Bredemeijer, G., and Heinen, W. (1967). The decay of cutin and cuticular components by soil microorganisms in their natural environment. *Acta Botanica Neerlandica* https://doi.org/10.1111/j.1438-8677.1967.tb00040.x.
68 Katayama, A. and Kuwatsuka, S. (1991). Effect of pesticides on cellulose degradation in soil under upland and flooded conditions. *Soil Science and Plant Nutrition* https://doi.org/10.1080/00380768.1991.10415003.
69 Lu, H., Madbouly, S.A., Schrader, J.A. et al. (2014). Biodegradation behavior of poly(lactic acid) (PLA)/distiller's dried grains with solubles (DDGS)

composites. *ACS Sustainable Chemistry & Engineering* 2 (12): 2699–2706. https://doi.org/10.1021/sc500440q.

70 Yew, S.P., Tang, H.Y., and Sudesh, K. (2006). Photocatalytic activity and biodegradation of polyhydroxybutyrate films containing titanium dioxide. *Polymer Degradation and Stability* https://doi.org/10.1016/j.polymdegradstab.2005.11.011.

71 Rock, C., Yang, W., Goodrich-Schneider, R., and Feng, H. (2012). Conventional and alternative methods for tomato peeling. *Food Engineering Reviews* https://doi.org/10.1007/s12393-011-9047-3.

72 Al-Wandawi, H., Abdul-Rahman, M., and Al-Shaikhly, K. (1985). Tomato processing wastes as essential raw materials source. *Journal of Agricultural and Food Chemistry* https://doi.org/10.1021/jf00065a009.

73 Liadakis, G.N., Tzia, C., Oreopoulou, V., and Thomopoulos, C.D. (1995). Protein isolation from tomato seed meal, extraction optimization. *Journal of Food Science* https://doi.org/10.1111/j.1365-2621.1995.tb09807.x.

74 Cifarelli, A., Cigognini, I., Bolzoni, L., and Montanari, A. (2014). Cutin isolated from tomato processing by-products : extraction methods and characterization. Uest.Ntua.Gr.

75 Taofiq, O., González-Paramás, A.M., Barreiro, M.F. et al. (2017). Hydroxycinnamic acids and their derivatives: cosmeceutical significance, challenges and future perspectives, a review. *Molecules* https://doi.org/10.3390/molecules22020281.

76 Holloway, P.J. and Baker, E.A. (1968). Isolation of plant cuticles with zinc chloride-hydrochloric acid solution. *Plant Physiology* https://doi.org/10.1104/pp.\ignorespaces43.11.1878.

77 Huelin, F.E. and Gallop, R.A. (1951). Studies in the natural coating of apples I. Preparation and properties of fractions. *Australian Journal of Biological Sciences* 4 (4): 526–532.

78 Orgell, W.H. (1955). The isolation of plant cuticle with pectic enzymes. *Plant Physiology* https://doi.org/10.1104/pp.\ignorespaces30.1.78.

79 Pacchiano, R.A., Sohn, W., Chlanda, V.L. et al. (1993). Isolation and spectral characterization of plant-cuticle polyesters. *Journal of Agricultural and Food Chemistry* https://doi.org/10.1021/jf00025a017.

80 Kolattukudy, P.E. (2001). Polyesters in higher plants. In: *Biopolyesters. Advances in Biochemical Engineering/Biotechnology*, vol. 71 (eds. W. Babel and A. Steinbüchel). Berlin, Heidelberg: Springer.

81 Phu, P.Q., Tri, T.M., Hien, L.P.T., and Ha, L.T.T. (2017). Effect of different parameters on extrication performance and tensile properties of cutin membrane from Basella alba cuticle. *Proceedings of the 2017 International Symposium on Materials Science and Technology*, pp. 31–39.

82 Benitez, J.J., Heredia-Guerrero, J.A., Bayer, I.S., and Athanassiou, A. (2016). A polyester from tomato fruit processing residues obtained by non-catalyzed melt-polycondensation in air. ES201630194, February 2016.

83 Gómez-Patiño, M.B., Cassani, J., Jaramillo-Flores, M.E. et al. (2013). Oligomerization of 10,16-dihydroxyhexadecanoic acid and methyl 10,16-dihydroxyhexadecanoate catalyzed by lipases. *Molecules* https://doi.org/10.3390/molecules18089317.

84 Gómez-Patiño, M.B., Gutiérrez-Salgado, D.Y., García-Hernández, E. et al. (2015). Polymerization of 10,16-dihydroxyhexadecanoic acid, main monomer of tomato cuticle, using the lewis acidic ionic liquid choline chloride·$2ZnCl_2$. *Frontiers in Materials* https://doi.org/10.3389/fmats.2015.00067.

85 Guzman-Puyol, S., Benítez, J.J., Domínguez, E. et al. (2015). Pectin-lipid self-assembly: influence on the formation of polyhydroxy fatty acids nanoparticles. *PLoS One* https://doi.org/10.1371/journal.pone.0124639.

86 Heredia-Guerrero, J.A., San-Miguel, M.A., Sansom, M.S.P. et al. (2009). Chemical reactions in 2D: self-assembly and self-esterification of 9(10), 16-dihydroxypalmitic acid on mica surface. *Langmuir* https://doi.org/10.1021/la9001412.

87 Heredia-Guerrero, J.A., Benítez, J.J., and Heredia, A. (2008). Self-assembled polyhydroxy fatty acids vesicles: a mechanism for plant cutin synthesis. *BioEssays* https://doi.org/10.1002/bies.20716.

88 Heredia-Guerrero, J.A., Domínguez, E., Luna, M. et al. (2010). Structural characterization of polyhydroxy fatty acid nanoparticles related to plant lipid biopolyesters. *Chemistry and Physics of Lipids* https://doi.org/10.1016/j.chemphyslip.2010.01.006.

89 San-Miguel, M.A., Oviedo, J., Heredia-Guerrero, J.A. et al. (2014). Biomimetic polymers of plant cutin: an approach from molecular modeling. *Journal of Molecular Modeling* https://doi.org/10.1007/s00894-014-2329-y.

90 Benítez, J.J., Heredia-Guerrero, J.A., Cruz-Carrillo, M.A. et al. (2017). Insolubilization and thermal stabilization of a long-chain polyester by noncatalyzed melt-polycondensation synthesis in air. *Journal of Applied Polymer Science* https://doi.org/10.1002/app.44350.

91 Benítez, J.J., Heredia-Guerrero, J.A., Guzmán-Puyol, S. et al. (2015). Long-chain polyhydroxyesters from natural occurring aleuritic acid as potential material for food packaging. *Soft Materials* https://doi.org/10.1080/1539445X.2014.993476.

92 Benítez, J.J., Heredia-Guerrero, J.A., Guzmán-Puyol, S. et al. (2015). Polyester films obtained by noncatalyzed melt-condensation polymerization of aleuritic (9,10,16-trihydroxyhexadecanoic) acid in air. *Journal of Applied Polymer Science* https://doi.org/10.1002/app.41328.

93 Cataldi, P., Heredia-Guerrero, J.A., Guzman-Puyol, S. et al. (2018). Sustainable electronics: sustainable electronics based on crop plant extracts and graphene: a “bioadvantaged” approach (Advanced Sustainable Systems 11/2018). *Advanced Sustainable Systems* https://doi.org/10.1002/adsu.201870042.

94 Heredia-Guerrero, J.A., Caputo, G., Guzman-Puyol, S. et al. (2019). Sustainable polycondensation of multifunctional fatty acids from tomato pomace agro-waste catalyzed by tin(II) 2-ethylhexanoate. *Materials Today Sustainability* https://doi.org/10.1016/j.mtsust.2018.12.001.

95 Tedeschi, G., Benitez, J.J., Ceseracciu, L. et al. (2018). Sustainable fabrication of plant cuticle-like packaging films from tomato pomace agro-waste,

Beeswax, and Alginate. *ACS Sustainable Chemistry & Engineering* https://doi.org/10.1021/acssuschemeng.8b03450.

96 Benítez, J.J., Heredia-Guerrero, J.A., Guzmán-Puyol, S. et al. (2015). Polyhydroxyester films obtained by non-catalyzed melt-polycondensation of natural occurring fatty polyhydroxyacids. *Frontiers in Materials* https://doi.org/10.3389/fmats.2015.00059.

97 Sutton, G.J. and Tighe, B.J. (1973). Poly- alpha-ester degradation studies - 1. Introduction: design and construction of equipment. *Journal of Polymer Science Part A: Polymer Chemistry* 11: 1069–1077.

98 Benítez, J.J., Guzman-Puyol, S., Cruz-Carrillo, M.A. et al. (2019). Insoluble and thermostable polyhydroxyesters from a renewable natural occurring polyhydroxylated fatty acid. *Frontiers in Chemistry* 7: 643. https://doi.org/10.3389/fchem.2019.00643.

99 Robertson, G.L. (2013). *Food Packaging: Principles and Practice*, 3e. Taylor & Francis Group.

100 Arvanitoyannis, I.S. and Kotsanopoulos, K.V. (2014). Migration phenomenon in food packaging. food-package interactions, mechanisms, types of migrants, testing and relative legislation-a review. *Food and Bioprocess Technology* https://doi.org/10.1007/s11947-013-1106-8.

101 Barnes, K., Sinclair, R., and Watson, D. (2006). *Chemical Migration and Food Contact Materials*. https://doi.org/10.1533/9781845692094.

102 Brody, A.L., Bugusu, B., Han, J.H. et al. (2008). Innovative food packaging solutions. *Journal of Food Science* https://doi.org/10.1111/j.1750-3841.2008.00933.x.

103 Montanari, A., Pezzani, A., Cassarà, A. et al. (1996). Quality of organic coatings for food cans: evaluation techniques and prospects of improvement. *Progress in Organic Coatings* https://doi.org/10.1016/S0300-9440(96)00625-X.

104 Prokop, Z., Hanková, L., and Jeřábek, K. (2004). Bisphenol A synthesis - modeling of industrial reactor and catalyst deactivation. *Reactive and Functional Polymers* https://doi.org/10.1016/j.reactfunctpolym.2004.02.013.

105 Arnich, N., Canivenc-Lavier, M.C., Kolf-Clauw, M. et al. (2011). Conclusions of the French Food Safety Agency on the toxicity of bisphenol A. *International Journal of Hygiene and Environmental Health* https://doi.org/10.1016/j.ijheh.2010.12.002.

106 Shelby, M.D. (2008). NTP-CERHR monograph on the potential human reproductive and developmental effects of bisphenol A. *NTP CERHR MON*.

107 Carwile, J.L., Ye, X., Zhou, X. et al. (2011). Canned soup consumption and urinary bisphenol A: a randomized crossover trial. *JAMA : The Journal of the American Medical Association* https://doi.org/10.1001/jama.2011.1721.

108 Lakind, J.S. and Naiman, D.Q. (2011). Daily intake of bisphenol A and potential sources of exposure: 2005–2006 National Health and Nutrition Examination Survey. *Journal of Exposure Science & Environmental Epidemiology* https://doi.org/10.1038/jes.2010.9.

109 Rudel, R.A., Gray, J.M., Engel, C.L. et al. (2011). Food packaging and bisphenol A and bis(2-ethyhexyl) phthalate exposure: findings from a dietary

intervention. *Environmental Health Perspectives* https://doi.org/10.1289/ehp.1003170.

110 Vandenberg, L.N., Hauser, R., Marcus, M. et al. (2007). Human exposure to bisphenol A (BPA). *Reproductive Toxicology* https://doi.org/10.1016/j.reprotox.2007.07.010.

111 Noonan, G.O., Ackerman, L.K., and Begley, T.H. (2011). Concentration of bisphenol a in highly consumed canned foods on the U.S. market. *Journal of Agricultural and Food Chemistry* https://doi.org/10.1021/jf201076f.

112 LaKind, J.S. (2013). Can coatings for foods and beverages: Issues and options. *International Journal of Technology, Policy and Management* https://doi.org/10.1504/IJTPM.2013.050999.

113 Bensaude-Vincent, B. (2007). Reconfiguring nature through syntheses : from plastics to biomimetics. In: *The Artificial and the Natural: An Evolving Polarity* (eds. B. Bensaude-Vincent and W.R. Newman), 293–312. Cambridge, Massachusetts, London, England: The MIT Press.

114 Altiokka, M.R. and Çitak, A. (2003). Kinetics study of esterification of acetic acid with isobutanol in the presence of amberlite catalyst. *Applied Catalysis A: General* https://doi.org/10.1016/S0926-860X(02)00381-2.

115 Bikiaris, D.N. and Achilias, D.S. (2008). Synthesis of poly(alkylene succinate) biodegradable polyesters, Part II: Mathematical modelling of the polycondensation reaction. *Polymer* https://doi.org/10.1016/j.polymer.2008.06.026.

116 Tesser, R., Di Serio, M., Guida, M. et al. (2005). Kinetics of oleic acid esterification with methanol in the presence of triglycerides. *Industrial and Engineering Chemistry Research* https://doi.org/10.1021/ie050588o.

117 Trinh, H., Yusup, S., and Uemura, Y. (2018). Optimization and kinetic study of ultrasonic assisted esterification process from rubber seed oil. *Bioresource Technology* https://doi.org/10.1016/j.biortech.2017.09.075.

118 Chan, C.M., Vandi, L.J., Pratt, S. et al. (2018). Composites of wood and biodegradable thermoplastics: a review. *Polymer Reviews* https://doi.org/10.1080/15583724.2017.1380039.

119 Bayer, I.S., Guzman-Puyol, S., Heredia-Guerrero, J.A. et al. (2014). Direct transformation of edible vegetable waste into bioplastics. *Macromolecules* https://doi.org/10.1021/ma5008557.

120 Hyppölä, R., Husson, I., and Sundholm, F. (1996). Evaluation of physical properties of plasticized ethyl cellulose films cast from ethanol solution Part I. *International Journal of Pharmaceutics* https://doi.org/10.1016/0378-5173(96)04436-5.

121 Espinoza-Herrera, N., Pedroza-Islas, R., Martín-Martinez, E.S. et al. (2011). Thermal, mechanical and microstructures properties of cellulose derivatives films: a comparative study. *Food Biophysics* https://doi.org/10.1007/s11483-010-9181-0.

122 Wang, H., Tong, D., Wang, L. et al. (2017). A facile strategy for fabricating PCL/PEG block copolymer with excellent enzymatic degradation. *Polymer Degradation and Stability* https://doi.org/10.1016/j.polymdegradstab.2017.04.015.

123 Townsend, K.J., Busse, K., Kressler, J., and Scholz, C. (2005). Contact angle, WAXS, and SAXS analysis of poly(β-hydroxybutyrate) and poly(ethylene glycol) block copolymers obtained via Azotobacter vinelandii UWD. *Biotechnology Progress* https://doi.org/10.1021/bp050044h.

124 Guzman-Puyol, S., Ceseracciu, L., Heredia-Guerrero, J.A. et al. (2015). Effect of trifluoroacetic acid on the properties of polyvinyl alcohol and polyvinyl alcohol-cellulose composites. *Chemical Engineering Journal* https://doi.org/10.1016/j.cej.2015.04.092.

125 Oh, E. and Luner, P.E. (1999). Surface free energy of ethylcellulose films and the influence of plasticizers. *International Journal of Pharmaceutics* https://doi.org/10.1016/S0378-5173(99)00224-0.

126 Zhao, Q., Qian, J., An, Q. et al. (2009). Pervaporation dehydration of isopropanol using homogeneous polyelectrolyte complex membranes of poly(diallyldimethylammonium chloride)/sodium carboxymethyl cellulose. *Journal of Membrane Science* https://doi.org/10.1016/j.memsci.2008.12.031.

127 Kamide, K. and Saito, M. (1985). Thermal analysis of cellulose acetate solids with total degrees of substitution of 0.49, 1.75, 2.46, and 2.92. *Polymer Journal* https://doi.org/10.1295/polymj.17.919.

128 Su, J.F., Huang, Z., Yuan, X.Y. et al. (2010). Structure and properties of carboxymethyl cellulose/soy protein isolate blend edible films crosslinked by Maillard reactions. *Carbohydrate Polymers* https://doi.org/10.1016/j.carbpol.2009.07.035.

7

Zein in Food Packaging

Ilker S. Bayer

Smart Materials, Istituto Italiano di Tecnologia, Via Morego 30, Genova 16163, Italy

7.1 Introduction

Zein is a general name attributed to plant-based alcohol-soluble proteins from the endosperm tissue of corn that is produced as a by-product during bioethanol production in industry [1]. John Gorham coined its name for the first time in 1821 [2]. Chemical amino acid composition analysis of zein shows that it has high amounts of hydrophobic amino acid residues which make it soluble in a number of organic solvents or blends and render it hydrophobic. In fact, more specifically, amino acid composition analyses show large amounts of hydrophobic residues such as leucine, proline, alanine, and phenylalanine [3]. The four main types of zein (denoted as α, β, γ, and δ) are classified according to their solubility properties. Although generally found as disulfide-bridged dimers, the individual α-zeins have molecular masses of 19 and 22 kDa, the size difference being from an amino acid insertion in the C-terminal region of the 22 kDa form [4]. Earlier studies proved that zein proteins contain repeat sequences and possess a rod-like structure with high helical content, and can self-aggregate, form fibers, and associate with membranes. A cluster of α-helices corresponding to the zein homologous repeat units and contained within it, a distorted cylindrical surface capped by polar glutamine residues would allow hydrogen-bonding patterns and hydrophobic van der Waals interactions to maintain zein structural integrity as well as intermolecular packing [3]. As per molecular weight, zein can assume different molecular weights, as mentioned above, but zein with 19 kDa molecular weight known as Z19 is a commonly studied protein.

It has a very complicated molecular structure that requires careful crystallinity analysis as well as molecular simulations [4]. Figure 7.1 demonstrates the physical morphology of typical zein powder as well as its (Z19) complicated chemical structure as dispersed in menthol (solvent) coupling triple-superhelix segments with nitrogen-terminal segment. Solvent methanol molecules are indicated by small lines. Lutein molecules (naturally occurring carotenoids) are shown in space-filling representation in the center but also separately (Figure 7.1c).

Zein is a film-forming protein when solvent-casted from ethanol solutions [5–7] and can also be extruded [8–10]. Hence, it is a very actively pursued

Sustainable Food Packaging Technology, First Edition. Edited by Athanassia Athanassiou.

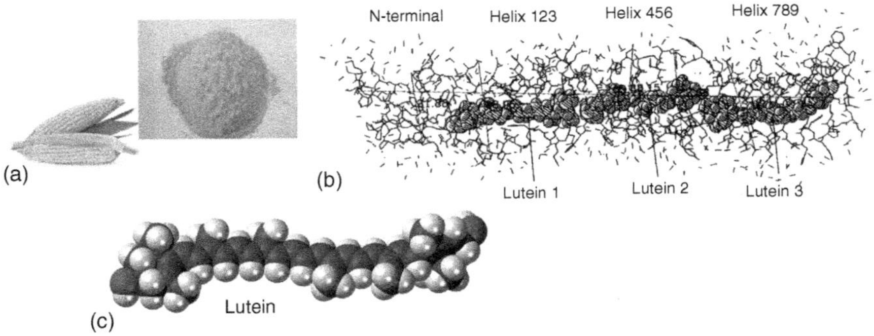

Figure 7.1 (a) Zein powder extracted from corn. (b) Complete structure of Z19 zein in methanol. Solvent methanol molecules are indicated as small lines. Lutein molecules are in the center space-filling representation. Source: Momany et al. [4]. © 2006, American Chemical Society. (c) Typical 3D structure of naturally occurring lutein molecule. Source: Modified from Momany et al. 2006 [4].

biopolymer for packaging [11]. Protein-based food packaging obtained from bio-based sources has advanced notably in recent years and become an alternative to conventional oil-derived polymer packaging because of its renewability and biodegradability. Compared to polysaccharides and lipids, plant-based proteins have a large amount of functional groups that can be reacted making it possible to meet the demand in green packaging materials. In film forming, α-zein's nine homologous repeating units are arranged in an antiparallel form by hydrogen bonds enabling it to easily form films through self-assembly [12]. Because of this (see Figure 7.2) and excellent vapor barrier property, zein has become a potential biodegradable material for food packaging and an attractive material for functional and smart sustainable materials [14]. As known by many, food packaging has been developed for the preservation of food quality after production until its end use. Today, synthetic polymers are extensively used since they offer versatile solutions for several needs such as softness, durability, transparency, barrier properties, recycling, etc. The key feature of protein-based biopolymers is their exceptional barrier to oxygen, comparable to that of ethylene vinyl alcohol (EVOH) or polyvinylidine chloride (PVDC) [15]. Although, corn zein (CZ) is a better barrier to moisture compared to other proteins and a very good barrier to oxygen, its use as standalone films in food packaging applications seems impossible due to poor mechanical properties. In addition to use of various plasticizers in order to increase flexibility of zein [16–18], many composite forming strategies were developed to utilize zein as an oxygen barrier material. These include lamination of zein with other proteins and lipids [19–22], blending with synthetic polymers or biopolymers [23–26], and coating polyolefin surfaces with corn zein [27]. Biopolymer coatings such as corn zein may be favorable for the environment, since identification and separation of layers for recycling of synthetic films are costly and such multilayer barrier packaging materials account for an important amount of plastic waste.

Since a majority of food packaging polymers and resins have been extruded, it is important to briefly review the status of extrusion of zein and its composites. This

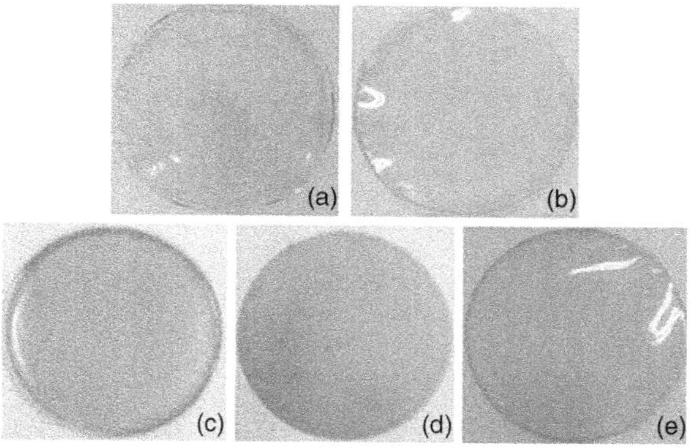

Figure 7.2 Biodegradable films produced from different proteins and their blends. (a) Hake protein isolate film, (b) wheat gluten film, (c) corn zein film, (d) zein/wheat gluten blend, (e) hake protein/wheat gluten blend film. Source: Nogueira and Martins [13]. © 2019, Springer Nature.

will be taken up in more detail in Section 7.4. In short, extrusion processing melts the polymer under high temperature and shear, forcing the polymer through a die, after which the extrudate is drawn and cooled to provide a fiber. The presence of degradation products (such as residual oxidants or residual monomers) within the extrudate has been shown to have a negative impact on a number of polymers. The use of extrusion techniques to process zein is fairly common [8, 9, 28]. It is generally extruded at around 90 °C. The effect of temperature on zein has been carried out using dynamic mechanical analysis (DMA) techniques at temperatures up to 166 °C [29]. It has been shown that extrusion can alter the properties of proteins [30]. Nonetheless, Figure 7.3 demonstrates that zein can be processed just like polyolefin polymers including blow molding [31]. Just like starch, zein is generally plasticized with agents such as polyethylene glycols before extrusion [32]. Extrusion forming of zein sheets and films with oleic acid through slit-die and blowing head were also demonstrated. Extruded zein sheets and films suffer from micro-voids left over by the evaporation of moisture. The number and size of the micro voids were affected by extrusion treatments and subsequent heat treatments [33].

Biodegradation of zein was studied in aerobic soil environment and compared with corn starch and dialdehyde starch for 180 days. The degradation process produced CO_2 equivalent to 63%, 64%, and 10% of theoretical carbon, respectively [34]. The biodegradable property of zein even *in vivo* was investigated thoroughly by some researches; *in vivo* the degradation was observed in rod-shaped rat implants within two weeks of surgical insertion [35].

The barrier properties of zein and other protein-, polysaccharide-, and lipid-based films have been widely investigated. The functionality of these materials can be enhanced by adding antimicrobial agents, antioxidants, nutrients, oxygen and carbon dioxide absorbers, enzymes, and other active compounds. For example, foods can spoil by surface growth of microorganisms, which can be controlled by edible coatings containing antimicrobial agents. Hence, in this chapter, we will review different processes and methods of forming zein films and their properties related to potential food packaging applications. Some examples, such as films made by solvent casting, extrusion, and blow molding, will be discussed in detail. We will present the state-of-the-art related to zein

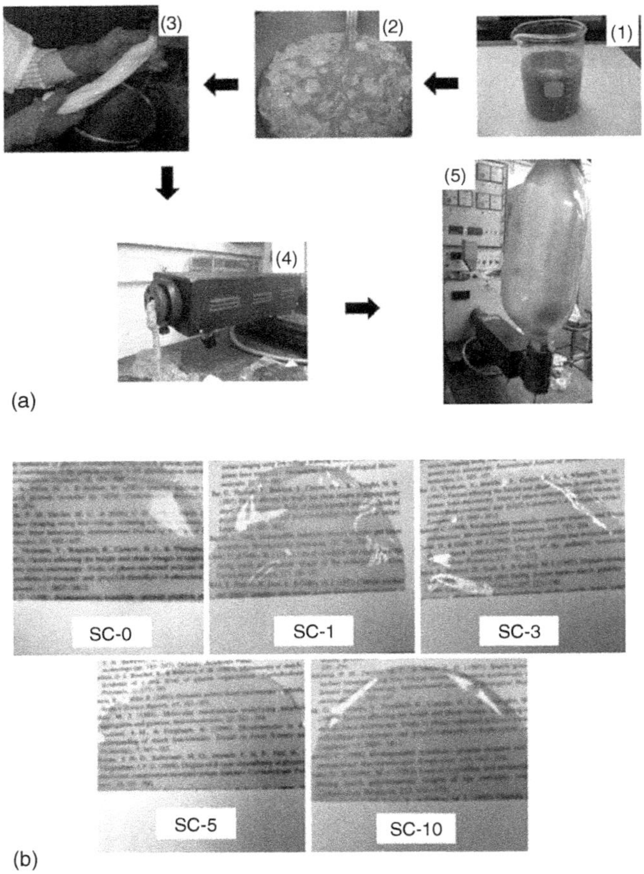

Figure 7.3 (a) The steps of the blown extrusion technique including (1) zein montmorillonite (MMT) solution, (2) precipitation, (3) resin formation, (4) cold extrusion, and (5) balloon formation. (b) Photographic pictures of zein-nanoclay (MMT) nanocomposite films prepared from solvent casting method (SC, number = MMT content [wt%]). Source: Luecha et al. [31]. © 2010, Springer Nature.

polymer blends as well as the effect of aforementioned functional additives on the performance of zein.

7.2 Solvent Cast Zein Films

Many different primary and binary solvents have been suggested and used for zein. The major work that identified solvents for zein was conducted by Evans and coworkers in the early 1940s [36–38]. Primary organic solvents for zein are a few such as hydroxyl compounds, amines, amides, or acids. The stability of zein in single solvents is, in general, much better than in binary solvents containing water.

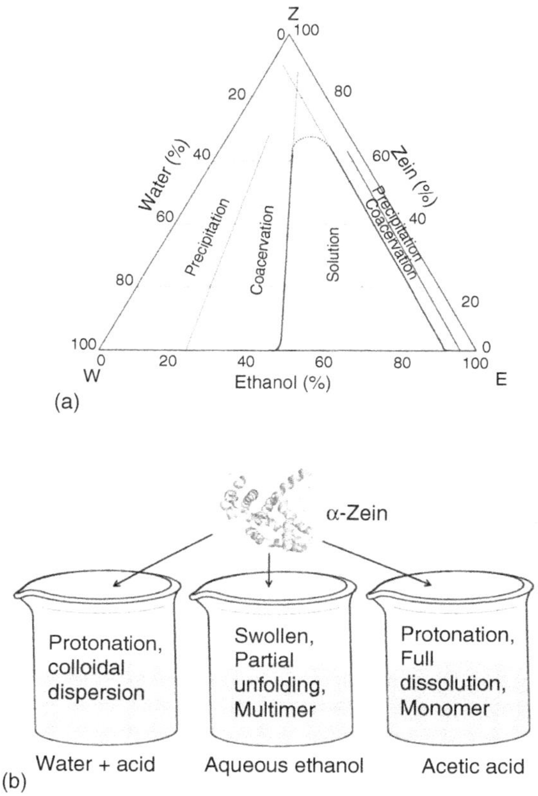

Figure 7.4 (a) Ternary phase diagram for the solubility of zein in ethanol and water. Source: Shukla and Cheryan [12]. © 2011, Elsevier. (b) Schematic diagram of α-zein in three different solvents: (left) water + acid; (middle) aqueous ethanol; (right) acetic acid. Source: Modified from Zhang et al. [39]. © 2015, Elsevier.

Evans and coworkers' early studies showed that 20 wt% solutions of zein in diethylene glycol, triethylene glycol, propylene glycol, and dipropylene glycol remained clear and fluid for two years when stored at 27 °C. They also showed that a similar concentration of zein in 85% aqueous ethyl alcohol stored at the same temperature gelled in about 30 days. The authors used the term "peptization" to indicate critical cloud points. Peptization (solution clouding upon cooling) describes the effect of temperature on the turbidity of protein solutions. In other words, the cloud point is the temperature at which the solution turns cloudy upon cooling. For a single substance to be a good solvent for zein, the molecule needs to have a proper balance between polar and nonpolar groups, for instance propylene glycol is a good solvent for zein but absolute propanol is not [38]. The solubility behavior of zein is shown in the form of a ternary phase diagram in Figure 7.4a. At constant temperature, the solubility of zein varies between 2 and 60 wt% (w/w), depending on the ethanol concentration. At lower (<40%) and higher (>90%) concentrations of ethanol, two liquid phases appear, containing zein, water, and ethanol. It corresponds to a transition state between complete solubilization and precipitation of zein. Data in Figure 7.4a indicates that it not possible to prepare solutions containing >65% zein without raising the temperature. Heating a zein solution to evaporate the ethanol gives an increasingly concentrated solution without discontinuity. The solution transforms from a thick transparent syrupy state to a glassy state in a thin coat, allowing zein to form glaze-like films with unique properties [12].

Solvents are particularly attractive for electrospinning zein and also for making drug-carrying nano or microparticles [39]. However, these two production processes are beyond the scope of this chapter. In a recent study [40], small-angle X-ray scattering (SAXS), static light scattering (SLS), and rheology were combined to investigate the structure and protein–solvent interaction of α-zein in both acetic acid and aqueous ethanol solutions. Zein was found to hold an elongated conformation (i.e. prolate ellipsoid) in all solutions, as revealed from SAXS data. The "aging effect" for zein in aqueous ethanol solutions was measured by the transition of Newtonian rheological profiles for fresh zein solutions to the non-Newtonian shear-thinning behavior for zein solutions after storage at room temperature for 24 hours. Shear-thinning behavior became more pronounced for zein solutions at higher concentrations. The SLS measurements indicated that acetic acid is a better solvent to dissolve zein than aqueous ethanol solution. This was shown to be caused by the protonation of the protein with the addition of acids. Ionic liquids have also been used to solubilize zein [41]. A systematic study measured solubility of zein gravimetrically from 30 to 60 °C in seven imidazolium-based ionic liquids and three conventional solvents, acetic acid, 70 vol.% aqueous ethanol, and 1-methylimidazole [41]. The study showed that at elevated temperatures the solubility of zein in six of the ionic liquids exceeded that of the current standard, 70 vol.% aqueous ethanol solution. The ionic liquids 1-methylimidazolium acetate and 1-ethyl-3-methylimidazolium dicyanamide dissolved 2.5 times more zein at 60 °C than 70 vol.% aqueous ethanol. This was related to the nature of some solvents that are able to dissolve more zein such that they have small molecular volumes, low polarities, and act as weak hydrogen bond acceptors.

Zein can also be gelled or cross-linked in solutions to modify its properties and resistance to acidic or basic media [42, 43]. For instance, it can be modified using polyethylene maleic anhydride (PEMA) in DMF solution to provide films with improved properties [43]. The films had improved resistance to dissolution by acetic acid with the amount of modified zein film that dissolved on exposure to AcOH ranging from 100% for control to 21% after treatment with 6% PEMA. The films also had increased tensile strength (32–42 MPa) with increased elongation (6–18%).

Zein has been shown to be dissolved in dimethyl sulfoxide (DMSO) and remain stable for at least two weeks [44]. Solution behavior and structure of zein in DMSO studied by rheology and SAXS indicated that zein adopted an elongated conformation in DMSO. Zein films could be casted by evaporation of DMSO at 70 °C. The final zein films contained approximately 15 wt% residual DMSO that could adsorb water from the environment that plasticized the films. Zein films equilibrated in a 76% RH environment showed an extremely large breaking/elongation strain (320.6%) along with a considerable break stress (1.9 MPa) [44]. This allowed zein films to be molded into different shapes and the shapes could be fixed in low humidity environment (RH < 52%), as shown in Figure 7.5.

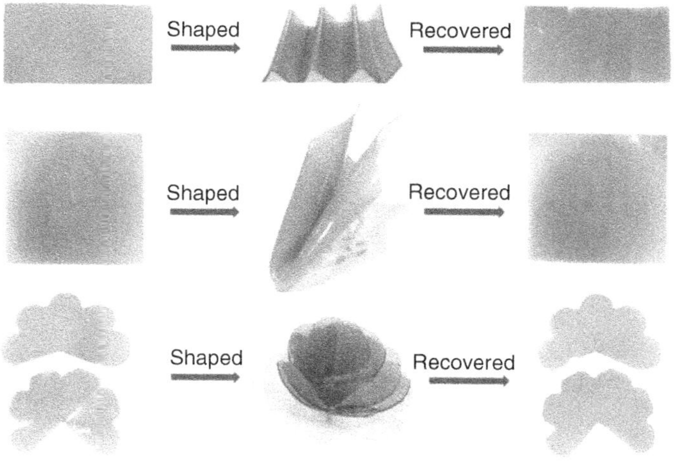

Figure 7.5 Different types of zein films solvent cast from DMSO and their shape recovery due to plasticizing effects of residual DMSO and ambient humidity. Source: Wei et al. [44]. © 2019, American Chemical Society.

To improve the mechanical properties of zein films, zein was reacted in glacial acetic acid with glutaraldehyde (GDA) from 0.5% to 16.0% by weight based on the weight of zein [45]. Above 4% GDA-modified zein generated a gel. That gel was insoluble in organic solvents known to solubilize zein. Films could be casted from the gelled solutions. Those films from 8% GDA-modified zein had a water vapor permeability of 5.9×10^{-8} g cm/Pa s m^2. GDA reaction with zein generated films with improved tensile strength, ductility, and stiffness when compared with respective zein control films that retained their integrity when subjected to either boiling water or extensive soaking, all important for food packaging [45]. The mechanical and barrier properties of zein films depend on many factors such as formulation, processing technique as well as the state of film-forming solution at a particular temperature and interaction with water. Understanding the phase behavior of zein films is very important to control their functionality at their end usage. A recent study [45] focused on this issue both at macro-scales and at nano-level polymer organizations. Films were investigated for their phase behavior as a function of water activity using differential scanning calorimetry (DSC) and the results were compared to the adhesive force measurements obtained from atomic force microscopy (AFM).

Finally, intelligent and biodegradable devices based on zein films were developed that could be integrated into food packaging materials [45]. Zein films can be used to reliably replicate micro-scale features (e.g. channels, posts, chambers), and strongly bond to glass slides. Zein films were bonded easily and quickly to different kinds of materials without requiring expensive equipment such as oxygen plasma generator. As such, zein microfluidic devices showed distinct permeability to small molecules enabling diffusive exchange between fluid flows and bulk zein microfluidic channels. Zein microfluidics with micro mixing channels (see Figure 7.6) has the potential to be used as fluid manipulator that can be coupled

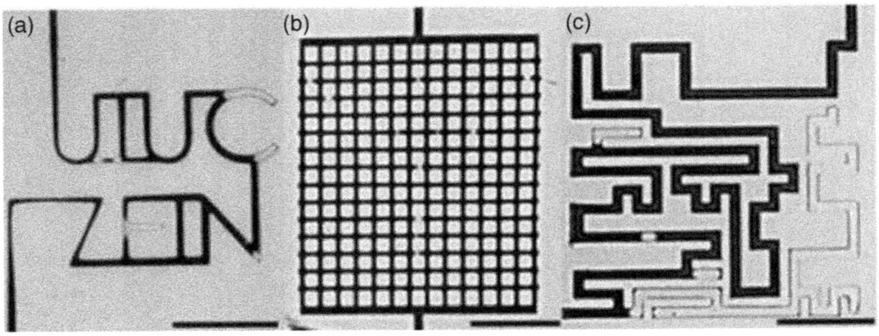

Figure 7.6 Photographs of zein-glass microfluidic devices with complex fluidic pathways. (a) Interconnected letters composed of continuous microfluidic channels, (b) a microfluidic network with channels and chambers, (c) a solved microfluidic maize maze with multiple false paths. Blue food dye was used for visual aid. All scale bars are 5 mm. Source: Luecha et al. [45]. © 2011, Royal Society of Chemistry.

with other analytical components for complex micro total analysis in packaging applications [45].

7.3 Chemical Characteristics of Solvent-Cast Zein Films

Fourier transform Raman (FT-Raman) spectroscopy was used to study solvent-cast zein films. Being free from fluorescence interference as well as free from pollutant chemical usage, FT-Raman was effectively used to characterize the molecular structure of zein films and for determining the thickness of zein coatings/films. Based on the FT-Raman results, the secondary structure of zein protein is mainly in α-helical form and is rarely stable during film formation. This is attributed to physical entrapment or hydrophobic interactions as well as disulfide bonding interactions within zein protein. The Table 7.1, adapted from [46], demonstrates typical wavenumbers of Raman bands and general assignments in zein protein structure.

Chemical and morphological state of zein films depend on the type of solvent used as well as the acidity of the solvents [47]. In a recent study, alternation of the solvents between ethanol and acetic acid to spin-cast zein films led to a distinct surface morphology and hydrophilicity [47]. Zein films prepared from ethanol solutions were more hydrophilic due to their higher percentage of polar functional groups on the surface. The different surface morphologies for zein films cast from different solvents originated from the difference in the interactions between zein molecules and Si wafers, resulting in different organization patterns of zein molecules during the film formation process. In the same work, the authors also oxidized zein films using UV/ozone treatment. Their results showed that UV/ozone treatment can efficiently control the surface hydrophilicity by converting some of the –C–H groups into –COO– groups (see Table 7.2 for details [47]). With controllable surface properties, many potentials of zein in packaging applications can be readily exploited.

Table 7.1 Typical wavenumbers of Raman bands and general assignments in protein structure.

Origin	Wavenumber	Assignment	Structural Information
Cystine	510	S–S stretch	Presence of gauche-gauche-gauche conformation; band broadening and/or shifts may indicate conformational heterogeneity of cystine residues
	525	S–S stretch	Gauche–gauche-trans conformation
	545	S–S stretch	Trans-gauche-trans conformation
	630–670	C–S stretch	Gauche conformation
	700–745	C–S stretch	Ttrans conformation
	2550–2580	S–H stretch	Presence of thiol of cysteine residues
Tyrosine	850/830	Fermi resonance between ring fundamental and overtone	State of phenolic OH group (exposed or buried, hydrogen-bond donor or acceptor)
Tryptophan	760, 880, 1360	Indole ring	Sharp intense band indicates buried residues; sensitive to environment polarity
Phenylalanine	1006	Ring breathe	Conformation insensitive; useful as an internal intensity standard
Aliphatic residues	1450, 1465	C–H bending	Microenvironment, polarity
	2800–3000	C–H stretching	Microenvironment, polarity
Amide I	1655 ± 5	Amide C=O stretch, N—H wag	α-helix
	1670 ± 3	Amide C=O stretch, N—H wag	Antiparallel β-sheet
	1665 ± 3	Amide C=O stretch, N—H wag	Disordered structure (solvated)
	1685	Amide C=O stretch, N—H wag	Disordered structure (non-hydrogen bonded)
Amide III	>1275	N–H in-plane bend, C—N stretch	α-helix
	1235 ± 5 (sharp)	N–H in-plane bend, C—N stretch	Antiparallel β-sheet
	1245 ± 4 (broad)	N–H in-plane bend, C—N stretch	Disordered structure
	1235	N–H in-plane bend, C—N stretch	Disordered structure (non-hydrogen bonded)

Source: Hsu et al. [46]. © 2015, American Chemical Society.

Table 7.2 Surface elemental compositions of zein films prepared from ethanol/water mixtures and acetic acid/water solutions obtained from XPS Survey Scans.

Zein film	Elemental composition				rms roughness (nm)	Water contact angle (°)
	C (%)	N (%)	O (%)	O/C		
A60	56.8	13.7	29.5	0.519	0.95 ± 0.07	73.7 ± 3.0
A100	63.1	13.4	23.5	0.372	1.51 ± 0.20	79.4 ± 2.7
E60	27.2	5.94	66.9	2.455	19.2 ± 0.4	40.9 ± 2.0
E95	57.4	14.4	28.2	0.491	32.1 ± 0.3	73.2 ± 0.7

Source: Shi et al. [47]. © 2009, American Chemical Society.

A study used sodium dodecyl sulfate-polyacrylamide gel electrophoresis(SDS-PAGE), which is a variant of polyacrylamide gel electrophoresis, Fourier transform infrared (FTIR), NMR, and SAX measurements simultaneously to resolve zein film structures [48]. SDS-PAGE is an analytical method in biochemistry for the separation of charged molecules in mixtures by their molecular masses in an electric field. It uses sodium dodecyl sulfate (SDS) molecules to help identify and isolate protein molecules. Zein showed two bands in SDS-PAGE, called Z19 and Z22 (19 and 22 kDa). Following this, conformation of the Z19 fraction was inspected by FTIR, nuclear magnetic resonance, and SAXS. The infrared results showed that Z19 had 46% of β helix and 22% of α sheet. The fast N–H to N–D exchange measured by 1H NMR spectroscopy showed that Z19 was not a compact structure. The scattering measurements indicated an extended structure with 12 by 130 Å. With these data, the authors had modeled the Z19 structure as a hairpin, composed of helical, sheet, turns, and secondary structures, folded back on itself [48].

Is zein denatured by alcohol, its most common solvent? To answer this question, secondary structure of α-zeins prepared with and without solubilization in alcohol (corn gluten meal [CGM] and protein bodies with high concentrations of α-zeins and traces of β-zeins) were analyzed by FTIR spectroscopy in KBr pellets and solid-state 13C-NMR spectroscopy [49]. The proportion of secondary structures obtained by FTIR of α-zeins prepared with and without solubilization in alcohol produced almost identical amounts of α helices and β sheets. Gluten and protein body samples with high β-zein content showed higher β sheet and lower α helix proportions than those obtained for α-zein preparations. The solid-state 13C-NMR spectra showed the carbonyl peak for the α-zeins at δ 176 and for the sample rich in β-zeins at δ 172 demonstrating the presence of a high content of α helices and β sheets, respectively (see Figure 7.7). As such it was confirmed that alcohol solubilization does not affect the conformation of α-zeins, validating the secondary structure measurements in solution.

An excellent review by turana and Kokini can be consulted for further advances in understanding the molecular structures and functionalities of zein using spectroscopic techniques [50].

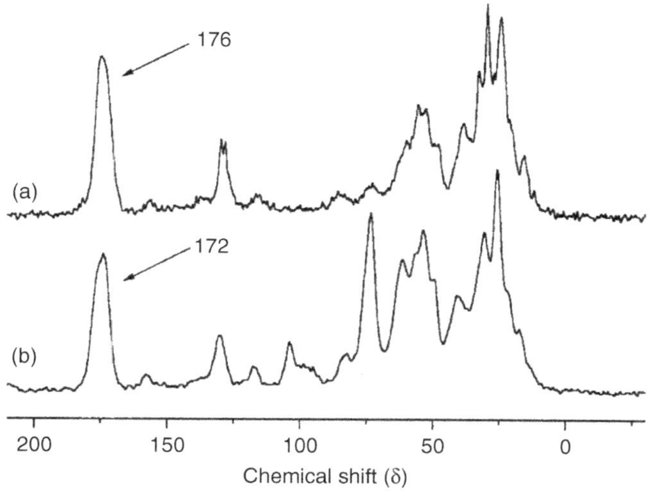

Figure 7.7 The solid-state ^{13}C-CP-MAS-NMR spectra of zeins (Z1) obtained by alcohol extract (spectrum A) and corn gluten meal (spectrum B). Source: Forato et al. [49]. © 2003, John Wiley & Sons.

7.4 Extrusion of Zein

Extrusion processing melts the polymer under high temperature and shear, forcing the polymer through a die, after which the extrudate is drawn and cooled to provide a fiber. The presence of thermal degradation products within the extrudate has a negative impact on the final product in general. Extruding zein to obtain films, fiber, or objects is fairly established nowadays. A common extrusion temperature for zein melt processing is around 90 °C. Selling [8] demonstrated that the extrusion processing of zein may have large changes on the structure of zein and resulting physical properties. Changes begin to occur at 120 °C where protein cross-linking begins. Chain cleavage begins to dominate above 180 °C. Significant color changes occur above temperatures of 160 °C. Protein structural changes slowly occur over the range from 100 to 220 °C. At 220 °C, the tertiary structure begins to rapidly change, and then at 240 °C, the secondary structure begins to rapidly change. Similarly, changes in the amide I and amide II peaks occur slowly until extrusion temperatures greater than 220 °C are reached, after which changes accelerate.

A study focused on such microstructural changes and dispersal patterns of zeins that occur during extrusion processing [51]. Therein, model systems were used that consisted of commercial zein and CGM, combined with corn starch. Commercial zein is not encased in protein bodies, whereas CGM is comprised of zeins encapsulated in protein bodies. Mixtures were extruded under conditions that resulted in different specific mechanical energy (SME). The microstructure of the resulting extrudates was analyzed using transmission electron microscopy (TEM) with immunogold labeling of the α-zein storage protein. The micrographs showed that the greatest zein dispersal occurred in CGM/starch extrudates,

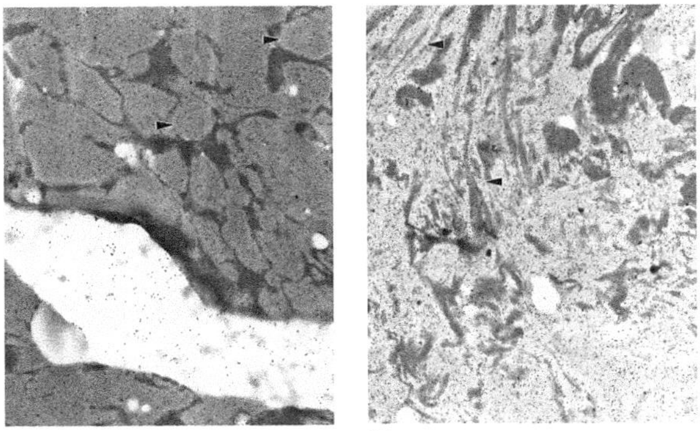

(a) (b)

Figure 7.8 (a) TEM micrograph showing protein in corn gluten meal/starch extrudate at low specific mechanical energy (SME 62 kJ/kg). Protein bodies are visible as round, spherical organelles with α-zeins immunolabeled with colloidal gold (black dots) contained within the central portion. Horizontal field width 7 mm = 0.5 μm. (b) TEM micrograph showing protein in corn gluten meal/starch extrudate at high specific mechanical energy (SME 220 kJ/kg). The spherical shapes of protein bodies are no longer visible, but rather they appear disrupted and have fused together forming what appears as fibrils. Source: Batterman-Azcona et al. [51]. © 1999, John Wiley & Sons.

especially under high SME conditions. The micrographs also showed that, at both mild and harsh extrusion conditions, commercial zein/starch extrudates contained large protein aggregates [51]. Under mild extrusion conditions, protein bodies in the CGM/starch extrudates partially retained their spherical shapes, but during harsher processing, protein bodies were disrupted and dispersed zein appeared to form fibrils, as shown in Figure 7.8a,b.

Reactive extrusion has also been used to successfully derivatize and cross-link zein in a recent work [52]. Formulations using 2% or less glyoxal (GLY) could be readily injection molded. All samples could be compression molded. Samples produced by either injection molding or compression molding were found to be resistant to dissolution by acetic acid. As the amount of GLY increased beyond 1.25%, the amount of material that was dissolved by acetic acid decreased significantly. At GLY levels of at least 1.75%, the samples were swollen on exposure to acetic acid. Samples produced by compression molding had higher tensile strength than those produced by injection molding. Incorporation of GLY did not improve tensile strength relative to control zein [52].

With successive passes through an extruder at 145 °C, zein underwent disulfide bond formation giving higher molecular weight [9]. The primary structure and chemical moieties present in the material remained substantially unchanged. Multiple passes through the extruder changed the secondary and tertiary structures of the protein and caused increased color change. The increase in molecular weight inflicted higher torque on the extruder and required increased compression molding temperature in order to produce specimens for physical property testing. There was a slight increase in modulus with successive extrusion passes with other properties remaining substantially unchanged. The authors concluded

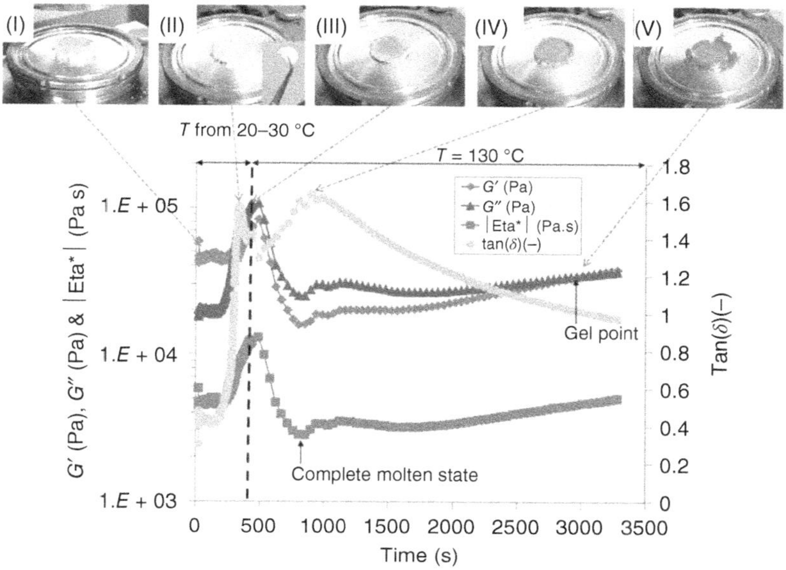

Figure 7.9 Rheological properties of powdery zein plasticized with 20 wt% glycerol at ambient conditions, during heating and at 130 °C. Photographs of the samples collected during dynamic rheometer testing (heating at 20 °C/min and followed by a plateau at 130 °C): initial wet powder (step I), at tan(δ) = 1 for $G'' > G'$ during heating (step II), at the beginning of the plateau at 130 °C (step III), when obtaining the complete molten state at the minimal viscosity and peak tan(δ) (step IV), and at gel point (after about 40 minutes at 130 °C, for tan(δ) = 1, step V). Source: From Chaunier et al. [54]. © 2017, Springer Nature.

that if physical properties were the most important attribute, then zein could be successfully recycled through an extruder [9].

Similarly, in a follow-up work [53], chemical and microscopic studies were conducted on corn flour that had been extruded under mild to extreme conditions to determine the SME required to break apart protein aggregates and release α-zein. TEM analysis revealed that starch granules and protein bodies remained intact under mild processing conditions (SME = 35–40 kJ/kg), but under harsher conditions, proteins transformed into α-zein. At SME ≈ 100 kJ/kg, protein bodies appeared highly deformed and fused together with the α-zein released, whereas at higher SME, protein bodies were completely disrupted and α-zein was dispersed and could form protein fibrils.

A recent work [54] on melt-processed zein plasticized with glycerol indicated a shear-thinning behavior in the 120–150 °C temperature range. Zein was also found to reach its gel point after 40 minutes at 130 °C (see Figure 7.9). The authors demonstrated that the structure of the network formed during melt processing involves various types of non-covalent and covalent cross-links. The small amounts of cysteine residues in α-zein amino acid sequences (<1 mol%) were shown to play a significant role in gelation kinetics. Plasticized zein behaved like a non-entangled polymer melt before gelation. The impact of the thermal stiffening (gelling) of zein-based melts can be limited by short time extrusion, leading to non-cross-linked materials. These results may be interesting for fast extrusion of zein-based thermoplastics and also to obtain feedstock materials that could be 3D printed. The authors argued that 3D printing of zein would allow one to construct edible or bioresorbing parts, with potential applications in the biomedical, food, and pharmaceutical sectors [54].

7.5 Zein Laminates with Various Packaging Films

Functional active packaging materials were also developed by forming laminates of zein with other polymeric films [21]. For instance, antioxidant laden corn-zein-laminated linear low-density polyethylene (LLDPE) films with minimum effective concentrations of the active compounds (for example, thymol, carvacrol, eugenol) were laminated to develop functional corn-zein-laminated films [55]. The release rate of antioxidant agents in gas and liquid media were determined along with the mechanical and water barrier properties of the films containing these compounds. Tensile strength and percentage elongation at break were reduced in the corn-zein-laminated LLDPE films when compared to typical LLDPE film. Furthermore, the ability of the corn-zein-laminated films to repel moisture decreased by approximately 12.2%, but was improved by incorporating hydrophobic antioxidant compounds in the corn-zein layer. The films were used for fresh ground beef packaging, and effectively inhibited lipid oxidation with a positive effect on the color stability of beef during storage. Similarly, zein coatings on polypropylene (PP) films were made as an alternative to commercial synthetic coatings (see Figure 7.10a) [56]. The study proposed a film structure for food packaging applications, which enabled easy recycling due to simple separation of the coating from the base plastic. The mechanical behavior of the PP films (tensile strength and elongation at break) was significantly improved by zein coating formulations developed therein. The plasticization degree of the coating solutions increased considerably with elongation of the coated PP films, and the tensile strength decreased with increasing plasticizer content used in the coating formulations [56].

Another highly cited work [57] attempted to prepare edible composite films using high-amylose corn starch (HACS) and zein and to determine their physical properties. HACS was gelatinized using a high-pressure container. HACS film containing 20 g sorbitol/100 g as a plasticizer with suitable physical properties was prepared. The HACS film coated with corn zein containing 20 g oleic acid/100 g had moderate physical properties and barrier properties (see Figure 7.10b) [57].

Zein was also applied on paper-based packaging materials to improve grease resistance [58] as a function of coating level, plasticizer addition, and time. Overall grease resistance was normalized by time and expressed as percent area stained per hour (% AS/h). For a 2.2 kg coating per ream of paper, % AS/h value was greater than both unplasticized and plasticized 4.4 and 6.6 kg/ream coating levels ($p < 0.001$). There was no difference in % AS/h values among 4.4 and 6.6 kg/ream zein-coated and commercial polyethylene laminated paper.

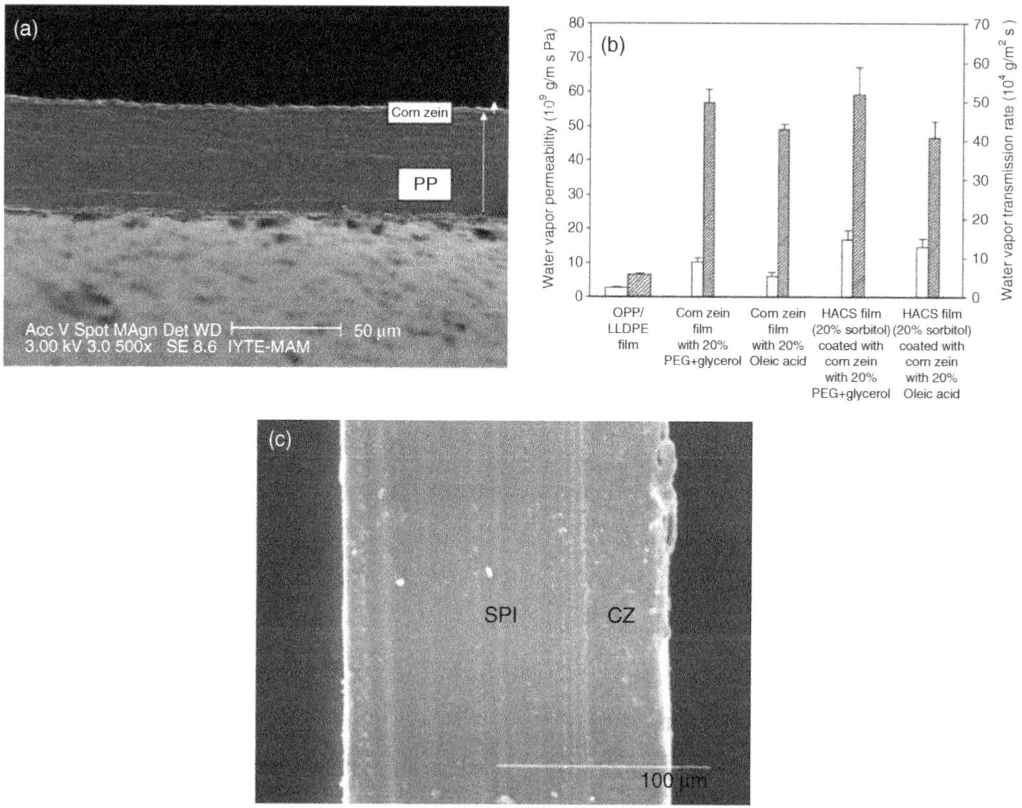

Figure 7.10 (a) SEM micrograph of the corn–zein-coated PP film for thickness determination. Source: Tihminlioglu et al. [56]. © 2011, John Wiley & Sons. (b) Water vapor permeability (white bar) and water vapor transmission rate values of (black bar) of OPP/LLDPE film and various edible films. HCAS stand for high-amylose corn starch. OPP/LLDPE stands for oriented polypropylene/linear low density polyethylene laminate film. Source: Ryu et al. [57]. © 2002, Elsevier. (c) SEM micrograph of the cross-section of the CZ/SPI bilayer film (CZ/SPI: corn zein/soy protein isolate). Source: Cho et al. [22]. © 2010, Elsevier.

Plasticizer increased ($p < 0.001$) and creasing had no effect ($p = 0.71$) on % AS/h value. Electron microscopy showed discontinuities in zein coating at the 2.2 kg/ream level. The 4.4 and 6.6 kg/ream levels showed uniform coatings. Literature review indicates that PP has been the main choice of packaging material to be treated or coated with proteins like zein. For instance, in a systematic study [59], three different plant protein coatings (soy protein isolate: SPI, whey protein isolate: WPI, zein) with various plasticizers were applied on common PP film. Among the proteins used, zein was classified as a non-suitable coating material for forming a composite structure with PP film. However, WPI as well as SPI were suggested to be used as coating materials to produce composite films when a proper plasticizer was applied. It was also suggested that WPI or SPI coatings on PP films would act as an efficient carrier for bioactive compounds in active food packaging systems. A recent study reported pouches to pack olive oil using bilayer films of zein and SPI [22]. The heat sealability of SPI film was improved by laminating the film with an additional layer of corn zein film (see Figure 7.10c). The authors claimed that the packaging of olive oil in edible and water disintegrable zein/SPI bilayer film pouches would increase the practicality and sustainability of packaged oil consumption since the film pouch disintegrated during cooking. In addition, the authors argued that the high moisture barrier properties of the CZ/SPI bilayer film under dry and intermediate humidity could extend the shelf life of olive oil by preventing oxidative rancidity during storage [22].

7.6 Zein Blend Films with Other Biopolymers

Zein can be blended in solution or during extrusion with several other plant proteins or other biopolymers including thermoplastic starch. For instance, even though WPI and zein possess different physicochemical properties (such as transparency, solubility, heat sealability, etc.), these proteins can be dispersed in a common solvent to form blend films. Their findings showed that while zein is not water soluble and WPI film has weak heat-sealing capability, the WPI/zein films were water soluble and heat-sealable [60]. Similarly, konjac glucomannan (KGM)/zein blend films were prepared with zein as the minor component 0–30% [61]. The hydrophobicity of blend films was better than pure KGM film with lower swelling and solubility properties, and moisture absorption. FTIR spectroscopy studies indicated that hydrogen bond interactions and Maillard reaction took place between KGM and zein molecules. After some optimization, a mixing ratio of KGM/zein blend ratio of 9/1 was found to display best film properties in terms of mechanical properties and morphological consistency [61].

Poly(vinyl alcohol) (PVA) and zein can be blended although they are not compatible. For instance, the effect of gamma irradiation and graft copolymerization with different ratios of acrylic acid monomer (AAc) on improving the compatibility between PVA and zein up to 50% were studied [62]. The FTIR analysis indicated that grafted AAc into zein/PVA matrix was stabilized by hydrogen bonding. The stress–strain curves showed that pure zein films were brittle, whereas pure PVA and zein/PVA blends films were tough materials either before or after gamma irradiation. The SEM micrographs indicated the formation of multilayers during the blending of zein and PVA, and these layers dispersed within one another after gamma irradiation and grafting with AAc [62].

To soften zein polymer, other polymers were also compounded such as block copolymers [63]. For instance, triblock copolymer Pluronic F127 was compounded with zein to alleviate its natural brittleness [63]. At low F127 loadings (0–35%), the plasticizing effect was dominant, and the elongation at break of zein composite film containing 35% F127 was about eightfold higher than that of the zein film with 10% F127. At high F127 loadings (50–100%), a large number of lamellae crystals were formed in the film matrix as verified by a combination of DSC and AFM (see Figure 7.11). The F127 crystallization surpassed its plasticizing effect, resulting in an increased brittleness of zein film with the further increase of F127 loading. Compared with the flat-on lamellae of pure F127, F127 chains folded into branch-like lamellar structures in the zein composite film containing 50% F127 due to the confinement of amorphous zein. The optimized F127 loading in zein/F127 composite film with a good overall performance was determined to be at around 35% [63]. In another study, polycaprolactone (PCL), zein and PCL/zein blends in the proportions 75/25, 50/50, and 25/75 wt% were mixed with glycerol at 5 wt% [64]. The mixtures were processed in a batch mixer equipped with roller rotors at 130 °C and 50 rpm for six minutes, to yield thermoplastic material. Sheets were prepared by a molding compression method using a molding temperature of 130 °C under 5 tons for five minutes. Blends of PCL/zein showed reduced tensile strength and elongation at break values, but increased Young's modulus compared to the pure polymers. Their findings indicated that PCL and zein were incompatible. Thermal studies showed that the thermal stability was enhanced by the addition of zein to PCL, whereas microscopy analysis showed a poor interfacial interaction between the polymers [64].

Effects of morphology on mechanical behavior of thermo-molded starch/zein blends with a constant zein concentration of 15% (w/w) were investigated in another study [65]. Microscopy analysis showed that different initial zein sizes (<100, <250, and >500 μm) had a strong influence on the morphology and thermo-mechanical properties for both blends. Small zein size enabled continuous morphology for amorphous starch in the blends. On the contrary, the zein network became continuous when zein size increased from 100 to 500 μm in native starch blends. It was also shown that the stress at rupture values of zein/blends decreased when the zein sizes increased from 100 to 500 μm, indicating that compact zein network impaired the stress distribution of corn starch matrix (see Figure 7.12 for details). On the contrary, the variations of the stress at rupture values of amorphous starch/zein blends did not show any significant differences, due to heterogeneous morphology of zein dispersion in the amorphous corn starch matrix [65].

Zein and chisotsan blend films with glycerol as plasticizer were prepared by solution mixing [66]. Rougher, more elastic, and softer film structures with better water vapor barrier properties were obtained using higher zein concentrations. Raman spectroscopy revealed chemical interactions such as the disappearance

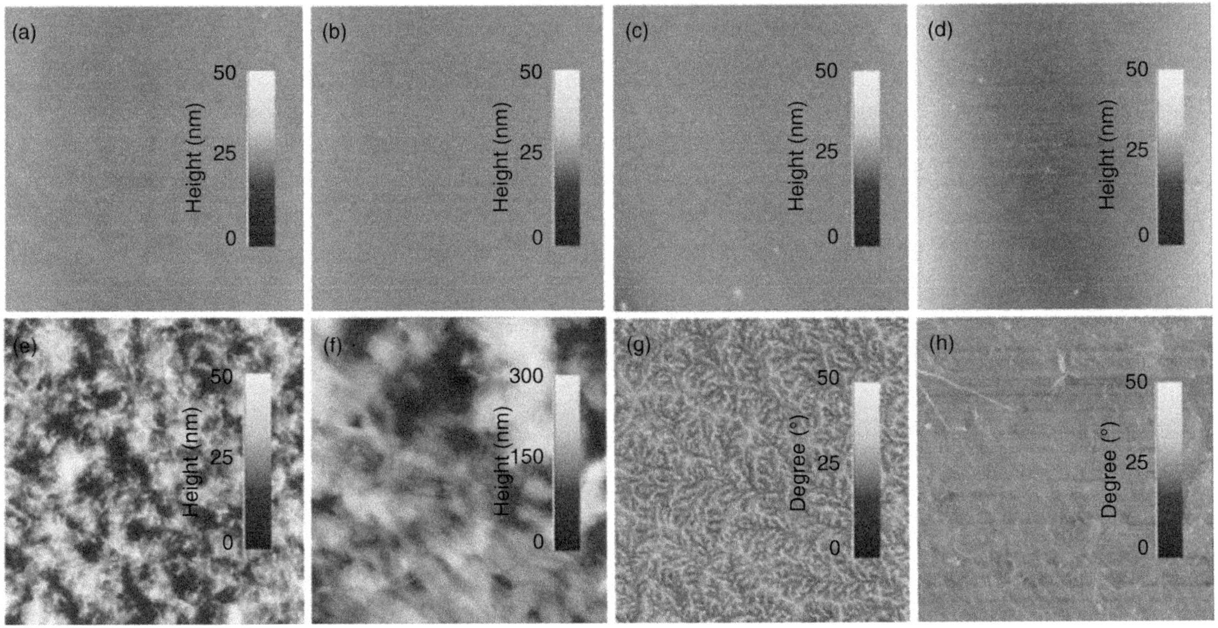

Figure 7.11 Tapping mode atomic force microscopy (TPAFM) images of zein/F127 composite films with different F127 loadings. Height images: (a) 0% F127; (b) 10% F127; (c) 20% F127; (d) 35% F127; (e) 50% F127; (f) 100% F127. Phase images: (g) 50% F127; (h) 100% F127. The scan size is fixed at 10 μm × 10 μm. Source: Woods et al. [62]. © 2010, American Chemical Society.

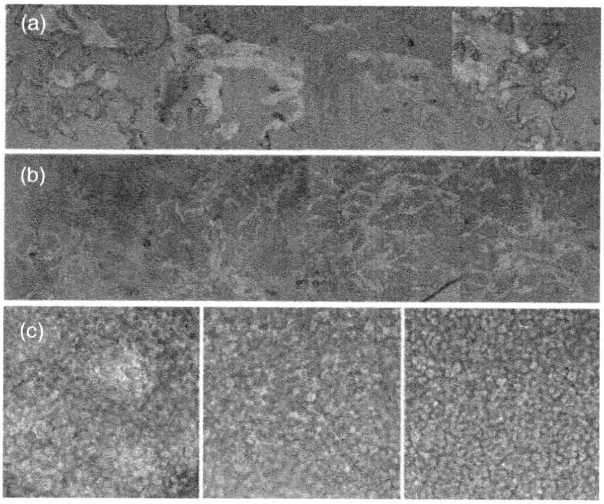

Figure 7.12 (a) Confocal light scanning microscopy images of different parts of cross-sections for amorphous starch/zein film with zein size of >500 μm. The red and green parts present zein and corn starch components, respectively. The corresponding image scale (each square panel) is 1272.5 μm × 1272.5 μm. (b) Confocal light scanning microscopy images of different parts of cross-sections for amorphous starch/zein film with zein size of <100 μm. The red and green parts present zein and corn starch components, respectively. The corresponding image (each square, four squares in total) scale is 1272.5 μm × 1272.5 μm. (c) Confocal light scanning microscopy images cross-sections for native starch/zein films with zein size of >500 μm (left), <250 μm (middle), <100 μm (right). The red and green parts present zein and corn starch components, respectively. The corresponding image scale is 319.4 μm × 319.4 μm (for each square panel). Source: Zeng et al. [65]. © 2011, Elsevier.

of the thiol groups of cysteine in the zein films and the appearance of O=S=O and C—O—S groups in the blended materials in conjunction with the disappearance of zein ε-amino and NH_2 functional groups in the zein film samples. The blend films had better barrier and mechanical properties than single pure zein or chitosan [66].

Table 7.3 summarizes recently published works on zein blend films with other bio-based polymers or proteins. In most cases, use of plasticizers has been reported and only few blends feature chemical interactions in the form of covalent bonding or hydrogen bond interactions within the blends. Both solvent and melt compounding approaches have been reported to date. It is evident from Table 7.3 that more and different plasticizers should be used to soften zein and its blends as glycerol is not waterproof and has no functionality such as antioxidant or antiseptic affects. Certain essential oils can be used to plasticize zein and its blends, particularly for solvent processing and casting films.

7.7 Outlook and Future Directions

Many materials scientists and polymer researchers today believe in the potential of edible packaging as a viable plastics replacement. This is justified by the

Table 7.3 Summary of zein blend films with other bio-based or biodegradable polymers.

Blending process	Biopolymer type	Biopolymer	Motivation	Plasticizer	Interaction	References
Thermoform	Polysaccharide	Starch	Edible packaging	None	Physical	[65]
Solvent	Polysaccharide	Chitosan	Edible-barrier packaging	Glycerol	Chemical	[66]
Solvent	Poloxamer	Pluronic F127	Flexible packaging	None	Physical	[63]
Melt compounding	Polysaccharide	TPS	General packaging	Glycerol	Physical	[67]
Solvent	Protein	Whey protein	Heat-sealable packaging	None	Physical	[60]
Solvent	Polysaccharide	Carrageenan	Packaging shelf-life extension	Glycerol	Physical	[68]
Solvent	Protein	Gelatin	Release packaging	None	Physical	[69]
Melt compounding	Polyester	PCL	Flexible packaging	Glycerol	Physical	[64]
Solvent	Vinyl polymer	PVA	Barrier packaging	Acrylic acid	Chemical	[62]
Solvent	Vinyl polymer	PVP	Soluble packaging	Triethylene glycol	Physical	[70]
Solvent	Polysaccharide	Pectin	Phobic/philic packaging	None	Chemical	[70]

number of new companies established to develop edible packaging for various food industries including hamburgers, fine waffles, and cheese to name a few. In fact, these packaging materials even supply intrepid consumers with extra fiber, minerals, and vitamins. The United States alone consumes a staggering 40 billion pieces of plastic cutlery a year, while the entire world's consumption multiplies this by 16 times. But today, even edible spoons from rice flour, wheat, and sorghum can be manufactured. Even liquid drinks can be packed in natural polymers that constitute edible packaging that lasts exactly as long as its contents. For instance, commercial examples of smoothie containers that come in jelly-like algae-plus-water packaging have been partially commercialized in Scandinavia.

As evidenced by this review chapter, the growing commercial importance of zein is reflected in the increasing number of articles appearing in the literature, particularly in the large number of patents issued the past few years. The film-forming properties of zein have been recognized for decades and are the basis for most commercial uses of zein. Even though zein is a protein, it presents unusually high resistance to water. Zein films are brittle but efficient plasticizers have been implemented and more therein should be done. Zein-based films show great potential for uses in edible coatings and bio-based packaging. For instance, clays can reduce water vapor permeability of zein by 50%. In some studies, zein was blended with other proteins, such as gluten and soy protein. These were reported to exhibit enhanced mechanical performance as a result of blending.

Zein also offers limitless number of opportunities in food, nutrition, pharmacy, and medicine. In other words, among biopolymers, zein is unique for fabrication of blends, due to its hydrophobic and solvent-friendly nature and hence mechanical and moisture barrier properties of zein can be improved significantly compared to other biopolymers. Zein as a matrix can be combined with various functional agents possessing antioxidant or flavoring properties to fabricate an active packaging. Note that it has been approved for oral use by Food and Drug Administration, FDA. Preliminary literature data indicates that applications of zein food packaging result in longer shelf life of foods with a better quality in comparison with various other polysaccharides and plant proteins.

Advances in the use of more green solvents, innovative extrusion techniques, and new developments in compounding zein with other natural products along with the unique characteristics and properties of zein open large area of investigations for zein-based packaging materials and their further applications in food and nutrition industries. Zein can also be used to fabricate devices or tools or microfluidic systems with a potential toward the development of biosensor elements to detect food and nutrient allergens and food/nutrient conditions or deficiencies during storage. Barrier properties of zein and its compounds must be improved and should remain stable for a long period of time to ensure long shelf life of packed products.

It appears that more studies and research efforts should be devoted to extrusion of zein with other thermoplastic biopolymers. As a result, more robust and flexible packaging materials with desired barrier properties can be developed. It is encouraging to see that zein can be extruded at relatively high temperatures (>100 °C) without undergoing thermal degradation. For instance, compounding with polylactic acids should be extensively undertaken and advanced. There exists almost no works on the compounding of zein with polyhydroxyalkanoate (PHA) polymers to date, as well. Recent advances in making nanoparticles or nanofibers of zein can be evaluated and used as inspirational frameworks for creating new thermoplastic zein composites for next generation food packaging.

7.8 Conclusions

The food industry is evolving toward a state where wasting food with or without its containers in any form will no longer be tolerated. In this context, use of food byproducts in the form of edible materials for food packaging and storage opens up new possibility of developing futuristic packing materials with smarter properties and added health benefits while being free from synthetic and bioaccumulating ingredients or constituents. As a result, identifying new edible materials along with exploring new ways of creating novel structures and materials from zein for food packaging have become a rapidly growing research area. Some of the latest research trends in the area of food packaging with zein including solvent casting, extrusion, and blending and laminating have been comprehensively reviewed in this chapter. Although the strong focus of this chapter is on zein processing for food packaging, the information presented in this chapter could be of significant interest to a broader community of biopolymer and biomaterials scientists and engineers working at the interface of biocomposites and polymer chemistry.

References

1 Huo, W., Wei, D., Zhu, W. et al. (2018). High-elongation zein films for flexible packaging by synergistic plasticization: preparation, structure and properties. *Journal of Cereal Science* 79: 354–361.
2 Gorham, J. (1820). Chemical analysis of Indian corn. *The New England Journal of Medicine, Surgery and Collateral Branches of Science* 9 (4): 320–328.
3 Argos, P., Pedersen, K., Marks, M.D., and Larkins, B.A. (1982). A structural model for maize zein proteins. *Journal of Biological Chemistry* 257 (17): 9984–9990.
4 Momany, F.A., Sessa, D.J., Lawton, J.W. et al. (2006). Structural characterization of α-zein. *Journal of Agricultural and Food Chemistry* 54 (2): 543–547.
5 Lai, H.M. and Padua, G.W. (1997). Properties and microstructure of plasticized zein films. *Cereal Chemistry* 74 (6): 771–775.
6 Parris, N. and Coffin, D.R. (1997). Composition factors affecting the water vapor permeability and tensile properties of hydrophilic zein films. *Journal of Agricultural and Food Chemistry* 45 (5): 1596–1599.
7 Yamada, K., Takahashi, H., and Noguchi, A. (1995). Improved water resistance in edible zein films and composites for biodegradable food packaging. *International Journal of Food Science & Technology* 30 (5): 599–608.
8 Selling, G.W. (2010). The effect of extrusion processing on Zein. *Polymer Degradation and Stability* 95 (12): 2241–2249.
9 Selling, G.W. and Utt, K.D. (2013). Effect of multiple extrusion passes on zein. *Polymer Degradation and Stability* 98 (1): 184–189.
10 Chaunier, L., Leroy, E., Valle, G.D. et al. (2017). 3D printing of maize protein by fused deposition modeling. In: *AIP Conference Proceedings*, vol. 1914, No. 1, p. 190003. AIP Publishing.
11 Corradini, E., Curti, P., Meniqueti, A. et al. (2014). Recent advances in food-packing, pharmaceutical and biomedical applications of zein and zein-based materials. *International Journal of Molecular Sciences* 15 (12): 22438–22470.
12 Shukla, R. and Cheryan, M. (2001). Zein: the industrial protein from corn. *Industrial Crops and Products* 13 (3): 171–192.
13 Nogueira, D. and Martins, V.G. (2019). Use of different proteins to produce biodegradable films and blends. *Journal of Polymers and the Environment* 27: 1–13.

14 Cataldi, P., Heredia-Guerrero, J.A., Guzman-Puyol, S. et al. (2018). Sustainable electronics based on crop plant extracts and graphene: a "bioadvantaged" approach. *Advanced Sustainable Systems* 2 (11): 1800069.
15 Ozcalik, O. and Tihminlioglu, F. (2013). Barrier properties of corn zein nanocomposite coated polypropylene films for food packaging applications. *Journal of Food Engineering* 114 (4): 505–513.
16 Lawton, J.W. (2004). Plasticizers for zein: their effect on tensile properties and water absorption of zein films. *Cereal Chemistry* 81 (1): 1–5.
17 Arcan, I. and Yemenicioğlu, A. (2011). Incorporating phenolic compounds opens a new perspective to use zein films as flexible bioactive packaging materials. *Food Research International* 44 (2): 550–556.
18 Ghanbarzadeh, B., Musavi, M., Oromiehie, A.R. et al. (2007). Effect of plasticizing sugars on water vapor permeability, surface energy and microstructure properties of zein films. *LWT - Food Science and Technology* 40 (7): 1191–1197.
19 Cho, S.Y., Park, J.W., and Rhee, C. (2002). Properties of laminated films from whey powder and sodium caseinate mixtures and zein layers. *LWT - Food Science and Technology* 35 (2): 135–139.
20 Rakotonirainy, A.M. and Padua, G.W. (2001). Effects of lamination and coating with drying oils on tensile and barrier properties of zein films. *Journal of Agricultural and Food Chemistry* 49 (6): 2860–2863.
21 Ghanbarzadeh, B. and Oromiehi, A.R. (2009). Thermal and mechanical behavior of laminated protein films. *Journal of Food Engineering* 90 (4): 517–524.
22 Cho, S.Y., Lee, S.Y., and Rhee, C. (2010). Edible oxygen barrier bilayer film pouches from corn zein and soy protein isolate for olive oil packaging. *LWT - Food Science and Technology* 43 (8): 1234–1239.
23 Vahedikia, N., Garavand, F., Tajeddin, B. et al. (2019). Biodegradable zein film composites reinforced with chitosan nanoparticles and cinnamon essential oil: physical, mechanical, structural and antimicrobial attributes. *Colloids and Surfaces B: Biointerfaces* 177: 25–32.
24 Yemenicioğlu, A. (2016). Zein and its composites and blends with natural active compounds: development of antimicrobial films for food packaging. In: *Antimicrobial Food Packaging* (ed. J. Barros-Velázquez), 503–513. Academic Press.
25 Arcan, I. and Yemenicioğlu, A. (2013). Development of flexible zein–wax composite and zein–fatty acid blend films for controlled release of lysozyme. *Food Research International* 51 (1): 208–216.
26 Gu, L. and Wang, M. (2013). Effects of protein interactions on properties and microstructure of zein–gliadin composite films. *Journal of Food Engineering* 119 (2): 288–298.
27 Tihminlioglu, F., Atik, İ.D., and Özen, B. (2010). Water vapor and oxygen-barrier performance of corn–zein coated polypropylene films. *Journal of Food Engineering* 96 (3): 342–347.
28 Bouman, J., Belton, P., Venema, P. et al. (2015). The development of direct extrusion-injection moulded zein matrices as novel oral controlled drug delivery systems. *Pharmaceutical Research* 32 (8): 2775–2786.

29 Zheng, X.Q., Liu, X.L., Yu, S.F. et al. (2014). Effects of extrusion and starch removal pretreatment on zein proteins extracted from corn gluten meal. *Cereal Chemistry* 91 (5): 496–501.

30 Verbeek, C.J. and van den Berg, L.E. (2010). Extrusion processing and properties of protein-based thermoplastics. *Macromolecular Materials and Engineering* 295 (1): 10–21.

31 Luecha, J., Sozer, N., and Kokini, J.L. (2010). Synthesis and properties of corn zein/montmorillonite nanocomposite films. *Journal of Materials Science* 45 (13): 3529–3537.

32 Oliviero, M., Verdolotti, L., Di Maio, E. et al. (2011). Effect of supramolecular structures on thermoplastic Zein–lignin bionanocomposites. *Journal of Agricultural and Food Chemistry* 59 (18): 10062–10070.

33 Wang, Y. and Padua, G.W. (2003). Tensile properties of extruded zein sheets and extrusion blown films. *Macromolecular Materials and Engineering* 288 (11): 886–893.

34 Spence, K.E., Jane, J.L., and Pometto, A.L. (1995). Dialdehyde starch and zein plastic: mechanical properties and biodegradability. *Journal of Environmental Polymer Degradation* 3 (2): 69–74.

35 Lin, T., Lu, C., Zhu, L., and Lu, T. (2011). The biodegradation of zein in vitro and in vivo and its application in implants. *AAPS PharmSciTech* 12 (1): 172–176.

36 Evans, C.D. and Manley, R.H. (1941). Solvents for zein. Primary solvents. *Industrial and Engineering Chemistry* 33 (11): 1416–1417.

37 Manley, R.H. and Evans, C.D. (1943). Binary solvents for zein. *Industrial and Engineering Chemistry* 35 (6): 661–665.

38 Lawton, J.W. (2002). Zein: a history of processing and use. *Cereal Chemistry* 79 (1): 1–18.

39 Zhang, Y., Cui, L., Che, X. et al. (2015). Zein-based films and their usage for controlled delivery: origin, classes and current landscape. *Journal of Controlled Release* 206: 206–219.

40 Li, Y., Li, J., Xia, Q. et al. (2012). Understanding the dissolution of α-zein in aqueous ethanol and acetic acid solutions. *The Journal of Physical Chemistry B* 116 (39): 12057–12064.

41 Tomlinson, S.R., Kehr, C.W., Lopez, M.S. et al. (2014). Solubility of the corn protein zein in imidazolium-based ionic liquids. *Industrial & Engineering Chemistry Research* 53 (6): 2293–2298.

42 Taylor, J., Anyango, J.O., and Taylor, J.R. (2013). Developments in the science of zein, kafirin, and gluten protein bioplastic materials. *Cereal Chemistry* 90 (4): 344–357.

43 Selling, G.W. and Utt, K.D. (2014). Improved zein films using polyethylene-maleic anhydride. *Journal of Applied Polymer Science* 131 (8): 40122.

44 Wei, Y., Hu, L., Yao, J. et al. (2019). Facile dissolution of zein using a common solvent dimethyl sulfoxide. *Langmuir* 35 (20): 6640–6649.

45 Luecha, J., Hsiao, A., Brodsky, S. et al. (2011). Green microfluidic devices made of corn proteins. *Lab on a Chip* 11 (20): 3419–3425.

46 Hsu, B.L., Weng, Y.M., Liao, Y.H., and Chen, W. (2005). Structural investigation of edible zein films/coatings and directly determining their thickness by

FT-Raman spectroscopy. *Journal of Agricultural and Food Chemistry* 53 (13): 5089–5095.

47 Shi, K., Kokini, J.L., and Huang, Q. (2009). Engineering zein films with controlled surface morphology and hydrophilicity. *Journal of Agricultural and Food Chemistry* 57 (6): 2186–2192.

48 Forato, L.A., Doriguetto, A.C., Fischer, H. et al. (2004). Conformation of the Z19 prolamin by FTIR, NMR, and SAXS. *Journal of Agricultural and Food Chemistry* 52 (8): 2382–2385.

49 Forato, L.A., Bicudo, T.D.C., and Colnago, L.A. (2003). Conformation of α zeins in solid state by Fourier transform IR. *Biopolymers* 72 (6): 421–426.

50 Turasan, H. and Kokini, J.L. (2017). Advances in understanding the molecular structures and functionalities of biodegradable zein-based materials using spectroscopic techniques: a review. *Biomacromolecules* 18 (2): 331–354.

51 Batterman-Azcona, S.J., Lawton, J.W., and Hamaker, B.R. (1999). Microstructural changes in zein proteins during extrusion. *Scanning* 21 (3): 212–216.

52 Selling, G.W., Woods, K.K., Biswas, A., and Willett, J.L. (2009). Reactive extrusion of zein with glyoxal. *Journal of Applied Polymer Science* 113 (3): 1828–1835.

53 Batterman-Azcona, S.J., Lawton, J.W., and Hamaker, B.R. (1999). Effect of specific mechanical energy on protein bodies and α-zeins in corn flour extrudates. *Cereal Chemistry* 76 (2): 316–320.

54 Chaunier, L., Della Valle, G., Dalgalarrondo, M. et al. (2017). Rheology and structural changes of plasticized zeins in the molten state. *Rheologica Acta* 56 (11): 941–953.

55 Park, H.Y., Kim, S.J., Kim, K.M. et al. (2012). Development of antioxidant packaging material by applying corn-zein to LLDPE film in combination with phenolic compounds. *Journal of Food Science* 77 (10): E273–E279.

56 Tihminlioglu, F., Atik, İ.D., and Özen, B. (2011). Effect of corn-zein coating on the mechanical properties of polypropylene packaging films. *Journal of Applied Polymer Science* 119 (1): 235–241.

57 Ryu, S.Y., Rhim, J.W., Roh, H.J., and Kim, S.S. (2002). Preparation and physical properties of zein-coated high-amylose corn starch film. *LWT - Food Science and Technology* 35 (8): 680–686.

58 Trezza, T.A. and Vergano, P.J. (1994). Grease resistance of corn zein coated paper. *Journal of Food Science* 59 (4): 912–915.

59 Lee, J.W., Son, S.M., and Hong, S.I. (2008). Characterization of protein-coated polypropylene films as a novel composite structure for active food packaging application. *Journal of Food Engineering* 86 (4): 484–493.

60 Tsai, M.J. and Weng, Y.M. (2019). Novel edible composite films fabricated with whey protein isolate and zein: preparation and physicochemical property evaluation. *LWT - Food Science and Technology* 101: 567–574.

61 Wang, K., Wu, K., Xiao, M. et al. (2017). Structural characterization and properties of konjac glucomannan and zein blend films. *International Journal of Biological Macromolecules* 105: 1096–1104.

62 Woods, M.M., Salmieri, S., El-Naggar, A.W. et al. (2010). Improving the compatibility of zein/poly(vinyl alcohol) blends by gamma irradiation and graft

copolymerization of acrylic acid. *Journal of Agricultural and Food Chemistry* 58 (7): 4470–4476.

63 Li, J., Li, Y., Lee, T.C., and Huang, Q. (2013). Structure and physical properties of zein/pluronic F127 composite films. *Journal of Agricultural and Food Chemistry* 61 (6): 1309–1318.

64 Corradini, E., Mattoso, L.H.C., Guedes, C.G.F., and Rosa, D.S. (2004). Mechanical, thermal and morphological properties of poly(ε-caprolactone)/zein blends. *Polymers for Advanced Technologies* 15 (6): 340–345.

65 Zeng, M., Huang, Y., Lu, L. et al. (2011). Effects of filler-matrix morphology on mechanical properties of corn starch–zein thermo-moulded films. *Carbohydrate Polymers* 84 (1): 323–328.

66 Escamilla-García, M., Calderon-Dominguez, G., Chanona-Perez, J.J. et al. (2013). Physical and structural characterisation of zein and chitosan edible films using nanotechnology tools. *International Journal of Biological Macromolecules* 61: 196–203.

67 Corradini, E., Carvalho, A.J.F.D., Curvelo, A.A.D.S. et al. (2007). Preparation and characterization of thermoplastic starch/zein blends. *Materials Research* 10 (3): 227–231.

68 Sanchez-Garcia, M.D., Hilliou, L., and Lagaron, J.M. (2010). Nanobiocomposites of carrageenan, zein, and mica of interest in food packaging and coating applications. *Journal of Agricultural and Food Chemistry* 58 (11): 6884–6894.

69 Deng, L., Kang, X., Liu, Y. et al. (2018). Characterization of gelatin/zein films fabricated by electrospinning vs solvent casting. *Food Hydrocolloids* 74: 324–332.

70 Woods, K.K., Selling, G.W., and Cooke, P.H. (2009). Compatible blends of zein and polyvinylpyrrolidone. *Journal of Polymers and the Environment* 17 (2): 115–122.

Part III

Biocomposites of Cellulose and Biopolymers in Food Packaging

8

Cellulose-Reinforced Biocomposites Based on PHB and PHBV for Food Packaging Applications

Estefania L. Sanchez-Safont, Luis Cabedo, and Jose Gamez-Perez

Universitat Jaume I, Polymers and Advanced Materials Group (PIMA), Avinguda de Vicent Sos Baynat, Castellón de la Plana, Valencia 12071, Spain

8.1 Introduction to Bioplastics

In the last few decades there has been a growing interest in the development of bioplastics as potential substitutes for conventional plastics. In this sense, there is a great interest in the development of new formulations of bio-based polymers that reduce dependence on petroleum-based plastics [1, 2] but also on the production of biodegradable plastics.

At this point, it is interesting to recall the subtleness of the terminology of bioplastics, biopolymers, bio-based polymers, biodegradable polymers, and compostable plastics.

Bioplastics represent a large group of plastics based on polymers that can be obtained from renewable sources, fossil resources, or a mixture of both. Therefore, a bioplastic can be either biologically based or biodegradable or both [3]. The IUPAC defines biopolymers as "Macromolecules produced by living organisms (including proteins, nucleic acids, and polysaccharides)." According to the IUPAC, biodegradable polymers are defined as "polymers susceptible to degradation by biological activity (enzyme, bacteria, fungi, algae) with the degradation accompanied by a lowering of its molar mass." [4] For practical purposes, in order for a plastic to be considered biodegradable in a given environment or under composting conditions, it must meet a series of specific requirements according to established regulations. In the specific case of "compostable" plastics, according to ASTM-D883, they are defined as "plastics that undergo biological degradation during composting to give rise to carbon dioxide, water, inorganic compounds, and biomass at a speed comparable to other known compostable materials and without leaving visually detectable or toxic residues." Considering this definition, a plastic will be compostable if it is biodisintegrable and biodegradable at a similar rate as food and crop residues and the products of its degradation are nontoxic [2, 5].

Sustainable Food Packaging Technology, First Edition. Edited by Athanassia Athanassiou.

There are different standards that allow evaluating these requirements and that establish the maximum periods of time and the minimum percentages of bio-disintegration and biodegradation for a material to be considered compostable. Thus, the ISO 17088 or ASTM 6400 standards establish the specifications for compostable plastics, the ISO 14855-1, ISO 14855-2, and ASTM D5338 standards allow evaluation of biodegradability in composting and ISO 20200 and ISO 16929 standards allow evaluation of disintegrability under composting conditions. The standard EN 13432 defines the criteria for a container to be labeled as biodegradable or compostable and contemplates the evaluation of the quality and toxicity of the resulting compost [5].

Finally, the biologically based polymers are those synthesized from molecules derived, in whole or in part, from biomass (e.g. bio-polyolefins, bio-polyterephthalate ethylene, bio-polyurethane, etc.); however, these do not necessarily have to be biodegradable [6, 7]. Therefore, although all these terms are sometimes used interchangeably, it is important to emphasize that a biopolymer may be biological based but not biodegradable [8] or, at the opposite extreme, it can derive from fossil resources but being biodegradable (e.g. poly(ε-caprolactone), PCL) [9].

Taking into account these considerations, bioplastics are all the polymers that can be classified in the following categories [3, 10–12]:

Biodegradable polymers – They can be divided depending on their synthesis as follows:
- (a) *Obtained directly from biomass* – this group includes carbohydrates such as cellulose, starch, chitosan, etc.; proteins of vegetable origin such as corn protein or glue; proteins of animal origin such as gelatin, collagen, etc.; and lipids such as waxes or fatty acids.
- (b) *Obtained by microbes* – polymers obtained by microbial fermentation such as polyhydroxyalkanoates (PHAs) or microbial polysaccharides such as pullulan or curdlan
- (c) Obtained by chemical synthesis
  1. *From biomass* – biopolymers obtained by conventional chemical synthesis from monomers from biomass such as polylactic acid (PLA)
  2. *From fossil resources* – polymers obtained by conventional chemical synthesis from monomers and polymers from fossil resources such as poly(glycolic acid) (PGA), poly(ε-caprolactone) (PCL), poly(butylene succinate) (PBS), polyvinyl alcohol (PVA), etc.

Non biodegradable biological based polymers – In this group are non biodegradable polymers in which at least one of their monomers comes from renewable sources such as bio-polyethylene, bio-polypropylene, bio-polyamides, bio-polyurethanes, etc.

If we focus on biodegradable bio-based polymers, several reviews [3, 10, 12, 13] highlight that, despite their enormous potential due to their renewable origin and their biodegradability, certain key aspects still exist limiting their application.

In general, these limitations are related to mechanical, thermal, and barrier properties lower than conventional polymers. These polymers usually present a high sensitivity to water, and a low stability due to aging and various processing problems. Also, most of them still have very high production costs compared to traditional polymers. Therefore, the main challenges to improve their competitiveness are focused on improving the stability of production, their processability, and their properties. [6]

8.2 PHB and PHBV: a SWOT (Strength, Weakness, Opportunity, and Threat) Analysis

8.2.1 Polyhydroxyalkanoates (PHA): Poly-3-hydroxybutyrate (PHB) and Poly-3-hydroxybutyrate-*co*-3-hydroxyvalerate (PHBV)

PHA are linear polyesters formed by hydroxyalkanoic acid monomers linked by ester linkages. The bond occurs between the carboxyl groups of a monomer with the hydroxyl group of the adjacent monomer [14]. The general structure of the PHA is shown in Figure 8.1.

PHAs are natural biopolymers synthesized and accumulated in the form of intracellular granules as a carbon and energy reserve by different prokaryotic microorganisms when there is a limitation of essential nutrients (nitrogen, oxygen, or phosphorus) in the presence of an excess carbon source [16]. This type of granular structures were first observed by Beijerink in 1888 inside Rhizobium cells, although he could not define their function or composition [14]. In 1923, the French scientist Maurice Lemoinge reported the production of 3-oxybutyric acid in *Bacillus subtilis* [17]. In 1927, he was able to extract and isolate the substance with chloroform from *Bacillus megaterium* cells and proved that it was a polymer, poly-3-hydroxybutyrate (PHB) [18]. Lemoige et al. reported different accumulations of PHB varying the growth medium and were the first to describe a method to quantify it [19]. Macrae and Wilkinson (1958) showed that the function of the PHB in the cell was to serve as an energy reserve. They were the first to propose a biosynthetic route to observe that the accumulation took place especially when the ratio of glucose to nitrogen in the medium of growth was high [14, 20]. Some bacteria are even able to synthesize PHAs without nutritional stress, such as recombinant *Escherichia coli* [21].

It was not until the early 1960s that the potential of the PHB began to be recognized and its production began to be explored. The first commercialization attempts of PHB were carried out by the company W.R. Grace & Co (USA), where the works of Baptist and Weber gave rise to various patents [22–24]. However, due to the low efficiency of the synthesis process, the low extraction performance, problems with the purity of the polymer, and the high cost, the company had to close.

Several years later, in 1974, Wallen and Rohwedder, documented the extraction of a polymer containing units of hydroxybutyrate (HB), hydroxyvalerate (HV), hydroxyhexanoate (HHx), and hydroxyheptanoate (HHp) in activated sewage sludges [25]. Since then, around 150 PHA monomers have been identified with

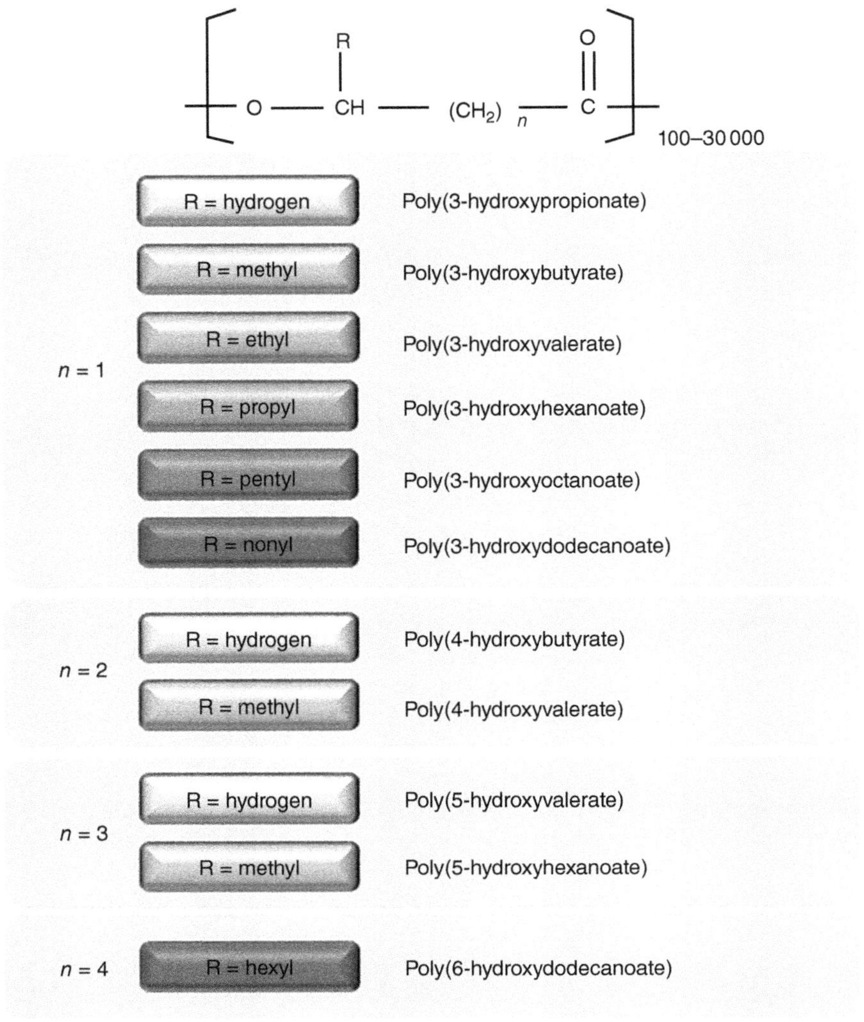

Figure 8.1 General structure of PHAs. Source: Adapted from Albuquerque and Malafaia 2018 [15].

molecular masses between 50 000 and 1 000 000 Da and more than 300 bacteria belonging to around 90 genera of microbial species capable of synthesizing these polymers [16, 26, 27].

The British company Imperial Chemical Industries (ICI) began to commercialize the poly-3-hydroxybutyrate-*co*-3-hydroxyvalerate (PHBV) copolymer under the name of BIOPOL® that was distributed by Monsanto and later by Metabolix [28, 29]. Currently, there are different companies that produce and market PHA, among which we can highlight Biomer Inc. (Germany), Goodfellow Cambridge Ltd. (United Kingdom), Kaneka (Singapore), Meredian (USA), Metabolix (USA), Mitsubishi Gas Chemicals (Japan), PHB Industrial S/A (Brazil), Tianan Biological

Materials (China), and Tianjin Green Biosciences (China). Even though growing, the production capacity of these companies is still low (around 1000–20 000 tons per year) and the price of these polymers exceeds €5 per kg in most cases [29, 30] (despite it has been substantially reduced in recent years).

The PHA production process comprises different stages that, in a very general way, can be described as follows: bacterial culture, fermentation in the medium (carbon source), synthesis and accumulation of PHA, extraction, and purification. The type of polymer or copolymer synthesized depends, fundamentally, on the selected carbon source and cell culture. The costs of this process are still high and are attributed to the cost of the raw materials used as a carbon source and also to the extraction and purification stages [15, 31].

PHA can be classified into three groups, depending on the number of carbons present in the monomer/monomers that form the polymer: (i) short chain length (scl-PHA: 3–5 C [carbon atoms]), (ii) medium chain length (mcl-HA, 6–14 C), and (iii) long chain length (lcl-PHA, >15 C) [32]. Therefore, PHAs comprise a broad family of polymers whose characteristics and physical and chemical properties such as melting and glass transition temperatures, crystallinity, etc., strongly depend on the structure of the constituent monomers, which can be more than 150 different ones [27].

In general, scl-PHA [33] are rigid and brittle thermoplastics. They possess a high crystallinity, low glass transition temperature, and high melting temperature [34]. The mcl- and lcl-PHA are thermoplastic elastomers with low crystallinity, low tensile strength, and high elongation at break, and have glass transition temperatures and fusion lower than scl-PHA [21].

Some general characteristics of PHAs are their intrinsic biodegradability due to their microbial origin, their biocompatibility, and non-toxicity. PHAs are soluble in chloroform and other chlorinated hydrocarbons. They are hydrophobic, insoluble in water, and relatively resistant to hydrolytic degradation. They are inert and stable in air. They have good ultraviolet resistance and low resistance to acids and alkali [29, 35].

8.2.2 PHB and PHBV: Strengths

PHB is a homopolymer of 3-HB whose molecular weight can vary between 10 000 and 3 000 000 Da depending on the microorganism that produces it, the carbon source, the concentration of available nutrients, and the growth conditions such as the pH or the temperature [27, 28]. PHBV is a random copolymer of 3-HB and 3-HV whose properties depend on the HV content. Thus, for very low HV contents their properties are close to those of PHB. At higher molar content of HV, the impact resistance and the elongation at break increases, while other properties such as glass transition and melting temperatures, crystallinity, rigidity, the tensile strength, and barrier properties decrease with respect to PHB [13, 35].

Both PHB and PHBV are stiff, biodegradable, biocompatible, and nontoxic thermoplastics. They are hydrophobic, insoluble in water, and relatively resistant to hydrolytic degradation. They are soluble in chloroform and other chlorinated hydrocarbons, showing good ultraviolet resistance. Nevertheless, their specific properties are closely related to their chemical structure, molecular mass, and crystalline structure.

PHB and low valerate content PHBV have a glass transition temperature around 5–10 °C, relatively high melting temperatures (~170 °C), and low melt viscosity [14, 29]. Although, semicrystalline polymers with relatively low glass transition temperatures usually have a mechanical behavior with low thermal resistance, the high crystallinity of PHB and PHBV (>60%) causes them to have a relatively high thermal resistance. Their heat deflection temperatures (HDTs) are higher than 100 °C, values well above those corresponding to other biopolymers such as amorphous PLA (HDT ~ 50 °C) [36, 37]. Due to their high crystallinity, they have good barrier properties, with low permeability to oxygen, water, and aromas [38, 39]. Regarding its mechanical properties, PHB is a rigid polymer, with low elongation at break and low impact strength [29, 40].

One of the properties that makes PHB and PHBV especially attractive for short-life applications (such as packaging or the manufacture of disposable products) is their biodegradability. In this sense, it is especially interesting to note that they are biodegradable in different environments, particularly in industrial composting. In many potential cases, the residues derived from these products could be discarded together with the organic fraction of the trash, so composting could then be considered as a viable option for waste treatment. The biodegradability of PHB and PHBV in different environments including marine, as well as lab-scale, industrial, or home composting has been widely studied and demonstrated [41–44].

8.2.3 PHB and PHBV: Weaknesses

Both PHB and the low hydroxyvalerate content PHBV have relatively high values of elastic modulus and tensile strength, but low elongation at break (<5%) that gives them a fragile character. Therefore, they are not very tough, and in addition, their fragility has a marked dependence over time [29, 34, 45]. This high fragility as well as the progressive embrittlement over time observed are strongly determined by the microstructure, which also depends largely on the crystallization conditions of the material during its processing.

In general, this mechanical behavior is attributed in the literature to the intrinsic fragility of spherulites that generally show large internal fissures, due to low nucleation density [46–48] (Figure 8.2). On the other hand, progressive embrittlement is also attributed to two phenomena: a secondary crystallization process [46, 49, 50] and a physical aging of the amorphous phase [51–53]. As highlighted by Crétois et al. [51], there is not a clear consensus to explain the fragility of PHB and PHBV over time, although it is widely accepted that a combination of both phenomena (crystallization and physical aging) occurs [51, 54, 55]. In the first case, small crystallites generated in secondary crystallization bind to the lamellae and act as points of union between lamellae, producing a constriction of the amorphous phase disturbing chain mobility [49].

The physical aging of the amorphous phase in PHB can be approached by the three-phase model proposed by Struik [56, 57]. In this model, two different types

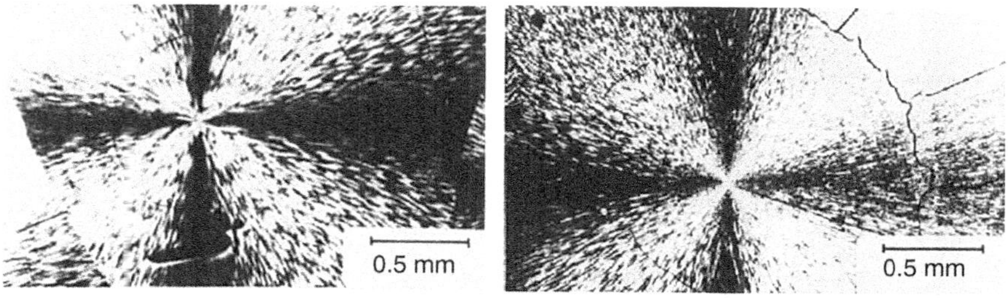

Figure 8.2 Optical micrographs of PHB spherulites showing cracks and fissures. Source: Barham and Keller [47]. © 1986, John Wiley & Sons.

of amorphous phase along with the crystalline phase are presented in a semicrystalline polymer: an unaltered (mobile) amorphous phase (MAF), and an altered amorphous phase with restricted mobility (RAF) adjacent to the crystals. Struik defined for the model two glassy transitions, one corresponding to the MAF and another corresponding to the devitrification of the RAF. Hence, the three-phase model contemplates the existence of the crystalline phase, the RAF, and the MAF [58–61]. The phenomenon of physical aging occurs, according to this model, due to the progressive densification of the MAF, resulting in a higher proportion of RAF. However, as Esposito et al. pointed out [53], this model is somehow controversial to explain the development of the RAF, which depends strongly on the crystallization conditions, as well as the structural reorganizations that occur both in the RAF and in the MAF [54, 62, 63]. Alternatively, the studies carried out by Esposito et al. [53] suggested that a complex two-phase model called "continuum of mobility" could more satisfactorily explain the mobility of the amorphous phase in PHBV under cold crystallization conditions (Figure 8.3).

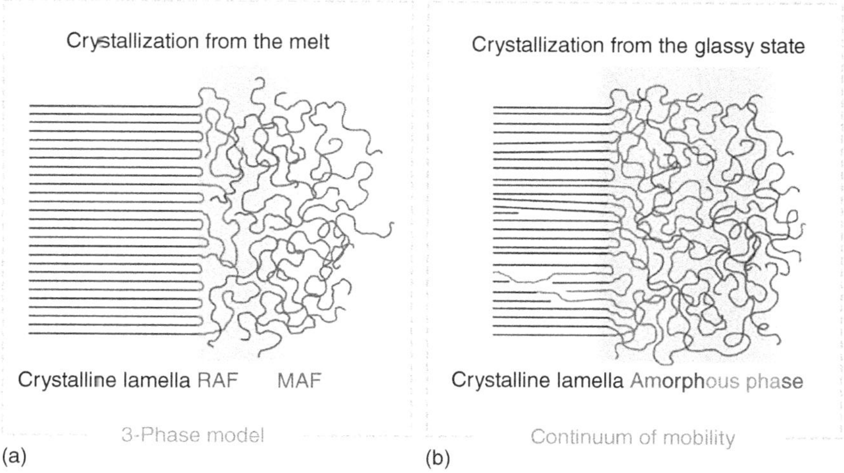

Figure 8.3 Schematic representation of the three-phase model and the "continuum of mobility" model. Source: From Esposito et al. 2016 [53].

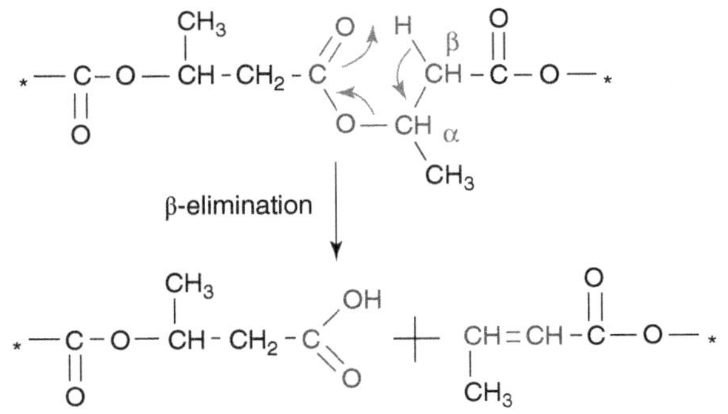

Figure 8.4 Mechanism of thermal degradation of PHAs by cis-elimination. The symbol "*" in artwork represents the polymer chain. Source: Adapted from Kopinke and Mackenzie 1997 [64].

Physical aging, as shown by Crétois et al. [51], has a reversible character without melting the material, by applying a mild annealing treatment. However, the reversibility of physical aging is only temporary. Indeed, this phenomenon is more relevant on the PHB embrittlement than secondary crystallization [45, 51, 55].

Regarding the thermal stability, PHB and PHBV are very sensitive to thermal degradation. PHAs thermally decompose at temperatures close to their melting temperature (generally above 190 °C). Thermal degradation of these polymers occurs by a mechanism of cleavage of random chains by cis-elimination, also known as McLafferty rearrangement [64] (Figure 8.4).

This degradation process is accompanied by the production of degradation by-products such as crotonic acid and other oligomers that, in turn, catalyze the degradation reaction as pointed out by Kopinke and Mackenzie [64]. This degradation process results in a sharp drop in molecular mass and a marked reduction in melt viscosity [29].

Although PHB (and PHBV) can be processed by conventional processing techniques, it shows several drawbacks that lead to narrow processing windows, such as the dependence of the crystallization rate on temperature, the high crystallinity, and high melting temperature in combination with its high sensitivity to thermal degradation [29, 31, 38, 44, 65, 66].

As mentioned, these materials are particularly attractive for short shelf life applications such as food packaging. It is worth mentioning that thermoforming, one of the most used techniques in the packaging industry, is particularly difficult to use with PHB and PHBV. Thermoforming consists of heating a plastic film until it reaches a semi-plastic state and the subsequent forming of a mold that can be positive or negative by stretching the film by applying pressure or vacuum. In amorphous polymers, the temperatures used for films are slightly above their glass transition temperature, while in the case of semicrystalline polymers the thermoforming temperatures are in the immediate vicinity of the melting temperature [67]. In the particular case of PHB and PHBV, this technique is especially difficult. Due to its high crystallinity, the material maintains its rigidity up to temperatures very close to that of melting and once this temperature is reached, a sharp drop in viscosity occurs. Indeed, the melting temperature is relatively close to their thermal degradation temperature, narrowing the processing window. This lack of sheet stability translates into defects associated with poor mold reproducibility and poor thickness distribution. In addition, during cooling, defects due to contractions caused by the recrystallization of the material may occur. The literature related to the thermoforming of PHB or PHBV is, anyhow, limited [65, 68].

8.2.4 PHB and PHBV: Opportunities

Some of the strategies described to improve the mechanical properties of PHAs are also applicable to the improvement of their processability. Thus, for example, the use of nucleating agents or the addition of nanofillers favors nucleation and crystallization rate and can contribute to improving thermal stability [10, 12, 13]. The use of plasticizers or mixing with other polymers can expand the processing window by reducing crystallization or melting temperatures [67, 69]. Thermal stability can be improved by adding chain extenders [70, 71]. The melt viscosity can be modified both by incorporating a second rigid phase, mixing with other polymers or plasticizers or using chain extenders. Therefore, by an appropriate design of the compositions and a suitable selection of the additives, it is possible to improve the processability of PHA.

In addition to a proper design of the composition, the adjustment of the operating parameters as well as the conditioning steps, like drying the materials before processing, is essential to prevent thermal degradation [72]. Mixing temperatures and speeds must be carefully chosen to avoid high temperature zones where thermal degradation can begin, especially if the material is going to be subjected to several thermal cycles. For instance, in extrusion processing, because the crystallization rate of these polymers is faster at temperatures between 60 and 130 °C [72], it may be convenient to reduce the temperature in the nozzle and even cool the polymer in a water bath to avoid sticking problems. Similarly, in injection molding by correctly adjusting the temperature of the mold, the rapid crystallization of the polymer can be favored, avoiding problems of bonding in the mold and favoring dimensional stability by preventing contractions due to secondary crystallization.

Regarding costs, first analysis by Choi & Lee [73] pointed out an estimate around US$5.6 per kg in 1997 , but in their analysis, they considered that by using carbon sources from wastes the price could be lowered to US$4.75 per kg. Although nowadays the consumer price of PHB is around €6 per kg, [29] the advances in technology using mixed cultures, as well as scaling savings, have decreased the production costs. It is expected that the prices will continue to decrease, reaching values of €2.5–3 per kg [14, 74]. One of the key factors to lower the price of PHAs is the ability to use waste as carbon source, which decreases the cost of the raw materials [75, 76]. Indeed, it has to be kept in mind that production of PHAs could be delocalized, not being necessary to transport tons of raw materials, which is also an opportunity to produce near the end users.

With respect to regulatory policies, the use of plastics is starting to be banned in some products (Directive [EU] 2019/904 of the European Parliament and of the Council of 5 June 2019 on the reduction of the impact of certain plastic products on the environment). Thus, the concern of pollution by plastic residues is expected to change the present scenario. Hence, there will be some markets that will be open to biodegradable biopolymers, regardless of the price difference, especially those associated with food hygiene, food safety, good hygiene practices, good manufacturing practices, consumer information, or traceability.

8.2.5 PHB and PHBV: Threats

The market of PHA products is still at an early stage. One of the threats to the development of new products based on PHB and PHBV is the limited availability of these polymers, which are not growing as fast as predicted a few years ago. In general terms, the growth of the market of biopolymers went from 10% between 2012 and 2014 to 4% from 2015 to 2018 [77]. Nevertheless, PHAs are the fastest growing biopolymers. Indeed, the growth rate of PHAs can be affected by several factors that have been identified [78, 79]:

(a) Biodegradable polyesters from chemical synthesis may prove to be more attractive economically, especially with low oil prices. Indeed, they may be considered by engineers as materials that are easier to modify in order to adapt them to the future market needs.
(b) The demand may not grow as fast as the production ability, due to the limitations in performance or higher cost of the raw materials.
(c) Even though the political framework is favorable to biodegradable materials, the plastics may be finally replaced by other biomaterials such as cellulose and other derivatives.
(d) There are also some societal concerns in the adoption of new bio-products.

As a final remark, considering the balance of the strength, weakness, opportunity, and threat (SWOT) analysis, PHAs can be certainly considered as very promising materials. They can boost in short time the sector of biodegradable polymers, since these new materials can be adapted to the current transformation technologies and not the other way around. Price reductions, encouraging policies, and new restrictions to polluting plastics may play a major role in increasing the number of plastic products based on PHAs.

8.3 Cellulose Biocomposites

The use of vegetable or lignocellulosic fibers as reinforcement in polymeric matrix composites has generated enormous interest in recent decades. This growing interest shown in the use of these materials is motivated not only by technical issues but also by socio-economic and environmental issues. From a technical point of view, these materials are suitable as reinforcements due to their low density, good strength, and high specific stiffness. From an economic point of view, since they are very abundant and cheap materials, they can help in reducing product costs. Finally, from the environmental point of view, the use of lignocellulosic fibers in biocomposites is attractive due to their natural origin and biodegradability and the possibility of recovery of residues of plant origin. Indeed, the development of composite materials based on lignocellulosic fibers combined with biodegradable and bio-based polymers, such as PHA, is in line with a product concept aimed at promoting sustainable development and circular economy [80–83].

8.3.1 Structure, Composition, and General Properties of Lignocellulosic fibers

Vegetable or lignocellulosic fibers are found in nature as part of the plant tissues of trees, plants, fruits, seeds, etc. Therefore, although they share certain general characteristics, their composition and specific properties depend on numerous factors such as the source of origin, the location in the plant, the age of the plant, the time of the year, the extraction methods used, etc. [84, 85].

These materials can be obtained from primary sources, that is, from plants grown for their high fiber content; from secondary sources, those in which fiber is obtained as a by-product; or, even tertiary, in which fiber is obtained from residues of plant origin as a form of recovery thereof [83, 86]. Depending on the location in the plant, plant fibers can be grouped into coarse or libero fibers (hemp, flax, jute), leaf fibers (sisal, abaca, agave), seeds or fruits (cotton, coconut, husks), grass, cane (bamboo), stem (wheat, rice, or corn straw) or wood (hard or soft) [82, 85–87]. In the case of plant residues, the classification within one of the aforementioned groups can be complicated since, often, they can be formed by mixtures, for example, of shells, stems and herbs.

Vegetable fibers can essentially be considered composite materials. Broadly speaking, they are formed by semicrystalline cellulose fibers embedded in an amorphous matrix composed mainly of hemicelluloses and lignin and other minor components such as pectin, waxes or fatty acids. Hence, they are also called lignocellulosic fibers [36, 37, 88]. These elementary fibers are joined together forming "packages" to configure different plant tissues with different functionalities within the plant [82]. In Figure 8.5, the characteristic structure of elementary plant fibers is shown.

Plant cells have three different parts: a central channel called lumen (responsible for the transport of nutrients and other substances to the plant), the cell wall and the intercellular layer or middle lamella. The cell wall is formed, in turn, by different layers composed of cellulose microfibrils arranged in different orientations depending on the layer to which they belong and the type of plant tissue. In the primary cell wall, cellulose microfibers are randomly arranged in a matrix of pectin, hemicellulose, lignin, and proteins. In the secondary cell wall, which is divided into three layers (internal, middle, and external), cellulose microfibers are arranged in a spiral or helical orientation characterized by a certain angle. This particular orientation of cellulose microfibrils has a marked influence on the mechanical properties of the fiber. The intercellular layer or middle lamella is composed mainly of pectin that acts as a binding material between different fibers [80, 82, 89]. Other constituents of lignocellulosic fibers are, to a lesser extent, waxes, fatty acid proteins, and other components and extracts depending on the type of plant.

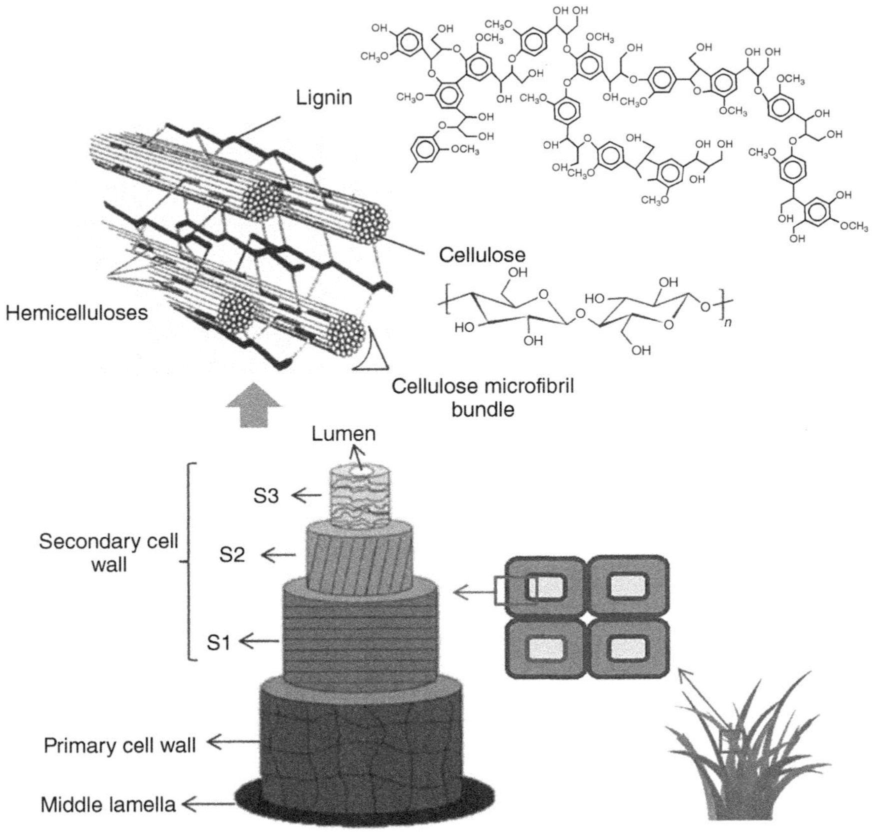

Figure 8.5 Schematic representation of the structure of lignocellulosic fibers.

Generally, cellulose represents about 40% or 50% of the dry mass of the fibers [90] and can even reach up to 80%, for example, in cotton fibers [91]. It is a long-chain linear hydrophilic polysaccharide formed by units of β-D-glucose linked by β-(1→4)-glycosidic bonds [80, 89]. Each monomer unit has three hydroxyl groups that can establish hydrogen bonds with other hydroxyl groups within the same chain or with adjacent chains to form cellulose microfibers. There are six polymorphic forms of cellulose. They are I, II, III_I, III_{II}, IV_I, and IV_{II}. Type I cellulose is the one found naturally in plants and can have two sub-allomorphs, form I_α and form I_β. In both cases, the chains are aligned in parallel and they differ only in the arrangement of hydrogen bonds [89, 90]. Cellulose fibrils have areas of high crystallinity and amorphous regions. In general, the cellulose present in wood or plants has a crystallinity of about 60–70% [89, 90].

Hemicelluloses are short-chain amorphous polysaccharides, including glucose, xylose, mannose, galactose, or arabinose [36, 37, 88, 90]. These components can represent up to 25–40% of the dry mass of the fibers [90].

Lignin acts as a support component in plants. It is a complex three-dimensional amorphous polymer consisting of both aliphatic and aromatic units [89]. It is formed by the dehydrogenation of phenylpropane molecules and their subsequent binding by ester, ether, or carbon–carbon bonds [90, 92]. The main monomer units of lignin are coumaryl, coniferyl and sinapyl alcohol [89]. The exact composition of lignin depends on the plant and even differs in the different tissues within it [90]. The lignin content depends on the plant species, the type of tissue, and the age of the plant and can vary greatly from one fiber to another from about 5% content, for example in flax fibers, to contents close to 50%, for example in coconut fibers [83].

Pectins are the main component of the middle lamina of the cell wall. They are branched heteropolysaccharides capable of forming semirigid gels in the presence of water and calcium ions. These branched structures provide rigidity to the plant [92]. The outer part of the plant tissues is covered by a bioprotective cuticle formed by waxes and other lipophilic components. In addition to these components, each plant has its own constituents depending on the species in question [90].

The properties of lignocellulosic fibers depend on both their microstructural characteristics and their composition (cellulose content, lignin, hemicelluloses, etc.), which in turn are a function of the type of plant, the type of plant tissue, the age of the plant, plant harvest time, weather conditions, storage, fiber extraction method, etc. [81, 93]. Therefore, taking into account the large number of plant species that exist and all the variables that influence, the properties of lignocellulosic fibers vary significantly from case to case. However, they have some common properties. They are biodegradable fibers, hydrophilic in nature with high water absorption capacity, have low density and high specific mechanical properties, and have a relatively low thermal stability [80, 82, 93].

In general, the mechanical properties of the fibers, that is stiffness and mechanical strength, are closely related to their composition, microstructure, and presence of defects. Mainly the cellulose content, microfibrillar angle, crystallinity or fibril diameter have a greater influence. Although as noted by Mohanty et al. [84] it is difficult to establish a direct relationship due to the complex microstructure of the fibers, in general, fibers with small diameter, high cellulose content, high crystallinity, and low microfibrillar angle have high mechanical strength and stiffness. The lignin content also has a marked influence on the microstructure and morphology of the fibers and, therefore, on their mechanical properties [36, 84]. Waxes and other lipid substances largely determine the wettability and adhesive characteristics of the fibers [84].

The thermal stability of the fibers is also closely related to their composition. Yang et al. [94] studied the pyrolysis of the main fiber components by means of TGA-DSC. Hemicelluloses begin their thermal degradation at relatively low temperatures, around 220 °C, thermal degradation of cellulose occurs in a temperature range between 315 and 400 °C, while thermal degradation of lignin occurs slowly over a wide range of temperatures (150–900 °C).

The different fiber components also exhibit different behavior against photochemical degradation or biodegradation. Lignin is more sensitive to ultraviolet degradation, followed by hemicelluloses and cellulose. Biodegradation is faster in hemicelluloses, amorphous regions of cellulose and finally the crystalline regions of cellulose and lignin [84, 95].

8.3.2 Lignocellulosic Fibers in Polymer Composites

One of the most used strategies to improve polymer properties and/or lower costs is the use of fillers. Filling materials can be classified into two large groups, loads and reinforcements. The main objective of the load is to reduce the cost of the product by reducing the amount of resin needed. The function of reinforcement materials is to provide rigidity and mechanical strength to the polymer matrix [67, 96]. Therefore, the efficiency of said reinforcement will depend on the properties of the matrix, the properties of the reinforcement, and its geometric characteristics and the adhesion between both phases. The reinforcement materials can be long or short fibers, oriented or not, microfibers, particles, or nano reinforcements, and must have high rigidity and mechanical strength [96].

Due to the abundance and availability of lignocellulosic fibers, their low cost, and their excellent specific mechanical properties, this type of materials are suitable both as loads and as reinforcements, covering a wide range of possibilities. However, they have certain limitations, such as their relatively low thermal stability or high hydrophilicity [81, 83, 85, 95]. Table 8.1 summarizes some of the main advantages and disadvantages of lignocellulosic fibers in polymeric matrices.

The high hydrophilic nature of these fibers causes them to exhibit low compatibility with polymeric matrices, generally hydrophobic. Since the fiber–matrix interaction is a determining factor in the final properties of the compound, different strategies, aiming to improve them, have been documented in the literature (reviewed in [83, 89, 92, 97–100]). These strategies are focused on two main approaches: (i) fiber modification through physical, chemical, or biological methods to improve its wettability or favor physical and/or chemical anchoring with the polymer matrix and (ii) establish chemical bonds between fiber and matrix.

Table 8.1 Main advantages and disadvantages of lignocellulosic fibers in polymer matrix composite materials.

Advantages	Disadvantages
Renewable origin, abundance, and low cost	Quality and variable properties
Biodegradability/compostability, low density	Relatively low thermal stability
High rigidity and specific mechanical resistance	High moisture absorption
Recyclability	Relatively low durability (due to biological and/or environmental degradation)
Non-abrasive for processing equipment	

Source: Adapted from Faruk et al. 2014 [81].

8.3.2.1 Fiber Modification

Physical modification The physical methods intend to introduce structural and surface modifications to the fibers that allow them to modify their surface energy, thus improving compatibility with the matrix and/or modifying its topography, favoring mechanical anchoring with the polymer. Some of the physical fiber treatments include stretching, calendering, heat treatments, corona discharge, or plasma treatment.

Chemical modifications The pretreatment of the fibers can lead to cleaner surfaces or introduce chemical changes that favor adhesion with the polymer matrix [82]. Chemical fiber modification treatments can be applied to modify the surface tension and polarity of the fibers or to activate the hydroxyl groups of their surface or introduce new functional groups that allow anchoring with the matrix [101]. There are a large number of methods of chemical modification of fibers based on the purification or "washing" of the fibers (removal of waxes and other noncellulosic components, delignification), impregnation of the fibers or surface functionalization thereof [100].

One of the most popular treatments is alkaline treatment with sodium hydroxide. This treatment partially eliminates noncellulosic components of the fibers, mainly lignin and waxes, and promotes the disruption of hydrogen bonds in the cross-linked structure of the fibers by increasing the number of reactive sites on the fiber surface. It also introduces greater surface roughness and may favor defibrillation and reduction in fiber diameter [86, 102]. With the elimination of noncellulosic components, fibers with a higher cellulose content and crystallinity are obtained, showing better mechanical performance, greater homogeneity of properties, greater whiteness, and reduction in size. Indeed, this treatment offers the possibility of processing the fibers at higher temperatures, since the onset of thermal degradation shifts to higher temperatures after eliminating hemicelluloses [103].

Other popular methods of fiber purification or delignification include extraction with organic solvents, the use of potassium permanganate, sodium hypochlorite, acetic acid, or peracetic acid, among others [80]. The use of chlorinated compounds is quite common in the pretreatment of lignocellulosic fibers. However, it results in the generation of contaminated spills, so purification processes are moving toward the use of chlorine-free treatments [104]. Peracetic acid treatment has gained some popularity, as it has proven effective and environmentally respectful [105].

Regarding the methods of surface functionalization in the literature, various strategies based on functionalization with silanes, isocyanates, acetylation, esterification, acrylation, etc. [89, 106, 107] have been described.

Biological modifications An alternative to the physical and chemical methods of fiber modification is biological modification by fungi or enzymes. Some of the advantages of these methods are that they allow the elimination of pectic or hemicellulosic components selectively with low energy costs [82, 101].

8.3.2.2 Fiber-matrix Chemical Anchor

The chemical anchoring of the fiber to the matrix can be achieved by ex-situ techniques based mainly on copolymerization or grafting strategies of the fibers with one or more monomers (this group would also include some surface functionalization strategies described above), or in-situ compatibility directly with the matrix by reactive processing [89].

Reactive processing involves the addition of reactive agents during melt mixing that interact with both the matrix and the fiber to promote compatibility. This technique has aroused great interest because the compatibilization is carried out in a single stage, so it is a simple and economically effective technique. In addition, it does not imply an additional cost of energy nor does it entail the generation of spills or other waste.

The most common reactive agents are: peroxides (e.g. dicumyl peroxide), capable of generating intermediate radical species to generate a chemical bond between the polymer and the fiber [108, 109]; bifunctional or multifunctional reagents capable of reacting with the fiber and with the polymer such as, for example, isocyanates [110, 111], epoxies [112, 113], or different carboxylic anhydrides [110, 114, 115], among others.

The different treatments of fiber modification and/or chemical anchoring with the polymer aim at enhancing fiber-matrix adhesion and, therefore, the mechanical behavior of the compounds. However, these treatments can also affect the water absorption capacity of the fibers [116], the thermal stability of the compounds, or the biodegradability. For example, the removal of impurities and noncellulosic components can have a positive impact on the thermal stability of the fibers; however, in some cases the chemical treatment may deteriorate the fiber structure and thermal stability may be reduced [103]. Different compatibilizing or coupling agents can, in some cases, reduce the rate of biodegradation. Such behavior has been observed, for example, in compounds compatible with maleic anhydride [117] or acrylic acid [118], in which the degradation rate was reduced with respect to the noncompatible compounds.

8.4 PHA/Fiber Composites

8.4.1 PHB and PHBV/Cellulose Composites: Achievements and Limitations

The development of compounds based on PHA and lignocellulosic fibers has been intensely investigated in recent decades. Bhardwaj et al. [119] point out that the relative polar nature of PHAs and the presence of carbonyl groups (>C=O) capable of establishing hydrogen bonds with hydroxyl groups of lignocellulosic fibers make for better fiber-matrix compatibility compared to other polymers, such as polypropylene. In fact, the presence of hydrogen bonds between the fibers and the polymer matrix has been confirmed by infrared spectroscopy by Gunning and coworkers [120] in PHB compounds with jute fibers. Torres-Tello et al. [121] also observed some compatibility between PHB and PHBV and agave fibers. In this case, the incorporation of the fibers meant an increase in the elastic modulus, both flexural and tensile, of impact resistance and a slight improvement in thermal stability. Rossa et al. [122] observed an improvement in the elastic modulus and mechanical resistance, as well as an increase in the HDT in PHBV compounds with carauá fibers; however, the thermal stability of those composites was reduced with respect to neat PHBV.

Despite the fact that some chemical compatibility between the lignocellulosic fibers and the PHA may exist, a lack of adhesion between the untreated lignocellulosic fibers and the matrix is generally observed. However, the results vary from one work to another depending on the type of fiber, the type of composite (continuous or discontinuous fibers, fiber size, etc.) and the processing technique. In general, in composites based on PHA and lignocellulosic fibers not subjected to any pretreatment, a lack of adhesion between the fiber and the matrix is observed. In most cases, although a reinforcement effect characterized by an increase in the elastic modulus is observed, the mechanical strength and elongation at break are negatively affected due to this lack of adhesion, which causes the fibers to act as concentrators of tensions [123–126]. Other properties, such as thermal stability, are also negatively affected in most cases [127, 128], although it depends, to a large extent, on the type of fiber and the processing conditions.

Regarding the influence of the fibers on the kinetics of crystallization and the crystallinity of PHA, the results vary widely. In some cases, the presence of the fibers does not seem to have any effect on the crystallization and melting temperatures of the polymer or on the crystallinity index, while in others they are affected differently depending on the matrix, the fiber content, the type of fiber, its size, or its geometry. However, it is difficult to establish a correlation. For example, Torres-Tello et al. [121] did not observe variations in these parameters in PHB/agave compounds, while in PHBV/agave compounds they observed a reduction in crystallinity with fiber content and an increase in melting temperature. These authors attribute these differences to the fact that fibers can act as nucleating agents, promoting crystallization, and restricting the movement of chains. Berthet et al. [125] observed a reduction in the crystallization temperature together with a slight decrease in crystallinity in PHBV compounds (with wheat straw and olive pomace milling residues), both attributed to a restriction in the movement of polymer chains due to the presence of the fibers. However, Avella et al. [129] observed an increase in the crystallization rate in PHBV compounds with wheat straw due to a nucleating effect of the fibers. Krishnaprasad et al. [123] also observed a nucleating effect of the fibers in PHB compounds with bamboo microfibers. Scalioni et al. [130] also found a nucleating effect of the fibers in PHB compounds with caraua fibers. In this case, they observed the formation of transcrystalline layers around the fibers, an increase in lamellar thickness, a decrease in crystallite size, and a spherulite-axial transition of the crystals.

There are many documented strategies to improve the behavior of PHA/fiber compounds. These strategies are mainly aimed at improving fiber-matrix adhesion, either by modifying the fibers or the matrix, or by the use of additives or compatibilizing reactive agents.

Dufresne et al. [131] studied the effect of plasma modification of fibers from fir wood flour and olive pits on the properties of PHBV matrix. These authors observed a poor adhesion between the fibers and the matrix even in the plasma-treated fibers, so that said treatment did not produce an improvement of the mechanical properties. Ventura et al. [132], on the other hand, did report an improvement in mechanical properties, both in stiffness and mechanical strength, in PHB compounds with flax fibers treated by plasma, attributed to greater fiber-matrix adhesion.

There are several authors who have studied different chemical treatments of lignocellulosic fiber purification and/or superficial chemical modification (wax removal, delignification, acetylation, use of coupling agents such as silanes, etc.) and their influence on the properties of PHA based compounds. Melo et al. [133] studied the effect of different chemical pretreatments on carnauba fibers in PHB matrix. Such treatments included: alkaline treatment with NaOH, and different combined treatments of NaOH + potassium permanganate, NaOH + acetylation, and NaOH + hydrogen peroxide. The treatment with hydrogen peroxide proved to be more efficient resulting in a greater wettability of the fibers by the matrix, which resulted in an increase in the elastic modulus and mechanical strength. Moreover, SEM observation of the fibers revealed an increase in the surface roughness, compared to the untreated fibers. This roughness would also improve the adhesion to the matrix. Ahankari et al. [134] also observed a clear improvement in mechanical properties and fiber-matrix adhesion in PHBV compounds with wheat straw, soybean, and corn straw fibers treated with NaOH. Indeed, the elastic modulus and mechanical strength values of the compounds with treated fibers were even higher than those corresponding to virgin PHBV, although the elongation at break and impact resistance were lower than those of the virgin polymer. In addition, the treatment resulted in an improvement in the thermal stability of the compounds. These authors also observed a greater compatibility of the fibers with the PHBV than with the PP.

The use of coupling agents such as maleic anhydride [110, 114, 135, 136], different silanes [114, 136], or acrylic acid [118, 137], through fiber functionalization techniques or through matrix grafting techniques, has demonstrated, in general, an improvement in fiber-matrix compatibility. The use of other additives such as plasticizers to improve matrix fiber compatibility has also been explored [138, 139]. In these cases, although there is some improvement in the wettability of the fiber by the matrix due to the adsorption of the plasticizer on its surface, the plasticizer can migrate through the matrix affecting its properties.

One of the strategies that has aroused the greatest interest is reactive compatibilization since it is a technique that includes a single stage and, therefore, is simple, economically viable, and more environmentally friendly than other techniques in terms of consumption of additional energy or discharge generation [140, 141]. There are different compatibilizing or coupling reactive agents that have functional groups capable of reacting in situ during the melt mixing stage with the abundant hydroxyl groups present on the surface of the fibers and even with the terminal hydroxyl and carboxyl groups of the polymer, thus establishing links between the fibers and the matrix and improving the adhesion between both components. Some of the most common reactive agents contain maleic anhydride groups (PHA-g-MA) [110, 142, 143], epoxy groups [110], isocyanate groups [110, 144], or carboxyl-terminated polyesters [110]. The use of peroxides, such as dicumyl peroxide [109, 140], as reactive compatibilization initiators by generation of intermediate radical species that allow the formation of random bonds between the fiber and the matrix, is also a very common strategy. In Table 8.2, some examples of different works documented in the literature related to the reactive compatibilization of PHA/fiber compounds and the main results obtained are summarized.

8.4.2 New Trends in PHB and PHBV/Cellulose-Reinforced Biocomposites

Logically, the improvement of matrix-fiber adhesion has a positive impact on mechanical properties. However, the high fragility of PHA causes PHA/fiber compounds to exhibit, in general, low toughness. Although, in some cases, the incorporation of the fibers may result in a certain improvement in impact strength, the ductility of the compounds is low and is limited by the properties of the matrix. Therefore, the incorporation and compatibilization of lignocellulosic fibers can be combined with the different strategies to improve the properties of the matrix (described in Section 8.2). Thus, for example, Nagarajan et al. [144] studied the properties of PHBV/polybutylene adipate terephthalate (PBAT) mixtures with different contents of rod grass fiber (*Panicum virgatum*) and the effect of using poly(diphenylmethane diisocyanate) (pMDI) as a compatibilizer, observing a general improvement in mechanical behavior. These authors suggest that pMDI acts not only by making the fibers compatible with the matrix but also by improving the interaction of the two polymeric phases by chain extension reactions. Zhang et al. [147] studied the behavior of mixtures of PHBV/PBAT, an impact modifier (epoxidized natural rubber, ENR) and miscanthus fibers obtained by reactive extrusion with DCP. These authors observed that the DCP and the ENR act collaboratively in the compatibility of the different phases. The combined effect of the different components results in balanced properties in terms of stiffness, mechanical strength, toughness, and thermal resistance.

Sanchez-Safont et al. [148] studied PHBV compounds with an elastomeric phase (thermoplastic polyurethane, TPU) and cellulose fibers as a technical reasonably good environmental friendly alternative to nonbiodegradable oil-based polymers for injection molding applications. The results showed a homogeneous distribution of the TPU and cellulose, well dispersed within the PHBV matrix. The fairly good affinity between PHBV and TPU and between PHBV and cellulose indicated that these systems might be compatible enough to be processed without any compatibilization agents. TPU promotes plastic deformation and an increase in toughness, whereas cellulose increases the elastic modulus but yields premature break of the composites. Thus, adding cellulose to PHBV/TPU blends counteracts some of the loss of mechanical performance but maintaining the toughness improvements ascribed to TPU. In addition, the PHBV/TPU/cellulose systems proved to be biodisintegrable under standard composting conditions.

Table 8.2 Examples of reactive compatibilization in PHA/fiber composites.

Matrix	Fiber	Reactive agent	Results	References
PHBV	Miscanthus	Dicumyl peroxide (DCP)	Enhanced fiber-matrix adhesion. Grafting evidenced by FTIR. Improvement of mechanical strength and elongation at break. T_m reduction but X_c does not vary.	[109]
PHBV	Miscanthus	DCP	Improvement of fiber-matrix adhesion. Improvement of mechanical tensile and flexural strength. Reduction of the elastic modulus attributed to decrease of X_c. Reduction of thermal stability	[140]
PHB/PHBV	α-Cellulose	DCP	Enhanced fiber-matrix adhesion. Improvement of elastic modulus and mechanical resistance. X_c decrease. Thermal stability is not affected	[145]
PHB	Pine wood flour	pMDI (poly(diphenyl-methane diisocyanate)) Uralac (carboxyl-terminated polyester) DER (epoxy resin) PHB-g-MA	In general compatibility effect: pMDI > Uralac > DER ≈ PHB-g-MA	[110]
PHBV	Kenaf	PHBV-g-MA	Improvement of rigidity and mechanical resistance. Improvement of water resistance	[143]
PHBV	Bambu fiber	pMDI	X_c is not affected with pMDI but increases T_c and T_m (greater crystalline perfection)	[146]
PHBV	Pine wood flour (WF) and organo polysiloxane treated pine WF	pMDI PHBV-g-MA	Enhanced adhesion Improved stiffness, mechanical strength and impact resistance. Thermal and oxidative stability are not affected	[111]

Despite the affinity between the matrix and dispersed phases, the failure of the compounds was triggered by the limited bond strength at the interface of PHBV with the other phases. In a later work [149], Sanchez-Safont et al studied the compatibilization of the PHBV/TPU/cellulose ternary systems by reactive extrusion with three reactive agents: hexamethylene diisocyanate (HMDI), Joncryl ® and triglycidyl isocianurate (TGIC). The results indicated that the reactive agents play a main role as compatibilizers at the interphases of the PHBV/TPU/Cellulose compounds, without compromising their biodisintegrability in composting conditions. HMDI showed the highest ability to compatibilize the cellulose and the PHBV in the compounds, with the topmost values of deformation at break and static toughness. Joncryl® and TGIC, on the other hand, seemed to enhance the compatibility between both the fibers and the polymer matrix as well as the TPU within the PHBV.

The findings of these works point to a route to modify properties of PHAs through blending, with the addition of reactive agents, which can help to overcome some of the difficulties that were encountered in these materials, allowing them to be used in standard applications making them also very promising candidates for food packaging.

8.4.3 The Potential Use of PHA-Based Composites in the Food Packaging Sector

The food packaging industry is nowadays under public criticism and pressure by the regulators because of the huge environmental impact related to the amount of waste generated by this sector. There is, therefore, an urgent need to move toward more sustainable and eco-friendly packaging. The first and more obvious solution could be to reduce the amount of packaging. This strategy makes sense as long as only unnecessary packaging is eliminated. However, packaging cannot be completely removed as it is a preservation medium. Without packaging, products will need to be produced locally and consumed only within a few days, and the unconsumed food waste would increase considerably. Moreover, this goes against current trends where people shop once a week and do not have time to cook.

Assuming food packaging is necessary and sustainable, there is a need to confront the problem of the residues. The first approach is the recycling of the post-use plastics. However, packaging waste is complex to recycle due to several reasons: use of multilayer structures, wide variety of compositions (materials, additives, and fillers), high impurity content of the residue flow, and, most relevantly, the use of post-consumer recycled plastic material in food contact applications is forbidden (besides PET). Therefore, recycling may not be the most sustainable option.

The second approach, which is the one most chapters of this book are devoted to, is the use of bioplastics and, more specifically, biodegradable plastics. This approach would allow the treatment of the food packaging residue together with the organic matter, thus using composting as the industrial waste treatment strategy. This scheme would solve some of the problems derived from the plastic residues, while allowing the food industry to use the optimized processes and advanced technologies of the current packaging industry.

Within this framework, the PHA-based composites pose some advantages with respect to other plastic solutions for the food packaging sector (including other bioplastics). As stated earlier, PHA is a thermoplastic polyester and can be processed and conformed into a package by using the actual equipment without further change than adjusting the processing conditions. As discussed in Section 8.2 of the chapter, the performance of PHA and PHA-based composites for food packaging sector is similar to nonbiodegradable plastics currently used in the industry. However, PHA is among the very few thermoplastic polymers whose structure is naturally present and, therefore, is fully biodegradable in the environment. This fact will allow PHA-based products to be treated as organic matter in industrial composting processes, but also in the case that they end up released in the environment, they will naturally biodegrade by microorganisms within a reasonably short period of time (thus reducing the long-term influence to the environment and the production of microplastics).

Finally, PHA, unlike other bioplastics, such as PLA or thermoplastic starch, does not come from food resources, i.e. it is not food-competitive. PHA is produced from bacteria that can be also grown by using residues, for instance organic matter or PHA residues. This new approach is what is being called biocircular economy. This concept will considerably reduce the residues of the food industry while ensuring new and safe material for the packaging sector.

In this chapter, we are advocating for single-use food packaging PHA-based concepts which do not need to be recycled and may serve to create new packaging in a closed cycle.

8.5 Conclusions

After analysing the current state of art of composites based on PHAs and lignocellulosic fibers, it is possible to perform a new SWOT analysis:

Strengths

- The use of vegetal fibers in PHA-based composites can contribute to increase the green concept of these materials due to their natural origin and biodegradability.
- Vegetal fibers are cheap, available and abundant in nature, and therefore their use in PHA-based composites allows to reduce the final costs of the product.
- The possibility of reusing local vegetable wastes in PHA-based composites in both matrix and reinforcement helps to promote the circular economy.
- The use of PHA-based composites can help to reduce the environmental impact of the residues derived from food packaging applications.
- The use of PHA composites allow the production of multilayer structures without decreasing the sustainability of the packaging.

Weaknesses

- The compatibility between vegetal fibers and PHAs is limited. Depending on factors such as the type of fiber or purity, said lack of compatibility may lead to a reduction in mechanical performance.
- The use of vegetal fibers in PHA-based composites may lead to a reduction in thermal stability during processing.
- The cost of PHA is still very high for the packaging sector, while the production capacity is not ready to provide enough PHA to satisfy the needs of the packaging industry.

Opportunities

- Strategies such as reactive extrusion or blending with biodegradable elastomers can counteract the technical limitations (mechanical and thermal performance) of PHA/fiber composites without compromising their biodegradability.
- The expected price reduction of PHAs in the coming years together with the contribution of vegetal fibers to the cost reductions offers the opportunity to position PHA/cellulose composites in the market.
- The growing demand for biodegradable materials driven by the increasingly restrictive environmental policies (e.g. the ban on single-use products made of plastics) is considered a great market opportunity for these materials. In addition, these composites can be processed by conventional techniques without the need to adapt the existing infrastructure in the plastics sector.
- The foreseen changes in the legislation will enforce the use of biodegradable polymers for food packaging applications.
- The use of highly sustainable and biocircular PHA-based materials for food packaging applications may help the society to accept plastic packaging as a sustainable option.

Threats

- These materials have to compete with other biodegradable materials such as paper, paperboard, or other polymers.
- It is necessary to have a guarantee of supply, quality, and homogeneity of the fibers, especially those from waste.
- The changes in legislation may not discriminate environmentally biodegradable PHA-based plastics from other polymers.

As it has been discussed, the SWOT analysis of PHA-based materials concluded that these materials can be considered very promising currently. Indeed, the growth expectations for composites based on PHAs and cellulose or vegetal fibers are even more optimistic. While adding cellulosic fibers to a nonbiodegradable material does not imply a significant added value from an environmental point of view, this kind of reinforcement when added to a biodegradable matrix such as PHAs enhances its value in three directions: performance, economics, and environmental friendliness.

References

1 Rujnić-Sokele, M. and Pilipović, A. (2017). Challenges and opportunities of biodegradable plastics: a mini review. *Waste Management & Research* 35: 132–140. https://doi.org/10.1177/0734242X16683272.
2 Lambert, S. and Wagner, M. (2017). Environmental performance of bio-based and biodegradable plastics: the road ahead. *Chemical Society Reviews* 46: 6855–6871. https://doi.org/10.1039/c7cs00149e.
3 Reddy, M.M., Vivekanandhan, S., Misra, M. et al. (2013). Biobased plastics and bionanocomposites: current status and future opportunities. *Progress in Polymer Science* 38: 1653–1689. https://doi.org/10.1016/j.progpolymsci.2013.05.006.
4 IUPAC (1997). *Compendium of Chemical Terminology* 2 (the "Gold Book") by A.D. McNaught and A. Wilkinson. Oxford: Blackwell Scientific Publications. XML on-line corrected version: http://goldbook.iupac.org (2006) created by M. Nic, J. Jirat, and B. Kosata; upd, 2014. doi:http://dx.doi.org/10.1351/goldbook.I03352.
5 Kale, G., Kijchavengkul, T., Auras, R. et al. (2007). Compostability of bioplastic packaging materials: an overview. *Macromolecular Bioscience* 7: 255–277. https://doi.org/10.1002/mabi.200600168.
6 Nakajima, H., Dijkstra, P., and Loos, K. (2017). The recent developments in biobased polymers toward general and engineering applications: polymers that are upgraded from biodegradable polymers, analogous to petroleum-derived polymers, and newly developed. *Polymers (Basel)* 9: 523. https://doi.org/10.3390/polym9100523.
7 Vert, M., Doi, Y., Hellwich, K.H. et al. (2012). Terminology for biorelated polymers and applications (IUPAC recommendations 2012). *Pure and Applied Chemistry* 84: 377–410.
8 Steinbüchel, A. (2005). Non-biodegradable biopolymers from renewable resources: perspectives and impacts. *Current Opinion in Biotechnology* 16: 607–613. https://doi.org/10.1016/j.copbio.2005.10.011.
9 Funabashi, M., Ninomiya, F., and Kunioka, M. (2009). Biodegradability evaluation of polymers by ISO 14855-2. *International Journal of Molecular Sciences* 10: 3635–3654. https://doi.org/10.3390/ijms10083635.
10 Rhim, J.-W., Park, H.-M., and Ha, C.-S. (2013). Bio-nanocomposites for food packaging applications. *Progress in Polymer Science* 38: 1629–1652. https://doi.org/10.1016/j.progpolymsci.2013.05.008.
11 Bordes, P., Pollet, E., and Averous, L. (2009). Nano-biocomposites: biodegradable polyester/nanoclay systems. *Progress in Polymer Science* 34: 125–155. https://doi.org/10.1016/j.progpolymsci.2008.10.002.
12 Lagaron, J.M. and Lopez-Rubio, A. (2011). Nanotechnology for bioplastics: opportunities, challenges and strategies. *Trends in Food Science and Technology* 22: 611–617. https://doi.org/10.1016/j.tifs.2011.01.007.
13 Mousa, M.H., Dong, Y., and Davies, I.J. (2016). Recent advances in bionanocomposites: preparation, properties, and applications. *International Journal of Polymeric Materials and Polymeric Biomaterials* 65: 225–254. https://doi.org/10.1080/00914037.2015.1103240.

14 Możejko-Ciesielska, J. and Kiewisz, R. (2016). Bacterial polyhydroxyalkanoates: still fabulous? *Microbiological Research* 192: 271–282. https://doi.org/10.1016/j.micres.2016.07.010.
15 Albuquerque, P.B.S. and Malafaia, C.B. (2018). Perspectives on the production, structural characteristics and potential applications of bioplastics derived from polyhydroxyalkanoates. *International Journal of Biological Macromolecules* 107: 615–625. https://doi.org/10.1016/j.ijbiomac.2017.09.026.
16 Gumel, A.M., Annuar, M.S.M., and Chisti, Y. (2013). Recent advances in the production, recovery and applications of polyhydroxyalkanoates. *Journal of Polymers and the Environment* 21: 580–605. https://doi.org/10.1007/s10924-012-0527-1.
17 Lemoigne, M. (1923). Production d'acide β-oxybutyrique par certaines bactéries du groupe du *Bacillus subtilis. Comptes Rendus Hebdomadaires des Séances de l'Académie des Sciences* 176: 1761.
18 Lemoigne, M. (1927). Chemical origin of the products of dehydration and of polymerization of β-hydroxybutyric acid. Hydroxybutyric fermentation. *Bulletin de la Société de Chimie Biologique.*
19 Lemoigne, M., Grelet, N., and Croson, M. (1950). Sur l'origine des lipides beta hydroxybutyriques formes par processus. *Bulletin de la Société de Chimie Biologique* 32: 719–721.
20 Lenz, R.W. and Marchessault, R.H. (2005). Bacterial polyesters: biosynthesis, biodegradable plastics and biotechnology. *Biomacromolecules* 6: 1–8. https://doi.org/10.1021/bm049700c.
21 Akaraonye, E., Keshavarz, T., and Roy, I. (2010). Production of polyhydroxyalkanoates: the future green materials of choice. *Journal of Chemical Technology and Biotechnology* 85: 732–743. https://doi.org/10.1002/jctb.2392.
22 Baptist, J.N. and Werber, F.X. (1960). Molded product containing poly-beta-hydroxybutyric acid and method of making. US3107172A.
23 Baptist, J.N. and Werber, F.X. (1962). Plasticized poly-beta-hydroxybutyric acid and process. US3182036A.
24 Noel, B.J. (1960). Process for preparing poly-beta-hydroxybutyric acid. US3044942A.
25 Wallen, W.K. and Rohwedder, L.L. (1974). Poly-betahydroxyalkanoate from activated sludge. *Environmental Science & Technology* 8: 576–579.
26 Keshavarz, T. and Roy, I. (2010). Polyhydroxyalkanoates: bioplastics with a green agenda. *Current Opinion in Microbiology* 13: 321–326. https://doi.org/10.1016/j.mib.2010.02.006.
27 Anjum, A., Zuber, M., Zia, K.M. et al. (2016). Microbial production of polyhydroxyalkanoates (PHAs) and its copolymers: a review of recent advancements. *International Journal of Biological Macromolecules* 89: 161–174. https://doi.org/10.1016/j.ijbiomac.2016.04.069.
28 Philip, S., Keshavarz, T., and Roy, I. (2007). Polyhydroxyalkanoates: biodegradable polymers with a range of applications. *Journal of Chemical Technology & Biotechnology* 82: 233–247. https://doi.org/10.1002/jctb.1667.
29 Bugnicourt, E., Cinelli, P., Lazzeri, A., and Alvarez, V. (2014). Polyhydroxyalkanoate (PHA): review of synthesis, characteristics, processing and

potential applications in packaging. *Express Polymer Letters* 8: 791–808. https://doi.org/10.3144/expresspolymlett.2014.82.

30 Chanprateep, S. (2010). Current trends in biodegradable polyhydroxyalkanoates. *Journal of Bioscience and Bioengineering* 110: 621–632. https://doi.org/10.1016/j.jbiosc.2010.07.014.

31 Wang, Y., Yin, J., and Chen, G.Q. (2014). Polyhydroxyalkanoates, challenges and opportunities. *Current Opinion in Biotechnology* 30: 59–65. https://doi.org/10.1016/j.copbio.2014.06.001.

32 Singh, M., Kumar, P., Ray, S., and Kalia, V.C. (2015). Challenges and opportunities for customizing polyhydroxyalkanoates. *Indian Journal of Microbiology* 55: 235–249. https://doi.org/10.1007/s12088-015-0528-6.

33 Martin, D.P. and Williams, S.F. (2003). Medical applications of poly-4-hydroxybutyrate: a strong flexible absorbable biomaterial. *Biochemical Engineering Journal* 16: 97–105. https://doi.org/10.1016/S1369-703X(03)00040-8.

34 Wang, S., Chen, W., Xiang, H. et al. (2016). Modification and potential application of short-chain-length polyhydroxyalkanoate (SCL-PHA). *Polymers (Basel)* 8: 273. https://doi.org/10.3390/polym8080273.

35 Laycock, B., Halley, P., Pratt, S. et al. (2013). The chemomechanical properties of microbial polyhydroxyalkanoates. *Progress in Polymer Science* 38: 536–583. https://doi.org/10.1016/j.progpolymsci.2012.06.003.

36 Reddy, N. and Yang, Y. (2005). Biofibres from agricultural byproducts for industrial applications. *Trends in Biotechnology* 23: 22–27. https://doi.org/10.1016/j.tibtech.2004.11.002.

37 Heredia, A., Jiménez, A., and Guillén, R. (1995). Composition of plant cell walls. *Zeitschrift für Lebensmittel-Untersuchung und -Forschung* 200: 24–31. https://doi.org/10.1007/BF01192903.

38 Keskin, G., Kızıl, G., Bechelany, M. et al. (2017). Potential of polyhydroxyalkanoate (PHA) polymers family as substitutes of petroleum based polymers for packaging applications and solutions brought by their composites to form barrier materials. *Pure and Applied Chemistry* 89: 1841–1848. https://doi.org/10.1515/pac-2017-0401.

39 Cava, D., Gimenez, E., Gavara, R., and Lagaron, J.M. (2006). Comparative performance and barrier properties of biodegradable thermoplastics and nanobiocomposites versus PET for food packaging applications. *Journal of Plastic Film & Sheeting* 22: 265–274. https://doi.org/10.1177/8756087906071354.

40 Pilla, S. (2011). *Handbook of Bioplastics and Biocomposites Engineering Applications*. doi: https://doi.org/10.1002/9781118203699.

41 Arcos-Hernandez, M.V., Laycock, B., Pratt, S. et al. (2012). Biodegradation in a soil environment of activated sludge derived polyhydroxyalkanoate (PHBV). *Polymer Degradation and Stability* 97: 2301–2312. https://doi.org/10.1016/j.polymdegradstab. 2012.07.035.

42 Weng, Y.-X., Wang, Y., Wang, X.-L., and Wang, Y.-Z. (2010). Biodegradation behavior of PHBV films in a pilot-scale composting condition. *Polymer Testing* 29: 579–587. https://doi.org/10.1016/j.polymertesting.2010.04.002.

43 Thellen, C., Coyne, M., Froio, D. et al. (2008). A processing, characterization and marine biodegradation study of melt-extruded polyhydroxyalkanoate (PHA) films. *Journal of Polymers and the Environment* 16: 1–11. https://doi.org/10.1007/s10924-008-0079-6.

44 Hermann, B.G., Debeer, L., De Wilde, B. et al. (2011). To compost or not to compost: carbon and energy footprints of biodegradable materials' waste treatment. *Polymer Degradation and Stability* 96: 1159–1171. https://doi.org/10.1016/j.polymdegradstab. 2010.12.026.

45 Corre, Y.-M., Bruzaud, S., Audic, J.-L., and Grohens, Y. (2012). Morphology and functional properties of commercial polyhydroxyalkanoates: a comprehensive and comparative study. *Polymer Testing* 31: 226–235. https://doi.org/10.1016/j.polymertesting.2011.11.002.

46 de Koning, G.J.M. and Lemstra, P.J. (1993). Crystallization phenomena in bacterial poly[(R)-3-hydroxybutyrate]: 2. Embrittlement and rejuvenation. *Polymer (Guildf)* 34: 4089–4094. https://doi.org/10.1016/0032-3861(93)90671-V.

47 Barham, P.J. and Keller, A. (1986). The relationship between microstructure and mode of fracture in polyhydroxybutyrate. *Journal of Polymer Science Part B: Polymer Physics* 24 https://doi.org/10.1002/polb.1986.180240108.

48 Martinez-Salazar, J., Sanchez-Cuesta, M., Barham, P.J., and Keller, A. (1989). Thermal expansion and spherulite cracking in 3-hydroxybutyrate/3-hydroxyvalerate copolymers. *Journal of Materials Science Letters* 8: 490–492. https://doi.org/10.1007/BF00720717.

49 de Koning, G.J.M., Scheeren, A.H.C., Lemstra, P.J. et al. (1994). Crystallization phenomena in bacterial poly[(R)-3-hydroxybutyrate]: 3. Toughening via texture changes. *Polymer (Guildf)* 35: 4598–4605. https://doi.org/10.1016/0032-3861(94)90809-5.

50 Biddlestone, F., Harris, A., Hay, J.N., and Hammond, T. (1996). The physical ageing of amorphous poly(hydroxybutyrate). *Polymer International* 39: 221–229. https://doi.org/10.1002/(SICI)1097-0126(199603)39:3<221::AID-PI511>3.0.CO;2-O.

51 Crétois, R., Chenal, J.M., Sheibat-Othman, N. et al. (2016). Physical explanations about the improvement of PolyHydroxyButyrate ductility: hidden effect of plasticizer on physical ageing. *Polymer (United Kingdom)* 102: 176–182. https://doi.org/10.1016/j.polymer.2016.09.017.

52 Hurrell, B.L. and Cameron, R.E. (1998). Physical ageing and the embrittlement of poly(hydroxybutyrate) on storage: a time resolved small angle X-ray scattering study. *Polymer International* 45: 308–312. https://doi.org/10.1002/(SICI)1097-0126(199803)45:3<308::AID-PI934> 3.0.CO;2-V.

53 Esposito, A., Delpouve, N., Causin, V. et al. (2016). From a three-phase model to a continuous description of molecular mobility in semicrystalline poly(hydroxybutyrate-co-hydroxyvalerate). *Macromolecules* 49: 4850–4861. https://doi.org/10.1021/acs.macromol.6b00384.

54 Di Lorenzo, M.L. and Righetti, M.C. (2013). Evolution of crystal and amorphous fractions of poly[(R)-3-hydroxybutyrate] upon storage. *Journal of*

Thermal Analysis and Calorimetry 112: 1439–1446. https://doi.org/10.1007/s10973-012-2734-3.

55 Srubar, W.V., Wright, Z.C., Tsui, A. et al. (2012). Characterizing the effects of ambient aging on the mechanical and physical properties of two commercially available bacterial thermoplastics. *Polymer Degradation and Stability*: 1922–1929. https://doi.org/10.1016/j.polymdegradstab.2012.04.011.

56 Struik, L.C.E. (1987). The mechanical and physical ageing of semicrystalline polymers: 1. *Polymer (Guildf)* 28: 1521–1533. https://doi.org/10.1016/0032-3861(87)90353-3.

57 Struik, L.C.E. (1987). The mechanical behaviour and physical ageing of semicrystalline polymers: 2. *Polymer (Guildf)* 28: 1534–1542. https://doi.org/10.1016/0032-3861(87)90354-5.

58 Alsleben, M. and Schick, C. (1994). The melting of polymers — a three-phase approach. *Thermochimica Acta* 238: 203–227. https://doi.org/10.1016/S0040-6031(94)85211-1.

59 Sedighiamiri, A., Van Erp, T.B., Peters, G.W.M. et al. (2010). Micromechanical modeling of the elastic properties of semicrystalline polymers: a three-phase approach. *Journal of Polymer Science Part B: Polymer Physics* 48: 2173–2184. https://doi.org/10.1002/polb.22099.

60 Kim, H.-S., Yang, H.-S., Kim, H.-J., and Kattan, M. (2004). Relaxations IN amorphous and semi-crystalline polyesters. *Journal of Thermal Analysis and Calorimetry* 76: 379–394. https://doi.org/10.1023/B:JTAN.0000028019.88096.e3.

61 El-Taweel, S.H., Höhne, G.W.H., Mansour, A.A. et al. (2004). Glass transition and the rigid amorphous phase in semicrystalline blends of bacterial polyhydroxybutyrate PHB with low molecular mass atactic R, S-PHB-diol. *Polymer (Guildf)* 45: 983–992. https://doi.org/10.1016/J.POLYMER.2003.12.007.

62 Di Lorenzo, M.L., Gazzano, M., and Righetti, M.C. (2012). The role of the rigid amorphous fraction on cold crystallization of poly(3-hydroxybutyrate). *Macromolecules* 45: 5684–5691. https://doi.org/10.1021/ma3010907.

63 Cheng, M.-L. and Sun, Y.-M. (2009). Relationship between free volume properties and structure of poly(3-hydroxybutyrate-co-3-hydroxyvalerate) membranes via various crystallization conditions. *Polymer (Guildf)* 50: 5298–5307. https://doi.org/10.1016/J.POLYMER.2009.09.035.

64 Kopinke, F.D. and Mackenzie, K. (1997). Mechanistic aspects of the thermal degradation of poly(lactic acid) and poly(β-hydroxybutyric acid). *Journal of Analytical and Applied Pyrolysis* 40–41: 43–53. https://doi.org/10.1016/S0165-2370(97)00022-3.

65 González-Ausejo, J., Sanchez-Safont, E., Lagaron, J.M. et al. (2017). Assessing the thermoformability of poly(3-hydroxybutyrate-co-3-hydroxyvalerate)/poly(acid lactic) blends compatibilized with diisocyanates. *Polymer Testing* 62: 235–245. https://doi.org/10.1016/j.polymertesting.2017.06.026.

66 Liu, Q.-S., Zhu, M.-F., Wu, W.-H., and Qin, Z.-Y. (2009). Reducing the formation of six-membered ring ester during thermal degradation of

biodegradable PHBV to enhance its thermal stability. *Polymer Degradation and Stability* 94: 18–24. https://doi.org/10.1016/j.polymdegradstab.2008.10.016.

67 Osswald, T.A. and Menges, G. (2012). *Material Science of Polymers for Engineers*. doi:https://doi.org/10.3139/9781569905241.

68 Guinault, A., Nguyen, A.S., Miquelard-Garnier, G. et al. (2012). The effect of thermoforming of PLA-PHBV films on the morphology and gas barrier properties. *Key Engineering Materials* 504–506: 1135–1138. https://doi.org/10.4028/www.scientific.net/KEM.504-506.1135.

69 Mekonnen, T., Mussone, P., Khalil, H., and Bressler, D. (2013). Progress in bio-based plastics and plasticizing modifications. *Journal of Materials Chemistry A* 1: 13379. https://doi.org/10.1039/c3ta12555f.

70 Duangphet, S., Szegda, D., Song, J., and Tarverdi, K. (2013). The effect of chain extender on poly(3-hydroxybutyrate-co-3-hydroxyvalerate): thermal degradation, crystallization, and rheological behaviours. *Journal of Polymers and the Environment* 22: 1–8. https://doi.org/10.1007/s10924-012-0568-5.

71 Kolahchi, A.R. and Kontopoulou, M. (2015). Chain extended poly(3-hydroxybutyrate) with improved rheological properties and thermal stability, through reactive modification in the melt state. *Polymer Degradation and Stability* 121: 222–229. https://doi.org/10.1016/J.POLYMDEGRADSTAB.2015.09.008.

72 Muthuraj, R., Misra, M., and Mohanty, A.K. (2018). Biodegradable compatibilized polymer blends for packaging applications: a literature review. *Journal of Applied Polymer Science* 135: 45726. https://doi.org/10.1002/app.45726.

73 Il Choi, J. and Lee, S.Y. (1997). Process analysis and economic evaluation for poly(3-hydroxybutyrate) production by fermentation. *Bioprocess Engineering* 17: 335–342. https://doi.org/10.1007/s004490050394.

74 Van Wegen, R.J., Ling, Y., and Middelberg, A.P.J. (1998). Industrial production of Polyhydroxyalkanoates using *Escherichia coli*: an economic analysis. *Chemical Engineering Research and Design* 76: 417–426.

75 Reddy, M.V., Mawatari, Y., Onodera, R. et al. (2019). Bacterial conversion of waste into polyhydroxybutyrate (PHB): a new approach of bio-circular economy for treating waste and energy generation. *Bioresource Technology Reports* 7: 100246. https://doi.org/10.1016/J.BITEB.2019.100246.

76 Nielsen, C., Rahman, A., Rehman, A.U. et al. (2017). Food waste conversion to microbial polyhydroxyalkanoates. *Microbial Biotechnology* 10: 1338–1352. https://doi.org/10.1111/1751-7915.12776.

77 Chinthapalli, R., Skoczinski, P., Carus, M. et al. (2019). Bio-based building blocks and polymers-global capacities. *Production and Trends*: 2018–2023. http://www.bio-based.eu/reports.

78 Dietrich, K., Dumont, M.J., Del Rio, L.F., and Orsat, V. (2017). Producing PHAs in the bioeconomy — towards a sustainable bioplastic. *Sustainable Production and Consumption* https://doi.org/10.1016/j.spc.2016.09.001.

79 Vandi, L.J., Chan, C.M., Werker, A. et al. (2018). Wood-PHA composites: mapping opportunities. *Polymers (Basel)* 10: 751. https://doi.org/10.3390/polym10070751.

80 Pereira, P.H.F., de Freitas Rosa, M., Cioffi, M.O.H. et al. (2015). Vegetal fibres in polymeric composites: a review. *Polímeros* 25: 9–22. https://doi.org/10.1590/0104-1428.1722.

81 Faruk, O., Bledzki, A.K., Fink, H.-P., and Sain, M. (2014). Progress report on natural fibre reinforced composites. *Macromolecular Materials and Engineering* 299: 9–26. https://doi.org/10.1002/mame.201300008.

82 George, M., Chae, M., and Bressler, D.C. (2016). Composite materials with bast fibres: structural, technical, and environmental properties. *Progress in Materials Science* 83: 1–23. https://doi.org/10.1016/j.pmatsci.2016.04.002.

83 Väisänen, T., Haapala, A., Lappalainen, R., and Tomppo, L. (2016). Utilization of agricultural and forest industry waste and residues in natural fibre-polymer composites: a review. *Waste Management* 54: 62–73. https://doi.org/10.1016/j.wasman.2016.04.037.

84 Mohanty, K., Misra, M., and Hinrichsen, G. (2000). Biofibres, biodegradable polymers and biocomposites: an overview. *Macromolecular Materials and Engineering* 276–277: 1–24. https://doi.org/10.1002/(SICI)1439-2054(20000301)276:1<1::AID-MAME1>3.0.CO;2-W.

85 Ramesh, M., Palanikumar, K., and Reddy, K.H. (2017). Plant fibre based bio-composites: sustainable and renewable green materials. *Renewable and Sustainable Energy Reviews* 79: 558–584. https://doi.org/10.1016/J.RSER.2017.05.094.

86 Faruk, O., Bledzki, A.K., Fink, H.-P., and Sain, M. (2012). Biocomposites reinforced with natural fibres: 2000–2010. *Progress in Polymer Science* 37: 1552–1596. https://doi.org/10.1016/J.PROGPOLYMSCI.2012.04.003.

87 Alkbir, M.F.M., Sapuan, S.M., Nuraini, A.A., and Ishak, M.R. (2016). Fibre properties and crashworthiness parameters of natural fibre-reinforced composite structure: a literature review. *Composite Structures* 148: 59–73. https://doi.org/10.1016/j.compstruct.2016.01.098.

88 Gibson, L.J. (2012). The hierarchical structure and mechanics of plant materials. *Journal of the Royal Society, Interface* 9: 2749–2766. https://doi.org/10.1098/rsif.2012.0341.

89 Wei, L. and McDonald, A. (2016). A review on grafting of biofibres for biocomposites. *Materials (Basel)* 9: 303. https://doi.org/10.3390/ma9040303.

90 Jones, D., Ormondroyd, G.O., Curling, S.F. et al. (2017). Chemical compositions of natural fibres. In: *Advanced High Strength Natural Fibre Composites in Construction*, 23–58. Elsevier https://doi.org/10.1016/B978-0-08-100411-1.00002-9.

91 Cheremisinoff, N.P. (1997). *Handbook of Engineering Polymeric Materials.* Marcel Dekker.

92 Yu, L. (2009). *Biodegradable Polymer Blends and Composites from Renewable Resources.* Wiley https://doi.org/10.1002/9780470391501.

93 Berthet, M.-A., Angellier-Coussy, H., Guillard, V., and Gontard, N. (2016). Vegetal fibre-based biocomposites: which stakes for food packaging applications? *Journal of Applied Polymer Science* 133 https://doi.org/10.1002/app.42528.

94 Yang, H., Yan, R., Chen, H. et al. (2007). Characteristics of hemicellulose, cellulose and lignin pyrolysis. *Fuel* 86: 1781–1788. https://doi.org/10.1016/j.fuel.2006.12.013.

95 Fogorasi, M. and Barbu, I. (2017). The potential of natural fibres for automotive sector - *review. IOP Conference Series: Materials Science and Engineering* 252: 012044. https://doi.org/10.1088/1757-899X/252/1/012044.

96 Tobergte, D.R. and Curtis, S. (2013). *Biocomposites: Desing and Mechanical Performance.* Elsevier Inc. https://doi.org/10.1017/CBO9781107415324.004.

97 Liu, M., Thygesen, A., Summerscales, J., and Meyer, A.S. (2017). Targeted pre-treatment of hemp bast fibres for optimal performance in biocomposite materials: a review. *Industrial Crops and Products* 108: 660–683. https://doi.org/10.1016/J.INDCROP.2017.07.027.

98 Jayavani, S., Deka, H., Varghese, T.O., and Nayak, S.K. (2015). Recent development and future trends in coir fibre reinforced green polymer composites: review and evaluation. *Polymer Composites* 37: 1–14. https://doi.org/10.1002/pc.23529.

99 Gandini, A. and Belgacem, M.N. (2011). *Modifying Cellulose Fibre Surfaces in the Manufacture of Natural Fibre Composites.* Woodhead Publishing Limited https://doi.org/10.1533/9780857092281.1.3.

100 Mohanty, A.K., Misra, M., and Drzal, L.T. (2001). Surface modifications of natural fibres and performance of the resulting biocomposites: an overview. *Composite Interfaces* 8: 313–343. https://doi.org/10.1163/156855401753255422.

101 George, M., Mussone, P.G., and Bressler, D.C. (2014). Surface and thermal characterization of natural fibres treated with enzymes. *Industrial Crops and Products* 53: 365–373. https://doi.org/10.1016/j.indcrop.2013.12.037.

102 Kalia, S., Kaith, B.S., and Kaur, I. (2009). Pretreatments of natural fibres and their application as reinforcing material in polymer composites-a review. *Polymer Engineering and Science* 49: 1253–1272. https://doi.org/10.1002/pen.21328.

103 Ndazi, B.S., Karlsson, S., Tesha, J.V., and Nyahumwa, C.W. (2007). Chemical and physical modifications of rice husks for use as composite panels. *Composites. Part A, Applied Science and Manufacturing* 38: 925–935. https://doi.org/10.1016/j.compositesa.2006.07.004.

104 Tripathi, S., Sharma, N., Alam, I., and Bhardwaj, N.K. (2018). Effectiveness of different green chemistry approaches during mixed hardwood bamboo pulp bleaching and their impact on environment. *International journal of Environmental Science and Technology*: 1–12. https://doi.org/10.1007/s13762-018-1887-4.

105 Duan, L., Yu, W., and Li, Z. (2017). Analysis of structural changes in jute fibres after peracetic acid treatment. *Journal of Engineered Fibers and Fabrics* 12: 33–42.

106 Ly, B., Thielemans, W., Dufresne, A. et al. (2008). Surface functionalization of cellulose fibres and their incorporation in renewable polymeric matrices. *Composites Science and Technology* 68: 3193–3201. https://doi.org/10.1016/j.compscitech.2008.07.018.

107 Yu, H.-Y. and Qin, Z.-Y. (2014). Surface grafting of cellulose nanocrystals with poly(3-hydroxybutyrate-co-3-hydroxyvalerate). *Carbohydrate Polymers* 101: 471–478. https://doi.org/10.1016/j.carbpol.2013.09.048.

108 Muthuraj, R., Misra, M., and Mohanty, A.K. (2017). Biocomposite consisting of miscanthus fibre and biodegradable binary blend matrix: compatibilization and performance evaluation. *RSC Advances* 7: 27538–27548. https://doi.org/10.1039/C6RA27987B.

109 Rodi, E., Langlois, V., Renard, E. et al. (2018). Biocomposites based on poly(3-Hydroxybutyrate-co-3-hydroxyvalerate) (PHBHV) and miscanthus giganteus fibres with improved fibre/matrix interface. *Polymers (Basel)* 10: 509. https://doi.org/10.3390/polym10050509.

110 Anderson, S., Zhang, J., and Wolcott, M.P. (2013). Effect of interfacial modifiers on mechanical and physical properties of the PHB composite with high wood flour content. *Journal of Polymers and the Environment* 21: 631–639. https://doi.org/10.1007/s10924-013-0586-y.

111 Chan, C.M., Vandi, L.-J., Pratt, S. et al. (2018). Mechanical properties of poly(3-hydroxybutyrate-co-3-hydroxyvalerate)/wood flour composites: effect of interface modifiers. *Journal of Applied Polymer Science*: 46828. https://doi.org/10.1002/app.46828.

112 Hao, M., Wu, H., Qiu, F., and Wang, X. (2018). Interface bond improvement of sisal fibre reinforced polylactide composites with added epoxy oligomer. *Materials (Basel)* 11, 398 https://doi.org/10.3390/ma11030398.

113 P. Nanthananon, M. Seadan, S. Pivsa-Art, H. Hiroyuki, S. Suttiruengwong, Biodegradable polyesters reinforced with eucalyptus fibre: effect of reactive agents, 2017: p. 070012. doi:https://doi.org/10.1063/1.5016739.

114 Srubar, W.V., Pilla, S., Wright, Z.C. et al. (2012). Mechanisms and impact of fibre–matrix compatibilization techniques on the material characterization of PHBV/oak wood flour engineered biobased composites. *Composites Science and Technology* 72: 708–715. https://doi.org/10.1016/J.COMPSCITECH.2012.01.021.

115 Gunning, M.A., Geever, L.M., Killion, J.A. et al. (2014). Effect of compatibilizer content on the mechanical properties of bioplastic composites via hot melt extrusion. *Polymer - Plastics Technology and Engineering* 53: 1223–1235. https://doi.org/10.1080/03602559.2014.891235.

116 Sreekala, M.S. and Thomas, S. (2003). Effect of fibre surface modification on water-sorption characteristics of oil palm fibres. *Composites Science and Technology* 63: 861–869. https://doi.org/10.1016/S0266-3538(02)00270-1.

117 Wu, C.-S. (2013). Preparation, characterization and biodegradability of crosslinked tea plant-fibre-reinforced polyhydroxyalkanoate composites. *Polymer Degradation and Stability* 98: 1473–1480. https://doi.org/10.1016/j.polymdegradstab. 2013.04.013.

118 C. Wu, Preparation and characterization of polyhydroxyalkanoate bioplastic-based green renewable composites from rice husk, (2014) 384–392. doi:https://doi.org/10.1007/s10924-014-0662-y.

119 Bhardwaj, R., Mohanty, A.K., Drzal, L.T. et al. (2006). Renewable resource-based green composites from recycled cellulose fibre and

poly(3-hydroxybutyrate-co-3-hydroxyvalerate) bioplastic. *Biomacromolecules* 7: 2044–2051.

120 Michael, C.L.H., Gunning, A., Geever, L.M. et al. (2013). Mechanical and biodegradation performance of short natural fibre polyhydroxybutyrate composites. *Polymer Testing* 32: 1603–1611.

121 Torres-Tello, E.V., Robledo-Ortíz, J.R., González-García, Y. et al. (2017). Effect of agave fibre content in the thermal and mechanical properties of green composites based on polyhydroxybutyrate or poly(hydroxybutyrate-co-hydroxyvalerate). *Industrial Crops and Products* 99: 117–125. https://doi.org/10.1016/j.indcrop.2017.01.035.

122 Rossa, L.V., Scienza, L.C., and Zattera, A.J. (2013). Effect of curauá fibre content on the properties of poly(hydroxybutyrate-co-valerate) composites. *Polymer Composites* 34: 450–456. https://doi.org/10.1002/pc.22432.

123 Krishnaprasad, R., Veena, N.R., Maria, H.J. et al. (2009). Mechanical and thermal properties of bamboo microfibril reinforced polyhydroxybutyrate biocomposites. *Journal of Polymers and the Environment* 17: 109–114. https://doi.org/10.1007/s10924-009-0127-x.

124 Seggiani, M., Cinelli, P., Mallegni, N. et al. (2017). New bio-composites based on polyhydroxyalkanoates and posidonia oceanica fibres for applications in a marine environment. *Materials (Basel)* 10 https://doi.org/10.3390/ma10040326.

125 Berthet, M., Angellier-coussy, H., Machado, D. et al. (2015). Exploring the potentialities of using lignocellulosic fibres derived from three food by-products as constituents of biocomposites for food packaging. *Industrial Crops and Products* 69: 110–122. https://doi.org/10.1016/j.indcrop.2015.01.028.

126 Berthet, M.-A., Mayer-Laigle, C., Rouau, X. et al. (2017). Sorting natural fibres: a way to better understand the role of fibre size polydispersity on the mechanical properties of biocomposites. *Composites. Part A, Applied Science and Manufacturing* 95: 12–21. https://doi.org/10.1016/j.compositesa.2017.01.011.

127 Macedo, J.D.S., Costa, M.F., Tavares, M.I.B., and Thiré, R.M.S.M. (2010). Preparation and characterization of composites based on polyhydroxybutyrate and waste powder from coconut fibres processing. *Polymer Engineering and Science* 50: 1466–1475. https://doi.org/10.1002/pen.21669.

128 Batista, K.C., Silva, D.A.K., Coelho, L.A.F. et al. (2010). Soil biodegradation of PHBV/peach palm particles biocomposites. *Journal of Polymers and the Environment* 18: 346–354. https://doi.org/10.1007/s10924-010-0238-4.

129 Avella, M., La Rota, G., Martuscelli, E. et al. (2000). Poly(3-hydroxybutyrate-co-3-hydroxyvalerate) and wheat straw fibre composites: thermal, mechanical properties and biodegradation behaviour. *Journal of Materials Science* 35: 829–836. https://doi.org/10.1023/A:1004773603516.

130 Scalioni, L.V., Gutiérrez, M.C., and Felisberti, M.I. (2017). Green composites of poly(3-hydroxybutyrate) and curaua fibres: morphology and physical, thermal, and mechanical properties. 134 https://doi.org/10.1002/app.44676.

131 Dufresne, A., Dupeyre, D., and Paillet, M. (2003). Lignocellulosic flour-reinforced poly(hydroxybutyrate-co-valerate) composites. *Journal of Applied Polymer Science* 87: 1302–1315.

132 Ventura, H., Claramunt, J., Rodríguez-Pérez, M.A., and Ardanuy, M. (2017). Effects of hydrothermal aging on the water uptake and tensile properties of PHB/flax fabric biocomposites. *Polymer Degradation and Stability* 142: 129–138. https://doi.org/10.1016/j.polymdegradstab. 2017.06.003.

133 Melo, J.D.D., Carvalho, L.F.M., Medeiros, A.M. et al. (2012). A biodegradable composite material based on polyhydroxybutyrate (PHB) and carnauba fibres. *Composites. Part B, Engineering* 43: 2827–2835. https://doi.org/10.1016/j.compositesb.2012.04.046.

134 Ahankari, S.S., Mohanty, A.K., and Misra, M. (2011). Mechanical behaviour of agro-residue reinforced poly(3-hydroxybutyrate-co-3-hydroxyvalerate), (PHBV) green composites: a comparison with traditional polypropylene composites. *Composites Science and Technology* 71: 653–657. https://doi.org/10.1016/j.compscitech.2011.01.007.

135 Yatigala, N.S., Bajwa, D.S., and Bajwa, S.G. (2018). Compatibilization improves physico-mechanical properties of biodegradable biobased polymer composites. *Composites. Part A, Applied Science and Manufacturing* 107: 315–325. https://doi.org/10.1016/j.compositesa.2018.01.011.

136 Wu, C.S., Liao, H.T., and Cai, Y.X. (2017). Characterisation, biodegradability and application of palm fibre-reinforced polyhydroxyalkanoate composites. *Polymer Degradation and Stability* 140: 55–63. https://doi.org/10.1016/j.polymdegradstab. 2017.04.016.

137 Wang, B. and Sain, M. (2007). The effect of chemically coated nanofibre reinforcement on biopolymer based nanocomposites. *BioResources* 2: 371–388.

138 Wong, S., Shanks, R., and Hodzic, A. (2002). Properties of poly(3-hydroxybutyric acid) composites with flax fibres modified by plasticiser absorption. *Macromolecular Materials and Engineering* 287: 647–655. https://doi.org/10.1002/1439-2054(200211)287:10<647::AID-MAME647>3.0.CO;2-4.

139 Wong, S., Shanks, R.A., and Hodzic, A. (2007). Effect of additives on the interfacial strength of poly(L-lactic acid) and poly(3-hydroxy butyric acid)-flax fibre composites. *Composites Science and Technology* 67: 2478–2484. https://doi.org/10.1016/j.compscitech.2006.12.016.

140 Muthuraj, R., Misra, M., and Mohanty, A.K. (2017). Reactive compatibilization and performance evaluation of miscanthus biofibre reinforced poly(hydroxybutyrate-co-hydroxyvalerate) biocomposites. *Journal of Applied Polymer Science* 134 https://doi.org/10.1002/app.44860.

141 Hao, M. and Wu, H. (2017). Effect of in situ reactive interfacial compatibilization on structure and properties of polylactide/sisal fibre biocomposites. *Polymer Composites* 39: E174–E187. https://doi.org/10.1002/pc.24484.

142 Jiang, L., Huang, J., Qian, J. et al. (2008). Study of poly(3-hydroxybutyrate-co-3-hydroxyvalerate) (PHBV)/bamboo pulp fibre composites: effects of nucleation agent and compatibilizer. *Journal of Polymers and the Environment* 16: 83–93. https://doi.org/10.1007/s10924-008-0086-7.

143 Avella, M., Bogoeva-Gaceva, G., Buzõarovska, A. et al. (2007). Poly(3-hydroxybutyrate-co-3-hydroxyvalerate)-based biocomposites reinforced with kenaf fibres. *Journal of Applied Polymer Science* 104: 3192–3200. https://doi.org/10.1002/app.26057.

144 Nagarajan, V., Misra, M., and Mohanty, A.K. (2013). New engineered biocomposites from poly(3-hydroxybutyrate-co-3-hydroxyvalerate) (PHBV)/poly(butylene adipate-co-terephthalate) (PBAT) blends and switchgrass: Fabrication and performance evaluation. *Industrial Crops and Products* 42: 461–468. https://doi.org/10.1016/j.indcrop.2012.05.042.

145 L. Wei, N.M. Stark, A.G. Mcdonald, Interfacial improvements in biocomposites based on poly(3-hydroxybutyrate) and poly(3-hydroxybutyrate-co-3-hydroxyvalerate) bioplastics reinforced and grafted with α-cellulose fibres, (2015) 4800–4814. doi:https://doi.org/10.1039/c5gc01568e.

146 Jiang, L., Chen, F., Qian, J. et al. (2010). Reinforcing and toughening effects of bamboo pulp fibre on poly(3-hydroxybutyrate-co-3-hydroxyvalerate) fibre composites. *Industrial and Engineering Chemistry Research* 49: 572–577. https://doi.org/10.1021/ie900953z.

147 Zhang, K., Misra, M., and Mohanty, A.K. (2014). Toughened sustainable green composites from poly(3-hydroxybutyrate-co-3-hydroxyvalerate) based ternary blends and miscanthus biofibre. *ACS Sustainable Chemistry & Engineering* 2: 2345–2354. https://doi.org/10.1021/sc500353v.

148 Sánchez-Safont, E.L., Arrillaga, A., Anakabe, J. et al. (2019). PHBV/TPU/cellulose compounds for compostable injection molded parts with improved thermal and mechanical performance. *Journal of Applied Polymer Science* 136: 47257. https://doi.org/10.1002/app.47257.

149 Sánchez-Safont, E.L., Arrillaga, A., Anakabe, J. et al. (2018). Toughness enhancement of PHBV/TPU/cellulose compounds with reactive additives for compostable injected parts in industrial applications. *International Journal of Molecular Sciences* https://doi.org/10.3390/ijms19072102.

9

Poly-Paper: Cellulosic-Filled Eco-composite Material with Innovative Properties for Packaging

Romina Santi[1], *Silvia Farè*[2,3], *Alberto Cigada*[2,3], *and Barbara Del Curto*[2,3]

[1] *Politecnico di Milano, Department of Design, Via Durando 38/A, Milano, Italy*
[2] *Materials and Chemical Engineering "Giulio Natta" Pizza Leonardo da Vinci, 32, Department of Chemistry, Milano 20133, Italy*
[3] *National Interuniversity Consortium of Materials Science and Technology, Local Unit Politecnico di Milano, Milan, Italy*

9.1 Introduction

Going back to the observation of nature, the concept of sustainable eco-oriented and responsible packaging starts with the idea that the best packaging is the one that "does not exist." So it is, for the orange and its peel and for the other fruits, where the protective layer is edible together with the fruit in some cases, or easily reusable, in others. In recent years, the search for the ideal packaging material has increasingly taken into account some new requirements: fully organic, perishable in a short time, but sufficiently resistant in order to protect the product inside, with reduced emissions and energy consumption for the production. In addition to the measurable physical quantities, there are characteristics related to functionality, safety, transparency, informative, ergonomic, cognitive, and the ones related to sensorial perception that in the case of packaging determine the preselection of the product by users and the way in which they will dispose the packaging [1]. There are many types of packaging materials, since there are many functions that packaging materials must fulfill depending on the product that should be contained. Different classes of materials are often combined industrially in a single package: corrugated cardboard, printed-paper, polystyrene (PS) foam, low-density polyethylene, and various other plastic components [2]. Each of these materials follows a different path for collection, disposal, and recycling. If used jointly, in a single package, it is possible that they might not be distinguished by the consumer and conferred jointly to the same waste stream. In a sustainable design process, the different packaging materials must be separable, separated by the consumer, and collected in each specific recycling channel. Cellulosic packaging is a perfect example of circularity, representing the most active and constantly growing recycling channel. In fact, the high percentage of recycling rate from the "Istituto Superiore per la Protezione e la Ricerca Ambientale" ISPRA, Urban Waste Report was calculated at 80% in 2016 [3].

Sustainable Food Packaging Technology, First Edition. Edited by Athanassia Athanassiou.

Moreover, the societal appraisal of aesthetic, expressive sensorial and semantic aspects of materials are in continuous evolution [4, 5]. These, as well as circular economy criteria, market needs, and legislative issues can be considered as drivers for new materials development. Nowadays, sustainability is the main trend cluster, in which research is predominantly focused on bio-based, biodegradable [6], recyclable, and/or recycled materials, but also on the reuse of production scraps, food waste or closely linked to territorial surpluses. Other approaches to sustainable material design include biomimicry (i.e. getting inspiration from nature) and edible materials in packaging, which follows a disruptive approach to battle food packaging waste.

In the last 20 years, economic, environmental, social, and legislative forces have been the main drivers for the development of sustainable packaging materials. The use of natural fibers in bioplastic matrices has made possible the development of sustainable composites with enhanced properties and performance. Natural fibers have multiple advantages over synthetic fibers as they are renewable, have low density and high mechanical strength and elastic modulus, and are not abrasive for machinery and plants [7]. In particular, the advantages of cellulose fibers used as fillers in sustainable composites are hardness, flexibility, ease of processing, recyclability, and eco-compatibility. This green reinforcement has been utilized in various investigations in the packaging field in recent decades [8–11].

The development of cellulose fibers as reinforcement in composite materials began upon the discovery of cellulose in 1838 by A. Payen. In 1987, the first thermoplastic composites were formulated using polypropylene (PP), high density polyethylene (HDPE), and PS as matrices, reinforced with cellulosic pulp fiber [8–10]. Since then, cellulose fibers have served as reinforcement in several thermoplastic polymers, such as Plasticised Wheat Starch (PWS) [12], poly(lactic acid), PLA [13–15], and polyvinyl alcohol (PVA) [16]. Since 1999, research has increasingly focused on renewable and sustainable materials as possible replacements for fossil-based nonbiodegradable plastics [8]. In particular, the study by Graupner et al. [13] describes the mechanical characteristics of PLA composites reinforced with different natural fibers obtained by compression molding for different product design applications. Ummartyotin and Pechyen [17] studied the possibility of obtaining microwavable packaging using different percentages of cellulose extracted from cotton (5%, 10%, and 20% w/w) as filler in a PP matrix. Zhang et al. [16] formulated thermoplastic PVA composites with plasticizers, using mechanically–chemically activated cellulose fibers as reinforcement to obtain biodegradable high-performance composite materials. Other investigations exploit different characteristics of cellulose fibers such as biodegradability, mechanical performance, and thermal and acoustic insulation in various fields of application [13, 14, 18]. It is therefore necessary to investigate different production processes of natural fiber reinforced composites such as extrusion, injection, and compression molding and filament winding. Before moving toward a specific production process, we need to focus on different criteria including desired properties, size and shape of the resultant composites, the production rate, processing raw materials and the manufacturing [19].

The aim of this chapter is to describe the development of a sustainable composite material, named Poly-Paper, for packaging based on renewable cellulose fibers in a water-soluble and eco-compatible polymeric matrix. Poly-Paper was developed during two Master Degree Theses in Design Engineering carried out at the Department of Department of Chemistry, Materials and Chemical Engineering, Politecnico di Milano. First, by Barelli [20] ,who investigated the properties of 3D printing and then by Santi [1], who optimized the injection molding, other forming processes, and recyclability issues.

The composite was the subject of two co-owned patents [21, 22] by Politecnico di Milano and NextMaterials s.r.l., an academic spin-off. The project idea is a material suitable for fulfilling the functions of packaging: contain, protect, and transport, which will be verified by mechanical characterizations [3]. The composite formulation ensures the recyclability in water, typical of the paper industry. The combination of a polymeric matrix with cellulosic fillers is a requirement for packaging intended for the cellulosic disposal chain. Another objective is to establish possible manufacturing processes for the Poly-Paper composite, taking in mind that the optimal processes are those that guarantee large-scale production, typical of the packaging sector. This would allow the composite to replace traditional plastic packaging items. In such a way, it is possible to obtain Poly-Paper composites that are:

- recyclable;
- processable through the typical industrial processes of plastics;
- resistant, not fragile.

9.2 Materials

9.2.1 Matrix

The polymer selected as matrix for the composite is an experimental formulation based on water-soluble PVA, with degree of hydrolysis between 75% and 90%, average molecular weight (Mw) between 75 and 150 kDa and a polydispersity index (Mw/Mn) between 2.5 and 4.3. PVA has been investigated for use as a potential matrix for sustainable composites thanks to its biodegradability, bio and eco-compatibility, high tensile strength, excellent adhesive properties, chemical resistance, and gas barrier properties [16, 23, 24]. PVA, unlike other biodegradable matrices (e.g. PLA), has the characteristic of dissolving in water without toxic residues for humans and the environment.

Today, PVA is mainly used as a protective colloid in the production of polymer emulsions. It is also found in many other applications, including the bonding of pigments and fibers, removable protective coatings, the production of detergents and cleaning agents, adhesives, and water-based paints and films. Moreover, PVA can also match other requirements: in fact, a biodegradable polymer is generally not solely chosen based on its biodegradability criterion, but must compete as a material based on its price/property characteristics, considering biodegradability a further advantage.

Table 9.1 Poly-paper formulations (from 0% to 55% of cellulose fibers content w/w).

Denomination	PVA content (w/w)	Cellulose micro fibers content (w/w)
Poly-Paper 0%	100	0
Poly-Paper 10%	90	10
Poly-Paper 20%	80	20
Poly-Paper 30%	70	30
Poly-Paper 40%	60	40
Poly-Paper 50%	50	50
Poly-Paper 55%	45	55

9.2.2 Reinforcement

The reason for the selection of natural fibers, especially in nonstructural applications, is linked not just to the economic side, but also to the environmental one: these materials are renewable and therefore "unlimited" in quantity, biodegradable, and if thermally disposed of can be used in all waste-to-energy processes [11, 18].

Much research interest has been centered around cellulose fibers because they are a recyclable reinforcement and can be obtained from recycling. They are the subject of study for the formulation of the recyclable composite material with paper and cardboard. In particular, for the experimentation, the filler chosen is a powder of additive cellulose fibers produced in various qualities (fiber length, thickness, purity) for a wide field of industrial applications such as insulation, cements and paints. The essential parameters of these fibers are insolubility, which allows them to be recycled and separated from the matrix, and the temperatures of use, which, in the case of matrix mixing, must adapt and combine properly with it. Cellulose fibers are selected as reinforcement, acquired by JRS Arbocel® BE 600/30 PU Cellulose microfibers[1] (Rettenmaier, Germany). The fibers are smaller than 45 μm (>45 μm, 0–0.1%), and apparent density is about 232–248 g/cm^3.

9.2.3 Composite Formulations

For the composition, the intention is to obtain as high a fiber content in the mixture as possible, to improve the mechanical characteristics, to render the composite easily recyclable, and to reduce the PVA dispersion in water.

Seven different percentages of cellulose fibers content were investigated (0–55% w/w, Table 9.1). The dry blend formulations (PVA/cellulose microfibers) were conditioned in an oven at 80 °C for 30 minutes in order to eliminate possible residual moisture. These blends were properly mixed with a kneading machine for 10 minutes until homogeneous mixtures were obtained.

1 Technical datasheet on: https://www.jrs.eu/jrs_en/cellulose-fiber/arbocel.php.

9.2.4 Extrusion Process

The dry blend of each formulation (Table 9.1) was processed by extrusion through a co-rotating twin-screw extruder FSCM series (TSA Industriale, Cernobbio, CO, Italy), with a screw diameter of 20 mm and L/D ratio of 40. The extruder has eight heating zones (Z1–Z8), whose temperatures were previously optimized for each extruded composite as detailed in Figure 9.1. After the die zone (i.e. Z8), the material flows over a conveyor belt ($L = 150$ cm), and it cools to room temperature by an air conditioner.

The semifinished outputs obtained through extrusion have different cross-sections depending on the selected extrusion dies, as reported in Figure 9.2. In particular:

(a) circular cross-section semifinished (Ø = 3 mm) were pelletized and used for the injection molding process;
(b) circular cross-section semifinished (Ø = 1.75 mm) were used as 3D printing filaments, or sheared in cylindrical samples for water degradation tests and morphological analysis;
(c) flat bars (250 mm × 2 mm) were extruded for tensile tests, using a rectangular die, taking as reference specimen type 2 (BS EN ISO 527-3:1996 standard) [25]. Nine specimens were obtained for each percentage of cellulose (Table 9.1);
(d) Thin tapes (250.0 mm × 0.5 mm) were extruded for thermoforming tests.

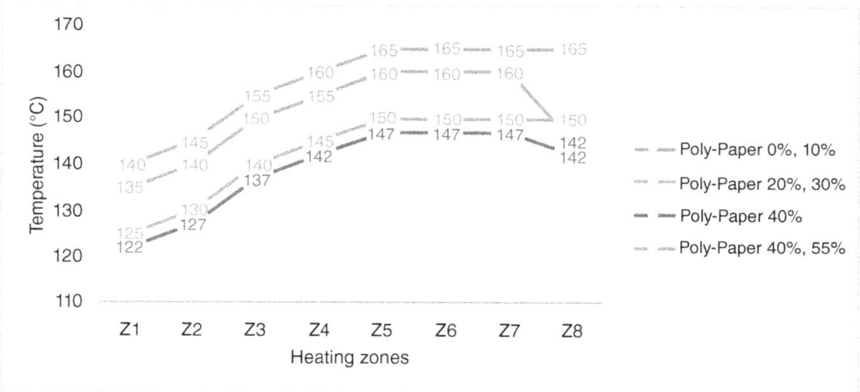

Figure 9.1 Process parameters (zone temperatures from Z1 to Z8) in relation to cellulose fiber content (w/w) in the PVA/Cellulose microfiber composites.

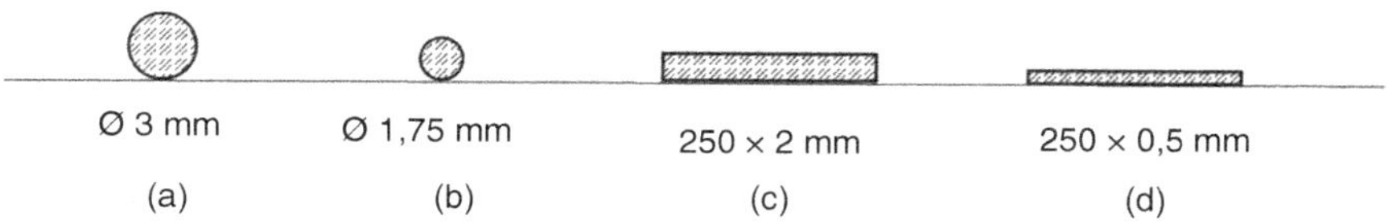

Figure 9.2 Cross-sections of the semifinished products obtained from extrusion.

Table 9.2 Mechanical parameters obtained from tensile mechanical tests performed on PVA/cellulose microfiber composites (0–55% w/w).

Cellulose microfiber content (w/w, %)	Stress at break (MPa)	Strain at break (%)	Modulus (MPa)
0	11.47 ± 0.55	239.70 ± 21.57	30.18 ± 1.83
10	10.23 ± 0.31	73.97 ± 20.59	97.55 ± 12.98
20	8.83 ± 0.25	67.31 ± 12.46	94.25 ± 8.13
30	10.37 ± 0.25	21.73 ± 7.34	175.84 ± 1.02
40	15.30 ± 0.79	15.86 ± 2.68	418.47 ± 24.25
50	29.23 ± 4.59	6.08 ± 1.11	1590.51 ± 227.14
55	30.33 ± 3.16	5.87 ± 1.28	1647.60 ± 244.10

9.3 Mechanical Properties

Mechanical tensile properties studied by Santi et al. [26] were evaluated using an electromechanical 1/MH (MTS, Eden Prairie, MN, USA) equipped with a 5 kN load cell. Tensile tests were performed in triplicate on extruded flat bar specimens (Figure 9.2c, length = 150 mm, cross-section = 50 mm^2), at 50 mm/min crosshead speed until break.

The tensile properties of PVA/Cellulose microfiber composites were compared with the results obtained for neat PVA. Increasing the percentage of cellulose microfibers used to reinforce the PVA matrix, a decrease in deformation can be observed (Table 9.2). In addition, by increasing fiber content up to 55% w/w, an increase in stress at break and stiffness of the composites was observed (Table 9.2). The elastic modulus (Figure 9.3) increases significantly ($p < 0.05$) between 0% and 55% w/w cellulose microfibers. Even the stress at break increases with higher cellulose content (i.e. up to +164% compared to the neat PVA, Table 9.2), demonstrating how the effects of the reinforcement are positive even when the relative fraction is higher than that of the matrix. At the same time, the strain at break significantly decreases due to the high percentage of cellulose microfiber filler that reduces the high deformability of PVA leading to an increasingly brittle behavior of the obtained composite (Table 9.2). In fact, the reinforcement at high percentage in the composite does not allow the sliding of the PVA macromolecules as possible in the neat PVA and at lower filler content.

9.4 Suitable Processes for Poly-Paper

The properties of composite materials are closely related to the properties of the materials from which they are formed, and to the reinforcement shape, size, concentration, distribution, and orientation. However, the mechanical behavior does not depend exclusively on the reinforcement material, but also on the

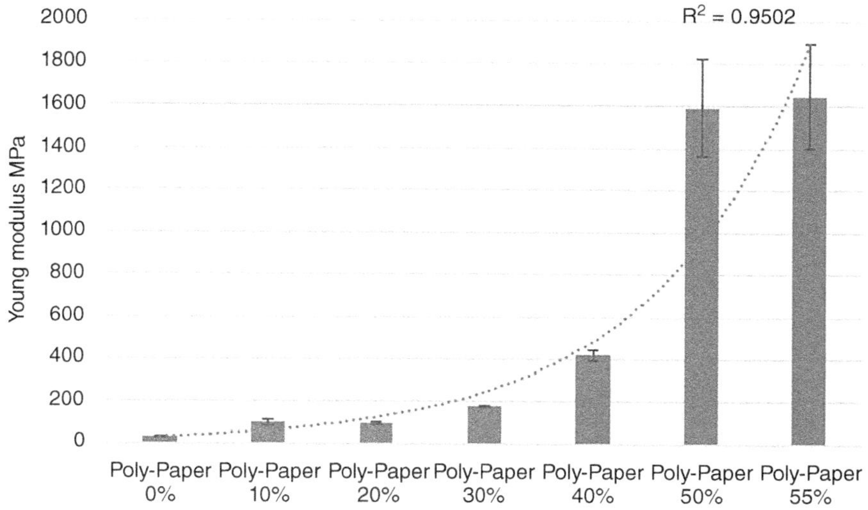

Figure 9.3 The effect of cellulose microfiber content on Young's modulus.

synergy between the reinforcement and the matrix. The mechanical strength and stiffness of the composite material depend largely on the reinforcing material and, in particular in the case of long or short fibers, on the orientation of fibers depending on the applied load direction.

The selected processes, which adhere to the characteristics of the matrix and of the fibers used in the composition of Poly-Paper, are based on the semifinished output obtained by extrusion, and are the following:

- injection molding;
- thermoforming;
- expansion.

Processability tests are particularly aimed at the typical processes of the food packaging industry. Injection molding and thermoforming are discussed in the next paragraphs (Sections 9.4.1 and 9.4.2) together with expansion (Section 9.4.3), which involves the use of blowing agent during the extrusion process.

9.4.1 Injection Molding

Poly-Paper pellets (diameter = 3 mm, Figure 9.2a) were used for injection molding. The feasibility tests were carried out with a Ray Ran Test sample pneumatic injection press apparatus (Ray-Ran Test Equipment Ltd., UK), with a pressure of 8 bar. All Poly-Paper formulations (Table 9.1) were tested with the injection molding press. Based on the tests carried out, it is possible to state that the optimum process temperatures are 180 °C for the injection cylinder and about 70 °C for the mold block. No problems relating to mold filling and fluidity were detected during the experimental tests even though the press did not have a mixing screw

(a) (b)

Figure 9.4 (a) Poly-Paper pellets and (b) Poly-Paper filled mold. Source: Romina Santi.

(Figure 9.4). Therefore, it is concluded that the material is processable also at industrial level. In addition, using an industrial machine with a mixing screw, the process temperatures could be lower and the entire process faster. Utilizing a packaging injection-molding plant it is possible to obtain not only various shapes, but also small thicknesses that contribute together with the characteristics of the composite material to a weight reduction compared to the traditional packaging. Through injection molding it is possible to identify applications such as trays, cups, caps, co-injections, closure accessories, or preforms for subsequent processing.

9.4.2 Thermoforming

For the thermoforming test, the Formec 450DT, Vacuum Thermoforming machine was used. To perform the test, several layers of thin Poly-Paper tape (250.0 mm × 0.5 mm) were welded together with water (since the composite is water-soluble) in order to form thin sheets (20 cm × 20 cm) for further characterization. The thermoforming tests were carried out using a hemispherical matrix as mold (diameter = 30 mm). The performed tests gave satisfactory results regarding the material thermo-formability. The obtained results suggest that the use of a smaller spherical mold could be advantageous in order to increase the ratio between the useful area and the deformable area (Figure 9.5). The tests have shown a good result in terms of detail in the final plates when 30%, 40%, and 50% w/w cellulose microfiber reinforcement were used. Thanks to thermoforming process optimization, applications such as trays, containers, hollow bodies, cups and lids, and so on can be envisaged.

9.4.3 Poly-Paper Expansion

The expansion tests were carried out on the composite with 30% w/w cellulose microfibers in PVA matrix, adding an increasing percentage of oxybisenzene-sulfonylhydrazide (OBSH) expanding agent, an endothermic foaming agent which is suitable for food contact applications.[2] As can be seen in Table 9.3,

2 TRAMACO Chemical Foaming Agents Datasheet. Available at: https://www.tramaco.de/fileadmin/user_upload/tramaco/Downloads/TRAMACOfoaming_agents_en.pdf.

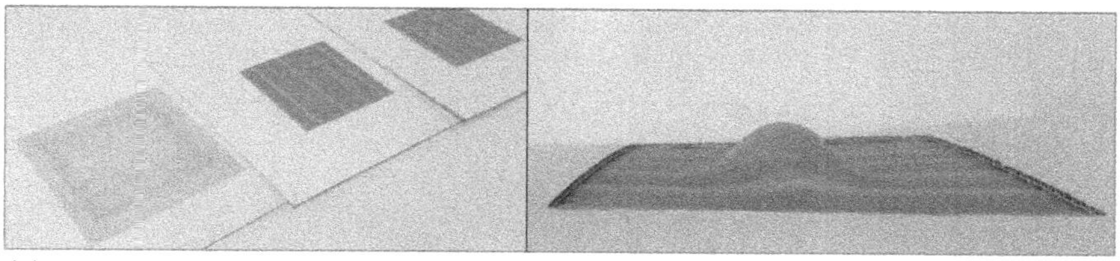
(a) (b)

Figure 9.5 Starting slabs (a) and result thermoformed Poly-Paper (b) [27]. Source: Bruno Testa.

Table 9.3 Characteristics (density, expansion factor, and color) based on different percentages of expander agent (from 2% to 6% w/w) in 70 : 30 PVA: cellulose microfiber ratio Poly-Paper composites.

PVA: cellulose microfiber ratio (w/w, %)	Expanding agent (w/w, %)	Volumetric mass density (g/cm³)	Expansion factor (%)	Color
70 : 30	0	0.70	0	Light brown
70 : 30	1	0.71	−1.3	Whitish
70 : 30	2	0.44	39.7	Whitish
70 : 30	4	0.47	35.8	Whitish
70 : 30	6	0.49	32.1	Whitish

the addition of the expanding agent has the greatest effect with an addition of 2% w/w (with respect to the total mass of the composite), decreasing for lower (1% w/w) and higher (4% and 6%) percentages. This apparently unpredictable behavior is attributed to the fact that small percentages of expanding agent tend to form a large number of small air pores, while larger expander percentages tend to form fewer larger pores.

Subsequently, various PVA : cellulose microfiber ratios were extruded with a fixed value of 2% w/w expanding agent. Table 9.4 shows the investigated compositions and the obtained results. In the case of pure PVA, the expanding agent has little effect, while it has an important effect in the presence of cellulose microfiber content of 30% and 40% w/w. By expansion, it is possible to reach an apparent density of about 0.45 g/cm^3 with an expansion of 40% compared to the same composite mixture extruder without an expander agent. Furthermore, the addition of an expander agent causes the granules to change color, which turns from light brown to whitish (Table 9.3).

Due to the lightness acquired, the products find interesting applications in food packaging. Further tests are moving toward thermal insulation tests. Furthermore, the expansion can significantly increase the ability to absorb mechanical shocks, a property that is particularly interesting for foods that tend to break (i.e. eggs) or for food delivery.

Table 9.4 Characteristics (density, expansion factor, and color) for 100 : 0, 70 : 30, 60 : 40 PVA : cellulose microfiber ratios compared to the non-expanded Poly-Paper formulations.

PVA: cellulose microfiber ratio (w/w, %)	Expanding agent (w/w, %)	Volumetric mass density (g/cm^3)	Expansion factor (%)	Color
100 : 0	0	0.74	0	Translucent brown
100 : 0	2	0.73	1.3	Translucent white
70 : 30	0	0.70	0	Light brown
70 : 30	2	0.44	39.7	Whitish
60 : 40	0	0.70	0	Brown
60 : 40	2	0.46	34.3	Whitish

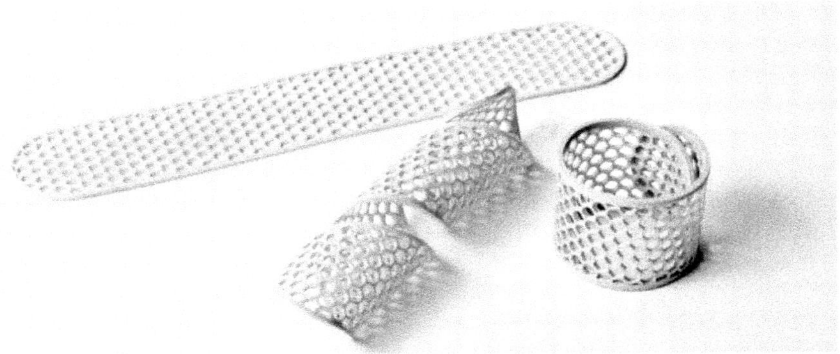

Figure 9.6 Poly-Paper 3D printed samples, flat printed and shaped via heat treatment. Source: Alberto Cigadaok.

9.5 Additional Properties of Poly-Paper

9.5.1 Shape Memory Forming

Products made of Poly-Paper can be formed and stabilized in the desired shape by heat treatment. In Figure 9.6, it is shown how flat specimens obtained by 3D printing can assume various configurations.

To quantify this effect and to assess whether this is correlated to the percentage of cellulose and/or varies based on the heat treatment, an appropriate test was set up.

Extruded filaments (Figure 9.2b) of the composite material with variable percentages of cellulose fiber (0%, 25%, 33%, and 50% w/w) were produced. Then, the filaments were wrapped on a 45 mm diameter mandrel to form springs of 10 turns. The springs were thermally stabilized at two different conditions: (a) 110 °C for 15 minutes, (b) 140 °C for 60 minutes. Subsequently, the springs were hung at one end and the length after forming was measured, immediately after hanging them and over time up to 14 days. The results are shown in the curves reported in Figure 9.7.

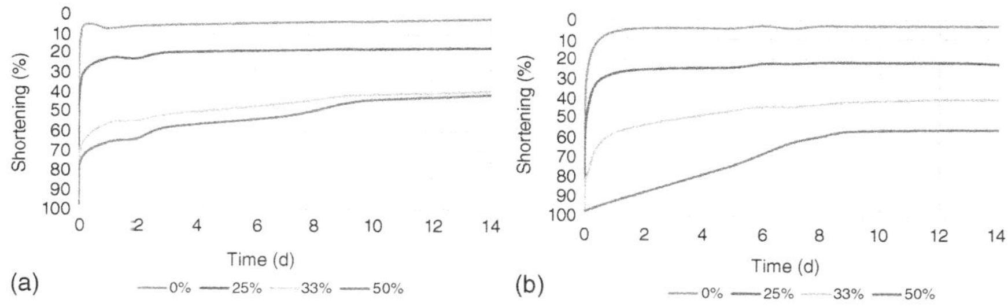

Figure 9.7 Shortening (%) vs. time (days) of different formulations of Poly-Paper heat treated at 110 °C for 15 minutes (a); and at 140 °C for 60 minutes (b).

By examining the results, some observations can be drawn:

- the material with a percentage of cellulose equal to 0% and 25% w/w relaxes almost completely, returning to a shape close to the initial one after just a few hours;
- for the material with 33% and 50% w/w cellulose fibers, recovery takes place at least after seven days, while maintaining a shape variation of about 50% over time;

The effectiveness of the maintenance of the form under load (i.e. the load is represented by fiber's weight) increases with the temperature and the treatment time.

It can be concluded that the property of Shape Memory Forming for Poly-Paper becomes quantitatively important only in the presence of a percentage of cellulose microfibers equal to 30% w/w.

The property was of primary importance for the subsequent thermoforming tests. Furthermore, this feature can be exploited to extend the life of the packaging product by remolding with a heat treatment, rendering it usable in different household and industrial applications.

9.5.2 Self-Healing by Water

During the various tests, it was verified that by simply wetting the surface of Poly-Paper with water it is possible to weld it strongly to itself, wood, and corrugated cardboard. To quantify self-healing effects, tensile tests were carried out following ISO 4587 Standard "Adhesives – Determination of tensile lap-shear strength of rigid-to-rigid bonded assemblies."

Poly-Paper samples (100 mm × 25 mm × 2 mm), obtained from the flat extruded bar and similar specimens of wood (thickness = 1 mm) and corrugated board BC wave (double wall thickness = 7 mm), were prepared. A 25 mm stretch at the end of the Poly-Paper specimens was wetted and placed next to a surface of the same size such as itself, wood, and cardboard, resulting in a gluing area of 625 mm^2 (Figure 9.8).

The two ends of the assembled specimen were then gripped in the jaws of the testing machine and subjected to tensile tests, measuring the force necessary to detach the two components of the whole specimen; the value was then

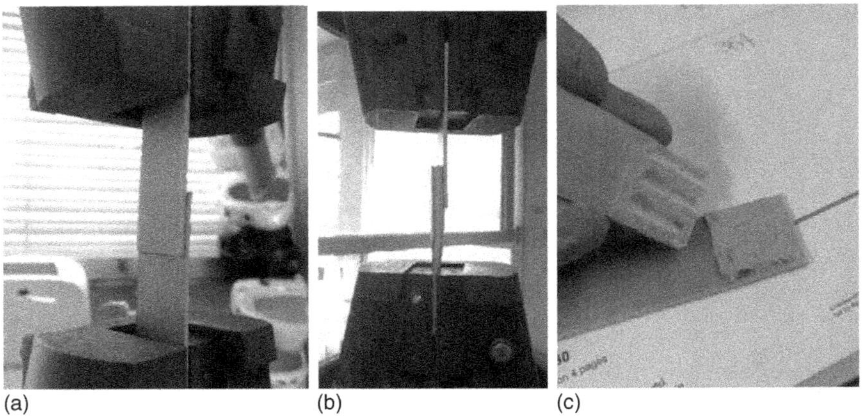

Figure 9.8 Adhesive determination of Poly-Paper samples to wood and cardboard. Source: Romina Santi.

Table 9.5 Adhesive property results of Poly-Paper 50% coupled with itself, peeled wood, and corrugated cardboard.

Coupling $A = 625\ mm^2$ to 0.15 mm/s	Force (N)	Shear stress (MPa)
Poly-Paper 50% – Poly-Paper 50%	815 ± 133	1.3 ± 0.21
Poly-Paper 50% – Peeled wood	798 ± 91	1.28 ± 0.14
Poly-Paper – Corrugated cardboard	347[a] ± 20	0.55[a] ± 0.03
Peeled wood – Poly-Paper 50% – Peeled wood	639 ± 56	1.02 ± 0.09
Corrugated cardboard – Poly-Paper – Corrugated cardboard	257[a] ± 19	0.49[a] ± 0.04

a) Breaking of corrugated cardboard.

normalized by the overlap area ($A = 625\ mm^2$). The tests were performed in triplicate; the results expressed as average and standard deviation values are reported in Table 9.5.

It should be emphasized that with the Poly-Paper-corrugated cardboard coupling, the corrugation of the cardboard always occurred before the separation of the two materials (Figure 9.8c).

The tests clearly show that the simple wetting with water of an area of 625 mm^2 is able to support a force of about 80 kg. Based on the obtained results, it is possible to hypothesize different applications that exploit this particular property. The welding by water allows the preparation of assemblies without using glues, rather only wetting with water, adding a thin layer of Poly-Paper in the bonding areas.

For example, the realization of integrated packaging composed of cardboard or paper pulp thermoformed products using a thin Poly-Paper film replacing petrochemical glues can be envisaged, so as to get fully recyclable packaging in the paper/cardboard chain after use. The use of water as a filler material for joining means that the resources used are bio-based and eco-compatible.

Moreover, assembly of trays made out in Poly-Paper, paper pulp, and cardboard can be detached and reattached by providing the user with additional thin films of Poly-Paper.

9.6 End-of-Life

The end-of-life describes the phase of the life cycle of a product or solution in which its active use concludes. For a packaging material, the disposal takes place very abruptly with respect to the insertion on the market, since the disposal of the packaging begins with the unpacking of each purchased product. For this reason, Poly-Paper application is intended for the food and nonfood packaging field, to introduce an environmental sensitivity, combined with the economic and social motivations that guided the design of the new material development. Considering the end-of-life of the materials and the possible user disposal errors, the intuition of having a mono-material entirely recyclable in a single chain has a significant importance. Packaging materials will therefore have to try to pursue the performance of current materials but with increased responsibility and uniqueness of their disposal at the end of their life.

Through injection molding, thermoforming, and expansion of Poly-Paper, the obtainable products, as well as the precise geometric design of the packaging, can be very easily adapted to the shapes and needs of the product. Thanks to the versatility of production, a wide range of products could be made of Poly-Paper, responding to the various packaging functions that, used together with paper and cardboard, can be recycled directly with paper, thereby avoiding disposal errors.

The end of the life of Poly-Paper is the most characteristic feature of the material: the whole material can be arranged together with the collection of paper and cardboard. Although PVA is a polymer of petrochemical origin, it is soluble in water, biodegradable, and considered nontoxic by the EFSA [28]. In addition, factors as temperature, pH and water turbulence influence the dissolution of the PVA matrix.

A preliminary dissolution test was carried out (Figure 9.9), investigating how the temperature and turbulence of the water increase the speed of dissolution of the material. The results of the solubility tests carried out have shown that the dissolution rate of Poly-Paper put into the pulp conditions makes it compatible with the paper recycling chain. In fact, the matrix is designed to dissolve, while the reinforcing cellulose fibers re-enter the recycling chain.

To have a precise view of the end-of-life of Poly-Paper, a preliminary test was conducted to analyze the recyclability level of cellulosic materials and products through the Aticelca test method mc 501: 2017 (Figure 9.10). The Aticelca method proposes a methodology for analyzing the level of recyclability in the paper industry and is applicable to all cellulose-based materials and products, not only in packaging field applications. This study is performed through a laboratory test that aims to simulate the paper industry's recycling phase.

Since the Poly-Paper cellulose microfibers have a dimension lower than 45 μm, the material was not tested alone but used as a layer added to corrugated

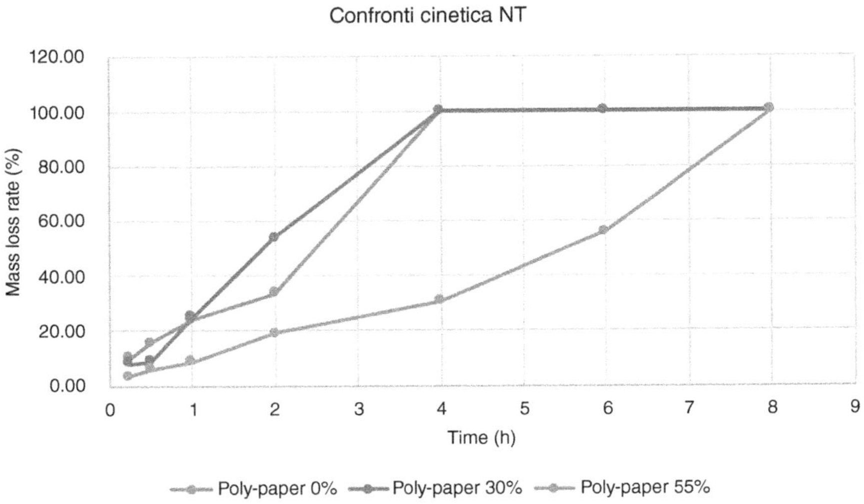

Figure 9.9 Water dissolution kinetics of Poly-Paper in three different formulations.

Figure 9.10 Preliminary test on Poly-Paper, Aticelca test method mc 501: 2017. Source: Romina Santi.

cardboard. Three samples were prepared using corrugated cardboard coated with a layer of Poly-Paper obtained by Fused Deposition Modeling (FDM) that constitutes 24.4% w/w of the whole sample.

The test carried out is to be considered only as a first indicator of the Poly-Paper behavior during the recycling phase in the paperboard supply chain. The test is not considered conclusive, but the starting point from which various paths can be considered.

Once this result has been obtained, tests could be performed using a different ratio between cellulose support (i.e. cardboard) and Poly-Paper, up to 50% w/w. In case these increasing tests are successful, the maximum percentage of Poly-Paper insertion in a product (packaging or not) in relation to the weight percentage of cellulosic material (paper or cardboard) can be established. The results, as shown in Figure 9.11 were encouraging. Poly-Paper has reached a level of recyclability "A" with the paper industry. All the results of the test would have ascertained an "A+" level but the fiber flakes, most likely due to the corrugated cardboard, slightly declassified the result.

It could also be investigated if the cellulose microfibers contained in a Poly-Paper composite tend to be dispersed in wastewater or if they tend to

Evaluation criteria	Recyclable with paper				Not recyclable with paper
	Level A+	Level A	Level B	Level C	Not recyclable with paper
Coarse scrap (%)	< 1.5	1.5 - 10.0	10.1 - 20.0	20.1 - 40.0	> 40.0
Area of adhesive particles Ø < 2000 µm (mm²/Kg)	< 2.500	2.500 - 10000	10.001 - 20.000	20.001 - 50.000	> 50.000
fiber flakes (%)	< 5.0	5.0 - 15.0	15.1 - 40.0	> 40.0	-
tackiness	none	none	none	none	present
Optical heterogeneity	Level 1	Level 2	Level 3	Level 3	-

Figure 9.11 Preliminary results of Aticelca test on Poly-Paper, test method mc 501: 2017.

interlace the structure of the cellulose and improve the mechanical characteristics of the recycled product.

At this stage, it may be desirable to use additives which retain the cellulose microfibers during the recycling phase. To prove this, a physical, morphological, and mechanical characterization of the obtained products will be necessary. In this way, Poly-Paper would not only be a product compatible with the paper mill, but it could prove to be an advantage for recycling thanks to the inclusion of microfibers in the structure of recycled paper.

9.7 Conclusions

Poly-Paper is a partly renewable composite that can be easily integrated with paper and cardboard packaging, being recyclable with them thanks to the easy separation of the two components: the PVA matrix is dissolved in water and the cellulose fibers extracted. Therefore, Poly-Paper combines the fundamental aspects of the two main packaging material families: the productive versatility from the polymers, and the conscientious recycling and the value of environmental responsibility from the cellulosic materials.

Poly-Paper has been developed for mono-material packaging solutions, as it can be processed through extrusion, injection molding, thermoforming, and expansion and can be disposed of with paper for waste collection. Poly-Paper behaves like thermoplastic polymers, so it can replace various plastic components in packaging having all the details and technical-aesthetic features thanks to the conventional thermoplastics processes [29]. The additional processes and finishes experimented have made it possible to hypothesize new applications of Poly-Paper for the development of expanded structures that could replace the expanded polystyrene in packaging. The innovation of this project concerns a conceptual rethinking before a technological and engineering one: the study of materials and processes, and management of circularity through the enhancement of natural resources and the minimization of waste production. The findings resulted in the filing of patents co-owned by the Politecnico di Milano and NextMaterials s.r.l.

References

1 Santi, R. (2016). *Polypaper. Un Materiale Composito per Packaging Ad Alta Sostenibilità Ambientale, Trasformabile Attraverso Le Tecnologie Tradizionali Di Lavorazione Dei Polimeri, Ottenuto Da Fibra Di Cellulosa Riciclata e Riciclabile Nella Filiera Carta, Cartone*. Politecnico di Milano.

2 Del Curto, B. (2016). *Packaging Naturalmente Tecnologico. Innovazioni Sostenibili per Il Food Packaging a Base Di Carta e Cartone*. Milano: Dativo Editore.

3 Santi, Romina, Agnese Piselli, Caterina Dastoli, Alberto Cigada, Barbara Del Curto, and Mauro Profaizer. 2018. "Poly-paper: an innovative thermoplastic eco-composite material recyclable with paper and cardboard." pp. 235-238. In TechConnect Briefs - ISBN:978-0-9975117-9-6 Vol. 2.

4 Bahrudin, Fadzli Irwan and Marco Aurisicchio. 2018. "The Appraisal of Sustainable Materials."In: International Design Conference (DESIGN 2018). ; 2018:2575-2584. doi:https://doi.org/10.21278/idc.2018.0455.

5 Karana, E. (2012). Characterization of 'natural' and 'high-quality' materials to improve perception of bio-plastics. *Journal of Cleaner Production*: 316–325.

6 European Bioplastics. 2017. What are bioplastics? Material types, terminology, and labels – an introduction bioplastics. https://docs.european-bioplastics.org/publications/fs/EuBP_FS_What_are_bioplastics.pdf (accessed 30 October 2019).

7 Cheng, Q., Wang, S., Rials, T.G., and Lee, S.H. (2007). Physical and mechanical properties of polyvinyl alcohol and polypropylene composite materials reinforced with fibril aggregates isolated from regenerated cellulose fibers. *Cellulose*: 593–602.

8 Abdul Khalil, H.P.S., Tye, Y.Y., Leh, C.P. et al. (2018). Cellulose reinforced biodegradable polymer composite film for packaging applications. In: *Bionanocomposites for Packaging Applications*, 49–69. Cham: Springer International Publishing.

9 Bledzki, A.K. and Gassan, J. (1999). Composites reinforced with cellulose based fibres. *Progress in Polymer Science* 24: 221–274.

10 Boldizar, A., Klason, C., Kubát, J. et al. (1987). Prehydrolyzed cellulose as reinforcing filler for thermoplastics. *International Journal of Polymeric Materials and Polymeric Biomaterials* 11 (4): 229–262.

11 Thakur, V.K. and Thakur, M.K. (2014). Processing and characterization of natural cellulose fibers/thermoset polymer composites. *Carbohydrate Polymers*: 102–117.

12 Avérous, L., Fringant, C., and Moro, L. (2001). Starch-Based Biodegradable Materials Suitable for Thermoforming Packaging. *Starch/Staerke* 53 (8): 368–371.

13 Graupner, N., Herrmann, A.S., and Müssig, J. (2009). Natural and man-made cellulose fibre-reinforced poly(lactic acid) (PLA) composites: an overview about mechanical characteristics and application areas. *Composites. Part A, Applied Science and Manufacturing*: 810–821.

14 Huda, M.S., Mohanty, A.K., Drzal, L.T. et al. (2005). 'Green' composites from recycled cellulose and poly(lactic acid): physico-mechanical and morphological properties evaluation. *Journal of Materials Science*: 4221–4229.
15 Mathew, A.P., Oksman, K., and Sain, M. (2005). *Mechanical properties of biodegradable composites from poly lactic acid (PLA) and microcrystalline cellulose (MCC). Journal of Applied Polymer Science* 97 (5): 2014–2025.
16 Zhang, W., Yang, X., Li, C. et al. (2011). Mechanochemical activation of cellulose and its thermoplastic polyvinyl alcohol ecocomposites with enhanced physicochemical properties. *Carbohydrate Polymers*: 257–263.
17 Ummartyotin, S. and Pechyen, C. (2016). Microcrystalline-cellulose and polypropylene based composite: a simple, selective and effective material for microwavable packaging. *Carbohydrate Polymers*: 133–140.
18 Curvelo, A.A.S., De Carvalho, A.J.F., and Agnelli, J.A.M. (2001). Thermoplastic starch-cellulosic fibers composites: preliminary results. *Carbohydrate Polymers*: 183–188.
19 Gurunathan, T., Mohanty, S., and Nayak, S.K. (2015). A review of the recent developments in biocomposites based on natural fibres and their application perspectives. *Composites. Part A, Applied Science and Manufacturing*: 1–25.
20 Barelli, N. (2015). *3D paper: A New Sustainable 3D Printing Material*. Politecnico di Milano.
21 Cigada, Alberto; Del Curto, Barbara; Fare', Silvia; Tanzi, Maria Cristina; Barelli, Nadia; Profaizer Mauro. 2015 Materiale composito ad alta sostenibilità ambientale. Italian patent 102015000028276, filed 26 June 2015 and issued 30 November 2017.
22 Highly environmentally sustainable composite material. PCT Application (UE, USA) IB2016/053777, filed 24 June 2016 and published EP 16745832.2 4 April 2018.
23 Chen, N., Li, L., and Wang, Q. (2007). New technology for thermal processing of poly(vinyl alcohol). *Plastics, Rubber and Composites* 36 (7-8): 283–290.
24 Ramaraj, B. (2006). Crosslinked poly(vinyl alcohol) and starch composite films. II. Physicomechanical, thermal properties and swelling studies. *Journal of Applied Polymer Science*: 909–916.
25 ISO 527–1: 2009. *Plastics – determination of tensile properties – Part 1: general principles*. International Standards Organization, Vernier, Geneva, Switzerland.
26 Santi, R., Cigada, A., Del Curto, B., and Farè, S. (2019). Modulable properties of PVA/cellulose fiber composites. *J. Appl. Biomater. Funct. Mater.*: 228080001983122.
27 Testa, Bruno. 2018. *"Designing with Novel Materials for the Circular Economy: The Case Study of Poly-Paper."*
28 EFSA Panel on Food Additives, Flavourings, Processing Aids and Materials in Contact with Food (2006). Opinion of the Scientific Panel on food additives, flavourings, processing aids and materials in contact with food (AFC) related to the use of polyvinyl alcohol as a coating agent for food supplements. *EFSA*

Journal 4 (2): 1–15. https://efsa.onlinelibrary.wiley.com/doi/epdf/10.2903/j.efsa.2006.294 (accessed 11 December 2017).

29 Del Curto, B., Barelli, N., Profaizer, M. et al. (2016). Poly-paper: a sustainable material for packaging, based on recycled paper and recyclable with paper. *Journal of Applied Biomaterials & Functional Materials*: 490–495.

10

Paper and Cardboard Reinforcement by Impregnation with Environmentally Friendly High-Performance Polymers for Food Packaging Applications

Uttam C. Paul[1] *and José A. Heredia-Guerrero*[2]

[1] *Smart Materials, Istituto Italiano di Tecnologia, Via Morego 30, Genova 16163, Italy*
[2] *Instituto de Hortofruticultura Subtropical y Mediterránea La Mayora, Universidad de Málaga - Consejo Superior de Investigaciones Científicas, Departamento de Mejora Genética y Biotecnología, Estación Experimental La Mayora, Avenida Dr. Wienberg, s/n, Algarrobo-Costa, Málaga E-29750, Spain*

10.1 Introduction

Packaging plays a vital role for the food industry since it protects the foods during distribution, storage, and sale, until consumption [1]. Currently, most of the used plastics for packaging are produced from fossil resources due to their low cost, lightness, flexibility, sealability, high strength, and excellent water barrier properties [2–4]. However, a massive use of these materials has led to major environmental problems mainly because of their lack of biodegradability [5]. On the other hand, biopolymers derived from biomass, especially the ones that are biodegradable, are realistic substitutes. They can be natural when they are directly extracted from the biomass (e.g. starch, proteins, cellulose) or synthetic (for instance, polylactic acid, PLA). Such biopolymers have the potential to replace the conventional oil-based plastics reducing their large impact on the environment by using renewable sources and decreasing CO_2 emissions.

In the same line, the use of paper and cardboard for packaging has gained a considerable amount of interest [6]. Paper and cardboard are preferred green packaging materials with good mechanical strength, flexibility, lightweight, recyclability, and low cost [2–4, 7, 8]. They consist mainly of cellulose, which is biodegradable and compostable. This polysaccharide is an attractive renewable, water-insoluble, and biodegradable biopolymer [1, 9] with an estimated annual biomass production between 10^{11} and 10^{12} tons [10]. It is mostly produced by plants and forms part of their cell walls as a structural component [11]. Cellulose is a polydispersed, linear, and syndiotactic polymer, whose basic monomeric unit is anhydroglucose [12]. Due to the hydrophilic nature and porous structure of papermaking cellulose, it is highly permeable to gases and water vapor, decreasing its strength with the presence of moisture/water, being these the main obstacles in its application as a packaging material [13–15]. Cellulose is usually laminated with aluminum or petroleum-based polymers such as polyethylene (PE), ethylene vinyl alcohol (EVOH), and polyvinylchloride (PVC)

Sustainable Food Packaging Technology, First Edition. Edited by Athanassia Athanassiou.

to improve its mechanical and barrier properties [8]. Examples include paper cups, plates, boxes, cartons, etc. In these cases, the paper and cardboard are used as structural materials and for printing, while the laminated plastic or aluminum layer provides the humidity barrier properties needed to prevent cellulose hydration and packaging failure. Since both the plastic films and the aluminum are problematic for sorting, recycling, and recovery, replacing one or more barrier layers with bio-based polymers is an interesting approach toward sustainable packaging. Replacing all the laminated layers with a single compostable biopolymer/cellulose composite, making the entire packaging material easier to produce, compostable, and environmentally friendly is even more interesting [16]. Several biopolymers such as polysaccharides (starch, cellulose, chitosan, and alginate), proteins (casein, whey, collagen, soya, and gluten), lipids (bees and carnauba waxes, and free fatty acids) and polyesters (polyhydroxyalkanoates (PHAs), poly-3-hydroxybutyrate (PHB), poly(3-hydroxybutyrate-*co*-3-hydroxyvalerate) (PHBV), and PLA) have already been used as coatings for paper and cardboard with moderate success [2, 6, 8, 13–21].

Since some of the most used biopolymers in industrial production of food containers are hydrophilic, the films produced from these materials are often hygroscopic, resulting in partial loss of their barrier properties at high humidity and reducing their performance as a packaging material in comparison to petroleum-based plastics [18]. Moreover, their molecular weight, rheological properties, and physical properties, such as time-dependent crystallization behavior, may cause problems and need to improve during industrial processing. Therefore, bio-based polymers are usually blended with other polymer matrices to improve their performance in terms of hydrophobicity, oleophobicity, and physio-mechanical properties [17, 18, 22]. Also in the case of cellulose-based paper and cardboard, to improve their water and oil barrier properties, treatment with external or internal coating layers, that reduce the interactions with water and protect the hydrophilic cellulose surface, is necessary. This is generally done by applying hydrophobic treatments on the paper or cardboard surface (e.g. surface coating or bulk fiber treatments) and is the focus of this chapter.

The first part of this chapter deals with the functionalization of cellulose networks by dip coating the paper and cardboard in environmentally friendly capstone (a polymer with C-6 fluorinated side chains) and ethyl cyanoacrylate (ECA) based solutions. In the second part, we present the fabrication, performance, and applications of several formulations based on silicone oil, starch, and polydimethylsiloxane (PDMS) to act as fillers and reinforcements of paper and cardboard matrices. The properties of the coated paper and cardboard such as water contact angle (WCA) and oil contact angle (OCA), moisture barrier properties, grease resistance test, and the tensile strength were determined.

10.2 Improving the Barrier Properties of Paper and Cardboard by Impregnation in Capstone and ECA Solutions

Papers are commonly protected against water by treating them with internal hydrophobic agents prior to sheet formation [7, 23]. The most common of them are rosin (a resin from pine trees), alkene ketene dimer (AKD) wax , and alkenyl succinic anhydride (ASA) [2]. In recent years, new hydrophobic products such as synthetic cationic or anionic polymers, fluorinated polyurethanes, fatty acids with different chain lengths, quaternary ammonium salts, fatty acid anhydrides, amines, ethylene copolymers, and surface-modified amphiphilic talcum have been introduced directly on the cellulose sheets [2, 17, 24, 25].

Fluorinated organic polymers have been comprehensively studied in the past for cellulose treatment because of their combined water and oil repellency properties [4, 26–28]. Fluoropolymers, particularly with ≥ C-8 fluorinated side chains exhibit very low surface free energy (≤18 dynes/cm). However, the US Environmental Protection Agency (EPA) has restricted the use of fluoropolymers having fluorinated side chains with C-8 or longer due to their various environmental concerns, as they decompose into perfluorooctanoic acid (PFOA) or perfluorooctanoic sulfonate (PFOS) that are persistent and bioaccumulate and, likely, carcinogenic [29, 30]. Hence, they were banned and a new generation of fluoropolymers with C-6 or lower fluorinated side chains that can degrade into environmentally safe perfluorohexanoic acid (PFHxA) with rapid bio-elimination rate were developed [26, 31]. In terms of water and oil repellency, C-6 fluorinated side chains perform worse than C-8 counterparts [31], although a proper application, chemical modification, and thermal treatment can be successfully used to create effective water- and oil/solvent-repellent coatings for many applications [26–28].

On the other hand, biocompatible cyanoacrylate monomers are extensively used in paper and cellulose functionalization and as tissue adhesives for the closure of skin wounds. Under ambient humidity and due to the presence of the hydroxyl groups of the cellulose, the ECA polymerizes around each cellulosic fiber forming a thin poly(ethyl cyanoacrylate) (PECA) shell [28, 32–34]. PECA interacts with the cellulose surface via hydrogen bonding as evidenced by Fourier transform infrared spectroscopy and thermogravimetric analysis measurements [34].

10.2.1 Preparation of the Samples

As mentioned above, the paper and cardboard were treated with capstone and ECA. We first prepared two formulations. They were solutions of polymer 1 (1–5 wt% Capstone ST-100 in acetone) and polymer 2 (1–5 wt% ECA in acetone). We deposited the solution onto the cardboard, using two methods: drop-casting and dip-coating. We found that the last procedure is more effective in reaching a good coating of paper fibers, without altering the macroscopic properties (i.e. the appearance) of the cardboard sheet. After soaking the cardboard for 30 seconds in the solutions, a coating around the fibers was formed without the use of complex curing instruments due to the material polymerization at ambient conditions, mainly driven by the humidity.

10.2.2 Morphological Characterization

Figure 10.1 shows the stereo-micro photograph of treated and untreated cardboard. The untreated cardboard showed a fibrous structure with cavities

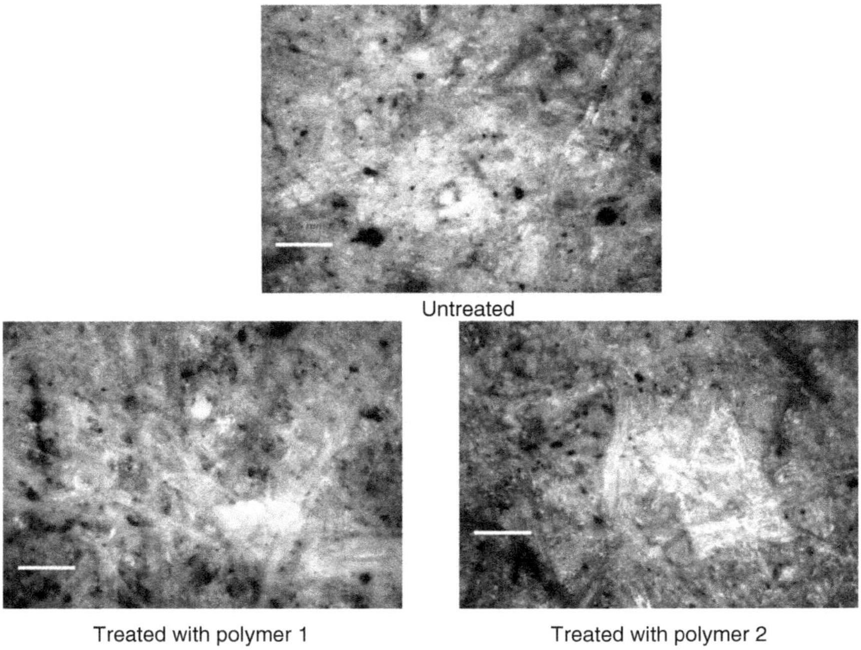

Figure 10.1 Stereo-micrograph of untreated and treated cardboard with polymer 1 (fluorinated polymer capstone) and polymer 2 (ECA). Scale bar 0.5 mm.

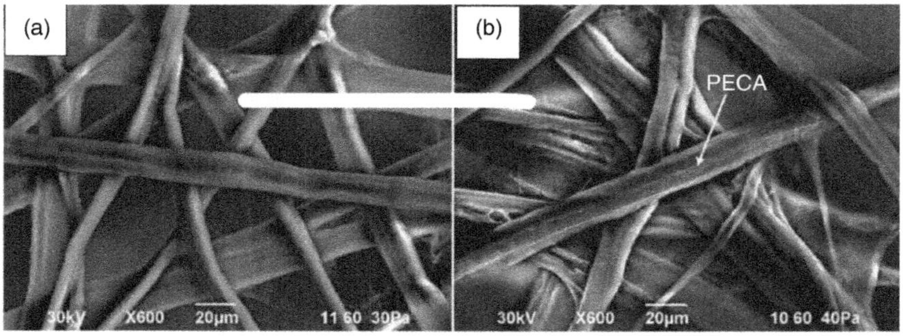

Figure 10.2 SEM images of (a) an untreated cellulosic sheet (35-μm thickness) and (b) treated sheet revealing the excellent preservation of the fiber network after treatment. Source: Bayer et al. [32]. © 2011, American Chemical Society.

from few to ~500 μm. The morphology of the treated cardboard is not changed, as there are no significant appearance changes between the treated and the untreated cardboard. In our previous studies [26, 32, 34], it was shown that this polymeric treatment forms a cladding structure around every single fiber, and not a unique, compact coating on the top the cardboard, Figure 10.2.

10.2.3 Chemical Characterization

XPS analysis confirmed the presence of both polymers in the cellulose matrix through the detection of characteristic peaks associated with capstone ($F \sim 28.2$ at.%) in one case and PECA ($N \sim 6.9$ at.%) in the other case, Figure 10.3.

To characterize the chemical interactions between cellulose and polymers (1 and 2), FTIR data were acquired. Figure 10.4 shows the FTIR spectrum of the raw (untreated) cellulose cardboard, where all the characteristic absorption bands of cellulose are observed. When cellulose cardboards are treated with the polymeric solutions of 1 and 2, the functional groups of both the cellulose and polymers can be identified in the FTIR spectra. For example, the characteristic absorption bands for the carbonyl C=O ester (1747 and 1735 cm^{-1} for polymers 1 and 2, respectively) are proofs of the intermolecular interactions by H-bonds between the hydroxyl groups of cellulose and the carbonyl groups of capstone and PECA [34].

10.2.4 Barrier Properties, Wettability, and Water Uptake

The treatments render the cardboards waterproof, as shown in Figure 10.5. In Figure 10.5, it is possible to observe a water drop deposited on the treated and untreated regions of the cardboard. The drop completely wets the untreated area and is absorbed by the cardboard in a few seconds. On the contrary, the water absorption is inhibited on the capstone or ECA-treated area for all the different used concentrations (1–5 wt%).

As shown in Figure 10.5, no water absorption is observed for the cardboard treated with both polymers, differently from the untreated one. This is true not only for the surface of the cardboard but also for the internal layers. We divided the cardboard sheet into two parts, peeling off the superficial layer, and placed a water drop between the two internal separated layers, Figure 10.6. Since the modified cardboard maintains its hydrophobic character also in the inner layers, we can assume that the treatment is not forming any type of film on the surface (i.e. a coating) but it is uniformly covering every single fiber in the entire volume of the cardboard.

We also tested the resistance of the treated cardboard when completely dipped into water. As presented in Figure 10.7, the untreated region becomes wet, while the treated one is waterproof, confirming the formation of a protecting polymeric covering around the cellulose fibers in the whole cardboard.

Figure 10.8 shows the WCA of treated and untreated cardboard. As expected, the hydrophilic cardboard turns hydrophobic after both polymer treatments. The WCA of polymer 1-treated cardboard is ~131° and for polymer 2–90°, while for the untreated cardboard, the value of WCA is below 20°, since the water drop is practically absorbed by the cellulose substrate. The increased WCA of treated cardboard is due to the hydrophobic nature of capstone and PECA protective layers that cover the hydrophilic cellulose [34].

Similarly, the oleophobicity of the samples was also analyzed. Figure 10.9 shows the oil repellency properties of treated and untreated cardboard by vegetable oil contact angle measurements. For polymer 1-treated cardboard the OCA is 111°.

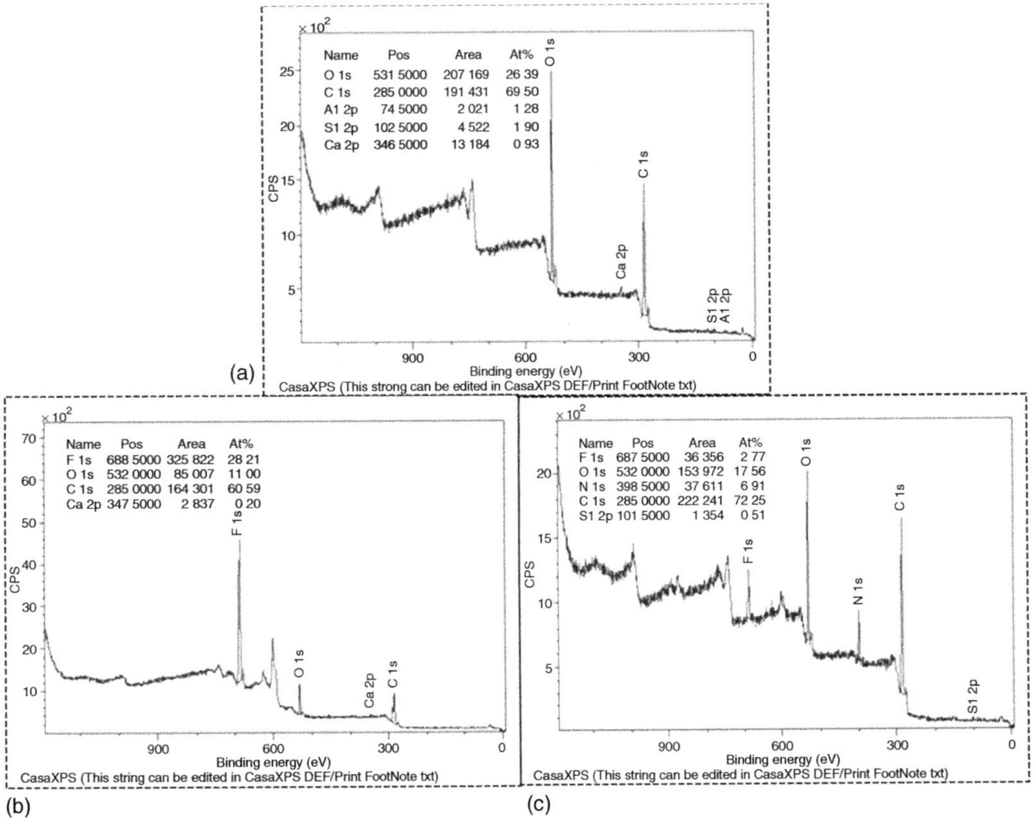

Figure 10.3 XPS analysis of untreated and treated cardboard (a) untreated, (b) treated with polymer 1 (fluorinated polymer capstone), and (c) treated with polymer 2 (ethyl cyanoacrylate, ECA).

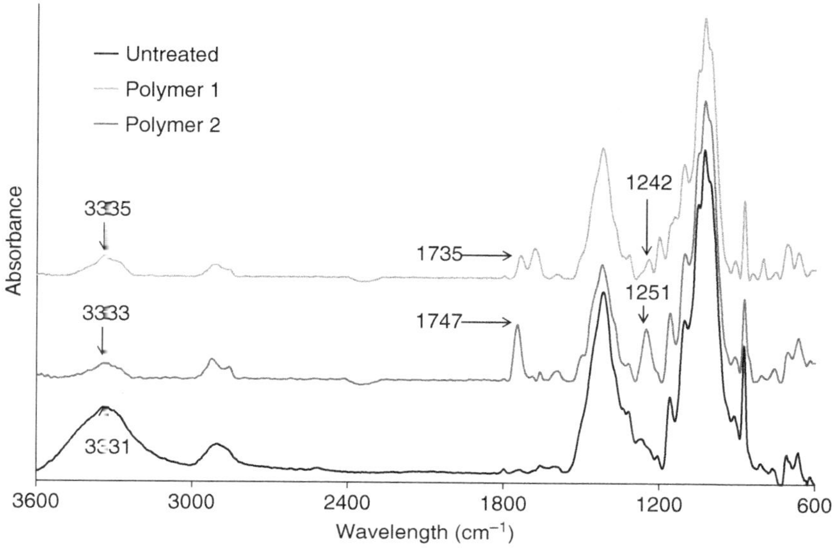

Figure 10.4 FTIR spectra of untreated cardboard, polymer 1 (capstone) treated cardboard, and polymer 2 (ECA) treated cardboard.

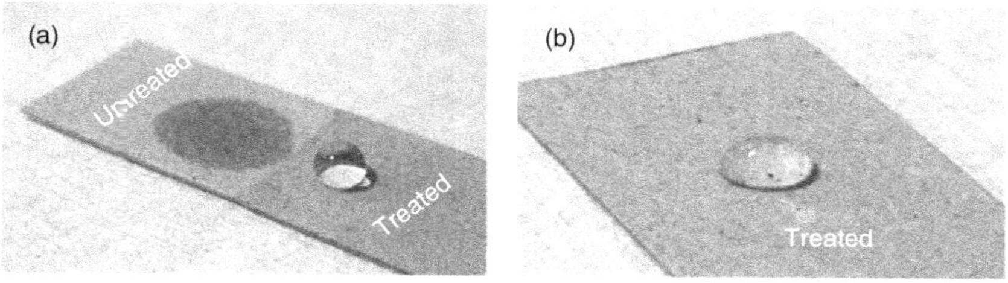

Figure 10.5 Water drop deposited onto the (a) untreated and fluorinated capstone (5 wt%) treated cardboard (cardboard weight 800–900 g/m^2), and (b) ECA (5 wt%) treated cardboard (cardboard weight 200–500 g/m^2).

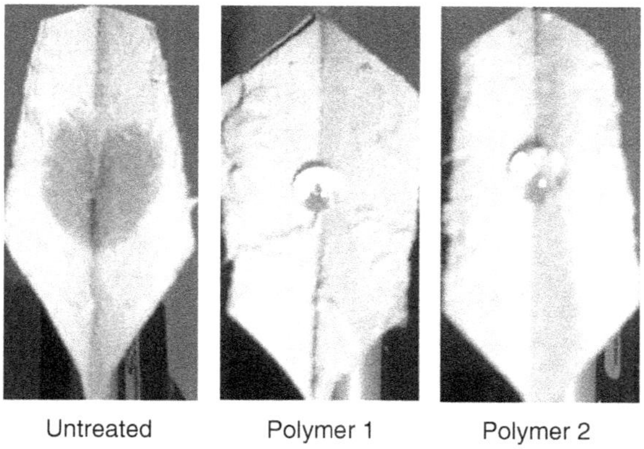

Figure 10.6 Water drop deposited onto the internal part of untreated and treated cardboard.

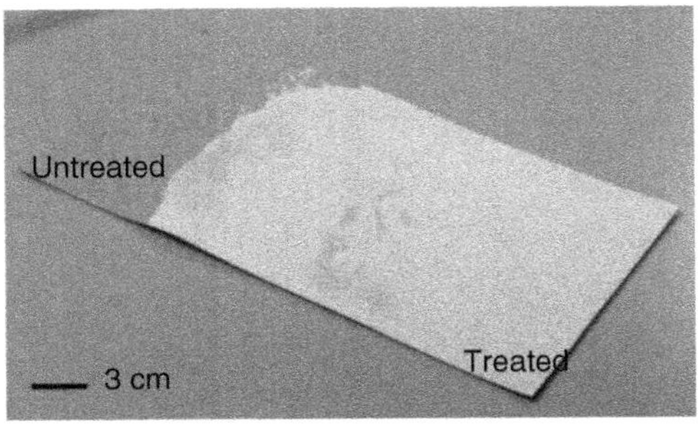

Figure 10.7 Image taken after complete dipping of the ECA (5 wt%) treated cardboard (cardboard weight 200–500 g/m^2) into water.

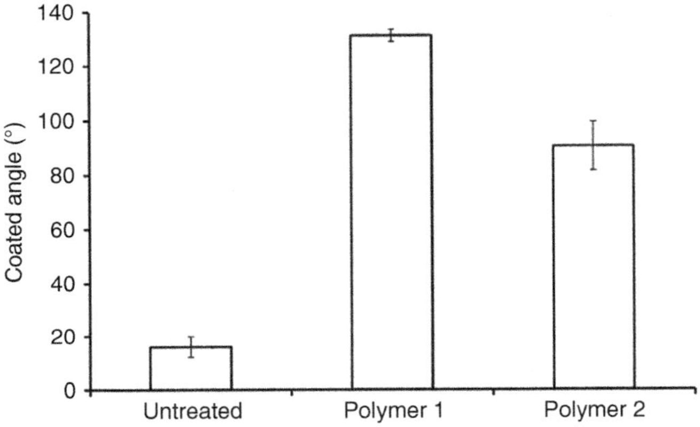

Figure 10.8 Water contact angle (WCA) of cardboard 1 (cardboard weight 800–900 g/m^2) untreated and treated with polymer 1 (capstone-fluorinated polymer) and polymer 2 (ECA).

On the other hand, the OCA of polymer 2-treated cardboard is only 50°, revealing the oleophilic character of the PECA polymer. Considering WCA and OCA results together, capstone-treated cardboards were both hydro- and oleophobic, while the PECA ones were only hydrophobic. Most probably, the inherent chemical composition of both polymers can explain this phenomenon. Thus, capstone wettability is similar to other fluoropolymers with good omniphobic behavior, whereas cyanoacrylate groups from PECA are hydrophobic but interact strongly with the oil molecules through dipole–dipole forces [16, 35, 36].

To evaluate the moisture barrier properties in realistic conditions, the WCAs of the treated cardboard 2 (weight 200–500 g/m^2) were measured at different relative humidity (RH) percentages (achieved by keeping the cardboard samples in closed chambers with oversaturated solutions of specific salts). Figures 10.10 and 10.11 show the measured WCA values for the different polymer formulations compared to the untreated cardboard 2 as a control sample. For the untreated cardboard, since the water drop is completely absorbed in a short time, we reported the initial WCA value (a few seconds after the deposition of the drop) at ambient conditions. In all cases, the contact angle of the different RHs treated cardboard is higher than 110°: for polymer 1 WCA is ~140°, while

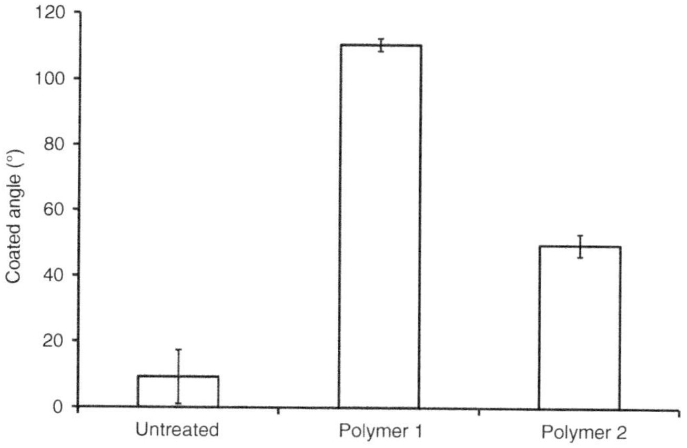

Figure 10.9 Oil contact angle (OCA) of cardboard 1 (cardboard weight 800–900 g/m^2) treated with different polymers. Polymer 1 (capstone) and polymer 2 (ECA).

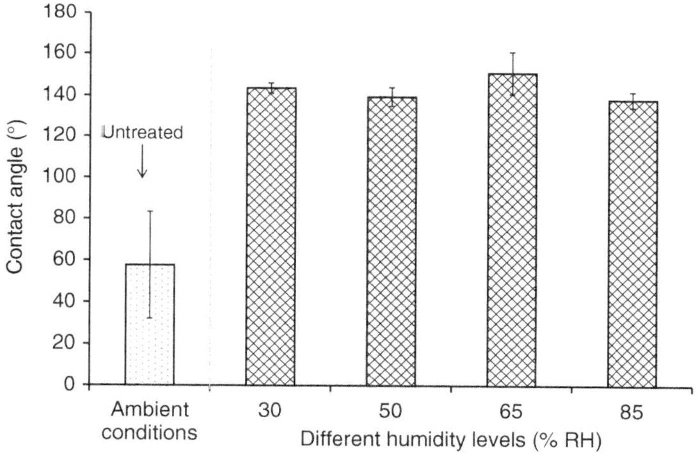

Figure 10.10 Water contact angle (WCA) of cardboard 2 (weight 200–500 g/m^2) treated with polymer at different relative humidity percentages.

for polymer 2–120°. This indicates that the cardboard is still hydrophobic, even if we keep it in different RHs, as required for packaging applications. Moreover, since no significant changes in WCA are observed for the different humidity conditions, we confirm the homogenous covering of the cellulose fibers. It should be mentioned here that the values of WCAs reported in Figures 10.10 and 10.11 are different from the ones reported in Figure 10.8 because the cardboards used for the tests are different.

The treated cardboards were also investigated for moisture barrier properties at different RH conditions. The moisture content (MC) in the samples stored at various RH conditions was evaluated by using the following equation:

$$\text{Moisture content } (\%) = (B - A)/A \text{ X}100 \tag{10.1}$$

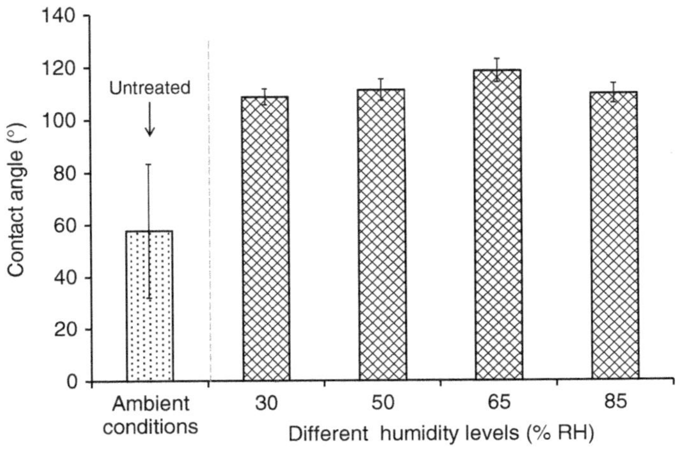

Figure 10.11 Water contact angle (WCA) of cardboard 2 (weight 200–500 g/m^2) treated with polymer 2 at different relative humidity percentages.

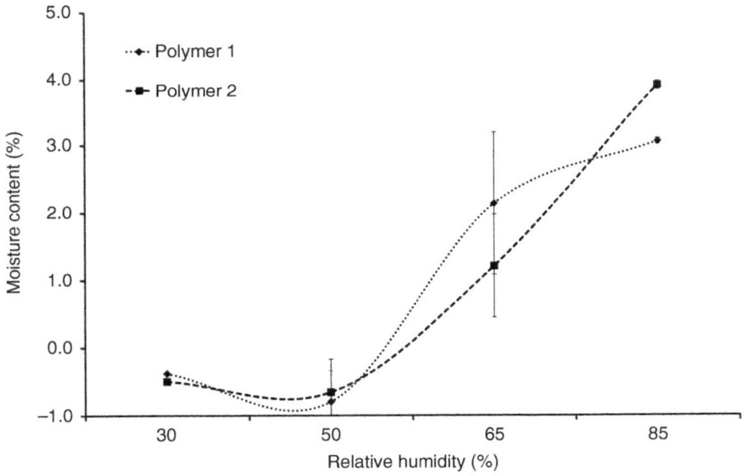

Figure 10.12 Moisture content for cardboard 1 (cardboard weight 800–900 g/m^2) treated with two formulations (polymer 1 and polymer 2) at different humidity levels.

where A is the weight of the cardboard sample before any conditioning in the chamber (room conditions) and B is the weight of the same cardboard sample at a specific humidity level.

Figures 10.12 and 10.13 show the moisture content on two types of treated and untreated cardboards at different humidity levels. The moisture content of the samples at 30% and 50% RH is lower than zero; this indicates that the cardboard stored in low relative humidity conditions can lose some inherent moisture, as in ambient conditions. Instead, at 65% and 85% RH, the MC is approximately 4.5% higher than in normal environmental conditions. This demonstrates that the

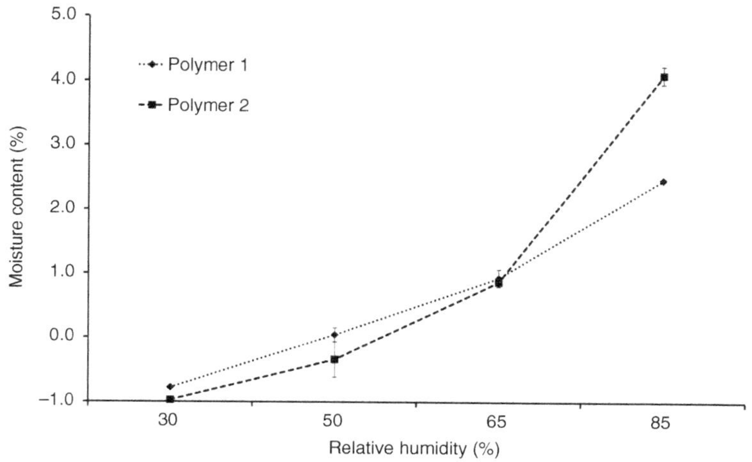

Figure 10.13 Moisture content for cardboard 2 (cardboard weight 200–500 g/m^2) treated with two formulations (polymer 1 and polymer 2) at different humidity levels.

polymer does not fill the cardboard porosity/free volume in the fiber networks of the treated cardboard, but at high humidity levels, this free volume can be occupied by moisture through diffusion process [37]. This is a further confirmation that the polymeric treatment does not create a thick coating but the network and porosity of fibers is maintained.

10.2.5 Mechanical Characterization

Both treatments protect the cardboard from the humidity absorption, so the mechanical properties are not compromised. Figures 10.14 and 10.15 show the tensile Young's modulus of two different cardboards treated with both formulations, respectively, calculated by DMA analysis at different humidity levels. There are no significant differences between the different treatment at the various RH, suggesting that mechanical properties are dominated by the rigid cellulose fibers instead of the ductile capstone and PECA coverings.

Water and oil repellency, mechanical properties, and moisture barrier properties of untreated cellulose cardboards and cardboards treated with two different polymeric formulations (i.e. capstone and ECA solutions in acetone) were studied as a function of relative humidity. Both treatments make the cardboard waterproof and increase the mechanical resistance, protecting the cardboards from the deterioration due to increased humidity, while the oil repellency properties are induced after treatment with polymer 1. We should point out that the treatment is done using exclusively polymers without the use of any functional filler. For this reason, the presented results are encouraging and indicate a promising route for the further improvement of the barrier properties. The incorporation of different types of functional fillers, known for their barrier properties, in the polymers is expected to further improve the performance of the paper and cardboard. In summary, we utilized C-6 fluorinated and PECA-based polymeric treatment

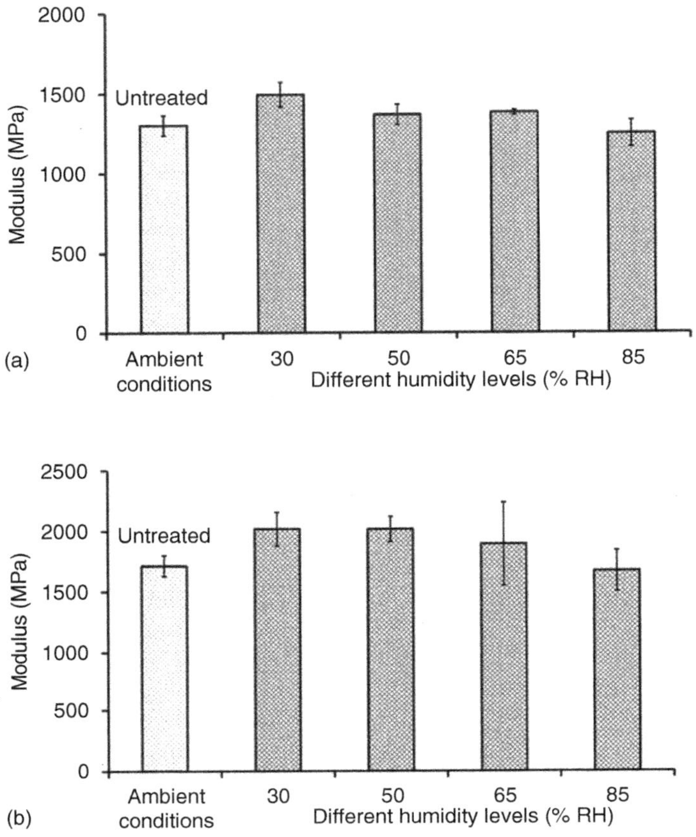

Figure 10.14 Mechanical properties of cardboard 1 (cardboard weight 800–900 g/m^2) treated with (a) polymer 1 and (b) polymer 2 at different humidity levels.

to make the cardboard hydrophobic, oleophobic, and prevent or minimize the mechanical damage by water or moisture of cardboard. Considering method simplicity and potential scalability, the proposed polymeric functionalization could be easily implemented in the large-scale production process of food packaging materials.

10.3 Water, Oil and Grease Resistance of Biocompatible Cellulose Food Containers

To turn cellulosic food storage or baking containers into water, oil, and grease resistant materials, the side of the paper packaging in contact with food is often coated with films of synthetic polymers [2, 38, 39]. However, the final cellulosic product loses its biodegradability as well as its recyclability. Moreover, the fabrication of the containers requires additional steps that complicate the process and increase the costs. Synthetic polymeric coatings on paper, such as polyethylene,

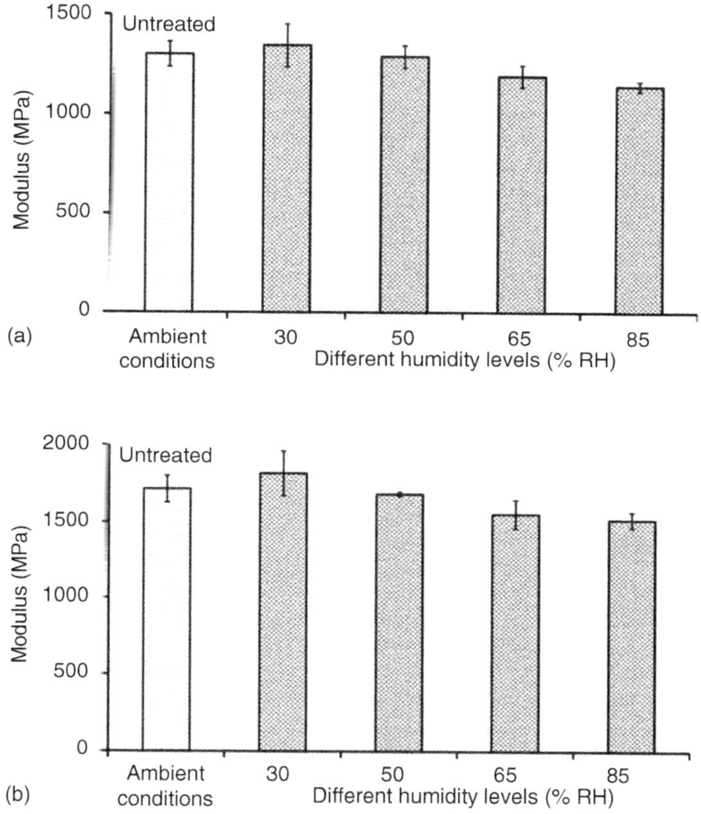

Figure 10 15 Mechanical properties of cardboard 2 (cardboard weight 200–500 g/m^2) treated with (a) polymer 1, and (b) polymer 2 at different humidity levels.

polyvinyl alcohol, rubber latex, and fluorocarbon polymers can be replaced by bio-based, biodegradable polymers to achieve the above-mentioned water, oil, and grease resistance for cellulose packaging materials [38, 40]. Indeed, recently, research interest has grown in the development of recyclable and biodegradable paper-based materials with good water, oil, and grease resistance properties for use in food packaging application [41]. For example, commercial printing paper and paper cups have been coated with melamine, an FDA approved-for-food contact materials and, subsequently, a PDMS-based water-repellent outer layer has been applied [41]. For the paper functionalization, various research approaches have been applied, such as paper sizing, chemical modification (layer-by-layer, grafting approach), and physiochemical modification (plasma etching, laser, etc.). However, most of these techniques are not applicable for practical large-scale production due to their time-consuming processes, and reliance on costly raw materials. Thus, the development of a facile, low-cost, safe-for-food contact, and scalable approach remains a critical practical challenge. The coating of paper with water- and oil-repellent materials using dip coating, rod coating, spraying, etc. is considered simple and commercially viable approach to develop biodegradable water and oil resistant paper. Typically, fluorinated polymers, polysiloxanes, or higher alkanes are applied to fabricate a hydrophobic and oleophobic coating for paper. Although fluorochemicals render excellent water and oil resistance properties to paper, as shown in Section 10.2.4, due to their potential toxicity and environmental concerns, demands have arisen for replacing these materials in paper coatings [38, 41] with environmentally friendly alternatives.

We will introduce here the combination of three alternatives, PDMS, silicone oil, and starch, which can be used for the successful functionalization of cellulosic substrates for food packaging applications: PDMS has attracted great attention due to its low surface tension, nontoxic nature, affordability, and environmental friendliness. PDMS is shown to enhance the hydrophobicity, mechanical properties, leakage of the fluid and gas, wear-resistance, water repellency, and heat-durability of treated paper [41–44]. Because of the two methyl groups on every silicon atom, PDMS is an oleophilic material that is readily swollen by oils and organic solvents [45]. On the other hand, silicone oil is a synthetic organic silicone polymer, which has (SiR_2–O) backbones with different polymerization degrees and chain lengths. The side group R is commonly methyl group. Silicone oils have low surface tension, high hydrophobicity, and can wet most surfaces. Silicone oils have also excellent lubricating and insulating functions as well as high thermal, weather, and chemical resistance. Due to these unique properties, silicone oils have been widely used in industrial and consumer products with other functional groups. For instance, hydrogen, alkoxy, amine, and higher number alkyl groups that can be introduced both to the side and to the end group of silicone oils in order to further enhance their properties and can be bonded to other polymeric materials [46].

Starch is a highly abundant, renewable, and biodegradable agricultural resource [47]. The molecular structure of starch consists of linear amylose and a branched macromolecular amylopectin molecule. Unmodified (granular) starch is used as a sustainable filler material for many polymeric materials [47–49]. However, the use of unmodified starch as a component for polymer compounding is quite challenging due to its large particle size, inherent hydrophilicity, and uptake of moisture. Therefore, blending starch granules with hydrophobic polymers requires complete particle drying and surface functionalization or modification in the form of grafting or coupling agents [47–50].

10.3.1 Preparation of the Samples

The treatment process relies on starch/PDMS, and silicone oil-functionalized starch/PDMS, which are affordable and biocompatible materials, that biodegrade into nontoxic silicate minerals and thus are safe for food processing [41, 47, 51]. The treatment was done on kraft white (65 g/m^2) paper sample with two polymeric formulations. Formulation 1: silicone oil (5 wt%)-functionalized potato starch (5 wt%) and PDMS (Syl-Off SL 160, Dow Corning) 90 wt%. Formulation 2: 5 wt% potato starch, and 95 wt% PDMS. To satisfy the need for water, oil, and grease resistant properties of paper, hydrophilic starch in formulation 1 was first functionalized with silicone oil and then mixed with PDMS. Formulations 1 and

2 were applied by a rod coater, and the treated paper was dried at 150 °C for 30 seconds.

10.3.2 Morphological Analysis

Scanning electron microscopy (SEM) analysis shows that the untreated kraft white paper (Figure 10.16a,b) has a porous structure typical of cellulose substrates. In case of polymeric formulation 1-treated paper (Figure 10.16c,d), the porosity of the paper was completely covered and few starch microparticles were visible on the paper surface. The formulation 2-treated paper (Figure 10.16e,f) also covered completely the porosity of the paper. However, more starch microparticles are visible on the top of the sample and appear aggregated, most likely because the silicon oil functionalization is missing.

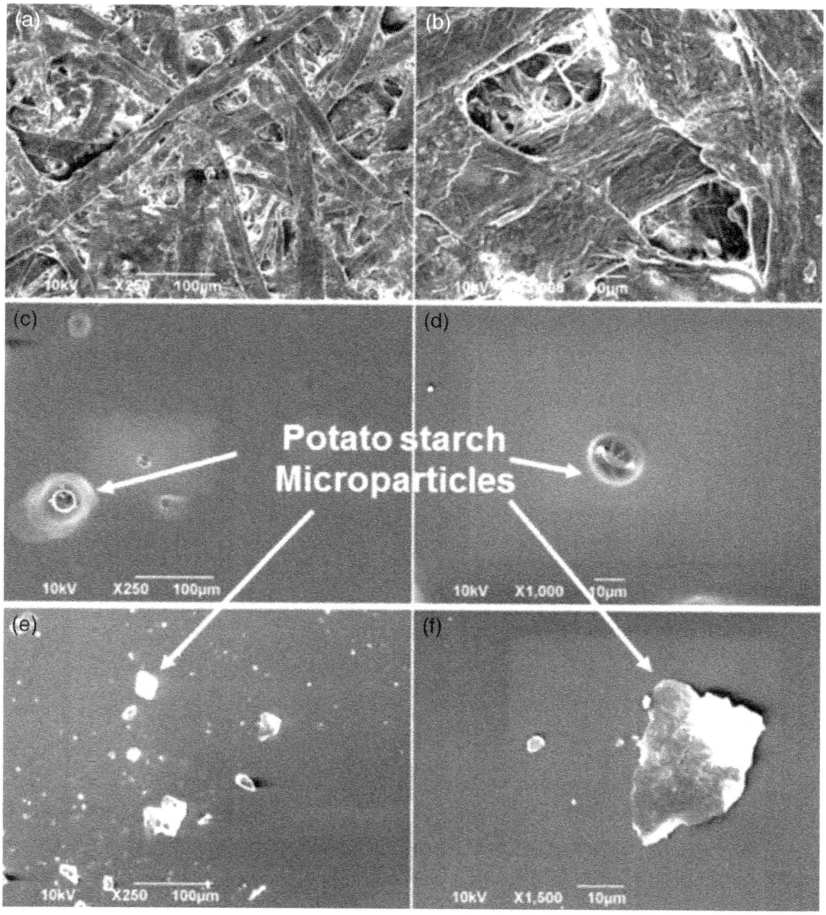

Figure 10.16 Scanning electron microscopy (SEM) analysis of (a, b) untreated paper surfaces (c, d) polymeric formulation 1-, and (e, f) polymeric formulation 2-treated paper, in two different magnifications.

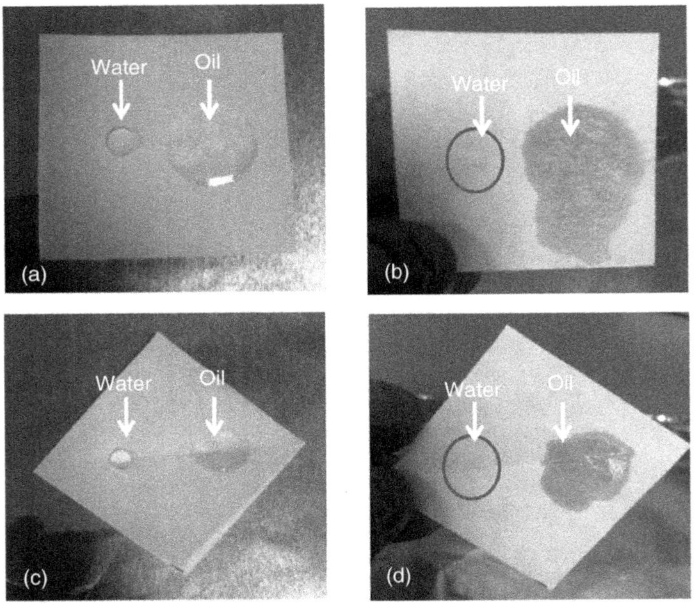

Figure 10.17 Water and oil droplets remaining trace onto the surface and bending area of the untreated paper. (a) water and oil droplets on surface in one to two minutes, (b) after removal of water and oil droplets, (c) water and oil droplets in bending point, and (d) after removal of water and oil from bending point.

10.3.3 Water and Oil Resistance Properties

As shown in Figure 10.17a,c, when water and oil drops, are placed on the surface and bending point of the untreated paper, respectively, they are easily absorbed and traces of them remain. See Figure 10.17b,d.

Instead, after the treatment, no water absorption is observed and the paper becomes waterproof. See Figure 10.18. We can assume that the treatment covers the cellulose surface uniformly, as confirmed previously by SEM. Moreover, differently from the untreated paper, the water drop slides on the surface of the treated paper without leaving a trace of water. On the other hand, for the treated papers, the oil is slightly spread on the surface/bending point. See Figure 10.18a,d, but it is not absorbed. There is not a visible oil spot when we wiped it out after 24 hours, indicating the oil resistance properties of treated papers.

10.3.4 Mechanical, Grease Resistance, and Barrier Properties of Treated Paper

Table 10.1 reports some of the physical characteristics and mechanical properties of untreated and treated paper. In general, the mechanical properties of all treated papers are superior to the untreated one, indicating a reinforcement of the paper fibrous structure.

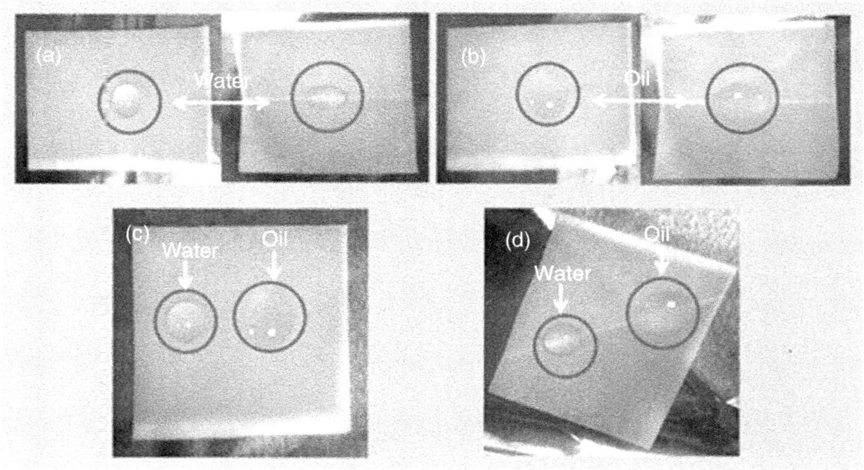

Figure 10.18 Water and oil resistance tests of treated paper. (a, b) Water and oil droplets trace onto the surface and bending point of the treated paper (treatment formulation 1). (c, d) Water and oil droplets trace onto the surface and bending point of the treated paper (treatment formulation 2).

Table 10.1 Physical and mechanical properties of untreated and treated papers.

	Paper weight (g/m^2)	Thickness (μm)	Water contact angle (°) after 10 min	Tensile strength (N)	Tensile strain (%)	Tensile stress (MPa)
Untreated	64.2 ± 0.5	51.0 ± 0.2	93.0 ± 1.5	7.1 ± 0.5	6.9 ± 0.2	22.9 ± 1.1
Treatment 1	100.0 ± 0.7	62.0 ± 0.3	105.0 ± 1.1	9.0 ± 0.4	7.9 ± 0.4	24.4 ± 0.1
Treatment 2	141.1 ± 2.5	96.0 ± 0.5	100.0 ± 2.2	9.7 ± 0.8	9.7 ± 0.3	20.1 ± 1.2

The grease resistance of a packaging material depends on its surface chemistry (i.e. hydrophilic or hydrophobic character), the barrier quality, density, and presence of pores and voids as well as on the coating thickness of the substrate [52]. The kit test is a common method for the evaluation of fat and grease resistance of paper. Grease resistance test was carried out according to the Tappi test method $T = 559\,\text{cm} - 12$ [53]. This test is a procedure for checking the degree of liquid repellency of the paper and cardboard. In this test, drops of liquids composed of oil and solvents in a specific proportion are placed on the surface of paper samples and the wetting behavior is observed. The highest numbered kit solution (kit solution no. 1–12) that remains on the surface of the paper without being absorbed is reported as higher "kit rating". If no penetration or wetting on the paper interface occurs after 15 seconds then the sample passes the test. The test continues by evaluating the next kit number until one of the test liquids shows obvious wetting. Table 10.2 shows the kit test results for treated and untreated paper.

Table 10.2 Comparison of grease resistance properties of paper coated with different biopolymers used in this study and in previous works.

Film/coating	Coating weight (g/m^2)	Grease resistance (kit number)[a]	References
Untreated paper	0	1	In this study
Formulation 1-treated paper	36	12+	In this study
Formulation 2-treated paper	77	8	In this study
cellulose fatty acid ester (C6) coated paper	10	6	[54]
cellulose fatty acid ester (C8) coated paper	11	10	[54]
cellulose fatty acid ester (C10) coated paper	9	7	[54]
cellulose fatty acid ester (C12) coated paper	6	9	[54]
cellulose fatty acid ester (C14) coated paper	7	11	[54]
cellulose fatty acid ester (C16) coated paper	12	5	[54]
Bio-LDPE coated paper	15	11	[54]
PLA coated paper	81	11	[54]
Sodium alginate (SA)/sodium carboxymethyl cellulose (CMC) coated paper	0.5	7	[54]
	0.7	8	[38]
	1.1	8	[38]
	1.3	8	[38]
	1.9	9	[38]
	2.4	9	[38]
	3	9	[38]
SA/propylene glycol alginate (PGA) coated paper	1.2	5	[38]
SA/propylene glycol alginate (SA/PGA 5 : 5) coated paper	2.4	4–5	[38]
Cellulose nanocrystal (CNC) coated paper	5	8	[55]
Cellulose nanofiber (CNF)/CNC coated multilayer paper	5	11	[55]
Virgin paper coated with chitosan	6	6	[52]
Virgin paper coated with alginate	6.1	12	[52]
Recycled paper coated with chitosan	6.1	5	[52]
Recycled paper coated with alginate	6	7	[52]

a) 0 = completely penetrate the oil/grease, and 12 = no penetration.

The higher is the kit number the better are the grease barrier properties of the paper. The maximum value obtainable is 12. Starch is a hydrophilic material and is able to close the voids and pores of the paper surface, thus meeting the initial criteria for a good grease barrier. Silicone oil-functionalized starch/PDMS (treatment formulation 1) applied on paper improved grease resistance, reached the maximum kit number of 12.0 and, thus, is classified as an excellent barrier material against grease. The non-functionalized starch/PDMS (treatment formulation 2) gives the kit value of 8, which could be of interest for some applications in the packaging industry because the kit values of 6 and 7 are acceptable values in most grease resistance applications [49]. The grease resistance properties of treated paper are also compared with other biopolymers film/coating, Table 10.2 [38, 50, 52, 54, 55], and it was found that formulation 1-treated paper has better grease resistance properties than others do. The grease resistance improvement factor indicates the effectiveness of this functional coating. From this study, we can see that the treatment formulation −1 shows the best results in case of water, oil and grease resistance properties as compared to untreated and other treatment processes.

Finally, the oxygen barrier properties of untreated and formulation 1-treated paper are measured and shown in Table 10.3. Table 10.3 shows the values of oxygen permeability of untreated paper and polymeric formulation 1-treated paper. The untreated paper showed a relatively high oxygen permeability of 1426 ml μm/m^2/d. The polymeric formulation 1-treated paper, however, showed lower permeability value of 33 (ml μm/(m^2 d), which corresponds to a reduction of ~97.7% with respect to the neat paper. The decrease in oxygen permeability may have been influenced by several factors, such as the ability of starch/silicon oil to fill the voids in PDMS and cover the surface of the paper matrix. The oxygen barrier properties of treated paper are also compared with other biopolymers film/coating and synthetic polymer (Table 10.3) [19, 56, 57], and found that formulation 1-treated paper has better oxygen barrier properties than others. The oxygen permeability

Table 10.3 Comparison of oxygen permeability (OP) of treated paper with other biopolymer and synthetic polymer films.

Film/coating	Conditions	OP (ml μm/m^2/d)	References
Untreated paper	50% RH, 23 °C	1426	In this study
Formulation 1-treated paper	50% RH, 23 °C	33	In this study
Chitosan treated paper	50% RH, 23 °C	78	[19]
Wheat glutin-treated paper	50% RH, 23 °C	413–825	[19]
Pectin-treated paper	50% RH, 23 °C	2400	[56]
Pectin + 10% nanoclay	50% RH, 23 °C	1200	[56]
Pectin + 20% nanoclay	50% RH, 23 °C	700	[56]
Pectin + 30% nanoclay	50% RH, 23 °C	2600	[56]
Cellulose nanofiber (CNF) bleach birch pulp-treated paper	50% RH, 23 °C	30	[56]
CNF bleach birch pulp (plasticized with glycerol)	50% RH, 23 °C	1700	[56]
Poly(ethylene terephthalate) (PET)	50% RH, 23 °C	1000–5000	[57]
Polypropylene (PP)	50% RH, 23 °C	50 000–100 000	[57]

Source: Vartiainen et al. 2014 [56]; Nanni et al. 2019 [57]; Gällstedt et al. 2005 [19].

improvement factor further indicates the effectiveness of this functional coating.

10.4 Conclusions

In summary, the results presented in this chapter indicate that paper and cardboard can be modified to become hydrophobic, oleophobic, and grease resistant by applying environmentally friendly polymeric coatings by impregnation or rod coating processes from the corresponding polymeric solutions. Sustainable, cost-effective, and ecologically compatible materials are needed for packaging applications because plastic packaging represents a large portion of land and ocean waste. New processes for modifying the paper and cardboard based on biocompatible and environmentally friendly materials open new opportunities for the development of bio-based packaging materials that can challenge the performance and cost of synthetic packaging materials. Review of the literature clearly shows that a certain hydrophobic agent is needed to make the cellulosic fibers suitable for packaging application. However, there are many inspiring new ways to make functional hydrophobic and oleophobic materials from cellulose and derivatives, and various practical applications have been demonstrated. Use of nontoxic, biocompatible, and environmentally friendly hydrophobic/oleophobic polymers and elastomers along with cellulose could be one interesting future direction, as such polymers do not pose alarming environmental concerns compared to more traditional harmful polymers used so far.

References

1 Andersson, C. (2008). New ways to enhance the functionality of paperboard by surface treatment – a review. *Packaging Technology and Science* https://doi.org/10.1002/pts.823.
2 Samyn, P. (2013). Wetting and hydrophobic modification of cellulose surfaces for paper applications. *Journal of Materials Science* 48: 6455–6498.
3 Zhang, D. and Xiao, H. (2013). Dual-functional beeswaxes on enhancing antimicrobial activity and water vapor barrier property of paper. *ACS Applied Materials & Interfaces* 5: 3464–3468.
4 Aulin, C., Gallstedt, M., and Lindstrom, T. (2010). Oxygen and oil barrier properties of microfibrillated cellulose films and coatings. *Cellulose* 17: 559–574.
5 Valdés, A., Burgos, N., Jiménez, A., and Garrigós, M.A. (2015). Natural pectin polysaccharides as edible coatings. 5: 865–886.
6 Bordenave, N., Grelier, S., Pichavant, F., and Coma, V. (2007). Water and moisture susceptibility of chitosan and paper-based materials: structure–property relationships. *Journal of Agricultural and Food Chemistry* 55: 9479–9488.
7 Shen, J., Fatehi, P., and Ni, Y. (2014). Biopolymers for surface engineering of paper-based products. *Cellulose* 21: 3145–3160.

8 Khwaldia, K., Arab-Tehrany, E., and Desobry, S. (2010). Biopolymer coatings on paper packaging materials. *Comprehensive Reviews in Food Science and Food Safety* 9: 82–91.

9 Kamel, S., Ali, N., Jahangir, K. et al. (2008). Pharmaceutical significance of cellulose: a review. *Express Polymer Letters* 2 (11): 758–778.

10 Klemm, D., Phillip, B., Heinze, T. et al. (1998). *Comprehensive Cellulose Chemistry*, vol. 2. Weinheim: Wiley.

11 Cosgrove, D.J. (2005). Growth of the plant cell wall. *Nature Reviews. Molecular Cell Biology* 6: 850–861.

12 Varshney, V.K. and Naithani, S. (2011). Chemical functionalization of cellulose derived from nonconventional sources. In: *Cellulose Fibers: Bio- and Nano-polymer Composites* (eds. S. Kalia, B.S. Kaith and I. Kaur). London: Springer.

13 Shankar, S. and Rhim, J.W. (2018). Antimicrobial wrapping paper coated with a ternary blend of carbohydrates (alginate, carboxymethyl cellulose, carrageenan) and grapefruit seed extract. *Carbohydrate Polymers* 196: 92–101.

14 Alouia, H., Khwaldia, K., Ben Slama, M., and Hamdi, M. (2011). Effect of glycerol and coating weight on functional properties of biopolymer-coated paper. *Carbohydrate Polymers* 86: 1063–1072.

15 Cataldi, P., Profaizer, M., and Bayer, I.S. (2019). Preventing water-induced mechanical deterioration of cardboard by a sequential polymer treatment. *Industrial and Engineering Chemistry Research* 58: 6456–6465.

16 Basu, S., Plucinski, A., and Catchmark, J.M. (2017). Sustainable barrier materials based on polysaccharide polyelectrolyte complexes. *Green Chemistry* 19: 4080–4092.

17 Ibrahim, M.M., Mobarak, F., El-Din, E.I.S. et al. (2009). Modified Egyptian talc as internal sizing agent for papermaking. *Carbohydrate Polymers* 75: 130–134.

18 Rastogi, V.K. and Samyn, P. (2015). Bio-based coatings for paper applications. *Coatings* 5: 887–930.

19 Gällstedt, M., Brottman, A., and Hedenqvist, M.S. (2005). Packaging-related properties of protein- and chitosan-coated paper. *Packaging Technology and Science* 18: 161–170.

20 Sangerlaub, S., Bruggemann, M., Rodler, N. et al. (2019). Extrusion coating of paper with poly(3-hydroxybutyrate-co-3-hydroxyvalerate) (PHBV) – packaging related functional properties. *Coatings* 9: 457.

21 Milionis, A., Ruffilli, R., and Bayer, I.S. (2014). Superhydrophobic nanocomposites from biodegradable thermoplastic starch composites (Mater-Bi®), hydrophobic nano-silica and lycopodium spores. *RSC Advances* 4: 34395.

22 Hult, E.L., Ropponen, J., Poppius-Levlin, K. et al. (2013). Enhancing the barrier properties of paperboard by a novel lignin coating. *Industrial Crops and Products* 50: 694–700.

23 Hubbe, M.A. and Heitmann, J.A. (2007). Review of factors affecting the release of water from cellulose fibers during paper manufacture. *BioResources* 2 (3): 500–533.

24 Trombetta, T., Iengo, P., and Turri, S. (2005). Fluorinated segmented polyurethane anionomers for water–oil repellent surface treatments of cellulosic substrates. *Journal of Applied Polymer Science* 98: 1364–1372.
25 Koskela, J.P. and Hormi, O.E.O. Long-chain fatty ammonium quaternaries in papermaking. *JAOCS* 79 (9): 1–11.
26 Zahid, M., Heredia-Guerrero, J.A., Athanassiou, A., and Bayer, I.S. (2017). Robust water repellent treatment for woven cotton fabrics with ecofriendly polymers. *Chemical Engineering Journal* 319: 321–332.
27 Masood, M.T., Heredia-Guerrero, J.A., Ceseracciu, L. et al. (2017). Superhydrophobic high impact polystyrene (HIPS) nanocomposites with wear abrasion resistance. *Chemical Engineering Journal* 322: 10–21.
28 Davis, A., Liakos, I., Genovese, M.E. et al. (2016). Water collection by sticky microislands on superomniphobic electrospun surfaces. *Advanced Materials Interfaces* 3: 1600606.
29 Schutzius, T.M., Bayer, I.S., Tiwari, M.K., and Megaridis, C.M. (2011). Novel fluoropolymer blends for the fabrication of sprayable multifunctional superhydrophobic nanostructured composites. *Industrial and Engineering Chemistry Research* 50: 11117–11123.
30 Jones, B. (2008). Fluoropolymers for coating applications. *Journal of Coatings Technology* 5 (9): 44–48.
31 Li, S.H., Huang, J.Y., Ge, M.Z. et al. (2015). Controlled grafting superhydrophobic cellulose surface with environmentally-friendly short fluoroalkyl chains by ATRP. *Materials and Design* 85: 815–822.
32 Bayer, I.S., Fragouli, D., Attanasio, A. et al. (2011). Water-repellent cellulose fiber networks with multifunctional properties. *ACS Applied Materials & Interfaces* 3 (10): 4024–4031.
33 Fragouli, D., Bayer, I.S., Di Corato, R. et al. (2012). Superparamagnetic cellulose fiber networks via nanocomposite functionalization. *Journal of Materials Chemistry* 22 (4): 1662–1666.
34 Ayadi, F., Bayer, I.S., Fragouli, D. et al. (2013). Mechanical reinforcement and water repellency induced to cellulose sheets by a polymer treatment. *Cellulose* 20: 1501–1509.
35 Sulaeva, I., Vejdovszky, P., Beaumont, M. et al. (2019). Fast approach to the hydrophobization of bacterial cellulose via the direct polymerization of ethyl 2-Cyanoacrylate. *Biomacromolecules* 20: 3142–3146.
36 Al-Refai, M., Ali, B.F., Said, A.B. et al. (2019). Synthesis, characterization, crystal structure and supramolecularity of ethyl(*E*)-2-cyano-3-(3-methylthiophen-2-yl)acrylate and a new polymorph of ethyl(*E*)-2-cyano-3-(thiophen-2-yl) acrylate. *Acta Cryst* 75: 1357–1361.
37 Prodduturi, S. et al. (2004). Water vapor sorption of hot-melt extruded hydroxypropyl cellulose films: effect on physico-mechanical properties, release characteristics, and stability. *Journal of Pharmaceutical Sciences* 93: 3047–3056.
38 Sheng, J., Li, L., and Zhao, L. (2019). Fabrication of grease resistant paper with non-fluorinated chemicals for food packaging. *Cellulose* 26: 6291–6302.

39 Perepelkin, K.E. (2005). Chemistry and technology of chemical fibers, (principles and method of modification of fibers and fiber materials): a review. *Fiber Chemistry* 37 (2): 123–140.
40 Khwaldia, K. (2010). Water vapor barrier and mechanical properties of paper-sodium caseinate and paper-sodium caseinate-paraffin wax films. *Journal of Food Biochemistry* 34: 998–1013.
41 Li, Z. and Rabnawaz, M. (2018). Fabrication of food-safe water-resistant paper coatings using a melamine primer and polysiloxane outer layer. *ACS Omega* 3: 11909–11916.
42 Wang, H.H., Yang, P.C., Liao, W.H., Yang, L.J. (2006). A new packaging method for pressure sensors by PDMS MEMS technology. Proceedings of the 1st IEEE International Conference on Nano/Micro Engineered and Molecular Systems, Zhuhai, China (18–21 January).
43 Xie, J., Hu, J., Lin, X. et al. (2018). Robust and anti-corrosive PDMS/SiO_2 superhydrophobic coatings fabricated on magnesium alloys with different-sized SiO_2 nanoparticles. *Applied Surface Science* 457: 870–880.
44 Mohamed, A.L., Er-Rafik, M., and Moller, M. (2013). Suitability of confocal Raman microscopy for monitoring the penetration of PDMS compounds into cotton fibres. *Carbohydrate Polymers* 96: 305–313.
45 Owen, M.J. (2017). Silicone hydrophobicity and oleophilicity. *SILICON* 9: 651–655.
46 Wang, Q. and Yang, Z. (2018). Quantification of Si in silicone Oils by ICP-OES. *SILICON* 10: 1–10.
47 Ceseracciu, L., Heredia-Guerrero, J.A., Dante, S. et al. (2015). Robust and biodegradable elastomers based on corn starch and polydimethylsiloxane (PDMS). *ACS Applied Materials & Interfaces* 7: 3742–3753.
48 Chandra, R. and Rustgi, R. (1997). Biodegradation of maleated linear low-density polyethylene and starch blends. *Polymer Degradation and Stability* 56: 185–202.
49 Kim, S. and Peterson, S.C. (2012). Development of degradable polymer composites from starch and poly(ethyl cyanoacrylate). *Polymer Composites* 33: 904–911.
50 De Buyl, F. (2001). Silicone sealants and structural adhesives. *International Journal of Adhesion and Adhesives* 21: 411–422.
51 Heredia-Guerrero, J.A., Ceseracciu, L., Guzman-Puyol, S. et al. (2018). Antimicrobial, antioxidant, and waterproof RTV silicone-ethyl cellulose composites containing clove essential oil. *Carbohydrate Polymers* 192: 150–158.
52 Kopacic, S., Walzl, A., Zankel, A. et al. (2018). Alginate and chitosan as a functional barrier for paper-based packaging materials. *Coatings* 8: 235.
53 Technical Association of the Pulp and Paper Industry (2012). *Grease Resistance Test for Paper and Cardboard*. TAPPI, T559 cm-12.
54 Willberg-Keyriläinen, P., Ropponen, J., Alakomi, H.-L., and Vartiainen, J. (2018). Cellulose fatty acid ester coated papers for stand-up pouch applications. *Journal of Applied Polymer Science*.
55 Tyagi, P., Lucia, L.A., Hubbe, M.A., and Pal, L. (2019). Nanocellulose-based multilayer barrier coatings for gas, oil, and grease resistance. *Carbohydrate Polymers* 206: 281–288.

56 Vartiainen, J., Vähä-Nissi, M., and Harlin, A. (2014). Biopolymer films and coatings in packaging applications – a review of recent developments. *Materials Sciences and Applications* 5: 708–718.

57 Nanni, G., Heredia-Guerrero, J.A., Paul, U.C. et al. (2019). Poly(furfuryl alcohol)-polycaprolactone blends. *Polymers* 11: 1069.

11

Nanocellulose-Based Multidimensional Structures for Food Packaging Technology

Saumya Chaturvedi[1], *Sadaf Afrin*[2], *Mohd S. Ansari*[3], *and Zoheb Karim*[4]

[1] *University of Delhi, Department of Food Technology, Shaheed Rajguru College of Applied Sciences for Women, 110096, Delhi, India*
[2] *Aligarh Muslim University, Department of Chemistry, Faculty of Sciences, 202002, Aligarh, India*
[3] *University of Delhi, Department of Biomedical Sciences, Shaheed Rajguru College of Applied Sciences for Women, 110096, Delhi, India*
[4] *MoRe Research Örnsköldsvik AB, SE-891 22, Örnsköldsvik, Sweden*

11.1 Introduction

Nanocellulose (NC) has potential applications across several industrial sectors and allows the development of innovative materials, as well as the enhancement of conventional material properties. Various processing routes have been followed for the production of NC-based multidimensional structures (Figure 11.1) for various applications such as biodegradable scaffolds for bone regeneration [1], structures (adsorbents, filters, composite membranes, functionally active frameworks) for water purification [2–5], flexible electronics [6, 7], etc.

There has been an explosive growth in the publication of peer-reviewed articles that combine key words related to "packaging" and "nanocellulose." Lots of significant publications could be seen in the field of NC-based packaging materials. Given such an acceleration of research around the globe, it might make sense to scrutinize the efforts made by researchers/scientists/product developers, etc. In principle, NC-based films, coatings, and intermediate layers, in addition to being light in weight, can provide benefits of renewability, recyclability, processability, and compatibility with health and the environment, and therefore application in the food packaging industry has been reported [1].

Cellulose nanostructures have been widely studied as components of materials for a variety of applications including food packaging [8]. They are usually incorporated as a reinforcement phase in nanocomposites. In other cases, cellulose nanostructures have been used as matrices for films production (bacterial cellulose deserving a special attention in this context, since it is produced as naturally nanostructured membranes).

Some novel properties required in produced structures (such as films, papers, coated materials, etc.) for food packaging have also been discussed in the literature. One example is the barrier properties, which are excellent for some

Sustainable Food Packaging Technology, First Edition. Edited by Athanassia Athanassiou.

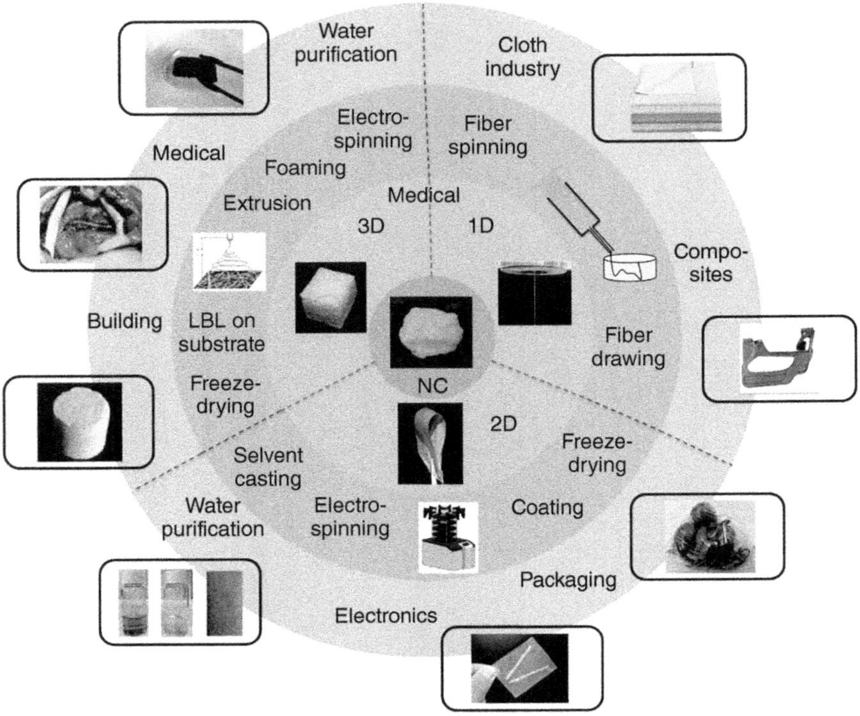

Figure 11.1 Various processing routes for the production of NC-based multidimensional structures. It is worth mentioning that the final properties of the produced structures are completely influenced by the route selected for processing.

plastic materials that show no or a very low permeability (diffusion) of gases (e.g. oxygen), vapors (e.g. water vapor) and aromas. On the other hand, the utilization of plastic for food packaging application is not very well appreciated by consumers due to sometimes-undefined barrier properties and nonbiodegradable nature. Therefore, applications of cellulosic nanomaterials in this field have attracted increasing attention in recent years. The possible utilization of NC as packaging materials depends on its barrier properties that will play an important role in determining its competitiveness with conventional materials [9, 10]. The good O_2 barrier properties of NC can be attributed to the dense network formed by nanofibrils with small and uniform dimensions [11]. Other properties that are required for a good packaging material include high tensile strength, moisture stability, high-pressure load, wet stability, biodegradability, etc. and interestingly we can tune NC-based structure to obtain these properties.

The use of NC for the production of food packaging materials (multidimensional structures) has been exploited recently [12–16]. The NC particles can be used as fillers, in composites manufacture, as coating and as self-standing thin films, achieving always very interesting and promising properties for food packaging. Very few of the several reviews that recently appeared on this topic in the scientific literature, however, summarized the potential of NC in packaging field [17, 18].

Therefore, this chapter expresses the potential of NC in food packaging. A general introduction of NC (types, isolation processes, related properties, etc.) is reported in the beginning of the chapter. Various processing routes used for the production of multidimensional packing materials are discussed and the effect of processing routes on the final properties of produced structures is summarized. Furthermore, surface chemistries and possible surface functionalization of NC for the introduction of desired properties are further explored. In the end, we have tried to explain the findings with the support of recently published articles/letters/scientific findings, etc.

11.2 Necessities in Food Packaging Industry

Packaging ensures food quality and safety, but has negative impact on the resources and the environment, and even leads to serious ecological problems. In this chapter, we take food packaging as the research object, analyze the necessity of food packaging and the negative impact of packaging on the environment, explore the current problems in food packaging system, and analyze the food packaging environmental pollution management approach. Nowadays, most materials used for food packaging are practically nondegradable, representing a serious global environmental problem. NC has been explored to develop biodegradable films in a large-scale effort to extend shelf life and improve quality of food while reducing packaging waste. One key point for the application of NC-based films is their selectivity toward gases (barrier properties) as discussed by Moon et al. [15].

Food packaging is a highly competitive and vibrant industry that is always on the research for new trends and innovations aiming to ease the life of consumers. Food packaging not only markets the product but also increases the shelf life of the products. The push for eco-friendliness has forced brands to ditch single-use plastics and provide biodegradable, recyclable, or reusable packaging. We urgently need to utilize paper packaging more than ever. A lot of research is also going into edible food packaging where the packaging itself can be consumed as a part of the food. NC-based packaging materials are still lacking in this field and we are hoping to read some interesting articles in future related to the NC-based materials for edible packaging. Nevertheless, research is going on in this direction and recently researchers in Tokyo University started mixing bacterial NC in ice cream for daily use.

Finally, time and temperature sensors are being embedded in beverage, food, and pharmaceutical packaging to ensure that the consumers get the product while it is still fresh. This also enables food items and other perishables to be sold online, like camel milk by desert farms. We can safely say that healthy food packaging trends largely incorporate technology.

Therefore, food packaging industry requires advanced reforms related to the utilization of green materials for the production of structures used for packaging. Moreover, such green packaging should have some extra advanced properties such as sensing, temperature controlling capability, nontoxicity, etc.

11.3 An Overview of NC

NC, prepared by a top-down approach from abundantly available wood biomass [19–21], has unique characteristic properties, such as high aspect ratio (>100), high crystallinity (65–95%), low ζ-potential (−65 mV), high tensile strength (200–300 MPa), and high elastic modulus (6–7 GPa) as reported by Castro et al. [19]. Furthermore, it can also be isolated from agricultural waste and microorganisms [4, 17]. It is a well-known fact that NC has tunable surface chemistry; therefore, it could be modified according to the required application [18, 22, 23]. This tunable surface chemistry could affect the surface charge density of available functional groups (−OH. −CHO, −COO) as discussed in our previous published article [24]. The introduced functional groups on the surface of NC not only affect the binding chemistry with matrix but alter the produced network too.

It is worth mentioning again here that food packaging materials require some unique properties as discussed in Section 11.2. It is not surprising that many studies have been reported to tune the surface of NC-based structures and NC itself to achieve these reported properties. For the achievement of specific surface properties of NC, two main routes are discussed in literature: (i) chemical functionalization and (ii) enzymatic modifications [25]. Various reinforcement materials such as inorganic fillers (clay, inorganic nanoparticles, etc.) have also been used to increase the mechanical properties of NC-based produced structures for food packaging, without compromising the gas permeability or the water vapor properties [22].

11.4 Cellulose Fibrils and Crystalline Cellulose

Cellulose is widely categorized into various types including microcrystalline cellulose (MCC), microfibrillated cellulose (MFC), nanofibrillated cellulose (NFC or CNF), nanocrystalline cellulose (NCC or CNC), nanorods, or cellulose whiskers [15]. The expression NC describes the cellulose fibril or crystallite containing at least one dimension in the nanoscale range, and it can be achieved by means of chemical or mechanical treatment of the wood pulp or plant cellulose and so on. The various types of NC are obtained by changing their source such as vegetable, lignocellulosic materials, bacterial cellulose, tunicate cellulose, algae cellulose, etc.

Nanocrystalline cellulose, also known as NCC or CNC, nanocrystals of cellulose, or cellulose nanowhiskers, is NC with high strength, which is usually extracted from cellulose fibrils by acid hydrolysis. It has short rod-like shape or whisker shape with 2–20 nm in diameter and 100–500 nm in length. Also, it contains 100% of cellulose chemical composition mainly in crystalline regions (high crystallinity around 54–88%). The amorphous parts are hydrolyzed and removed by acid.

NFC or CNF, also known as cellulose nanofibrils, or nanofibrillar cellulose, is the long, flexible, and entangled NC which can be extracted from cellulose fibrils by mechanical method. It has long fibril shape with 1–100 nm in diameter and 500–2000 nm in length. Also, it contains both crystalline and amorphous regions. Comparing with nanocrystalline cellulose, NFC has a longer length, higher aspect ratio (length to diameter), higher surface area, and higher amount of hydroxyl groups which are easily accessible for surface modification.

Bacterial NC is another kind of NC which is different from nanocrystalline cellulose and NFC. A detailed flowchart has been shown in Figure 11.2 for various NC types and isolation processes.

In the rest of the chapter, we have reported the utilization of these nanomaterials in the food packaging application. CNC- and CNF-related literature has been

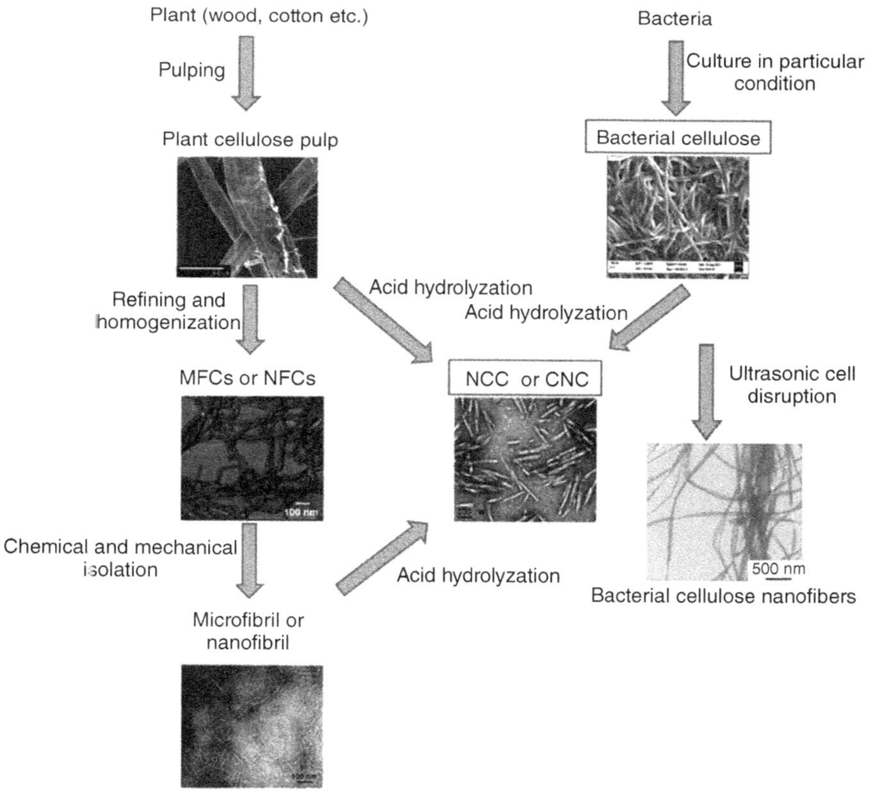

Figure 11.2 Various types of NC isolated from wood and bacterial culture. Acid hydrolysis is utilized for the production of cellulose nanocrystals (NCC or CNC), whereas, mechanical grinding is utilized for the isolation of cellulose nanofibrils (NFC). Bacterial cellulose is isolated by culturing. Source: Mishra et al. [26]. © 2018, Elsevier.

discussed and the effect of the structural properties on the packaging application has been summarized (gas permeability and water vapor transportation).

11.5 Why NC for Packaging?

NC has gained increasing attention due to the high strength, stiffness, biodegradability and renewability, and drastic increase in physical and functional properties when incorporated in polymeric composites. Because of the equatorial positions (C6) of the hydroxyls on the cellulose chain, they protrude laterally along the extended molecule and are readily available for hydrogen bonding. These hydrogen bonds cause the chains to group together in a highly ordered structure, which makes them mechanically stable. Good mechanical strength and rigidity can offer resistance to high-pressure environmental packaging application. NC has a highly crystalline lattice structure, which provides the possibility to perform well within different pH ranges, with various gases, environmental pressure, and moisture. Oksman et al. [11] reported the crystallinity of NC to be in the range of 70–80% depending on the process of isolation. Furthermore, the high crystallinity value makes it resistant to acid, fumes, etc. Sehaqui et al. [27] reported a surface area of 482 m^2/g for nanopaper prepared using supercritical CO_2 drying. However, it will be challenging to maintain this high surface area during large-scale processing.

With the presence of large number of chemical functionalities within their structure, NC building blocks provide a unique platform for significant surface modification through various chemistries [28]. The introduced functional groups after modification can be tailored for selectivity of gases. Volesky [29] summarized that functional groups can be added, such as carbonyl (ketone), carboxyl, sulfhydryl (thiol), sulfonate, thioether, amine and secondary amine, phosphate, phosphodiester, imidazole, etc. [30] has shown that the presence of COO^-, SO_3^-, PO_3^{2-} on the surface of NC makes them selective toward the charged species [30].

11.6 Effect on NCs on Networking

Apart from the abovementioned advanced properties, NC-based structures must be selectively non-permeable to targeted gases. In some recent articles, impermeability to various gases through the formed structures (usually film-like 2D structures) has been discussed in detail. The permeability of a particular gas completely depends on the networking behavior of the produced structures. As shown in Figure 11.3, the tight networking in a film of cellulose crystals produced using acid hydrolysis was responsible for the improved gas barrier properties compared to the film produced using cellulose nanofibers (CNFs). Indeed, the cellulose crystals-based network was highly dense compared to the fiber-based network [11, 15].

Furthermore, the oxygen transmittance rate of cellulose nanofibers at 0% relative humidity was found within the range 17–18 ml/m^2 d. The increase of relative

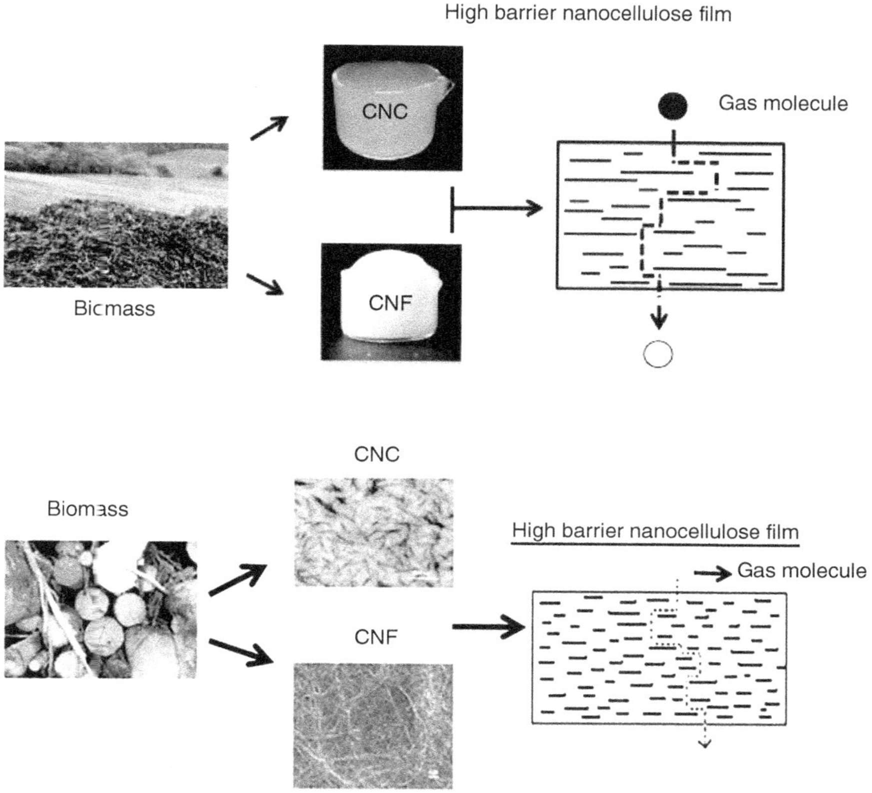

Figure 11.3 Gas permeability representation of NC-based films. Tighter networking of CNC produced high barrier property compared to CNF based films. In the extreme right, the possible migration of gas molecules through the produced film is schematically represented. Dense networking of film makes it difficult for the gas molecules to penetrate.

humidity enhances the oxygen permeability as a result of the constrained hydrogen bonding as well as loose networks as water molecules enter [11]. The water and oxygen permeability diminished for thicker films due to the insufficient interconnectivity of pores throughout the volume. It was additionally reported that the application of MFC, in the form of a surface layer on a paper sheet, decreased its oxygen permeability. This was ascribed to the dense configuration established by the microfibrils leading to decreased sheet porosity [18, 26].

Cellulose nanofibrils have the ability to build films with minimal moisture diffusivity because of a stiff fiber network and, consequently, possess outstanding barrier characteristics. The water vapor intake rate of cellulose nanofibrils films is 20% lesser compared to those composed of microfibers (in the range 20–30%) [9, 10]. Finally, CNCs films are anticipated to offer an excellent barrier to water molecules because CNCs films have an even more crystalline nature as compared to cellulose nanofibrils [8].

11.7 Migration Process of Molecules Through NC Dimensional Film

Migration of molecules between two adjacent volumes separated by a thin film of solid polymer or membrane occurs in three basic steps. In the first step, the molecules are adsorbed onto the sample surface. In the second step, the molecules diffuse through the film or membrane. Finally, the diffusing molecules exit the sample by desorbing from the surface. The gas permeability through an NC film mainly depends on the dissolution of gas and its rate of diffusion in the film. Thus, permeability of gas molecules through an NC film can be expressed as:

$$P = DS \tag{11.1}$$

where P is the permeability, D is the diffusion coefficient, and S is the solubility coefficient.

The permeability coefficient P is obtained from the application of Henry's law of solubility to Fick's law of diffusion,

$$P = DS = ql/At\Delta p \tag{11.2}$$

where q is the amount of material passing through the film, l is the thickness, A is the cross-sectional area, t is time, and Δp is the pressure difference between the two sides of film.

The gas molecules should be first dissolved in the membrane or film before diffusing. Even though the surface of films influences the permeating gas molecules, the most dominant factor in molecular migration is bulk flow, i.e. rate of molecule diffusion in the membrane or film [31]. The dense nanofibrils form more complex and smaller pores compared to pure cellulose fibers which are in micro scale. This complex dense network increases the tortuosity within the film and thereby decreases the permeability within the films [32]. Also, the high crystalline structure within the nanofibrils or whiskers contributes to the gas barrier properties [33].

11.8 Processing Routes of NC-based Multidimensional Structures for Packaging

Production of NC-based structures is an emerging topic. NC alone (neat) or with other polymers (composites) can be used for the production of 1D (electrospun fibers), 2D (films, membranes, nanopapers), and 3D (hydrogels, aerogels, Pickering emulsions) networks for various applications, including food packaging. Achieving one specific structure (1D, 2D, and 3D) depends on the processing route to obtain unique final properties required for a specific application.

Spinning – Research at Luleå University of Technology (LTU), Sweden, produced wood-based CNC porous fiber mats in chitosan using electrospinning. Randomly oriented networks of fibers with porous and highly open structures (porosity ≥ 93%) were observed using Scanning Electron Microscope analysis [23]. Furthermore, aligned NC in the field direction were also found. Kvien and Oksman [34] applied a magnetic field (7 T) during solution casting of a CNC suspension (2 wt% CNC), and they were able to lock in a preferential alignment. The CNC particle alignment was perpendicular to the field direction as determined from analysis of fracture surfaces that were ion etched to remove the matrix. These porous dimensional structures could not be used for the packaging of fresh food but literature shows that dry food could be easily packaged using these porous mat-like structures [32].

Solvent casting – Ideally, solvent casting is easy and scalable for the production of dense 2D structures for food packaging. Normally, NC alone or with a polymeric matrix (composites) are used for the production of sheet-like 2D or in a three-dimensional mold to produce 3D structures. Solvent-cast cellulose nanocomposites were first reported by the French research group Favier et al. [35, 36], and the first studies focused on tunicin-based nanocrystals in a latex matrix of poly(styrene-co-butyl acrylate), where an interesting reinforcing effect was demonstrated, especially above the polymer relaxation temperature [35, 36]. Generally, the literature concerning the preparation of cellulose sheet-like structures (nanocomposites) involves the solvent casting method as the main processing approach [37–40]. When water-soluble polymers are used as matrices, the interactions between the NCs and the respective matrix are strong due to the polar nature of both constituents.

Filtration – Vacuum-filtration is another approach for the production of a compact network, but not always; for example; multilayered (tri-layered) flat 2D dense structures of CNF and CNC were fabricated using vacuum-filtration followed by hot pressing. The final properties confirmed the presence of highly compacted dimensional structures (97 MPa, tensile strength and zero water flux) [2]. Pore sizes, determined by N_2 adsorption, were ~74 Å. Subsequently, less compacted layered flat sheet-like structures were produced using a filtration approach. Sludge of microfibers was used for the bottom layer, and CNC sludge was used for the top layer [3]. Low tensile strength (13 MPa) and high-water flux (900–4000 l/h m^2) were recorded through the flat sheet. Therefore, it is possible to use the same approach for the production of totally different structures having distinct final properties.

Coating – Coating is a very simple approach for the production of a sophisticated smooth surface deposited on various supports. NC was coated using spray coating, dipping, spinning, rod, or even hand laminations [41–43]. Coated two-dimensional structures have remarkable barrier properties, increased strength and smoothness compared to base materials. Coating of bacterial cellulose was performed to examine whether oil and water resistances changed after coating or not and to evaluate the nature of BC (to hydrophilic and hydrophobic materials). Results indicated an increase in oil, grease, oxygen, and water resistance after coating [42, 43]. Furthermore, coated 2D structures showed high hydrophilic properties compared to base paper [44].

Layer-by-layer approach – Recently, a robust and rapid method for the layer-by-layer (LbL) assembly of functional polymers and nanoparticles on cross-linked CNF aerogel with porosity close to 99%, high strength, and nanoscale shape integrity in water was developed. An LbL assembly of thin films of biomolecules, conducting polymers, and carbon nanotubes is able to provide electronic and mechanical properties. Furthermore, it has been shown that aerogels with dry strength and elastic wet strength enhancement, elastic mechano-responsive resistance, and supercapacitor properties with specific capacitance values exceeding 400 Fg^{-1} have been developed [45].

11.9 CNFs for Barrier Application

CNFs are a strong gas barrier material [19]. Saito and Isogai [46] showed that the degree of crystallinity varied from 78% to 91% for CNFs produced from (2,2,6,6-Tetramethylpiperidin-1-yl)oxyl (TEMPO) oxidation of cotton linter. Films made purely of mechanically fibrillated CNFs have very high air and oxygen barrier property. The oxygen transmission rates (OTRs) of CNF films with a thickness of 21 μm were as low as 17 ± 1 ml/m^2 d. These values are competitive with other best synthetic polymers such as ethylene vinyl alcohol (EVOH) (3–5 ml/m^2 d) and polyvinylidene chloride (PVdC) coated polyester films (9–15 ml/m^2 d) of approximately same thickness with respect to OTR [32]. Recently, Osterberg et al. [33] demonstrated a rapid method of making robust CNF films with high oxygen barrier property. The CNF solutions were first filtered followed by hot pressing at high pressure followed by air drying. At a relative humidity below 65%, the oxygen permeability of these films was below 0.6 cm^3 μm/m^2 d kPa. However, oxygen permeability of CNF films increases with the increase in relative humidity. This is mainly due to the plasticizing and swelling of nanofibrils through the adsorption of water molecules at high relative humidities.

CNFs also have tunable barrier properties. Rodionova et al. [47] showed that both pure and partially acetylated CNF films can be used for modified atmosphere packaging (MAP) with permeability below 10–20 ml/m^2 d. Similarly, carboxymethylated CNF films with thicknesses of 2.54 and 3.19 μm have very low oxygen permeability of 0.009 and 0.0006 cm^3 μm/m^2 d kPa, respectively [48]. High oxygen performance barrier films with permeability of 0.007 cm^3 μm/m^2 d kPa can be made by controlled thermal treatment at different temperatures. It was demonstrated that after thermal treatment at 175 °C, the oxygen permeability of the CNF films was reduced by 96% compared to the films without heat treatment [49]. Table 11.1 shows the oxygen permeability of films made from NC compared with those from commercially available petroleum-based materials and other polymers.

Also, CNFs are used to coat the surface of films to enhance the barrier properties. Fukuzumi et al. [52] prepared high oxygen barrier CNF films by TEMPO-mediated oxidation. The 0.1 μm thick CNF film when coated on a polylactic acid (PLA) film, reduced the oxygen permeability from 746 ml/m^2 d Pa for pure PLA film to 1 ml/m^2 d Pa for PLA film with CNF layer. The oxygen barrier of polyethylene terephthalate (PET) films was raised by more than 350 times through a 1 μm thick layer of the same TEMPO-oxidized CNFs [53].

Table 11.1 Oxygen permeability of NC film compared to those made from commercially available petroleum-based materials and other polymers.

Material	Oxygen permeability (cc μm/m² day kPa)	Conditions	References
CNF	0.6	At 65% RH	Saito and Isogai [46]
CNF (carboxymethylated)	0.006	At 0% RH	Aulin et al. [48]
CNF (carboxymethylated)	0.85	At 50% RH	Aulin et al. [48]
Cellophane	0.41	At 0% RH	Wu and Yuan [50]
PVdC	0.1–3	At 50% RH	Lange and Wyser [51]
Polyvinyl alcohol (PVOH)	0.20	At 0% RH	Lange and Wyser [51]
EVOH	0.01–0.1	At 0% RH	Lange and Wyser [51]

Table 11.2 WVTR of NC compared to commercially available petroleum-based materials and other polymers.

Material	Water vapor transmission rate (WVTR) (g/m² d)	Average thickness of the film (μm)	Conditions	References
CNF	234	42	50% RH	Rodionova et al. [47]
CNF (acetylated for 0.5 h)	167	46	50% RH	Rodionova et al. [47]
PVdC	3.07	12.7	100% RH	Steven and Hotchkiss [55]
Polyethylene (PE)	16.8	18.3	100% RH	Steven and Hotchkiss [55]
Plasticized polyvinyl chloride (PVC)	118.56	12.7	100% RH	Steven and Hotchkiss [55]

Similarly, Fujisawa et al. [54] found that the use of TEMPO-oxidized CNFs with sodium carboxylate groups improved the oxygen barrier of PET films more than when coated with TEMPO-oxidized CNFs with carboxyl groups. Furthermore, some water vapor transmission rates are also mentioned in Table 11.2 for modified and coated CNF.

11.10 CNCs for Barrier Application

Compared to CNFs, very few studies have been directed toward the study of 100% pure CNC films or treated CNC films. Belbekhouche et al. [31] compared the gas barrier properties between CNF and CNC films. They found that the films made of CNCs were more permeable to oxygen than those made of CNFs. The oxygen molecules penetrated much more slowly within CNF film due to the higher fibril entanglements within the film which increased the tortuosity factor.

CNCs, which have crystallinity greater than 60%, combined with their ability to form a dense hydrogen bonded network in principle can increase gas barrier property. Bacterial cellulose nanocrystals (BNCs) films present excellent oxygen barrier at low relative humidity, but their high moisture sensitivity results in dramatically decreased barrier when the relative humidity is higher than 70%. The oxygen permeability of 6.99 ± 10^{-22} m^3 m/m^2 s Pa at 0% humidity increased to 5.97 ± 10^{-18}m^3 m/m^2 s Pa at 80% humidity. However, this permeability was reduced by 97% and 74% when BNC films were coated with annealed PLA electrospun nanostructured fibers and 3-(aminopropyl)trimethoxysilane (APTS), respectively [56]. Herrera et al. [22] studied thin spin coated films made from CNCs prepared with sulfuric acid and hydrochloric acid. The hydrochloric acid-treated CNCs resulted in films with low permeability to oxygen, while the sulfuric acid-treated CNCs resulted in films with higher permeability.

CNCs have been studied as fillers for various natural polymers for enhancing the barrier properties. Saxena et al. [57] produced nanocomposite films with low oxygen permeability by casting an aqueous solution containing xylan, sorbitol, and nanocrystalline cellulose. Oxygen permeability of films prepared from xylan, sorbitol, and 50% by weight of sulfonated CNC exhibited a significantly reduced oxygen permeability of 0.1799 cm^3 m/m^2 d kPa compared to films prepared solely from xylan and sorbitol with an oxygen permeability of 189.1665 cm^3 m/m^2 d kPa. PLA nano-biocomposites containing 5 wt% of CNCs exhibited the highest oxygen barrier. The OTR for PLA nanocomposites with 5% w/w of unmodified CNCs was 17.4 ± 1.4 cm^3 mm/m^2 d, while that for modified CNCs with an acid phosphate ester of ethoxylated nonylphenol in a 1/4 (wt/wt) ratio was 15.8 ± 0.6 cm^3 mm/m^2 d [58]. Addition of 1 wt% of silver nanoparticles to these modified CNC-PLA composites further decreased the OTR to 12.6 ± 0.1 cm^3 mm/m d [59]. The OTR values of ternary systems consisting of PLA, poly hydroxybutyrate (PHB), and 5 wt% unmodified CNCs was 15.3 cm^{3}mm/m^2 d, while that for modified CNCs with an acid phosphate ester of ethoxylated nonylphenol in a 1/1 (wt/wt) ratio was 13 cm^3 mm/m^2 d. Water contact angle measurements showed that the ternary system had high hydrophobicity and the presence of sulfate groups with low polarity on the surface of CNCs increased the surface hydrophobicity of the final composite material [60].

11.11 Conclusion

In this chapter, an overview of a sustainable approach using NC as base material for the food packaging is discussed. NC-based multidimensional structures (films/papers/foams/membranes etc.) could be produced using various approaches such as coating, casting, vacuum-filtration, layer-by-layer approaches etc. and the used processing route has a direct influence on the properties of multidimensional structures. The water vapor and gas permeability through produced films/papers varies and depends on the networking behavior of produced structures.

Furthermore, CNFs and CNCs were widely used for the production of film-like 2D structures for food packaging. The networking behavior and arrangements of these tiny particles within the used matrices have a direct influence on the selectivity of gases. It has been further discussed that moisture content within the produced films has a direct impact on the permeability of gases and water. Therefore, some special packaging materials are required for the packaging of temperature-sensitive food or edible items.

It could be further concluded that NC is an emerging material in the field of food packaging and a lot of research is going on regarding the use of NC in packaging applications. We are still awaiting some commercial products made up of nanocellulose for food packaging. To sum up, NC has potential in food packaging but some time will be required to see it in the market.

References

1 Khalil, H.P.S., Chaturbhuj, K., Saurabh, A.S. et al. (2016). A review on chitosan-cellulose blends and nanocellulose reinforced chitosan biocomposites: Properties and their applications. *Carbohydrate Polymers* 150: 216–226.

2 Karim, Z., Claudpierre, S., Grahn, M. et al. (2016). Nanocellulose based functional membranes for water cleaning: tailoring of mechanical properties, porosity and metal ions capture. *Journal of Membrane Science* 514: 418–428.

3 Karim, Z., Mathew, A.P., Grahn, M. et al. (2014). Nanoporous membranes with cellulose nanocrystals as functional entity in chitosan: removal of dyes form water. *Carbohydrate Polymers* 112: 668–676.

4 Karim, Z., Afrin, S., Husain, Q. et al. (2017). Necessity of enzymatic hydrolysis for production and functionalization of nanocelluloses. *Critical Reviews in Biotechnology* 37: 355–370.

5 Mautner, A., Maples, H.A., Kobkeatthawin, T. et al. (2016). Phosphorylated nanocellulose papers for copper adsorption from aqueous solutions. *International journal of Environmental Science and Technology* 13: 1861–1872.

6 Chen, W., Yu, H., Lee, S.Y. et al. (2018). Nanocellulose: a promising nanomaterial for advanced electrochemical energy storage. *Chemical Society Reviews* 47: 2837.

7 Shin, Y., Blackwood, J.M., Bae, I.T. et al. (2007). Synthesis and stabilization of selenium nanoparticles on cellulose nanocrystal. *Materials Letters* 61: 4297–4300.

8 Azeredo, H.M.C., Rosa, M.F., and Mattoso, L.H.C. (2017). Nanocellulose in bio-based food packaging applications. *Industrial Crops and Products* 97: 664–671.

9 Kian, L.L., Jawaid, M., Afrrine, H. et al. (2019). Morphological, physiohemical and thermal properties of nanowhiskers from roselle fibers. *Cellulose* 26: 6599–6613.

10 Kian, L.L., Jawaid, M., Afrrine, H. et al. (2018). Isolation and characterization of nanocrystalline cellulose from Roselle derived microcrystalline cellulose. *International Journal of Biological Macromolecules* 114: 54–63.

11 Oksman, K., Etang, J.A., Mathew, A.P. et al. (2011). Cellulose nanowhiskers separated from a bio-residue from wood bioethanol production. *Biomass and Bioenergy* 35: 146–152.
12 Azizi Samir, M.A.S., Alloin, F., and Dufresne, A. (2005). Review of recent research into cellulosic whiskers, their properties and their application in nanocomposite field. *Biomacromolecules* 6: 612–626.
13 Duran, N., Lemes, A.P., and Seabra, A.B. (2012). Review of cellulose nanocrystals patents: preparation, composites and general applications. *Recent Patents on Nanotechnology* 6 (1): 16–28.
14 Lin, N., Huang, J., and Dufresne, A. (2012). Preparation, properties and applications of polysaccharide nanocrystals in advanced functional nanomaterials: a review. *Nanoscale* 4: 3274–3294.
15 Moon, R.J., Martini, A., Nairn, J. et al. (2011). Cellulose nanomaterials review: structure, properties and nanocomposites. *Chemical Society Reviews* 40 (7): 3941–3994.
16 Singh, S. and Ray, S.S. (2007). Polylactide based nanostructured biomaterials and their applications. *Journal of Nanoscience and Nanotechnology* 7 (8): 2596–2615.
17 Afrin, S. and Karim, Z. (2017). Isolation and surface modification of nanocellulose: necessity of enzymes over chemicals. *ChemBioEng Reviews* 4: 289–303.
18 Herrera, N., Mathew, A.P., and Oksman, K. (2015). Plasticized polylactic acid nanocomposite films with cellulose and chitin nanocrystals prepared using extrusion and compression molding with two cooling rates: effects on mechanical, thermal and optical properties. *Composites Science and Technology* 106: 149–155.
19 Castro, D.O., Karim, Z., Medina, L. et al. (2018). The use of a pilot-scale continuous paper process for fire retardant cellulose-kaolinite nanocomposites. *Composites Science and Technology* 162: 215–225.
20 Melvin Ng, H.K. and Leo, C.P. (2019). The coherence between TiO_2 nanoparticles and microfibrillated cellulose in thin film for enhanced dispersal and photodegradation of dye. *Progress in Organic Coatings* 132: 70–75.
21 Ogeda, T.L., Silva, I.B., Fidale, L.C. et al. (2012). Effect of cellulose physical characteristics, especially the water sorption value, on the efficiency of its hydrolysis catalyzed by free or immobilized cellulose. *Journal of Biotechnology* 157: 246–252.
22 Herrera, M.A., Mathew, A.P., and Oksman, K. (2014). Gas permeability and selectivity of cellulose nanocrystals films (layers) deposited by spin coating. *Carbohydrate Polymers* 112: 494–501.
23 Naseri, N., Mathew, A.P., Girandon, L. et al. (2015). Porous electrospun nanocomposite mats based on chitosan-cellulose nanocrystals for wound dressing: effect of surface characteristic of nanocrystals. *Cellulose* 22: 521–534.
24 Karim, Z., Sbedberg, A., Lee, K.Y. et al. (2019). Processing-structure-property correlation understanding of microfibrillated cellulose based dimensional structures for ferric ions removal. *Scientific Reports* 9: 10277.

25 Karim, Z., Hakalahti, M., Tammelin, T. et al. (2017). In situ TEMPO surface functionalization of nanocellulose membranes for enhanced adsorption of metal ions from aqueous medium. *RSC Advances* 7: 5232–5241.

26 Mishra, R.K., Sabu, A., and Tiwari, S.K. (2018). Materials chemistry and the futurist eco-friendly applications of nanocellulose: status and prospect. *Journal of Saudi Chemical Society* 22: 949–978.

27 Sehaqui, H., Zhou, Q., Ikkala, O. et al. (2011). Strong and tough cellulose nanopaper with high specific surface area and porosity. *Biomacromolecules* 12: 3638–3644.

28 Habibi, Y. (2014). Key advances in the chemical modification of nanocelluloses *Chemical Society Reviews* 21: 345–387.

29 Volesky, B. (2007). Biosorption and me. *Water Research* 41: 4017–4029.

30 Liu, P. (2014). Nanopolysaccharides for adsorption of heavy metals ions from water. Licentiate thesis. Lulea University of Technology, Sweden.

31 Belbekhouche, S., Bras, J., Siqueira, G. et al. (2011). Water sorption behaviour and gas barrier properties of cellulose whiskers and microfibrils films. *Carbohydrate Polymers* 83: 1740–1748.

32 Syverud, K. and Stenius, P. (2009). Strength and barrier properties of MFC films. *Cellulose* 2009 (16): 75–85. https://doi.org/10.1007/s10570-008-9244-2.

33 Osterberg, M., Vartiainen, J., Lucenius, J. et al. (2013). A fast method to produce strong NFC films as a platform for barrier and functional materials. *ACS Applied Materials and Interfaces* 5: 4640–4647.

34 Kvien, I. and Oksman, K. (2007). Orientation of cellulose nanowhiskers in polyvinyl alcohol. *Applied Physics A: Materials Science and Processing* 87: 641–643.

35 Favier, V., Canova, G.R., Cavaillé, J.Y. et al. (1995). Nanocomposite materials from latex and cellulose whiskers. *Polymers for Advanced Technologies* 6: 351–355.

36 Favier, V., Canova, G.R., Cavaillé, J.Y. et al. (1995). Polymer nanocomposites reinforced by cellulose whiskers. *Macromolecules* 28: 6365–6367.

37 Liu, H.Z. and Laborie, M.P.G. (2011). Bio-based nanocomposites by in situ cure of phenolic prepolymers with cellulose whiskers. *Cellulose* 18: 619–630.

38 Mathew, A.P. and Dufresne, A. (2002). Morphology investigation of nanocomposites from sorbitol plasticized starch and tunicin whiskers. *Biomacromolecules* 3: 609–617.

39 Mikkonen, K.S., Mathew, A.P., and Pirkkalainen, K. (2010). Glucomannan composite films with cellulose nanowhiskers. *Cellulose* 17: 69–81.

40 Siqueira, S., Bras, J., and Dufresne, A. (2008). Cellulose whiskers versus microfibrils: influence of the nature of the nanoparticles and its surface functionalization on the thermal and mechanical properties of nanocomposites. *Biomacromolecules* 10: 425–432.

41 Oksman, K., Aitomäki, Y., Mathew, A.P. et al. (2016). Review of the recent developments in cellulose nanocomposite processing. *Composites Part A: Applied Science and Manufacturing* 83: 2–18.

42 Tayagi, P., Hubbe, M.A., and Lucia, L. (2018). High performance nanocellulose-based composite coatings for oil and grease resistance. *Cellulose* 25: 3377–3391.

43 Tayagi, P., Hubbe, M.A., and Lucia, L. (2019). Nanocellulose-based multilayer barrier coatings for gas, oil, and grease resistance. *Carbohydrate Polymers* 206: 281–288.

44 Wang, Z., Ma, H., Chu, B. et al. (2017). Fabrication of cellulose nanofiber-based ultrafiltration membranes by spray coating approach. *Journal of Applied Polymer Science* 134: 44583.

45 Hamedi, M., Karabulut, E., Marais, A. et al. (2013). Nanocellulose aerogel functionalized by rapid layer-by-layer assembly for high charge storage and beyond. *Angewandte Chemie International Edition* 52: 12038–12042.

46 Saito, T. and Isogai, A. (2004). TEMPO-mediated oxidation of native cellulose. The effect of oxidation conditions on chemical and crystal structures of the water-insoluble fractions. *Biomacromolecules* 5: 1983–1989.

47 Rodionova, G., Lenes, M., Eriksen, O. et al. (2011). Surface chemical modification of microfibrillated cellulose: improvement of barrier properties for packaging applications. *Cellulose* 18: 127–134.

48 Aulin, C., Gallstedt, M., and Lindstrom, T. (2010). Oxygen and oil barrier properties of microfibrillated cellulose films and coatings. *Cellulose* 17: 559–574.

49 Sharma, S., Zhang, X., Nair, S.S. et al. (2014). Thermally enhanced high performance cellulose nano fibril barrier membranes. *RSC Advances* 4: 45136–45142.

50 Wu, J. and Yuan, Q. (2002). Gas permeability of a novel cellulose membrane. *Journal of Membrane Science* 204: 185–194.

51 Lange, J. and Wyser, Y. (2003). Recent innovations in barrier technologies for plastic packaging – a review. *Packaging Technology and Science* 16: 149–158.

52 Fukuzumi, H., Saito, T., Wata, T. et al. (2009). Transparent and high gas barrier films of cellulose nanofibers prepared by TEMPO-mediated oxidation. *Biomacromolecules* 10: 162–165.

53 Rodionova, G., Saito, T., Lenes, M. et al. (2012). Mechanical and oxygen barrier properties of films prepared from fibrillated dispersions of TEMPO-oxidized Norway spruce and Eucalyptus pulps. *Cellulose* 19: 705–711.

54 Fujisawa, S., Okita, Y., Fukuzumi, H. et al. (2011). Preparation and characterization of TEMPO-oxidized cellulose nanofibril films with free carboxyl groups. *Carbohydrate Polymers* 84: 579–583.

55 Steven, M.D. and Hotchkiss, J.H. (2002). Comparison of flat film to total package water vapor transmission rates for several commercial food wraps. *Packaging Technology and Science* 15: 17–27.

56 Martinez-Sanz, M., Lopez-Rubio, A., Lagaron, J.M. et al. (2013). High-barrier coated bacterial cellulose nanowhiskers with reduced moisture sensitivity. *Carbohydrate Polymers* 98: 1072–1082.

57 Saxena, A., Elder, T.J., Kenvin, J. et al. (2010). High oxygen nanocomposite barrier films based on xylan and nanocrystalline cellulose. *Nano-Micro Letters* 2: 235–241.

58 Fortunati, E., Peltzer, M., Armentano, I. et al. (2012). Effects of modified cellulose nanocrystals on the barrier and migration of PLA nano-composites. *Carbohydrate Polymers* 90: 948–956.

59 Fortunati, E., Peltzer, M., Armentano, I. et al. (2013). Combined effects of cellulose nanocrystals and silver nanoparticles on the barrier and migration properties of PLA nano-biocomposites. *Journal of Food Engneering* 118: 117–124.

60 Arrieta, M.P., Fortunati, E., Dominici, F. et al. (2014). PLA-PHB/cellulose based films: mechanical, barrier and disintegration properties. *Polymer Degradation and Stability* 107: 139–149.

Part IV

Natural Principles in Active and Intelligent Food Packaging for Enhanced Protection and Indication of Food Spoilange or Pollutant Presence

12

Sustainable Antimicrobial Packaging Technologies

Selçuk Yildirim and Bettina Röcker

Zurich University of Applied Sciences, Department of Life Sciences and Facility Management, Institute of Food and Beverage Innovation, Food Packaging Research Group, Campus Reidbach, Wädenswil 8820, Switzerland

12.1 Introduction

Packaging is of essential importance in the whole food supply chain. As a primary function, food packaging serves as a container enabling efficient transport within the supply chain. Thereby, packaging overtakes a passive role of food preservation as it protects the food from any spoilage by acting as a physical barrier between the food product and external conditions. This barrier function has been well optimized to achieve the desired shelf life for individual products. Over the last few decades, however, consumer demand has increased for high-quality food that does not contain any preservatives, is non-processed or minimally processed but is still safe and has an acceptable shelf life. To respond to this need, the protective function of packaging has been improved over the last few years leading to the development of innovative packaging technologies such as active packaging (AP) [1–4]. In the European regulation (EC) No. 450/2009 "active packaging" is defined as a packaging system that interacts with the food and is designed to "deliberately incorporate components that would release or absorb substances into or from the packaged food or the environment surrounding the food" [5]. In this context, exceptional but essential functions, going beyond the simple barrier function of a packaging, can be implemented leading to multiple benefits. AP can extend the shelf life of food products while ensuring their quality, safety, and integrity. This in turn can lead to a reduction of food waste and thereby contribute to a more sustainable handling of food.

AP systems can be divided into active scavenging systems (absorbers) removing undesired compounds from the food or its environment; and active releasing systems (emitters) adding compounds into the headspace or to the packaged food. Table 12.1 provides an overview of the main active packaging systems [1, 4, 6].

AP systems, such as antimicrobial or carbon dioxide releasers and oxygen, moisture, or ethylene absorbers, are often designed in the form of a sachet that has to be additionally added to the food packaging. However, such sachets have several drawbacks such as the risk of accidental breakage leading to an involuntary content consumption by the consumer, the need of an additional packaging

Sustainable Food Packaging Technology, First Edition. Edited by Athanassia Athanassiou.

Table 12.1 Main active packaging systems.

Active scavenging systems	Active releasing systems
Oxygen absorbers	Antimicrobial releasers
Moisture absorbers	Antioxidant releasers
Ethylene absorbers	Ethylene releasers
Carbon dioxide absorbers	Carbon dioxide releasers
Odor absorbers	Flavor releasers

operation step as well as the unsuitability to liquids and beverages [4, 7, 8]. Moreover, sachet-based applications are not well accepted by consumers in European countries, in contrast to Asia or the United States [9]. For this reason, latest developments have aimed to directly incorporate the active components into the polymer matrix of the packaging material [4].

Due to the ease of processing, so far research on active packaging systems has predominantly been carried out by the use of polymeric films, such as polyesters, produced from fossil-based resources. There is scientific evidence that the packagings relative environmental impact, compared to that of the food products itself (food production, transport emissions, product wastage, etc.), is rather low and a suitable packaging therefore clearly benefits from added value in terms of waste reduction [10–13]. Nevertheless, the next generation of active packaging materials should inevitably target a replacement of the petroleum-based commodity plastics by biodegradable materials arising from biological and renewable resources [10, 14, 15]. In this context, this chapter presents the latest developments in sustainable active packaging technologies for potential food application. Thereby, the focus is on bioactive substances that have been incorporated into biopolymer matrices derived from renewable resources and that have been shown to provide antimicrobial activity against foodborne pathogens and spoilage microorganisms. Special emphasis is placed on publications where the antimicrobial packaging has been successfully tested on food systems. Therefore, the functions of antimicrobial food packaging systems are elucidated in the beginning.

12.2 Antimicrobial Food Packaging

Antimicrobial packaging systems are innovative advancements designed to inhibit the growth of spoilage and pathogenic microorganisms in packaged foodstuffs [3, 4, 6, 7, 16]. Accompanying the physical protection, active antimicrobial packaging constitutes a promising complementary method to traditional preservation methods where antimicrobial substances are added directly into the food. A direct addition of bioactive substances to the food may result in an inhibition or loss of their activity due to interactions with the food matrix. Besides that, microbial growth and degradation on fresh and processed food mostly occurs at the surface only. Therefore, it might be more effective to release antimicrobial agents via active packaging to the surface of the food instead of distributing it in the whole bulk of the food. By this way, the amount of active substances required for the desired antimicrobial activity may be decreased as well [6].

To obtain antimicrobial activity in a packaging system, different approaches are possible. These are the addition of sachets or pads containing antimicrobial volatiles, immobilization or adsorption methods, the incorporation of antimicrobial agents to the packaging matrix or the surface of a film by co-extrusion, or the use of packaging raw materials that are inherently antimicrobial. In terms of its mode of action, antimicrobial packaging systems can be differentiated into three types: (i) Systems requiring direct food contact release antimicrobial agents, such as organic acids, from the packaging material to the food surface from where they further diffuse into the food. (ii) Systems that do not require direct food contact contain volatile antimicrobial agents, such as plant extracts. Released from the packaging material, such agents diffuse into the packaging headspace from where they are absorbed by the food. (iii) Non-migratory antimicrobial polymers consisting of an antimicrobial agent attached to the polymer backbone. Such systems also require direct food contact since the antimicrobial agent does not intentionally migrate out. This category includes surface-modified films with antimicrobial properties or polymer films that are antimicrobial by themselves such as chitosan [4, 6]. According to the European regulation (EC) No. 1333/2008, an active agent that is intentionally released to the food is considered as a food additive and has to be declared as such [17]. This is the case for systems (i) and (ii) described above.

In the last decade, huge efforts have been made in research and development of antimicrobial packaging systems, and various concepts have been developed [3, 4, 6, 7, 16]. Antimicrobial agents such as essential oils (EOs), organic acids, bacteriocins, enzymes, metals, and metal oxides as well as their nanoparticles have been incorporated into various polymer matrices. Such antimicrobial films were proven to be effective against a wide range of foodborne pathogens and spoilage microorganisms on laboratory media (*in vitro*) as well as on real food applications (*in vivo*) [4]. Overall, antimicrobial active packaging comes along with multiple benefits including the reduction of spoilage microorganisms, the extension of shelf life leading to food waste reduction, the inhibition of growth of pathogenic microorganisms increasing food safety, the maintenance of food quality in terms of sensory or color, stating an additional hurdle leading to easier food processing, and finally, a step toward a clean labeling by the replacement of artificial preservatives with natural ones, or use of less preservatives.

Nevertheless, the vast majority of the reported active antimicrobial packaging systems consist either of inorganic active agents with questionable biocompatibility, such as nanoparticles of metals or metal oxides; or natural active agents have been integrated into petroleum-based polyesters such as PET, PE, and LDPE. Due to environmental concerns and consumer acceptance, research trend is now moving toward the integration of natural antimicrobial agents into polymer matrices made from renewable resources [18, 19] and/or polymer matrices that are biodegradable [20]. Such developments are not only of advantage from an environmental, but also from a technological point of view. High thermal polymer processing methods such as injection molding or extrusion can negatively influence bioactive substances due to their thermal instability. Formation of biopolymer films, in contrast, may occur under mild conditions. Thus, the incorporation of bio-preservatives into biopolymer films may be more suitable than their incorporation into plastics [18, 21–24].

12.3 Natural Antimicrobial Agents

In this chapter, the most studied antimicrobial food packaging systems have been classified according to their bioactive substance/material: plant extracts such as EOs and phenolic compounds (Table 12.2); organic aids, their salts and anhydrides (Table 12.3); proteinaceous compounds such as bacteriocins (Table 12.4) and enzymes (Table 12.5); and the antimicrobial polymer chitosan (Table 12.6).

12.3.1 Plant Extracts

EOs are secondary metabolites and they play an important role in plant defense; thus, some of them express strong antimicrobial properties. Moreover, most of them are classified as GRAS [68] and have therefore been extensively studied against food spoilage and pathogenic microorganisms. As shown in Table 12.2, numerous EOs such as cinnamon, oregano, basil leaf, clove, marjoram, savory, or apricot kernel EOs have been incorporated into bio-based films and coatings and were successfully tested against a wide range of spoilage microorganisms including bacteria, fungi, and yeasts in both laboratory medium (*in vitro*) as well as on real food application.

Cinnamon essential oil is among the most studied EOs in active materials. Wen et al. [25] incorporated a cinnamon EO/β-cyclodextrine inclusion complex into electrospun polylactic acid (PLA) nanofibers. Application of the active film on pork meat extended the microbial shelf life by inhibiting the growth of *Escherichia coli* and *Staphylococcus aureus*. Higueras et al. [26] controlled the EO release by reversibly anchoring cinnamaldehyde, the major compound of cinnamon EO, to a chitosan polymer. *In vitro* effectiveness of the chitosan–cinnamaldehyde films was demonstrated against *E. coli* and *S. aureus*. Moreover, the active films were shown to inhibit the growth of *Listeria monocytogenes*, inoculated in milk. Marjoram EO and Savory EO have been incorporated into nanocomposite films based on sodium alginate–montmorillonite [27] and agar–cellulose [28], respectively. Both active films were shown to be effective against *L. monocytogenes*, *Bacillus cereus*, *S. aureus* and *E. coli* on laboratory media. Oregano EO has been successfully incorporated into pectin edible coatings [29] and into films from whey protein isolate (WPI) [30]. On tomatoes coated with Oregano EO-pectin films the inhibition of *Alternaria alternata*, one of the main fungus affecting tomatoes, was demonstrated. On beef cuts wrapped with the Oregano-EO-WPI films, a significant reduction of pseudomonads and total bacteria flora, and an entire inhibition of lactic acid bacteria was shown. Active biopolymer films containing basil leaf EO and clove EO were also shown to

Table 12.2 Examples of antimicrobial plant extracts incorporated into bio-based polymer matrices.

Antimicrobial agent	Matrix/packaging application	Microorganisms tested	Food application	References
Cinnamon EO + β-cyclodextrine	Polylactic acid electrospun nanofibers	*E. coli*, *S. aureus*	Fresh pork	[25]
Cinnamaldehyde	Reversible covalent immobilization on chitosan polymer	*S. aureus*, *E. coli*, *L. monocytogenes*	*In vitro*, milk	[26]
Marjoram EO	Sodium alginate–montmorillonite nanocomposite films	*L. monocytogenes*, *S. aureus*, *B. cereus*, *E. coli*	*In vitro*	[27]
Savory EO	Agar–cellulose nanocomposite films	*L. monocytogenes*, *S. aureus*, *B. cereus*, *E. coli*	*In vitro*	[28]
Oregano EO	Pectin edible coatings	Fungi (*Alternaria alternata*)	Tomatoes	[29]
	Whey protein isolate	TVC, Pseudomonas, LAB	Beef cuts	[30]
Basil leaf EO	Fish protein isolate/fish skin gelatin film	TVC, Psychrophilic bacteria, LAB, H_2S-producing bacteria, Pseudomonads, Enterobacteriaceae	Sea bass slices	[31]
Clove EO	Soy protein isolate-montmorillonite nanocomposite films	TVC, total aerobic mesophiles, Pseudomonas spp., LAB, Enterobacteriaceae	Muscle fillets of bluefin tuna (*Thunnus thynnus*)	[32]
Apricot kernel EO	Chitosan films	*B. subtilis*, *E. coli*, visible mold growth	*Bacteria: in vitro* Mold: bread slices	[33]
Cinnamon and Ginger EO	Chitosan	Total viable counts (TVC)	Pork slices	[34]

(*continued overleaf*)

Table 12.2 (Continued)

Antimicrobial agent	Matrix/packaging application	Microorganisms tested	Food application	References
Epigallocatechin-3-gallate, blueberry ash fruit, macadamia skin extracts	Pea starch-guar gum edible films	*Bacteria*: *Enterococcus faecalis, Bacillus subtilis, Staphylococcus epidermidis, Staphylococcus lugdunensis, Salmonella typhimurium, Pseudomonas fluorescence, Klebsiella pneumoniae, Enterobacter aerogenes, E. coli* *Fungi*: *Aspergillus niger, Penicillium italicum, Penicillium digitatum, Candida albicans, Geotrichum candidum, Mucor* sp., *Rhizopus* sp.	*In vitro*	[35]
Stem, leaf, and seed extracts of *Pistacia terebinthus*	Edible chitosan films	*E. coli, Proteus microbilis, Proteus vulgaris, P. aeruginosa, Enterobacter aerogenes, Salmonella typhimurium, Bacillus thuringiensis, S. aureus, Streptococcus mutans*	*In vitro*	[36]
Apple peel polyphenols	Chitosan film	*B. cereus, E. coli, S. typhymurium, S. aureus*	*In vitro*	[37]
Tannic acid	Chitosan, gelatin and methylcellulose films	*E. coli, S. aureus*	Cherry tomatoes, grapes	[38]
Vanillin	Starfish gelatin films	*L. monocytogenes*	Crab stick	[39]

Table 12.3 Examples of antimicrobial organic acids, their salts and anhydrides incorporated into bio-based polymer matrices.

Active substances	Matrix/packaging application	Microorganisms tested	Food application	References
Citric acid	Starch/PVA films	*L. monocytogenes*, *E. coli*	Fresh figs	[40]
Sorbic acid + Algae (*Fucus spiralis*)	PLA films	Total aerobics, psychrotrophs, *Enterobacteriaceae*	Flat fish megrim	[41]
Potassium sorbate and/or Sodium benzoate	Chitosan and/or methyl cellulose	*Penicillium notatum*, *Rhodotorula rubra*	*In vitro*	[42]
Potassium sorbate	Post-extrusion anti-stacking dip solution of recycled PET	*Botrytis cinerea*	Blueberries	[43]
	Sodium caseinate film	*L. monocytogenes*	*In vitro*	[44]
	Thermoplastic starch/PBAT films	Total coliforms, total viable counts, *E. coli*	Restructured chicken steaks (frozen storage)	[45]
Propionic acid	Propionic acid encapsulated in β-cyclodextrine, incorporated in carboxymethyl cellulose films	Fungi	Post-harvest wheat grains	[46]

extend the microbial shelf life of fish products. On sea bass slices, wrapped with fish protein isolate/fish skin gelatin films containing basil leaf EO, the growth of total flora (total viable count, TVC), psychrophiles, LAB, H_2S-producing bacteria, Pseudomonads, and Enterobacteriaceae was significantly inhibited [31]. Similarly, soy protein isolate-montmorillonite nanocomposite films containing clove EO were shown to inhibit the growth of TVC, total aerobic mesophiles, LAB, Pseudomonas spp. and Enterobacteriaceae on Muscle fillets of bluefin tuna (*Taunnus thynnus*) [32]. Apricot kernel EO extracted from kernels of bitter apricot (*Prunus armeniaca*) has been incorporated into chitosan films [33]. The active films were shown to express *in vitro* antibacterial activity against *Bacillus subtilis* and *E. coli* and when applied on packaged bread slices, mold growth was completely inhibited.

A combination of different EOs can also be an effective approach to reinforce their antimicrobial effects. In this context, Wang et al. [34] incorporated cinnamon and ginger EO into edible chitosan films for pork packaging purposes. Thereby, TVCs of bacteria on pork slices wrapped with the active films were successfully inhibited.

Table 12.4 Examples of bacteriocins incorporated into bio-based polymer matrices.

Active substances	Matrix/packaging application	Microorganisms tested	Food application	References
Nisin	Nisin/EDTA immobilized on the surface of the cellulose nanocrystal/chitosan nanocomposite films	Psychotrophic bacteria, Mesophilic bacteria, *L. monocytogenes*, *E. coli*	Boneless pork loin meat	[47]
	PHB/PCL with organo-clays	*Lactobacillus plantarum*	Sliced ham	[48]
	Embedded nano-liposome-encapsulated nisin peptides into biodegradable hydroxypropyl methylcellulose films	*L. monocytogenes*	*In vitro*	[49]
	Carboxylated cellulose nanofibers	*B. subtilis*, *S. aureus*	*In vitro*	[50]
Natamycin	Natamycin incorporated into cellulose polymeric films and laminated films	*P. roqueforti*	Gorgonzola cheese	[51]
Enterocins	Alginate, zein and polyvinyl alcohol films	*L. monocytogenes*	Cooked ham	[52]
Pediocin	Pediocin incorporated into cellulose acetate film	*L. innocua*, *Salmonella* spp.	Sliced ham	[53]
Lactocins	Lactocins coated on wheat gluten-based films	*L. plantarum*, *L. innocua*	Wieners	[54]

Since EOs are odor-intensive volatiles, their application on food products can be disadvantageous. However, undesirable organoleptic effects caused by the release of EOs can be minimized by the selection of EOs according to the type of food. For example, the combination of cinnamon and ginger EO are reported to be compatible with meat in terms of the Chinese habit of using cinnamon and ginger as spices to improve flavor and taste [34]. In terms of cinnamaldehyde released to pasteurized milk in the study of Higueras et al. [26], the effect was considered as positive by the sensory panel as the treated milk was preferred to the control sample.

Besides the well-known essential oils, plants often contain other bioactive substances such as phenolic compounds. Phenolics can inhibit bacterial growth by a number of mechanisms such as membrane distraction, development of complexes with cell walls, and suppression of bacterial adhesion or inactivation of bacterial enzyme systems [69, 70]. Apart from the application of phenolic compounds in their pure forms, there are more and more approaches reported incorporating whole plant extracts into biopolymer matrices serving as alternative food packaging materials.

Table 12.5 Examples of antimicrobial enzymes incorporated into bio-based polymer matrices.

Active substances	Matrix/packaging application	Microorganisms tested	Food application	References
Lysozyme	Cellulose acetate	*E. coli*, *B. amyloliquefaciens*	*In vitro*	[55]
	Zein	*B. subtilis*, *L. plantarum*	*In vitro*	[23]
	Chitosan	*E. coli*, *Enterococcus faecalis*	*In vitro*	[56]
Lysozyme, lactoferrin, lactoperoxidase	Whey protein isolate (WPI) film	*Salmonella enterica*, *E. coli*	*In vitro*	[57]
Lysozyme, lactoferrin	Cellulose/ carboxymethyl cellulose paper sheets	*E. coli*, *L. innocua*, total aerobic bacteria	*In vitro*, ready-to-eat thin-cut veal meat (raw)	[58]
Lactoferrin	Bacterial cellulose	*E. coli*, *S. aureus*	Fresh sausages	[59]
Lysozyme nanofibers	Pullulan	*S. aureus*	*In vitro*	[60]

Saberi et al. [35] incorporated epigallocatechin-3-gallate (EGCG) and two native Australian plants blueberry ash fruit (BBA) and macadamia skin (MAC) extracts into pea starch–guar gum edible films. The authors demonstrated that EGCG, BBA, and MAC are capable of preventing an extensive spectrum of food pathogenic and spoilage microorganisms. The edible films were shown to possess antimicrobial activities against most of the tested bacterial and fungal strains listed in Table 12.2. Thereby, the EGCG-films showed highest activity against all test strains. Kaya et al. [36] blended edible chitosan films with stem, leaf, and seed extracts of the turpentine tree, a species of Pistacia (*Pistacia terebinthus*) that is rich in flavonoids and polyphenols. *In vitro* tests clearly confirmed antimicrobial efficacy of the active films against foodborne pathogens, especially against gram-negative bacteria such as *E. coli*, *Proteus microbilis*, *Proteus vulgaris*, and *Pseudomonas aeruginosa*.

Apple fruit is another source of polyphenols. Chitosan films enriched with apple peel polyphenols where shown to have antibacterial activity against food-borne pathogens [37]. Whereas the plain chitosan-based films already slightly inhibited the growth of *B. cereus*, *E. coli*, *S. typhimurium*, and *S. aureus*, the inhibitory effect was shown to increase significantly with increasing polyphenol concentration for all the bacteria tested. Halim et al. [38] incorporated tannic acid (TA) into different biopolymer matrices such as gelatin, chitosan, and methylcellulose films for food packaging applications. Methylcellulose-TA, gelatin-TA, and chitosan films clearly demonstrated *in vitro* antimicrobial activity against *E. coli* and *S. aureus*. However, the highest inhibitory effect was observed with the combination of chitosan and tannic acid. This observed cumulative antibacterial effect may be reasoned by the addition of tannic acid

Table 12.6 Examples of chitosan-based antimicrobial films.

Active substances	Matrix/packaging application	Microorganisms tested	Food application	References
Chitosan	Paper sheets	*In vitro*: *L. innocua, E. coli, S. aureus, Pseudomonas putida.* *In vivo*: *L. innocua*	*In vitro* Sliced raw veal meat	[61]
Chitosan	Chitosan–montmorillonite (MMT) nanocomposites	*E. coli, S. aureus*	*In vitro*	[62]
Chitosan	Chitosan–PVA coatings	*E. coli, S. aureus , B. subtilis*	Whole and cut fresh tomatoes	[63]
Chito–oligosaccharide	Chitosan/PVA films	*S. aureus*	*In vitro*	[64]
Chitosan Chitosan + lignin	Lignin/chitosan films	*P. aeruginosa, B. subtilis*	*In vitro*	[65]
Chitosan chitosan + different organic acids	Coatings on PLA films	*L. innocua, L. monocytogenes, S. typhimurium*	Ready-to-eat turkey slices	[66]
Chitosan Chitosan + LAE	Chitosan/glycerol films	*Bacteria*: *S. aureus, L. monocytogenes, E. coli, S. enterica, Pseudomonas putida* *Yeasts*: *Candida utilis, S. cerevisiae, Torulopsis pinus* Fungi: *A. niger, Penicillum chrysogenum, Cladosporium cladosporioides*	*In vitro*, chicken breast fillet	[67]

into the chitosan biopolymer solution resulting in an enhanced acidity of the solution, leading to an increased number of cations [38]. The biopolymer-TA films were also evaluated for their ability to preserve fresh fruits. Compared to fruits wrapped with commercial cling film, cherry tomatoes and grapes wrapped in the active films were shown to better preserve the fruits, resulting in lower weight loss and lower browning indexes. However, antimicrobial activity was not evaluated on the food application.

Vanillin, obtained from the cured bean or pod of the tropical Vanilla orchid, has also been proposed as a bioactive antimicrobial agent. Vanillin is used as a food preservative and flavor enhancer as well as a cross-linking agent. Because of the aldehyde group in the chemical structure of vanillin, the growth of pathogenic microorganisms can be effectively suppressed [71]. Lee et al. [39] incorporated vanillin into starfish gelatin films and demonstrated its antimicrobial activity against *L. monocytogenes* on inoculated crab sticks.

12.3.2 Organic Acids, Their Salts and Anhydrides

Organic compounds such as selected organic acids (acetic, sorbic, citric, lactic, malic, propionic, and tartaric among others), their salts and anhydrides have been used in the food industry for many years for fungal and yeast spoilage inhibition and decontamination of spoilage bacteria [72–74]. Due to their powerful antimicrobial activity, organic acids have also been incorporated into bio-based packaging films (Table 12.3).

Ternary blend films from starch/polyvinyl alcohol (PVA)/citric acid showed high antimicrobial activity against *L. monocytogenes* and *E. coli* in laboratory media [40]. Moreover, freshness test of fresh figs packed into the active films were shown to prevent both the formation of condensed water on the surface of the film as well as the deterioration of figs during storage. A PLA biodegradable film including sorbic acid and lyophilized alga *Fucus spiralis* was applied to protect the flat fish megrim (*Lepidorhombus whiffiagonis*) from microbial growth [41]. The active films showed antimicrobial activity against psychrotrophs, aerobics, and Entrerobacteriaceae after seven days of storage. Although no significant antimicrobial effect could be demonstrated at the end of the shelf life (11 days), improved sensory properties (external odor, gill appearance, and odor) were reported for megrim packed with active films, while control samples were considered unacceptable from a sensorial point of view.

Potassium sorbate has been evaluated as an antimicrobial agent in various biopolymer matrices [42–45]. Chitosan/methyl cellulose films containing potassium sorbate, sodium benzoate, or a combination thereof were shown to express significant antifungal properties against *Penicillium notatum* and *Rhodotorula rubra* on laboratory medium [42]. Junqueira-Gonçalves et al. [43] suggested the use of potassium sorbate as an effective antifungal active packaging for berries. The addition of potassium sorbate to the post-extrusion anti-stacking dip solution of post-consumer recycled PET showed an inhibitory effect (2 log cycles) against *Botrytis cinerea* on packed blueberries. Incorporated into sorbitol-plasticized sodium caseinate films, potassium sorbate was also highly effective against *L. monocytogenes* [44]. Cestari et al. [45] incorporated 5% potassium sorbate into thermoplastic starch/PBAT blended films to prevent microbial growth in restructured chicken steaks during frozen storage. Whereas in the control samples 1.24 log CFU/g of *E. coli* was detected after the completion of 150 days frozen storage, *E. coli* (initial count 1.94 log CFU/g) was not detected anymore in samples packed with potassium sorbate films from 30 days up to the end of storage.

Propionic acid (PA) represents a powerful and volatile antifungal agent widely used for protection of dry agricultural products such as cereals and grains and it is generally recognized as safe (GRAS) [75]. However, its volatile nature leads to undesired drawbacks such as the need of multiple treatments, lack of homogenous dispersion, or undesired aftertaste in the food products. Thus, systems for controlled release of PA are more preferable [46, 75]. In this context, Rutenberg et al. [46] prepared active films based on biodegradable cellulose-derived hydrocolloids (CMC, HPMC, and MC) capable of controlled release of PA. Thereby, β-cyclodextrine, a cyclic oligosaccharide, was engaged to increase the PA loading capacity in the films. Antifungal tests with CMC-based films containing PA and β-cyclodextrine on post-harvest wheat grains confirmed that the active films were effective in preventing fungal growth.

12.3.3 Bacteriocins

Bacteriocins are antimicrobial peptides produced by bacteria and can inhibit the growth of other bacteria, mainly gram-positive bacteria. The most common bacteriocin is nisin which is heat-stable, non-toxic, commercially available, and has already been tested for many applications. It has a broad spectrum of antimicrobial activity including foodborne pathogens and spoilage bacteria. These properties have already been used successfully for the development of a variety of antimicrobial food packaging systems [4, 6]. As shown in Table 12.4, bacteriocins have also been incorporated into bio-based polymer matrices or coatings that can release the compound on contact with food.

Nisin and ethylenediaminetetraacetic acid (EDTA) were immobilized on the surface of cellulose nanocrystal/chitosan-films by using genipin as a cross-linking agent [47]. The bionanocomposite films restricted the growth of mesophiles, psychrotrophs, and Lactobacillus spp. in fresh pork loin meats, and a microbial shelf life extension by more than five weeks was achieved. Moreover, *E. coli* and *L. monocytogenes* counts in meat samples could be clearly reduced by the antimicrobial films. Nisin activated poly(hydroxybutyrate)/poly(caprolactone) (PHB/PCL) nanocomposite films were tested against *Lactobacillus plantarum* inoculated on sliced ham [48]. The obtained results confirmed the effectiveness of such films to extend the shelf life of vacuum-packed sliced cooked ham. A promising approach was made by Imran et al. [49] who embedded nano-liposome-encapsulated nisin peptides into hydroxypropyl methylcellulose films. These films exhibited antimicrobial activity against the foodborne pathogen *L. monocytogenes*. Saini et al. [50] developed a novel antimicrobial film with covalently linked nisin on the surface of TEMPO-oxidized carboxylated cellulose nanofibers (CNFs) for food packaging. The nisin grafted CNF displayed excellent antimicrobial activity against *B. subtilis* and *S. aureus* with significant 3.5 log reduction. Other bacteriocins such as natamycin [51], enterocins [52], pediocin [53], and lactocins [54] have been incorporated into biopolymer-based films or used as coatings. They reduced fungal (*Penicillium roqueforti*) or bacterial (*L. monocytogenes, Listeria innocua*, Salmonella spp. and *L. plantarum*) growth on Gorgonzola cheese, cooked and sliced ham, and Wieners, respectively.

In future, more research effort is still needed to evaluate the release of bacteriocins from diverse films and coatings into packaging, as well as diffusion to the surface of the food [4]. To date, only nisin (E234) and natamycin (E235) have been approved for the use as food preservative in Europe under Directive 95/2/EC on food additives [76] and were affirmed GRAS by the US Food and Drug Administration. Therefore, legislative issues regarding the use of bacteriocins as food preservatives may remain the main limitation for commercial exploitation. Nevertheless, application of bacteriocins combined with other preservation techniques can result in synergistic effects in food packaging and therefore, maintaining the quality and safety of fresh and minimally processed food products.

12.3.4 Enzymes

Enzymes are of interest for food packaging applications since they are naturally occurring bioactive compounds and can act as effective antimicrobial agents [57]. The low tolerance of enzymes toward high temperatures excludes their application from harsh film processing technologies such as extrusion or casting from solutions. Alternatively, enzymes can be immobilized by either physically entrapping in, or chemically bonding to packaging films. Finally, the activity of the enzymes that are released from the packaging to the food will further depend on intrinsic conditions such as temperature, pH, and the presence of salts [6]. A well-known antimicrobial enzyme is lysozyme, which is effective mainly against gram-positive bacteria and acts by splitting the bonds between *N*-acetylglucosamine and *N*-acetylmuramic acid of the peptidoglycan in their cell walls. On gram-negative bacteria, the outer membrane is surrounded by a peptidoglucan layer protecting them against the antibacterial activity of lysozyme. However, if lysozyme is combined with a chelating agent such as EDTA, the outer membranes of gram-negative bacteria are destabilized enhancing their susceptibility [23, 55]. So far, the antimicrobial activity of enzyme-containing packaging films was mainly tested under laboratory conditions. As shown in Table 12.5, however, there are also studies available that have proven the stability of the enzymatic activity in real food application.

Gemili et al. [55] incorporated lysozyme with Na_2EDTA into asymmetric and porous cellulose acetate films showing antimicrobial activity against both gram-positive (*Bacillus amyloliquefaciens*) and gram-negative (*E. coli*) bacteria on laboratory media. Similar results were obtained by Mecitoğlu et al. [23] who incorporated lysozyme into zein films showing antimicrobial activity against *B. subtilis* and *L. plantarum*. By the addition of Na_2EDTA, the films were also effective against *E. coli*. Another approach to enhance the inhibition efficacy of lysozyme against gram-negative bacteria was demonstrated by Park et al. [56] who developed lysozyme–chitosan composite films. Thereby, it is assumed that the outer membranes of bacteria were uncovered or disrupted by the polycationic nature of the chitosan molecules so that lysozyme was also effective against gram-negative bacteria. As a result, lysozyme incorporation enhanced the antibacterial activity of chitosan films against both *Enterococcus faecalis* and *E. coli*, with a log cycle reduction of 2.7 and 3.8, respectively, in liquid media.

Lactoferrin is an iron-binding glycoprotein that exhibits antimicrobial activity by making iron, an essential component, unavailable for microorganisms. It exhibits antimicrobial activity against gram-positive and gram-negative bacteria, as well as fungi and parasites [3, 57, 59]. Lactoferrin was incorporated into bacterial cellulose-based films and assed as edible antimicrobial packaging [59]. Bactericidal efficiency was observed against *E. coli* and *S. aureus* with a mean reduction of 69% and 97% in the films per se and 94% and 36% when inoculated on fresh sausages, respectively. Min et al. [57] studied the antimicrobial effects of lactoferrin and lysozyme in liquid media, and a lactoperoxidase system (LPOS) in edible WPI films. In this study, however, neither lactoferrin nor lysozyme alone showed antimicrobial activity against *Salmonella enterica* and *E. coli*. In contrast, the LPOS-WPI films showed strong bactericidal effects against these pathogens (4 log CFU/cm^2), since both were completely inhibited when contacted with the active films. Thus, LPOS-WPI films have potential for controlling *S. enterica* and *E. coli* when already present on food products, as well as for inhibiting their growth from contamination of film-wrapped or coated food products. Combinations of lysozyme and lactoferrin have also been studied [58]. Both enzymes were incorporated into paper sheets containing carboxymethyl cellulose, which allowed noncovalent binding of the positively charged proteins to the paper matrix. *In vitro* tests resulted in a 2 log/CFU/ml reduction showing a synergistic interaction of lysozyme and lactoferrin against *E. coli* and *L. innocua*. Food application was performed with veal "carpaccio" (thin slices of veal fillet, meant for consumption as raw meat) laid on paper sheets containing either or both antimicrobial proteins. Results indicated that lysozyme was most effective in preventing growth of total aerobic bacteria in the meat sample, giving 1 log cycle reduction with respect to the control.

Although lysozyme is known to exhibit antimicrobial activity particularly against gram-positive bacteria, this is not the case for the foodborne pathogen *S. aureus* since all pathogenic Staphylococcal strains which are O-acetylated are completely resistant to this cationic cell wall-lytic enzyme [77–80]. With the advent of nanotechnology, however, Silva et al. [60] demonstrated that nano-fibrillation of lysozyme confers antibacterial activity to lysozyme nanofibers against *S. aureus*. Thereby, pullulan films reinforced with lysozyme nanofibers inhibited the growth of *S. aureus* with about 3.2 log CFU reduction within 48 hours, whereas no inhibition was observed for the films containing native lysozyme and the plain pullulan films. Such promising results show the potential of proteinaceous nanofibers for the development of eco-friendly edible films for active food packaging.

In Europe, the use of lysozyme (E1105) falls under Directive 95/2/EC on food additives [76] and it has also GRAS status as an antimicrobial agent on casings for frankfurters and on cooked meat and poultry products [81]. Lactoferrin is accepted by the U.S. FDA as an ingredient in milk-based infant formula [82].

12.3.5 Chitosan

Polymers such as chitosan, along with their derivatives, are inherently antimicrobial and have therefore been proven to be beneficial for food packaging (Table 12.6). Chitosan, a deacetylated derivative of chitin, is a natural polysaccharide, extracted from marine sources such as crustacean shells [83] or obtained from fungi [84, 85]. After cellulose, chitin is the second most abundant natural biopolymer [86]. Due to its broad applicability such as in films, blends, composites or coatings, chitosan has attracted the attention of both the food and packaging industry and the scientific community because of its nontoxicity, biodegradability, and biocompatibility. Moreover, chitosan provides antifungal and antibacterial activity against a wide range of foodborne pathogens [83, 87, 88]. To date, the antimicrobial effect of chitosan is still unclear [4]; however, several mechanisms have been proposed. According to Fortunati [83], the most widely accepted theory is based on the fact that interactions between the negatively charged bacterial cell membranes and the positively charged chitosan amine groups may lead to the leakage of low-molecular-weight materials, nucleic acids, and proteins.

Chitosan has been used for the development of paper-based active packaging materials [61]. Its antimicrobial activity has been confirmed *in vitro* against *L. innocua*, *E. coli*, *S. aureus*, and *Pseudomonas putida* as well as on thinly sliced raw veal meat inoculated with *L. innocua*. Han et al. [62] reported an increased antimicrobial activity of chitosan films by the incorporation of sodium montmorillonite (Na^+-MMT) nanocomposites. The resulting chitosan-Na^+-MMT films exhibited significantly higher antimicrobial activity against *E. coli* and *S. aureus* compared to pure chitosan and Na^+-MMT. This synergistic effect has been explained by the strong electrostatic interaction of the cationic chitosan molecules with the anionic silicate layers. Antimicrobial films of chitosan and PVA were prepared by solution casting method with glutaraldehyde as cross-linker [63]. Coating application on whole and cut fresh tomatoes demonstrated antimicrobial activity of the chitosan–PVA films against the food pathogenic bacteria *E. coli*, *S. aureus*, and *B. subtilis*. In another study [64], a chitosan–PVA blend has been developed by immobilizing a chito-oligosaccharide (COS, degree of polymerization of 9) on PVA by cross-linking with the bi-functional compound *N*-methylolacrylamide. The PVA–COS films showed excellent antimicrobial activity against *S. aureus* on laboratory media. Bacterial growth was completely suppressed when PVA films with COS concentrations between 4.49% and 8.5% were applied. Biodegradable antimicrobial composites with lignin incorporated into chitosan films were tested against *B. subtilis* and *P. aeruginosa* [65]. Both chitosan and lignin exerted antibacterial activity individually. However, for their composite a synergistic antibacterial effect has been demonstrated *in vitro* with inhibition zone diameters almost double the size compared to that of the individual materials. The authors concluded that the enhanced antimicrobial properties caused by the addition of lignin are mainly due its hydrophobicity, lipophilicity, and capability to bind with proteins at acidic pH. Edible chitosan coatings on PLA films were shown to inhibit the growth of *L. innocua*, *L. monocytogenes*, and *S. typhimurium*. Chitosan coatings with incorporated organic acids (lauric arginate ester, sodium lactate, or sorbic acid) were also effective against the respective bacteria inoculated on the surface of ready-to-eat sliced turkey [66]. The meat had not been tested with the pure chitosan coatings; however, the *in vitro* antimicrobial effect of the pure chitosan coatings was shown to be very similar to that of those containing organic acids suggesting a similar antimicrobial effect *in vivo*. Higueras et al. [67] studied the combination of chitosan with the antimicrobial compound ethyl-N^{α}-dodecanoyl-L-arginate (LAE) by the production of cast films with 20% glycerol. In laboratory media, the chitosan–LAE films were effective against a wide range of selected microorganisms (Table 12.6). Particularly on yeasts, a synergistic effect of chitosan and LAE has been observed, both known to be yeast inhibitors. For both chitosan and chitosan–LAE films, antimicrobial activity has also been confirmed for food application. On fresh chicken breast fillet, the chitosan films expressed antimicrobial activity in the range of 0.47–2.96 log reductions and the chitosan–LAE films resulted in 1.78–5.81 log reductions, depending on time and microorganism studied (mesophiles, psychrophiles, Pseudomonas spp., coliforms, lactic acid bacteria, hydrogen sulfide-producing bacteria).

12.4 Conclusions and Perspectives

It has been shown that various organic compounds can provide antimicrobial properties to food packaging materials and that such bioactive compounds can be incorporated into biopolymer matrices. This offers a great potential to maintain or improve the quality of food products leading to a reduction of food waste and thereby contributing to a more sustainable handling of food. Concerning the environmental impact, the next generation of AP should inevitably target a replacement of the petroleum-based commodity plastics. To develop sustainable antimicrobial packaging systems, the end of life of the whole package should be considered.

If the AP is applied to biodegradable packaging materials, all the components of the AP including the active compound should be biodegradable. Here it is vital to consider the effect of antimicrobial AP on the biodegradability. The effect of antimicrobial bioactive compounds on the compostability of the materials has not been studied well so far [89]. In future, more recyclable materials will be used for food packaging. The foundation therefore has been laid by the "European Strategy for Plastics in a Circular Economy" [90]. The strategy proposes a new plastics economy where the design and production of plastic and plastic products should fully respect reuse, repair, and recycling needs and more sustainable materials should be developed and promoted. Therefore, AP systems applied to recycle packaging materials should not interfere with the recycling process. They should be either recycled with the whole package or should be easily separated during the recycling process.

For successful implementation of bioactive materials in food packaging, however, the cost has to correspond to the benefit gained by the particular food product, regulatory and legislative issues must be addressed, and broad consumer acceptance is required. In the Japanese market, active packaging had already gained high popularity in the mid-1970s [9]; in Europe however, AP is only now beginning to increase. High cost, low consumer acceptance, and strict legislations are the main points hindering a diffusion of such systems in the European market. The benefit of AP to develop clean label products or to reduce or eliminate the use of preservatives should be communicated to increase the acceptance by consumers and to justify the cost. Understanding the technological, social, and marketing challenges of AP technologies and strengthening the collaborative research activities between the research institutes and the food and packaging companies to overcome such challenges will pave the way for future commercial applications.

References

1 Arvanitoyannis, I.S. and Oikonomou, G. (2012). Active and intelligent packaging. In: *Modified Atmosphere and Active Packaging Technologies* (ed. I.S. Arvanitoyannis), 628–654. Boca Raton, FL: CRC Press.

2 Brockgreitens, J. and Abbas, A. (2016). Responsive food packaging: recent progress and technological prospects. *Comprehensive Reviews in Food Science and Food Safety* 15 (1): 3–15. https://doi.org/10.1111/1541-4337.12174

3 Vilela, C., Kurek, M., Hayouka, Z., et al. (2018). A concise guide to active agents for active food packaging. *Trends in Food Science & Technology* 80: 212–222. https://doi.org/10.1016/j.tifs.2018.08.006

4 Yildirim, S., Röcker, B., Pettersen, M.K., et al. (2018). Active packaging applications for food. *Comprehensive Reviews in Food Science and Food Safety* 17 (1): 165–199. https://doi.org/10.1111/1541-4337.12322

5 European Comission (2009). EU Guidance to the Commission Regulation (EC) No 450/2009 of 29 May 2009 on active and intelligent materials and articles intended to come into the contact with food (version 1.0). Retrieved from https://ec.europa.eu/food/safety/chemical_safety/food_contact_materials_en (accessed 3 July 2020).

6 Yildirim, S. (2011). Active packaging for food biopreservation. In: *Protective Cultures, Antimicrobial Metabolites and Bacteriophages for Food and Beverage Biopreservation* (ed. C. Lacroix), 460–489. Cambridge: Woodhead Publishing Ltd.

7 Pereira de Abreu, D.A., Cruz, J.M., and Losada, P.P. (2012). Active and intelligent packaging for the food industry. *Food Reviews International* 28 (2): 146–187. https://doi.org/10.1080/87559129.2011.595022

8 Suppakul, P. (2015). Active and intelligent packaging. In: *Polymers for Packaging Applications* (eds. S. Alavi, S. Thomas, K.P. Sandeep, et al.), 395–399. Toronto: Apple Academic Press.

9 Restuccia, D., Spizzirri, U.G., Parisi, O.I., et al. (2010). New EU regulation aspects and global market of active and intelligent packaging for food industry applications. *Food Control* 21 (11): 1425–1435. https://doi.org/10.1016/j.foodcont.2010.04.028

10 Guillard, V., Gaucel, S., Fornaciari, C., et al. (2018). The next generation of sustainable food packaging to preserve our environment in a circular economy context. *Frontiers in Nutrition* 5: 13. https://doi.org/10.3389/fnut.2018.00121

11 Heller, M.C., Selke, S.E.M., and Keoleian, G.A. (2019). Mapping the influence of food waste in food packaging environmental performance assessments. *Journal of Industrial Ecology* 23 (2): 480–495. https://doi.org/10.1111/jiec.12743

12 Licciardello, F. (2017). Packaging, blessing in disguise. Review on its diverse contribution to food sustainability. *Trends in Food Science & Technology* 65: 32–39. https://doi.org/10.1016/j.tifs.2017.05.003

13 Svanes, E., Oestergaard, S., and Hanssen, O.J. (2019). Effects of packaging and food waste prevention by consumers on the environmental impact of

production and consumption of bread in Norway. *Sustainability* 11 (1): 20. https://doi.org/10.3390/su11010043
14 Díez-Pascual, A.M. (2019). Synthesis and applications of biopolymer composites. *International Journal of Molecular Sciences* 20 (9): 2321. https://doi.org/10.3390/ijms20092321
15 Helanto, K., Matikainen, L., Talja, R., Rojas, O. J (2019). Bio-based polymers for sustainable packaging and biobarriers: a critical review. *Bioresources* 14 (2): 4902–4951
16 Realini, C.E. and Marcos, B. (2014). Active and intelligent packaging systems for a modern society. *Meat Science* 98 (3): 404–419. https://doi.org/10.1016/j.meatsci.2014.06.031
17 European Union (2008). Regulation (EC) No 1333/2008 of the European Parliament and of the council of 16 December 2008 on food additives (Text with EEA relevance). Retrieved from https://ec.europa.eu/food/safety/chemical_safety/food_contact_materials_en (accessed 3 July 2020).
18 Gan, I. and Chow, W.S. (2018). Antimicrobial poly(lactic acid)/cellulose bionanocomposite for food packaging application: a review. *Food Packaging and Shelf Life* 17: 150–161. https://doi.org/10.1016/j.fpsl.2018.06.012
19 Raquel, R., Maria, V., Lorena, A., et al. (2018). Biopolymers carrying essential oils, or their compounds, for food antimicrobial packaging. *Current Organic Chemistry* 22 (12): 1141–1156. http://dx.doi.org/10.2174/1385272822666171129153011
20 Muthuraj, R., Misra, M., and Mohanty, A.K. (2018). Biodegradable compatibilized polymer blends for packaging applications: a literature review. *Journal of Applied Polymer Science* 135 (24): 45726. https://doi.org/10.1002/app.45726
21 Appendini, P. and Hotchkiss, J.H. (2002). Review of antimicrobial food packaging. *Innovative Food Science & Emerging Technologies* 3 (2): 113–126. https://doi.org/10.1016/S1466-8564(02)00012-7
22 Han, J.H. (2000). Antimicrobial food packaging. *Food Technology* 54 (3): 56–65
23 Mecitoğlu, Ç., Yemenicioğlu, A., Arslanoğlu, A., et al. (2006). Incorporation of partially purified hen egg white lysozyme into zein films for antimicrobial food packaging. *Food Research International* 39 (1): 12–21. https://doi.org/10.1016/j.foodres.2005.05.007
24 Suppakul, P., Miltz, J., Sonneveld, K., et al. (2003). Active packaging technologies with an emphasis on antimicrobial packaging and its applications. *Journal of Food Science* 68 (2): 408–420. https://doi.org/10.1111/j.1365-2621.2003.tb05687.x
25 Wen, P., Zhu, D.-H., Feng, K., et al. (2016). Fabrication of electrospun polylactic acid nanofilm incorporating cinnamon essential oil/β-cyclodextrin inclusion complex for antimicrobial packaging. *Food Chemistry* 196: 996–1004. https://doi.org/10.1016/j.foodchem.2015.10.043
26 Higueras, L., López-Carballo, G., Gavara, R., Pilar Hernández-Muñoz (2015). Reversible covalent immobilization of cinnamaldehyde on chitosan films via Schiff Base formation and their application in active food packaging. *Food and Bioprocess Technology* 8 (3): 526–538. https://doi.org/10.1007/s11947-014-1421-8

27 Alboofetileh, M., Rezaei, M., Hosseini, H., Mehdi Abdollahi (2018). Morphological, physico-mechanical, and antimicrobial properties of sodium alginate-montmorillonite nanocomposite films incorporated with marjoram essential oil. *Journal of Food Processing and Preservation* 42 (5): e13596. https://doi.org/10.1111/jfpp.13596

28 Atef, M., Rezaei, M., and Behrooz, R. (2015). Characterization of physical, mechanical, and antibacterial properties of agar-cellulose bionanocomposite films incorporated with savory essential oil. *Food Hydrocolloids* 45: 150–157. https://doi.org/10.1016/j.foodhyd.2014.09.037

29 Rodriguez-Garcia, I., Cruz-Valenzuela, M.R., Silva-Espinoza, B.A., et al. (2016). Oregano (Lippia graveolens) essential oil added within pectin edible coatings prevents fungal decay and increases the antioxidant capacity of treated tomatoes. *Journal of the Science of Food and Agriculture* 96 (11): 3772–3778. https://doi.org/10.1002/jsfa.7568

30 Zinoviadou, K.G., Koutsoumanis, K.P., and Biliaderis, C.G. (2009). Physico-chemical properties of whey protein isolate films containing oregano oil and their antimicrobial action against spoilage flora of fresh beef. *Meat Science* 82 (3): 338–345. https://doi.org/10.1016/j.meatsci.2009.02.004

31 Arfat, Y.A., Benjakul, S., Vongkamjan, K., et al. (2015). Shelf-life extension of refrigerated sea bass slices wrapped with fish protein isolate/fish skin gelatin-ZnO nanocomposite film incorporated with basil leaf essential oil. *Journal of Food Science and Technology* 52 (10): 6182–6193. https://doi.org/10.1007/s13197-014-1706-y

32 Echeverría, I., López-Caballero, M.E., Gómez-Guillén, M.C., et al. (2018). Active nanocomposite films based on soy proteins-montmorillonite-clove essential oil for the preservation of refrigerated bluefin tuna (Thunnus thynnus) fillets. *International Journal of Food Microbiology* 266: 142–149. https://doi.org/10.1016/j.ijfoodmicro.2017.10.003

33 Priyadarshi, R., Sauraj, Kumar, B., et al. (2018). Chitosan films incorporated with apricot (*Prunus armeniaca*) kernel essential oil as active food packaging material. *Food Hydrocolloids* 85: 158–166. https://doi.org/10.1016/j.foodhyd.2018.07.003

34 Wang, Y., Xia, Y., Zhang, P., et al. (2017). Physical characterization and pork packaging application of chitosan films incorporated with combined essential oils of cinnamon and ginger. *Food and Bioprocess Technology* 10 (3): 503–511. https://doi.org/10.1007/s11947-016-1833-8

35 Saberi, B., Chockchaisawasdee, S., Golding, J.B., et al. (2017). Characterization of pea starch-guar gum biocomposite edible films enriched by natural antimicrobial agents for active food packaging. *Food and Bioproducts Processing* 105: 51–63. https://doi.org/10.1016/j.fbp.2017.06.003

36 Kaya, M., Khadem, S., Cakmak, Y.S., et al. (2018). Antioxidative and antimicrobial edible chitosan films blended with stem, leaf and seed extracts of *Pistacia terebinthus* for active food packaging. *RSC Advances* 8 (8): 3941–3950. https://doi.org/10.1039/C7RA12070B

37 Riaz, A., Lei, S., Akhtar, H.M.S., et al. (2018). Preparation and characterization of chitosan-based antimicrobial active food packaging film incorporated

with apple peel polyphenols. *International Journal of Biological Macromolecules* 114: 547–555. https://doi.org/10.1016/j.ijbiomac.2018.03.126
38 Halim, A.L.A., Kamari, A., and Phillip, E. (2018). Chitosan, gelatin and methylcellulose films incorporated with tannic acid for food packaging. *International Journal of Biological Macromolecules* 120: 1119–1126. https://doi.org/10.1016/j.ijbiomac.2018.08.169
39 Lee, K.-Y., Lee, J.-H., Yang, H.-J., et al. (2016). Characterization of a starfish gelatin film containing vanillin and its application in the packaging of crab stick. *Food Science and Biotechnology* 25 (4): 1023–1028. https://doi.org/10.1007/s10068-016-0165-9
40 Wu, Z., Wu, J., Peng, T., et al. (2017). Preparation and application of starch/polyvinyl alcohol/citric acid ternary blend antimicrobial functional food packaging films. *Polymers* 9 (3): 102. https://doi.org/10.3390/polym9030102
41 García-Soto, B., Miranda, J.M., Rodríguez-Bernaldo de Quirós, A., et al. (2015). Effect of biodegradable film (lyophilised alga *Fucus spiralis* and sorbic acid) on quality properties of refrigerated megrim (*Lepidorhombus whiffiagonis*). *International Journal of Food Science & Technology* 50 (8): 1891–1900. https://doi.org/10.1111/ijfs.12821
42 Chen, M.-C., Yeh, G.H.-C., and Chiang, B.-H. (1996). Antimicrobial and physicochemical properties of methylcellulose and chitosan films containing a preservative. *Journal of Food Processing and Preservation* 20 (5): 379–390. https://doi.org/10.1111/j.1745-4549.1996.tb00754.x
43 Junqueira-Gonçalves, M.P., Alarcón, E., and Niranjan, K. (2014). Post-consumer recycled PET packaging for fresh berries: a comparative study between incorporating an antifungal agent superficially and into the main body of the packaging. *Food and Bioprocess Technology* 7 (9): 2610–2617. https://doi.org/10.1007/s11947-014-1262-5
44 Kristo, E., Koutsoumanis, K.P., and Biliaderis, C.G. (2008). Thermal, mechanical and water vapor barrier properties of sodium caseinate films containing antimicrobials and their inhibitory action on *Listeria monocytogenes*. *Food Hydrocolloids* 22 (3): 373–386. https://doi.org/10.1016/j.foodhyd.2006.12.003
45 Cestari, L.A., Gaiotto, R.C., Antigo, J.L., et al. (2015). Effect of active packaging on low-sodium restructured chicken steaks. *Journal of Food Science and Technology* 52 (6): 3376–3382. https://doi.org/10.1007/s13197-014-1357-z
46 Rutenberg, R., Bernstein, S., Paster, N., et al. (2016). Antimicrobial films based on cellulose-derived hydrocolloids. A synergetic effect of host–guest interactions on quality and functionality. *Colloids and Surfaces B: Biointerfaces* 137: 138–145. https://doi.org/10.1016/j.colsurfb.2015.06.022
47 Khan, A., Gallah, H., Riedl, B., et al. (2016). Genipin cross-linked antimicrobial nanocomposite films and gamma irradiation to prevent the surface growth of bacteria in fresh meats. *Innovative Food Science & Emerging Technologies* 35: 96–102. https://doi.org/10.1016/j.ifset.2016.03.011
48 Correa, J.P., Molina, V., Sanchez, M., et al. (2017). Improving ham shelf life with a polyhydroxybutyrate/polycaprolactone biodegradable film activated with nisin. *Food Packaging and Shelf Life* 11: 31–39. https://doi.org/10.1016/j.fpsl.2016.11.004

49 Imran, M., Revol-Junelles, A.-M., René, N., et al. (2012). Microstructure and physico-chemical evaluation of nano-emulsion-based antimicrobial peptides embedded in bioactive packaging films. *Food Hydrocolloids* 29 (2): 407–419. https://doi.org/10.1016/j.foodhyd.2012.04.010

50 Saini, S., Sillard, C., Naceur Belgacem, M., et al. (2016). Nisin anchored cellulose nanofibers for long term antimicrobial active food packaging. *RSC Advances* 6 (15): 12422–12430. https://doi.org/10.1039/C5RA22748H

51 de Oliveira, T.M., de Fátima Ferreira Soares, N., Pereira, R.M., Kátia de Freitas Fraga (2007). Development and evaluation of antimicrobial natamycin-incorporated film in gorgonzola cheese conservation. *Packaging Technology and Science* 20 (2): 147–153. https://doi.org/10.1002/pts.756

52 Marcos, B., Aymerich, T., Monfort, J.M., et al. (2007). Use of antimicrobial biodegradable packaging to control *Listeria monocytogenes* during storage of cooked ham. *International Journal of Food Microbiology* 120 (1–2): 152–158. https://doi.org/10.1016/j.ijfoodmicro.2007.06.003

53 Santiago-Silva, P., Soares, N.F.F., Nóbrega, J.E., et al. (2009). Antimicrobial efficiency of film incorporated with pediocin (ALTA® 2351) on preservation of sliced ham. *Food Control* 20 (1): 85–89. https://doi.org/10.1016/j.foodcont.2008.02.006

54 Massani, M.B., Molina, V., Sanchez, M., et al. (2014). Active polymers containing Lactobacillus curvatus CRL705 bacteriocins: effectiveness assessment in wieners. *International Journal of Food Microbiology* 178: 7–12. https://doi.org/10.1016/j.ijfoodmicro.2014.02.013

55 Gemili, S., Yemenicioğlu, A., and Altınkaya, S.A. (2009). Development of cellulose acetate based antimicrobial food packaging materials for controlled release of lysozyme. *Journal of Food Engineering* 90 (4): 453–462. https://doi.org/10.1016/j.jfoodeng.2008.07.014

56 Park, S.-I., Daeschel, M.A., and Zhao, Y. (2004). Functional properties of antimicrobial lysozyme-chitosan composite films. *Journal of Food Science* 69 (8): M215-M221. https://doi.org/10.1111/j.1365-2621.2004.tb09890.x

57 Min, S., Harris, L.J., and Krochta, J.M. (2005). Antimicrobial effects of lactoferrin, lysozyme, and the Lactoperoxidase system and edible whey protein films incorporating the lactoperoxidase system against *Salmonella enterica* and Escherichia coli O157:H7. *Journal of Food Science* 70 (7): m332-m338. https://doi.org/10.1111/j.1365-2621.2005.tb11476.x

58 Barbiroli, A., Bonomi, F., Capretti, G., et al. (2012). Antimicrobial activity of lysozyme and lactoferrin incorporated in cellulose-based food packaging. *Food Control* 26 (2): 387–392. https://doi.org/10.1016/j.foodcont.2012.01.046

59 Padrão, J., Gonçalves, S., Silva, J.P., et al. (2016). Bacterial cellulose-lactoferrin as an antimicrobial edible packaging. *Food Hydrocolloids* 58: 126–140. https://doi.org/10.1016/j.foodhyd.2016.02.019

60 Silva, N.H.C.S., Vilela, C., Almeida, A., et al. (2018). Pullulan-based nanocomposite films for functional food packaging: exploiting lysozyme nanofibers as antibacterial and antioxidant reinforcing additives. *Food Hydrocolloids* 77: 921–930. https://doi.org/10.1016/j.foodhyd.2017.11.039

61 Rollini, M., Mascheroni, E., Capretti, G., et al. (2017). Propolis and chitosan as antimicrobial and polyphenols retainer for the development of paper based

active packaging materials. *Food Packaging and Shelf Life* 14: 75–82. https://doi.org/10.1016/j.fpsl.2017.08.011

62 Han, Y.-S., Lee, S.-H., Choi, K.H., In Park (2010). Preparation and characterization of chitosan–clay nanocomposites with antimicrobial activity. *Journal of Physics and Chemistry of Solids* 71 (4): 464–467. https://doi.org/10.1016/j.jpcs.2009.12.012

63 Tripathi, S., Mehrotra, G.K., and Dutta, P.K. (2009). Physicochemical and bioactivity of cross-linked chitosan–PVA film for food packaging applications. *International Journal of Biological Macromolecules* 45 (4): 372–376. https://doi.org/10.1016/j.ijbiomac.2009.07.006

64 Cho, Y.-W., Han, S.-S., and Ko, S.-W. (2000). PVA containing chito-oligosaccharide side chain. *Polymer* 41 (6): 2033–2039. https://doi.org/10.1016/S0032-3861(99)00351-1

65 Rai, S., Dutta, P., and Mehrotra, G. (2017). Lignin incorporated antimicrobial chitosan film for food packaging application. *Journal of Polymer Materials* 34 (1): 171

66 Guo, M., Jin, T.Z., and Yang, R. (2014). Antimicrobial polylactic acid packaging films against listeria and salmonella in culture medium and on ready-to-eat meat. *Food and Bioprocess Technology* 7 (11): 3293–3307. https://doi.org/10.1007/s11947-014-1322-x

67 Higueras, L., López-Carballo, G., Hernández-Muñoz, P., et al. (2013). Development of a novel antimicrobial film based on chitosan with LAE (ethyl-N α-dodecanoyl-L-arginate) and its application to fresh chicken. *International Journal of Food Microbiology* 165 (3): 339–345. https://doi.org/10.1016/j.ijfoodmicro.2013.06.003

68 Ruiz-Navajas, Y., Viuda-Martos, M., Sendra, E., et al. (2013). In vitro antioxidant and antifungal properties of essential oils obtained from aromatic herbs endemic to the southeast of Spain. *Journal of Food Protection* 76 (7): 1218–1225. https://doi.org/10.4315/0362-028X.JFP-12-554

69 Alkan, D. and Yemenicioğlu, A. (2016). Potential application of natural phenolic antimicrobials and edible film technology against bacterial plant pathogens. *Food Hydrocolloids* 55: 1–10. https://doi.org/10.1016/j.foodhyd.2015.10.025

70 Cowan, M.M. (1999). Plant products as antimicrobial agents. *Clinical Microbiology Reviews* 12 (4): 564–582. https://doi.org/10.1128/cmr.12.4.564

71 Stroescu, M., Stoica-Guzun, A., Isopencu, G., et al. (2015). Chitosan-vanillin composites with antimicrobial properties. *Food Hydrocolloids* 48: 62–71. https://doi.org/10.1016/j.foodhyd.2015.02.008

72 Brul, S. and Coote, P. (1999). Preservative agents in foods: mode of action and microbial resistance mechanisms. *International Journal of Food Microbiology* 50 (1): 1–17. https://doi.org/10.1016/S0168-1605(99)00072-0

73 Mani-López, E., García, H.S., and López-Malo, A. (2012). Organic acids as antimicrobials to control Salmonella in meat and poultry products. *Food Research International* 45 (2): 713–721. https://doi.org/10.1016/j.foodres.2011.04.043

74 Paster, N. (1979). A commercial scale study of the efficiency of propionic acid and calcium propionate as fungistats in poultry feed. *Poultry Science* 58 (3): 572–576. https://doi.org/10.3382/ps.0580572
75 Poverenov, E., Granit, R., and Gabai, S. (2013). Encapsulation and controlled release of antifungal propionic acid utilizing biodegradable active films based on natural polymers. *European Food Research and Technology* 237 (1): 19–26. https://doi.org/10.1007/s00217-013-2011-0
76 European Union (1995). European Parliament and Council Directive No 95/2/EC of 20 February 1995 on food additives other than colours and sweeteners.
77 Bera, A., Biswas, R., Herbert, S., Friedrich Götz (2006). The presence of peptidoglycan O-acetyltransferase in various staphylococcal species correlates with lysozyme resistance and pathogenicity. *Infection and Immunity* 74 (8): 4598–4604. https://doi.org/10.1128/iai.00301-06
78 Bera, A., Biswas, R., Herbert, S., et al. (2007). Influence of wall Teichoic acid on lysozyme resistance in *Staphylococcus aureus*. *Journal of Bacteriology* 189 (1): 280–283. https://doi.org/10.1128/jb.01221-06
79 Bera, A., Herbert, S., Jakob, A., et al. (2005). Why are pathogenic staphylococci so lysozyme resistant? The peptidoglycan O-acetyltransferase OatA is the major determinant for lysozyme resistance of *Staphylococcus aureus*. *Molecular Microbiology* 55 (3): 778–787. https://doi.org/10.1111/j.1365-2958.2004.04446.x
80 Pushkaran, A.C., Nataraj, N., Nair, N., et al. (2015). Understanding the structure–function relationship of lysozyme resistance in *Staphylococcus aureus* by peptidoglycan O-acetylation using molecular docking, dynamics, and Lysis assay. *Journal of Chemical Information and Modeling* 55 (4): 760–770. https://doi.org/10.1021/ci500734k
81 FDA (2001). Office of Premarket Approval, Agency Response Letter GRAS Notice No. GRN 64: Lysozyme. Retrieved 21.05, 2019, from https://www.accessdata.fda.gov/scripts/fdcc/?set=GRASNotices&id=64.
82 FDA (2017). Office of Premarket Approval, Agency Response Letter GRAS Notice No. GRN 669: Cow's milk-derived lactoferrin. Retrieved 21.05, 2019, from https://www.accessdata.fda.gov/scripts/fdcc/index.cfm?set=GRASNotices&id=669.
83 Fortunati, E. (2016). Multifunctional films, blends, and nanocomposites based on chitosan: use in antimicrobial packaging. In: *Antimicrobial Food Packaging* (ed. J. Barros-Velazquez), 467–477. San Diego, CA: Academic Press.
84 Peter, M.G. (2005). Chitin and chitosan in fungi. In: *Biopolymers Online*: A. Steinbüchel, Wiley-VCH.
85 Pochanavanich, P. and Suntornsuk, W. (2002). Fungal chitosan production and its characterization. *Letters in Applied Microbiology* 35 (1): 17–21. https://doi.org/10.1046/j.1472-765X.2002.01118.x
86 Ghadi, A., Mahjoub, S., Tabandeh, F., Farid Talebnia (2014). Synthesis and optimization of chitosan nanoparticles: potential applications in nanomedicine and biomedical engineering. *Caspian Journal of Internal Medicine* 5 (3): 156–161

87 Elsabee, M.Z. and Abdou, E.S. (2013). Chitosan based edible films and coatings: a review. *Materials Science and Engineering: C* 33 (4): 1819–1841. https://doi.org/10.1016/j.msec.2013.01.010
88 Srinivasa, P.C. and Tharanathan, R.N. (2007). Chitin/chitosan — safe, ecofriendly packaging materials with multiple potential uses. *Food Reviews International* 23 (1): 53–72. https://doi.org/10.1080/87559120600998163
89 Balaguer, M.P., Villanova, J., Cesar, G., et al. (2015). Compostable properties of antimicrobial bioplastics based on cinnamaldehyde cross-linked gliadins. *Chemical Engineering Journal* 262: 447–455. https://doi.org/10.1016/j.cej.2014.09.099
90 European Comission (2018). A European Strategy for Plastics in a Circular Economy. Communication from the commission to the European parliament, the council, the European economic and social committee and the committee of the regions. Retrieved from https://eur-lex.europa.eu/legal-content/EN/TXT/HTML/?uri=CELEX:52018DC0028&from=EN (accessed 3 July 2020).

13

Active Antioxidant Additives in Sustainable Food Packaging

Thi-Nga Tran

Smart Materials Group, Istituto Italiano di Tecnologia, Via Morego 30, 16163 Genoa, Italy

13.1 Introduction

Due to a fast-growing world population which is predicted to reach 9.8 billion in 2050 and 11.2 billion in 2100 [1], food demand emerges to be among the major issues of global food security and sustainability in the twenty-first century. In order to provide the needed amount of food to the whole increasing population, the transition to healthy diets from sustainable food production systems is an urgent global scientific target aiming at improving human health and environmental sustainability [2, 3]. To achieve this challenge, one logical solution seems to be increasing the capability to produce more food products, but this might lead to a high stressful workflow in agriculture and food industry along with a risk of exhausted resources and climate change. For these reasons, one of the most important strategies is to significantly reduce the food losses and wastes along with the improvement of food quality and shelf life. The extension of food lifetime by preventing its spoilage represents a promising approach to tackle this problematic issue. Indeed, up to 50% of the total food weight produced is lost or wasted, for instance, as unconsumed foods in households, restaurants, and supermarkets or as damaged and spoilt foods during inappropriate storage and transportation [3]. The high amount of discarded food has a negative impact on the food demand/producing capacity balance as well as on the environment. To this end, much effort has been devoted to develop the so-called innovative active food packaging with several achievements so far. Generally, the primary function of food packaging material is to protect the food during transportation, storage, and consumption. The conventional packaging materials only prevent the food from the physical damage that can be caused through cuts, stress conditions, extreme conditions such as high temperature, low temperature, pathogen and pest attacks from outside environment, or they are simply used for the marketing of foods [3, 4]. Thus, the possibility to prolong food lifetime by preventing intrinsic spoilage and contaminations using conventional packaging is limited. As a result, the development and use of active food packaging materials to progressively replace the conventional ones become a topic of global interest. As being indicated in the European Union legislation for food packaging,

Sustainable Food Packaging Technology, First Edition. Edited by Athanassia Athanassiou.

"active food contact materials and articles are designed to deliberately incorporate 'active' components that are capable of scavenging or eradicating the spoiling factors which are released from the food" [5, 6]. As such, this novel type of packaging materials is expected to maintain or prolong the shelf life of food products while ensuring their quality, safety, and integrity. Unlike the case in which active components are supplemented directly into the food as additives, their integration into packaging materials is much more advantageous. This is mainly due to the fact that a controlled release of active components by the gradually diffusion of nonvolatile agents from the packaging to the food surface or the emission of volatile compounds into the atmosphere surrounding the food can be achieved. In addition, several benefits employing this approach can be also highlighted such as: (i) controlled and long-term activity thanks to the gradual migration of active components from the package to the food matrix; (ii) reduction in cost of food processing, and (iii) conservation of the food structure due to the retrenchment of mixing, immersion, or spraying steps of the active substances into food. These features make active food packaging outstanding in comparison to other approaches. In fact, some active food packaging materials have been already commercialized and are available in various structures, sizes, and shapes depending on the food to be packed, such as self-adhesive labels, sachets, films, trays, and monolayer or multilayer packaging sheets [7, 8]. They can be used in a wide range of packed food, including oil, nuts, butter, fresh meat and seafood, processed meat, bakery products, fruits, and vegetables.

Even though food spoilage can be induced by several factors and reasons, oxidation stands out as the most common origin. Due to the complicated compositions of food, the generation of decaying factors in food occurs continuously and it is getting even more predominant in contact with the water and oxygen in food surrounding environment. Lipid oxidation is the main mechanism causing food spoilage since food lipids are the most prone to oxidation food components. It results in changes in the appearance, texture, taste, and/or odor of the food as well as on the reduction of its nutritional quality [9, 10]. During the initial phase of lipid oxidation, numerous free radicals, mainly oxo, hydroxyl, and superoxide species are generated [9, 11]. Subsequently, they get actively involved in the deterioration of food through chemical oxidation in which the reactive components in food get oxidized leading to the destruction of food composition and, eventually, its spoilage. Thus, active packaging with the potential to scavenge such free radicals (antioxidant activity) is of great interest. It offers the opportunity to prevent or delay the initiation or the propagation of lipid oxidation, retard the development of off-flavors, and improve the color and sensory stability [4, 11]. Using active antioxidant packaging, the lipid oxidation can be prevented and suspended by eliminating those radicals as soon as they are formed and thus significantly delaying further oxidation. Obviously, the use of antioxidant active packaging also bestows the possibility to eliminate the need of both high-barrier and vacuum packaging materials. To tackle this issue, the initial studies started with a very straightforward and simple approach in which synthetic antioxidants such as butylated hydroxytoluene, butylated hydroxyanisole, and *tert*-butylhydroquinone [8], were integrated into packaging materials. However, there were then the concerns regarding the safety of active compounds from

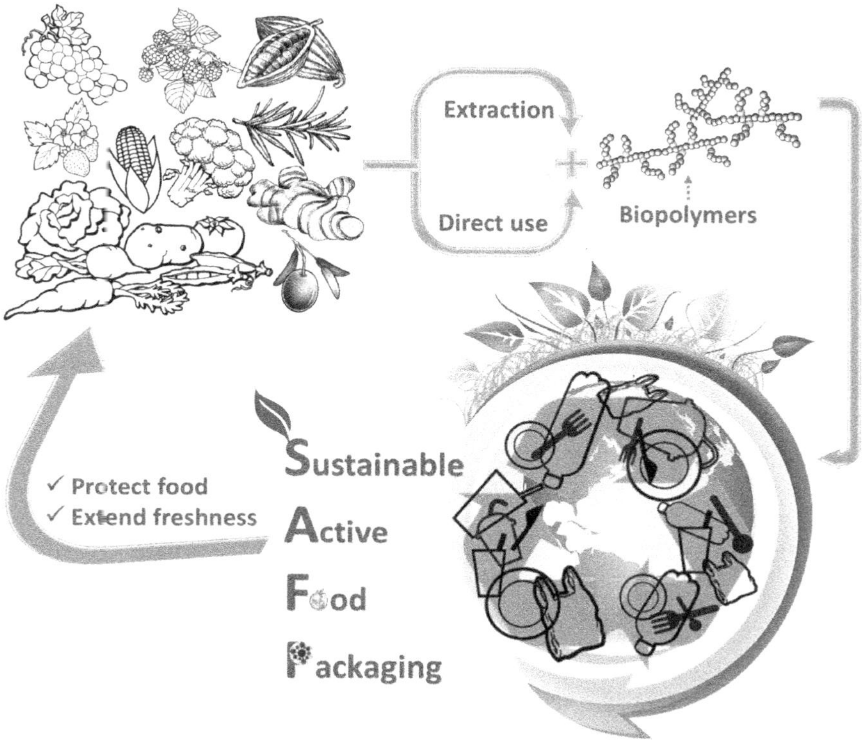

Figure 13.1 Circular economy idea of developing active food packaging from natural antioxidant additives in biopolymers.

synthetic source. In addition, the issues of marketing benefits and especially consumer demands had a great impact on the shift to the use of ecofriendly/natural origin antioxidants in form of either extracts or raw materials. The concept of using active natural components and polymers from natural resources to protect food follows perfectly the idea of circular economy and is illustrated schematically in Figure 13.1. In the last 15 years, this novel concept has drawn the interest both of the scientific community and the economic sectors, and thus, the exploration of natural antioxidants and polymers in active packaging holds a great promise to substitute conventional packaging.

The majority of plastics used for packing materials are synthetic and nonbiodegradable, and are nowadays a notorious problem worldwide. The annual production of plastics worldwide increases rapidly and continuously, namely from 335 million tons in 2016 to 348 million tons in 2017. About 40% of that massive amount of annual plastic production, equally to 139 million tons, is produced for packaging application [12]. Packaging plastics become waste at the end of their service life (usually less than one year, and only weeks or days in case of fresh food packages). Thirty percent of the waste plastic is disposed by landfilling. After their disposal, they cause serious damages on the environment and ecosystem, such as soil and water pollution [13, 14]. Hence, the use of active food packaging not only prolongs the food lifetime but can also reduce the demand on plastic production. Moreover, using renewable biopolymers, instead of petroleum-based, nonbiodegradable plastics, especially for short-term applications like food packaging, can further reduce the damage.

Biopolymers include the proteins and polysaccharides extracted directly from the plants and living organisms, such as zein, cellulose, chitin, chitosan, glycogen, arabinoxylans, pectins, poly(hydroxyl alkanoate)s and hyaluronic acid as well as their derivatives [4, 15–18]. Besides, polymers obtained from the polymerizations of natural monomer derivatives such as lactic acid, glycolic acid, and lactones are also considered as biopolymers [19, 20]. Owing to their natural origin, biopolymers usually offer several advantages over the synthetic ones, including renewability, biodegradability, biocompatibility, recyclability, reusability, and environmental benignness. For these reasons, biopolymers are expected to replace the non-sustainable synthetic ones progressively in packaging applications and in many other sectors of daily life [8, 21]. Therefore, to meet the demand of extending the shelf life of food and secure a sustainable environment for the future, both scientific research and the food industry have been concentrating on the development of new packaging systems that are based on the combination of biopolymers and natural antioxidant substances. Such combination holds a great potential to solve one of the most challenging issues worldwide. This chapter aims at providing an overview on materials developed for sustainable food packaging, consisting of natural plant-based antioxidants and biopolymers, especially their composition, fabrication pathways, mechanical and barrier properties, and antioxidant activity. Furthermore, the future perspectives will be discussed in terms of the performance and sustainability.

13.2 Antioxidant Capacities of Plant-Based Food Packaging Materials

Plants are the most abundant organisms and among the most important components of the ecosystem and life on Earth. In addition, they are also the most predominant source of food for the living systems in general and humans in particular. Apart from their importance of being a vital nutrient source for carbohydrates, minerals, vitamins and water, plants in general, and herbals as well as spices in particular, are also rich in antioxidant, antimicrobial, and anti-inflammatory active compounds that provide huge additional health benefits [22–25]. In fact, from the ancient ages until now, numerous herbal and spice plants have been used as traditional medicines, wound treatments, energy drinks, coloring materials, and food preservatives [26–28]. Nowadays, they are highly desirable components to produce cosmetic and beauty care products for daily use. For active food packaging applications, the antioxidants in plants are of particular interest due to their natural origin, abundancy, and high activity. These antioxidants can be found in all the parts of the plants including their grains, fruits, nuts, seeds, leaves, roots, arils, and barks [25, 29, 30]. To integrate antioxidants derived from natural sources into biopolymer matrices and achieve the so-called active packaging biomaterials, two main fabricating methods have been employed so far. In the first approach, the antioxidants are extracted from the plants using an aqueous-based/organic solvent media and then integrated in the biopolymers using solvent casting technique. The second approach focuses on the use of raw materials derived from the plants as an antioxidant source to be embedded in biopolymer matrices by either solvent casting or melting extrusion technique, which is more relevant to the industry scale.

13.2.1 Antioxidant Natural Extracts in Food Packaging

Generally, the main components of natural plant-based antioxidants are phenolic compounds that consist of tocopherols, flavonoids, and phenolic acid families, with strong hydrogen donating activity, and thus possessing high free radical scavenging activity. The solubility and antioxidant activity of such compounds depend on their chemical structures including the molecular skeleton along with the number and location of free aromatic hydroxyl (–OH) groups [31–33]. These compounds can be extracted from the plants by conventional organic solvents, or aqueous-based media with or without the assistance of microwave and/or high pressure. Solvent systems, temperature, and time are the key parameters that determine the yield of the extraction process and the quality (antioxidant activity) of the extract. Generally, extraction using organic solvents results in a better yield and quality of the extract while using aqueous-based media offers a much lower cost and greener technique. However, higher temperature and longer extraction time are typically needed in this case. To incorporate such active extracts into biopolymers, mostly in the form of thin films, solvent casting represents the most conventional technique. In this step, the extracts and biopolymers are dissolved in the proper solvent or multi-solvents systems in order to obtain homogenous mixtures. Successively, the mixtures are casted into a mold and the free-standing films are collected after solvent evaporation. Following this procedure, a variety of active materials have been prepared and reported to show several interesting features, which are desirable for active food packaging applications. It has been demonstrated that the addition of plants or fruits extracts to biopolymer matrices results in great improvement not only of the antioxidant capacity but also of some physicochemical properties such as mechanical properties, and ultra violet (UV) light and water vapor barrier properties [16, 34, 35].

Tea is one of the most popular beverages in daily life due to its pleasant taste and several health benefits. For a long time, tea has been used as an idealistic anti-aging component and functional food due to its high antioxidant activity. Thanks to a high amount of catechins, theaflavins, thearubigins, and flavonols [27, 36], tea leaves exhibit a very high capability to scavenge free radicals. In particular, in comparison to other extracts, namely grape seed, basil, ginger, gingko leaf, tea extracts exhibit more efficient and faster antioxidant activity [37–39]. Along with the ease to be extracted by the simple and modest hot water immersion, the extract from dried tea leaves represents the most attractive antioxidant source to be applied in the active packaging field. Indeed, several tea extracts including the ones from green, black, and oolong tea have been introduced to various biopolymer matrices, such as chitosan, starch, alginate, and gelatin [15, 23, 37–43]. In general, once being incorporated alone or along with other synthetic antioxidants, tea extracts retain their antioxidant activity and thus confer to the resulting composite films enhanced capability to eradicate the free radicals in their surrounding environment. A slow release of antioxidants from polymer matrices also results in a sustainable and long-term radical scavenging capacity in most of the cases. In addition, the antioxidant activity can be tuned by simply changing the percentage of the extract used in the films. As a proof of concept, Yang et al. demonstrated that biofilms composed of tea extract and grain protein could protect fresh beef and pork during 10 days of storage at 4 °C against lipid oxidation and discoloration [39]. Besides antioxidant activity, the films made of tea extract and several biopolymers such as chitosan, gelatin, potato starch, myofibrillar protein and alginate also show a significant improvement in their mechanical properties in terms of tensile strength which could be assigned to the formation of hydrogen bonding between the active compounds in the extract with the polymer matrices. Nevertheless, it is necessary not to rule out the possibility that the tea extract might play the role of a plasticizer. Given its intrinsic color, tea extract is also used as a natural colorant which is able to promote UV and water vapor barrier properties of potato starch and gelatin films [23, 41]. Particularly, the extract of green tea also helps to improve the antimicrobial activity of chitosan films which is represented by a higher potential for shelf life extension of pork sausages [44].

Apart from tea leaves, the extract of berry family is also another antioxidant source of interest. Berry in general is well-known for its high content of antioxidants. Berry contains a high amount of phenolic acids, anthocyanins, ellagitannins, flavanols, and procyanidins [45] which are among the strongest radical scavenging substances. This makes the berry family not only a precious antiaging food but also an interesting candidate to derive materials for active packaging. The first example regarding this class of antioxidant source is blueberry. It was shown that soy protein isolate films incorporating blueberry extract exhibit enhanced mechanical and barrier properties as well as modest antioxidant activity [18]. Later, it was reported that the incorporation of raspberry extract into soy protein isolate led to a significant improvement in mechanical properties (tensile strength and elongation at break) along with a high antioxidant activity of the resulting films [34]. A similar effect was also observed in the study in which Chilean berry was used as an antioxidant source and methyl cellulose was used as the polymer matrix [46]. The improvement in the physiochemical properties of the resulting composite films can be also explained with a similar mechanism discussed for the case of tea extract. Table 13.1 provides a more detailed information of the materials based on tea and berry extracts.

Besides tea leaves and berries, the extracts obtained from herbs and spices also represent interesting candidates to be incorporated in active packing materials. Herbs and spices have been used by humans since the ancient ages in traditional medicine remedies and in cooking as seasoning. They are now becoming vital ingredients in cosmetic and beauty care products industry [33]. Similar to berry and tea leaves, herbs and spices extracts also show strong antioxidant activity. It is well known that herbs and spices contain a significant amount of phenolic acids such as caffeic, coumaric, ferulic, and neochlorogenic; along with flavonoids

Table 13.1 Active bioplastic films containing tea and berry extracts.

Plant extracts	Extraction condition	Biopolymer matrices	Features	References
Black tea, green tea	Ethanol	Chitosan	↑ Light barrier, ↑ protection against oxidative processes, ↑ moisture content, ↑ tensile strength	[17]
Green tea	n/a	Myofibrillar protein	↑ Resistance against water solubility, ↑ water vapor barrier, ↓ elongation at break, ↑ tensile strength, ↑ thermal stability and moisture content	[48]
Green tea	Water (90 °C)	Chitosan	↑ Mechanical and water vapor barrier properties, ↑ polyphenolic content and antioxidant activity	[44]
Green tea	n/a	Potato starch	↑ Protection against lipid oxidation and ↑ color stability of fresh beef stored at 4 °C and at room temperature for 10 days, ↑ water vapor transmission rate, ↑ elastic modulus, ↓ elongation at break	[23]
Green tea, black tea	n/a	Chitosan	↑ Water vapor barrier, ↑ antioxidant ability	[42]
Green tea	n/a	Alginate	↑ Antioxidant capacity, ↑ water barrier, green tea extract is more efficient than grape seed	[37]
Green, black, and oolong tea	Water (80 °C)	Grain protein	↑ Protection against lipid oxidation of fresh pork during 10 days stored at 4 °C, green tea has greatest antioxidant activity	[39]
Tea	n/a	Gelatin, sodium alginate	↑ Tensile strength, ↑ contact angle, ↓ elongation at break, ↑ water vapor barrier, ↑ antioxidant capacity	[40]
Tea	n/a	Chitosan	↑ Protection against lipid oxidation, ~acceptable sensory quality of pork meat patties stored at 4 ± 1 °C for 12 days	[15]
Blueberry	Ethanol, water	Soy protein isolate	↑ Mechanical and barrier properties, ↑ protection against oxidation and hydrolysis of packaged lard during the storage at 36 °C and relative humidity of 40% for five weeks	[18]
Chilean berry	Ethanol (40 °C)	Methylcellulose	↑ Mechanical properties, ↑ antioxidant activity	[46]

n/a: not available.

including quercetin, luteolin, apigenin, kaempferol, and isorhamnetin [29, 32]. This class of compounds, having unique chemical structures, exhibits several interesting biological activities, not only radical scavenging capability but also antimicrobial, antibacterial and antiseptic properties [29, 33]. Therefore, they also draw a great deal of attention for active food packaging application as outlined in Table 13.2. Among those, rosemary represents the most extensively studied probably due to its abundancy and popularity. Either water or ethanol extract of rosemary was used as active substance to integrate the antioxidant activity to gelatin (from bovine-hide, pork, and tuna skin), chitosan, protein (from whey) and cassava starch films [10, 50–52]. In addition, an improvement in mechanical properties in terms of tensile strength, elongation at break, and general flexibility was achieved in these composite films after the addition of the rosemary extract. Furthermore, the films made of cassava starch incorporating rosemary extract absorb more UV light than the neat ones, thus providing a much better UV barrier performance [50]. Outstandingly, in a practical model, the composite films composed of classical gelatin and rosemary extract were able to suppress the lipid oxidation and the growth of microbial organisms significantly of smoked sardines, as demonstrated by Gomez-Estaca et al. [10]. In the same study, it was also shown that the oregano extract offers a similar performance in comparison to the rosemary extract. Other numerous studies also focused on the utilization of different herbs and spice sources such as cinnamon, guarana, boldo-do-Chile, sage, ginseng, ginger, clove, basil, thyme as well as leaves of kenal, betel, and gingko [26, 47, 49, 53]. In these cases, several biopolymers have been employed as matrices, including gelatin from bovine-hide, tuna and pig skin [51], whey protein [52], polysaccharides such as starch from cassava, sago, and pea [17, 50, 54], alginate, [54] and chitosan [47]. These composite materials further verify the benefit of incorporating plant extracts in biopolymer films. In all cases, a remarkable increase in radical scavenging capability was observed and coupled with an enhancement of mechanical properties thanks to intermolecular hydrogen bonding network and/or plasticizer effect.

Other interesting sources to provide active antioxidant extracts are the plant-based food waste and by-products of food industry such as fruit seeds (i.e. from grape, grapefruit, and mango) and peel (i.e. from pomegranate) [55–58]. Like other edible parts of a plant such as leaves and fruits, these nonedible components are also rich in many biologically active substances in general and antioxidants in particular. Therefore, they have been also involved in the development of functional composite materials that show very good antioxidant capabilities. More details on this type of active materials are summarized in Table 13.3. Among the food wastes and by-products of food industry, fruit seeds represent a focal group for active packaging materials. They contain a high amount of phenolic compounds such as catechin, epicatechin, epicatechin gallate, quercetin, and gallic acid, which exhibit a strong capability to scavenge free radicals [31, 61]. Additionally, a significant amount of procyanidin, lutein, zeaxanthin, cryptoxanthin, β-carotene, and α-tocopherol is also identified in their composition [62]. Thanks to a unique interaction, the incorporation of fruit seed extract improves the mechanical properties of certain biopolymers films, such as gelatin, soy protein isolate, myofibrillar, and barley bran protein

Table 13.2 Different bioplastic films incorporated with extractions from herbs.

Plant extracts	Extraction condition	Biopolymer matrices	Features	References
Cinnamon, guarana, rosemary, boldo-do-chile	Ethanol	Pig skin gelatin and chitosan	Significantly ↑ antioxidant activity, ↑ elongation at break	[26]
Ginger, rosemary, sage, black tea, green tea, and kenaf leaves	Ethanol	Chitosan	↑ Light barrier, ↑ protection against oxidative processes, ↑ moisture content, ↑ tensile strength	[47]
Green tea and basil	Water (100 °C)	Cassava starch	Significant ↑ antioxidant activity, green tea is better than basil,↑ degradation in soil (less than two weeks), ↑ thermal stable up to 240 °C, ↑ water vapor barrier	[38]
Clove	n/a	Ethyl cellulose	↑ Flexibility and ↑ antioxidant capacity	[49]
Oregano, rosemary	Water (45 °C)	Gelatin	↑ Protection against lipid oxidation and microbial growth of smoked sardine, ↑ phenol content	[10]
Rosemary	Water (50 °C)	Cassava starch	↑ Antioxidant activity, ↑ UV light barrier, ↑ biodegradation after 14 days of composting	[50]
Rosemary, oregano	Water (45 °C)	Bovine-hide and tuna-skin gelatins	↑ Glass transition temperature, ↓ deformability, ↑ water solubility, ↑ opacity	[51]
Rosemary, thyme	Ethanol, methanol, acetone	Whey protein	↑ Flexibility, ↑ antioxidant activity	[52]
Thyme	Water (100 °C)	Chitosan, pea starch	Remarkable ↑ antioxidant activity, ↑ tensile behavior, ↑ opacity	[17]

n/a: not available.

Table 13.3 Active packaging compositions and the effect of natural additives on mechanical, barrier and other properties.

Plant extracts	Extraction condition	Biopolymer matrices	Features	References
Grape fruit seed	n/a	Agar	↑ Color, ↑ UV barrier, ↑ moisture content, ↑ water solubility and water vapor permeability, ↑ surface hydrophilicity, ↑ flexibility	[56]
Grape fruit seed	n/a	Chitosan	↑ Amorphous, ↓ tensile strength, ↑ elongation at break, ↑ protection of packed bread	[24]
Grape seed	n/a	Chitosan	↑ Water vapor transmission rate, ↑ surface wettability, ↑ total phenol content and antioxidant activity, ↓ tensile strength and elongation at break	[57]
Grape seed	Water	Soy protein isolate	↑ Tensile strength, ↑ color darkness	[59]
Grape seed and green tea	n/a	Myofibrillar protein	↓ Water solubility, ↑ water vapor barrier, ↑ tensile strength, ↑ thermal stability and moisture content	[48]
Grapefruit seed	n/a	Barley bran protein and gelatin	↑ Antioxidant activity, ↑ freshness of packed salmon after 15 days of storage	[60]
Grapefruit seed, green tea	n/a	Gelatin	↑ Tensile strength, ↑ water vapor permeability, ↑ quality of packed pork loins during storage	[30]
Green tea, grape seed	n/a	Alginate	↑ Antioxidant capacity, ↑ water barrier	[37]
Green tea, grape seed, ginger, gingko leaf	n/a	Gelatin	↑ Barrier to UV light and moisture, excellently ↑ antioxidant activity	[41]
Mango kernel	Ethanol	Soy protein isolate, fish gelatin	↑ Tensile strength, ↑ water vapor permeability, ↑ antioxidant activity	[55]
Pomegranate peel	Ethanol (40 °C)	Chitosan	Greatly ↑ UV light barrier, ↑ protection against moisture, ↓ water vapor permeability,↓ tensile strength and elongation at break, ↑ total phenol content and antioxidant activity	[35]
Propolis	Ethanol (50 °C)	Chitosan	↑ Tensile strength, ↑ elongation at break, ↑ total phenolic content, ↑ antioxidant activity, ↑ water vapor and oxygen barrier properties	[16]

n/a: not available.

[30, 48, 59, 60], while it deteriorates these properties of polysaccharides, such as agar and chitosan, once being integrated [56, 57]. A similar behavior was observed for chitosan films when pomegranate peel extract was used instead of fruit seed extract [58]. The unique extract which enables the improvement of tensile strength along with elongation at break of the resulting film is the one from propolis [16]. Due to its hydrophobicity, this extract also provides resulting films with better barrier properties against water vapor and oxygen. Interestingly, several proofs of concept have been reported regarding the capability of biopolymers combined with grapefruit seed extract to protect the food. Indeed, Tan et al. showed that active chitosan films with grape fruit seed extract help to prolong the shelf life of packed bread [24]. Meanwhile, Hong et al. and Song et al. demonstrated that the incorporation of grapefruit seed extract enables barley bran protein and gelatin films to prolong the freshness of packed salmon and improve the quality of packed pork loins during storage, respectively [30, 60].

13.2.2 Antioxidant Raw Materials Derived from Food Wastes and Agro-Industry by-Products

Packaging films made of biopolymers and different natural plant extracts have shown a great potential for use in the active food packaging field. Nevertheless, extraction of antioxidants from plant-based materials are associated with high solvent consumption, long extraction time, and a higher risk of the decay of antioxidant components due to the thermal degradation or hydrolysis during the extraction. These issues might lead to several problems during processing steps and thus a much higher cost in comparison to the conventional packaging materials could be anticipated. Alternatively, the use of raw dried plants powders (either the edible food wastes or agro-industry by-products) as fillers to increase the mechanical properties and confer sustainable antioxidant activity is also of great interest. Following this strategy, neither the extraction process nor the use of excess solvents is needed, thus resulting in the reduction of production cost and time. Indeed, the cost for the use of dried plant waste powders as antioxidant source is minimized. Additionally, such approach definitely enables their valorization along with a minimization of food industry wastes, and therefore, this makes them outstanding natural sources of antioxidants for active food packaging both from economical and sustainable points of view. In fact, the fabrication of active food packaging materials employing this strategy emerges to be a very active research field and has drawn an increasing interest of the scientific community in recent years. In this procedure, the raw plant materials are collected, dried, and finally grounded and/or micronized to obtain powders with difference size ranges [13, 25]. The as-prepared plant powders are incorporated into either biopolymer matrices or biodegradable synthetic polymers/elastomers via solvent casting process or industrial scalable extrusion technique.

Industrial food wastes such as vegetable peels, seeds, branches, and shells are practically discarded or a small part of them can be used as animal food or fertilizer [22, 63]. However, they own abundant quantity of strong antioxidant compounds that are useful for active food packaging. Among them, for instance, red beetroot, a common vegetable for cooking in daily life, contains a high amount of active compounds with several desirable biological activities. In particular, betanin present in red beetroot demonstrates exceptional antioxidant ability, strong free radical scavenging activity and effective antitumor property [4]. Inspired by these features, Tran et al. used red beetroot peel waste powder as a natural and low-cost antioxidant source to functionalize biodegradable starch-based bio-elastomers in order to obtain active biofilms [4]. The antioxidant bio-elastomers were produced by a simple solvent casting process. The release of the antioxidant molecule, predominantly betanin, was performed through diffusion process rather than polymer matrix disintegration and could be controlled by varying the amount of starch granules within the bio-elastomers. The released betanin retained its antioxidant activity, demonstrated by a high capability of scavenging free radicals. These bio-elastomers showed good hydrophobicity and resistance to disintegration in water. Fascinatingly, prolonged underwater immersion did not impair their hydrophobicity. The incorporation of red beetroot powder into starch-based bio-elastomers helped to increase the Young's modulus without compromising their elongation ability.

Another interesting antioxidant food waste comes from the chocolate industry. Today, due to continuously increasing consumption of chocolate and cacao-related products worldwide, the amount of cocoa bean shell waste generated from chocolate industry is also increasing and estimated at around 0.5 million tons per year [64]. Cocoa shell waste contains various valuable substances including lipids, lignin, and polysaccharides as well as polyphenols and other antioxidants [64, 65]. Cocoa shell is reported to effectively protect fatty foods against oxidative rancidity [64]. Taking advantage of the high antioxidant activity of cocoa shell waste, a novel active bio-elastomer with a capability of incorporating a significant amount of cocoa shell waste (up to 75 wt%) was developed by Tran et al. The micronized cocoa beans shell waste was directly incorporated into poly(dimethylsiloxane) matrix to achieve active bio-elastomers with antioxidant and several desirable properties. Cocoa bio-elastomers showed an enhanced Young's modulus, good hydrophobicity, and water barrier ability as well as a high potential to eradicate fee radical [25]. In another approach, cocoa beans shell waste has been also compounded with biodegradable poly(ε-caprolactone) using extrusion, a solvent-free and industrial scalable process, to produce active bio-filaments for 3D printing application [20]. It was shown that micronized cocoa shell waste can be incorporated up to 50 wt% without a significant alteration of the Young's modulus of the resulting bio-filaments with respect to the pure biopolymer. The developed cocoa bio-filaments were successfully employed for 3D printing using fused deposition modeling technique at a lower temperature compared to the one used for commercially available filaments, thus resulting in a lower energy consumption. The printed specimens displayed a well-defined structure with good adhesion between layers and good resolution. Such cocoa bio-filaments hold a great potential for use in the production of various active food packaging systems with different design, size, and shape employing 3D printing technique and/or other melting processes such as hot-press and injection molding.

Along with utilizing agro-industry by-products as fillers to fabricate active packaging materials, a direct transformation of edible agro-waste to active biofilms via the spontaneous dissolution of biopolymers and antioxidants also represents an interesting approach. This strategy relies on the fact that several biopolymers and active compounds are soluble in acidic solvents or media, for instance trifluoroacetic and diluted hydrochloric acids. Generally, the vegetable waste powders are dissolved in an acidic medium and then casted in a mold. After the evaporation of the solvent, freestanding bioactive plastic films are collected. Using trifluoroacetic acid solutions, 100% natural amorphous cellulose-based bioplastics from a wide range of industrially processed edible vegetable and cereal wastes including parsley, spinach stems, rice hulls, and cocoa pod husks were obtained [14, 22]. The antioxidant substances present in these plants offer the resulting bioplastics good biocompatible, biodegradable characteristics and exceptional thermo-physical properties, which are comparable with petroleum-based nonbiodegradable plastics. Trifluoroacetic acid was also exploited to dissolve red, green, and brown seaweeds to obtain marine-source-based biofilms [28]. Interestingly, the antioxidant activity of the biofilms increased significantly in comparison to the starting materials thanks to an efficient extraction process of their intrinsic antioxidants in trifluoroacetic acid. Furthermore, the active biocomposite films showed high biocompatibility and were fully biodegradable in seawater after one month.

Toward a more environmentally friendly fabrication pathway, Perotto et al. used diluted aqueous HCl to dissolve carrot, parsley, radicchio, and cauliflower vegetable wastes at room temperature [13]. Freestanding, flexible, and biodegradable bioplastic films were obtained by solvent casting method. It was also shown that the antioxidant capability and color of original vegetables in the bioplastics were preserved thanks to a mild, simple, and versatile extraction step. Given their flexibility and antioxidant activity, the films prepared following this technique are very promising for active food packaging applications.

13.3 Conclusions and Future Perspectives

Food demand and plastic waste are now considered as two of the most challenging issues for human beings to pursue a sustainable future. Despite the fact several solutions have been considered to tackle them, the application of bio-based renewable active materials in general and in particular antioxidant ones showing the capability to prolong food shelf life stands out as a flagship approach. Therefore, fabricating low-cost renewable materials that are able to prevent the food from being quickly spoilt and are completely biodegraded once being discarded in the environment without generating any toxic substances has become a new paradigm in material research. Indeed, numerous new generations of active packaging have been developed to meet these demands of consumers, environment, and food industry. A large number of scientific researches on the development of active packaging materials have realized the potential, effectiveness, and benefit of active packages in the view of economy and sustainability. Due to their abundancy and renewability, much attention has been focused on the antioxidant sources based on plants, either their extract or raw powder. The polymers used for the fabrication of such active packaging materials include natural ones such as polysaccharides and proteins as well as biodegradable synthetic ones, for instance poly(ε-caprolactone) and crosslinked bio-elastomers. Using either solvent casting or extrusion, several bio-based materials have been prepared and found to have an outstanding antioxidant activity, represented by the capability to inhibit and delay lipid oxidation and discoloration of food resulting in the extension of food shelf life. In addition, they also exhibit good physical properties which are comparable to the conventional packaging materials. Hence, the advantage of the utilization and valorization of natural antioxidant additives via their integration into the so-called active bio-packaging is visible. In addition, having a natural origin, such active packaging materials might quickly obtain the authorization from either the European Commission or the U.S.A. Food Drug Administration (FDA) in comparison to the synthetic-based ones.

Nowadays, despite the intense research as well as the increasing industrial manufacturing capacity worldwide, few natural antioxidant films or packaging are commercially available. Several issues remain elusive before the applications of natural active food packaging in daily life substitute the conventional ones. One of those is an intrinsic problem originating from the physiochemical properties of plant extracts or dried powder. Since they contain a significant amount of natural antioxidants, plant extracts and raw dried powders feature some distinguished colors which could interfere with the colors of the packed food leading to the misunderstanding of the customer. In addition, this coloration might be also problematic for the visual monitoring of the spoilage of the food. Furthermore, the chemical composition of the natural extracts and plants dried powders can vary remarkably depending on the plant characteristics, the climate, and the processing procedures. This leads to differences in antioxidant activity among the batches, thus resulting in a drastic variation of the quality of packaging materials. This also gives rise to the difficulty in comparison between various materials due to the use of different extracts, plants, packed food, and characterization techniques. Therefore, there will be a need for a standardized criterion in terms of free radical scavenging capability to have a better comparison and normalization between different types of antioxidant materials. In addition to that, the polymers of choice for preparing such active packaging materials also draw considerable attention. In spite of their natural origin, biocompatibility, and biodegradability, biopolymers such as polysaccharides and proteins suffer a limitation relating to their processing techniques. Due to poor thermal processability of natural biopolymers, in most of the cases, solvent casting technique has to be used to fabricate the active packaging films which is not desirable for industry scale, environmentally friendly and cost-effective demand. In this regard, the exploration of new biopolymer materials which can be incorporated into industry-relevant fabrication methodology such as melt extrusion or injection molding techniques is on a high demand. This problem might be addressed via either the improvement of already existing biopolymers or the incorporation of a sustainable amount of synthetic biodegradable polymers as co-matrices.

Another issue arises from possible interactions between active packaging and food in contact. Apart from the capability to eliminate malignant free radicals, active packaging materials might also interact with packed food in an uncontrolled manner. Therefore, several aspects regarding this issue have to be also addressed. In future works, there will be a need to (i) investigate thoroughly the migration of materials from the packaging films into the food, (ii) evaluate the effect of released substances on the color, flavor, and texture of packed food. On the other hand, a high amount of free radicals is generated during food processing, transportation, and storage. Hence, it is important to develop accurate and very sensitive analytical methods to quantify the generated amount of free radicals for each type of food in order to determine a needed amount of antioxidants to be integrated.

Once the mentioned above issues are sufficiently addressed, active packaging materials based on renewable sources can be quickly introduced into the daily life to replace the conventional ones in near future. Given their outstanding features and great potential, along with other strategies to achieve the so-called sustainable packaging technology, active food packaging materials will play an important role on the roadmap toward a sustainable future.

References

1 United Nations (2017). World Population Prospects: The 2017 Revision, Key Findings and Advance Tables. *ESA/P/WP/248*.

2 United Nations (2018). The Sustainable Development Goals Report. ISBN: 978-92-1-101390-0.

3 Willett, W., Rockström, J., Loken, B. et al. (2019). Food in the Anthropocene: the EAT–lancet commission on healthy diets from sustainable food systems. *The Lancet* 393 (10170): 447–492.

4 Tran, T.N., Athanassiou, A., Basit, A. et al. (2017a). Starch-based bio-elastomers functionalized with red beetroot natural antioxidant. *Food Chemistry* 216: 324–333.

5 European Commission (2009). *Commission regulation (EC) No. 450/2009 on active and intelligent materials and articles intended to come into contact with food. Off J Eur Comm L.* 135: 3–11.

6 European Commission (2004). *Regulation (EC) No. 1935/2004 of the European Parliament and of the council of 27 October 2004, on materials and articles intended to come into contact with food and repealing directives 80/509/EEC and 89/109/EEC. Official Journal of the European Union L.* 338: 4–17.

7 Pereira de Abreu, D., Cruz, J.M., Paseiro Losada, P. et al. (2012). Active and intelligent packaging for the food industry. *Food Reviews International* 28 (2): 146–187.

8 Vilela, C., Kurek, M., Hayouka, Z. et al. (2018). A concise guide to active agents for active food packaging. *Trends in Food Science & Technology* 80: 212–222.

9 Sanches-Silva, A., Costa, D., Albuquerque, T.G. et al. (2014). Trends in the use of natural antioxidants in active food packaging: a review. *Food Additives and Contaminants: Part A*: 374–395.

10 Gómez-Estaca, J., Montero, P., Giménez, B. et al. (2007). Effect of functional edible films and high pressure processing on microbial and oxidative spoilage in cold-smoked sardine (Sardina pilchardus). *Food Chemistry* 105 (2): 511–520.

11 Fang, Z., Zhao, Y., Warner, R.D. et al. (2017). Active and intelligent packaging in meat industry. *Trends in Food Science & Technology* 61: 60–71.

12 Plastics Europe (2018). Plastics – the Facts 2018.

13 Perotto, G., Ceseracciu, L., Simonutti, R. et al. (2018). Bioplastics from vegetable waste via an eco-friendly water-based process. *Green Chemistry* 20 (4): 894–902.

14 Tran, T.N., Paul, U., Heredia-Guerrero, J.A. et al. (2016). Transparent and flexible amorphous cellulose-acrylic hybrids. *Chemical Engineering Journal* 287: 196–204.

15 Qin, Y.-Y., Yang, J.-Y., Lu, H.-B. et al. (2013). Effect of chitosan film incorporated with tea polyphenol on quality and shelf life of pork meat patties. *International Journal of Biological Macromolecules* 61: 312–316.

16 Siripatrawan, U. and Vitchayakitti, W. (2016). Improving functional properties of chitosan films as active food packaging by incorporating with propolis. *Food Hydrocolloids* 61: 695–702.

17 Talón, E., Trifkovic, K.T., Vargas, M. et al. (2017). Release of polyphenols from starch-chitosan based films containing thyme extract. *Carbohydrate Polymers* 175: 122–130.

18 Zhang, C., Guo, K., Ma, Y. et al. (2010). Incorporations of blueberry extracts into soybean-protein-isolate film preserve qualities of packaged lard. *International Journal of Food Science & Technology* 45 (9): 1801–1806.

19 Battegazzore, D., Bocchini, S., Alongi, J. et al. (2014). Plasticizers, antioxidants and reinforcement fillers from hazelnut skin and cocoa by-products: extraction and use in PLA and PP. *Polymer Degradation and Stability* 108: 297–306.

20 Tran, T.N., Bayer, I.S., Heredia-Guerrero, J.A. et al. (2017b). Cocoa shell waste biofilaments for 3d printing applications. *Macromolecular Materials and Engineering* 302 (11): 1700219.

21 Samsudin, H., Soto-Valdez, H., Auras, R. et al. (2014). Poly(lactic acid) film incorporated with marigold flower extract (Tagetes erecta) intended for fatty-food application. *Food Control* 46: 55–66.

22 Bayer, I.S., Guzman-Puyol, S., Heredia-Guerrero, J.A. et al. (2014). Direct transformation of edible vegetable waste into bioplastics. *Macromolecules* 47 (15): 5135–5143.

23 Nisa, I., Ashwar, B.A., Shah, A. et al. (2015). Development of potato starch based active packaging films loaded with antioxidants and its effect on shelf life of beef. *Journal of Food Science and Technology* 52 (11): 7245–7253.

24 Tan, Y., Lim, S., Tay, B. et al. (2015). Functional chitosan-based grapefruit seed extract composite films for applications in food packaging technology. *Materials Research Bulletin* 69: 142–146.

25 Tran, T.N., Heredia-Guerrero, J.A., Mai, B.T. et al. (2017c). Bioelastomers based on cocoa shell waste with antioxidant ability. *Advanced Sustainable Systems* 1 (7): 1700002.

26 Bonilla, J. and Sobral, P.J. (2016). Investigation of the physicochemical, antimicrobial and antioxidant properties of gelatin-chitosan edible film mixed with plant ethanolic extracts. *Food Bioscience* 16: 17–25.

27 Cao, G., Sofic, E., Prior, R.L. et al. (1996). Antioxidant capacity of tea and common vegetables. *Journal of Agricultural and Food Chemistry* 44 (11): 3426–3431.

28 Guzman-Puyol, S., Russo, D., Penna, I. et al. (2017). Facile production of seaweed-based biomaterials with antioxidant and anti-inflammatory activities. *Algal Research* 27: 1–11.

29 Hinneburg, I., Dorman, H.D., Hiltunen, R. et al. (2006). Antioxidant activities of extracts from selected culinary herbs and spices. *Food Chemistry* 97 (1): 122–129.

30 Hong, Y.H., Lim, G.O., Song, K. et al. (2009). Physical properties of Gelidium corneum–gelatin blend films containing grapefruit seed extract or green tea extract and its application in the packaging of pork loins. *Journal of Food Science* 74 (1): C6–C10.

31 Parry, J., Su, L., Moore, J. et al. (2006). Chemical compositions, antioxidant capacities, and antiproliferative activities of selected fruit seed flours. *Journal of Agricultural and Food Chemistry* 54 (11): 3773–3778.

32 Wojdyło, A., Oszmiański, J., Czemerys, R. et al. (2007). Antioxidant activity and phenolic compounds in 32 selected herbs. *Food Chemistry* 105 (3): 940–949.

33 Yanishlieva, N.V.M., Emma, Pokorný, J. et al. (2006). Natural antioxidants from herbs and spices. *European Journal of Lipid Science and Technology* 108 (9): 776–793.

34 Wang, S., Marcone, M., Barbut, S. et al. (2012). The impact of anthocyanin-rich red raspberry extract (ARRE) on the properties of edible soy protein isolate (SPI) films. *Journal of Food Science* 77 (4): C497–C505.

35 Yuan, G., Lv, H., Yang, B. et al. (2015). Physical properties, antioxidant and antimicrobial activity of chitosan films containing carvacrol and pomegranate peel extract. *Molecules* 20 (6): 11034–11045.

36 Higdon, J.V., Balz, F. et al. (2003). Tea catechins and polyphenols: health effects, metabolism, and antioxidant functions. *Critical Reviews in Food Science and Nutrition* 43 (1): 89–143.

37 Fabra, M.J., Falcó, I., Randazzo, W. et al. (2018). Antiviral and antioxidant properties of active alginate edible films containing phenolic extracts. *Food Hydrocolloids* 81: 96–103.

38 Medina-Jaramillo, C., Ochoa-Yepes, O., Bernal, C. et al. (2017). Active and smart biodegradable packaging based on starch and natural extracts. *Carbohydrate Polymers* 176: 187–194.

39 Yang, H.-J., Lee, J.-H., Won, M. et al. (2016). Antioxidant activities of distiller dried grains with solubles as protein films containing tea extracts and their application in the packaging of pork meat. *Food Chemistry* 196: 174–179.

40 Dou, L., Li, B., Zhang, K. et al. (2018). Physical properties and antioxidant activity of gelatin-sodium alginate edible films with tea polyphenols. *International Journal of Biological Macromolecules* 118: 1377–1383.

41 Li, J.-H., Miao, J., Wu, J.-L. et al. (2014). Preparation and characterization of active gelatin-based films incorporated with natural antioxidants. *Food Hydrocolloids* 37: 166–173.

42 Peng, Y., Wu, Y., Li, Y. et al. (2013). Development of tea extracts and chitosan composite films for active packaging materials. *International Journal of Biological Macromolecules* 59: 282–289.

43 Siripatrawan, U. and Harte, B.R. (2010). Physical properties and antioxidant activity of an active film from chitosan incorporated with green tea extract. *Food Hydrocolloids* 24 (8): 770–775.

44 Siripatrawan, U., Suparat, N. et al. (2012). Active film from chitosan incorporating green tea extract for shelf life extension of pork sausages. *Food Hydrocolloids* 27 (1): 102–108.

45 Kähkönen, M.P., Hopia, A.I., Heinonen, M. et al. (2001). Berry phenolics and their antioxidant activity. *Journal of Agricultural and Food Chemistry* 49 (8): 4076–4082.

46 Dicastillo, C.L., Bustos, F., Guarda, A. et al. (2016). Cross-linked methyl cellulose films with murta fruit extract for antioxidant and antimicrobial active food packaging. *Food Hydrocolloids* 60: 335–344.

47 Souza, V.G.L., Fernando, A.L., Pires, J.R.A. et al. (2017). Physical properties of chitosan films incorporated with natural antioxidants. *Industrial Crops and Products* 107: 565–572.

48 Nie, X., Gong, Y., Wang, N. et al. (2015). Preparation and characterization of edible myofibrillar protein-based film incorporated with grape seed procyanidins and green tea polyphenol. *LWT - Food Science and Technology* 64 (2): 1042–1046.

49 Heredia-Guerrero, J.A., Ceseracciu, L., Guzman-Puyol, S. et al. (2018). Antimicrobial, antioxidant, and waterproof rtv silicone-ethyl cellulose composites containing clove essential oil. *Carbohydrate Polymers* 192: 150–158.

50 Piñeros-Hernandez, D., Medina-Jaramillo, C., López-Córdoba, A. et al. (2017). Edible cassava starch films carrying rosemary antioxidant extracts for potential use as active food packaging. *Food Hydrocolloids* 63: 488–495.

51 Gómez-Estaca, J., Montero, P., Fernández-Martín, F. et al. (2009). Physical and chemical properties of tuna-skin and bovine-hide gelatin films with added aqueous oregano and rosemary extracts. *Food Hydrocolloids* 23 (5): 1334–1341.

52 Andrade, M.A., Ribeiro-Santos, R., Bonito, M.C.C. et al. (2018). Characterization of rosemary and thyme extracts for incorporation into a whey protein based film. *LWT* 92: 497–508.

53 Norajit, K., Kim, K.M., Ryu, G.H. et al. (2010). Comparative studies on the characterization and antioxidant properties of biodegradable alginate films containing ginseng extract. *Journal of Food Engineering* 98 (3): 377–384.

54 Nouri, L. and Nafchi, A.M. (2014). Antibacterial, mechanical, and barrier properties of sago starch film incorporated with betel leaves extract. *International Journal of Biological Macromolecules* 66: 254–259.

55 Adilah, Z.M., Jamilah, B., Hanani, Z.N. et al. (2018). Functional and antioxidant properties of protein-based films incorporated with mango kernel extract for active packaging. *Food Hydrocolloids* 74: 207–218.

56 Kanmani, P. and Rhim, J.-W. (2014). Antimicrobial and physical-mechanical properties of agar-based films incorporated with grapefruit seed extract. *Carbohydrate Polymers* 102: 708–716.

57 Moradi, M., Tajik, H., Rohani, S.M.R. et al. (2012). Characterization of antioxidant chitosan film incorporated with Zataria multiflora Boiss essential oil and grape seed extract. *LWT - Food Science and Technology* 46 (2): 477–484.

58 Yuan, G., Lv, H., Tang, W. et al. (2016). Effect of chitosan coating combined with pomegranate peel extract on the quality of Pacific white shrimp during iced storage. *Food Control* 59: 818–823.

59 Sivarooban, T., Hettiarachchy, N., Johnson, M. et al. (2008). Physical and antimicrobial properties of grape seed extract, nisin, and EDTA incorporated soy protein edible films. *Food Research International* 41 (8): 781–785.

60 Song, H.Y., Shin, Y.J., Song, K.B. et al. (2012). Preparation of a barley bran protein–gelatin composite film containing grapefruit seed extract and its application in salmon packaging. *Journal of Food Engineering* 113 (4): 541–547.

61 Lutterodt, H., Slavin, M., Whent, M. et al. (2011). Fatty acid composition, oxidative stability, antioxidant and antiproliferative properties of selected cold-pressed grape seed oils and flours. *Food Chemistry* 128 (2): 391–399.

62 Kennedy, J.A., Matthews, M.A., Waterhouse, A.L. et al. (2000). Changes in grape seed polyphenols during fruit ripening. *Phytochemistry* 55 (1): 77–85.

63 Moraes Crizel, T., de Oliveira Rios, A., Alves, V.D. et al. (2018). Active food packaging prepared with chitosan and olive pomace. *Food Hydrocolloids* 74: 139–150.

64 Sarriá, B., Mateos, R., Sierra-Cinos, J.L. et al. (2012). Hypotensive, hypoglycaemic and antioxidant effects of consuming a cocoa product in moderately hypercholesterolemic humans. *Food & Function* 3 (8): 867–874.

65 Sarriá, B., Martínez-López, S., Sierra-Cinos, J.L. et al. (2015). Effects of bioactive constituents in functional cocoa products on cardiovascular health in humans. *Food Chemistry* 174: 214–218.

14

Natura and Biocompatible Optical Indicators for Food Spoilage Detection

Maria E. Genovese, Jasim Zia, and Despina Fragouli

Smart Materials, Istituto Italiano di Tecnologia, Via Morego 30, 16163, Genova, Italy

The quality of food products is subject to changes during their cycle since foods are perishable by nature. The properties and the related intrinsic quality attributes of food commodities are susceptible to various spoilage processes and quality defects due to different mechanisms, which depend on the type of food product and on its processing, distribution, storage, and handling conditions. Moreover, agriculture products are often heterogeneous with respect to diverse quality parameters, due to many variables such as seasonal and regional differences. On the other hand, over the last few years, consumers have requested more convenient, fresh, and organic food products and, at the same time, stricter regulations related to full traceability, the adoption of higher quality standards, and the prediction of food shelf life have been imposed on producers and suppliers by regulatory agencies. This complexity necessitates a capillary, reliable, and real-time quality control of food products applicable on a large scale throughout their entire processing and distribution cycle. Smart packaging and portable sensing technologies can be used as a tool to monitor the established quality requirements as well as to provide information on the food state and freshness in supply chain management and to customers and final consumers. These technologies rely on the interaction between functional components that can be embedded in the packaging material and specific by-products originating from deterioration reactions in the food that give rise to a change in one or more measurable physicochemical properties (e.g. electric conductivity, color, spectral absorption, fluorescence) of the sensing material. In particular, the use of natural or biocompatible optical and colorimetric sensing platforms constitutes the most effective approach for the detection of food spoilage processes, as it enables the real-time wireless communication of the output data in a user-friendly manner through a spectral/color analysis and direct naked eye visualization and, at the same time, it allows a safe on-site (on-package) integration eliminating the toxicity risks associated to the contact of the sensing device with food. In this way, the criteria for more sustainable and eco-friendly smart packaging solutions can be achieved.

In this chapter, the most recent challenges and technological solutions addressing the optical and colorimetric detection of food spoilage through biocompatible

Sustainable Food Packaging Technology, First Edition. Edited by Athanassia Athanassiou.

and natural indicators will be discussed. In particular, in Section 14.1, a preliminary introduction on food spoilage processes with regard to the challenges, key aspects, and factors that determine them will be presented. In Section 14.2, the traditional and most recent technologies available for the detection of food spoilage and the main advantages and limitations of each type will be outlined. In Section 14.3, relevant chemically active and stimuli-responsive indicators based on natural and biocompatible materials will be described, and will be categorized by the type of the analytes detected. In Section 14.4, the main conclusions, challenges, and future trends in the research and development of sustainable optical and colorimetric indicators for food spoilage detection will be briefly summarized.

14.1 Food Spoilage

14.1.1 Food Spoilage: A Never-ending Challenge

Many food products are highly perishable and subject to deterioration due to the inherent susceptibility of their natural biomacromolecules to microbial and physicochemical degradation. Despite the availability of both conventional and more recently developed methods for food preservation such as salting, curing, refrigeration, freezing, freeze-drying, heat treatments, ultrafiltration, use of preservatives, irradiation, and high hydrostatic pressure, food spoilage remains inevitable, constituting a serious financial burden and raising safety concerns. Food spoilage refers to an ensemble of complex and often interconnected intrinsic and extrinsic factors associated with microbial, chemical, and physical processes that render a food product undesirable or unacceptable for human consumption. As shown in Figure 14.1, the manifestations of food spoilage can vary and may become evident through a sensory assessment of the food appearance (discoloration, slime production, colony formation, breakdown of structure, blowing of container), smell (off-odors), and taste (off-flavors) [1].

14.1.2 Microbial Spoilage

Microbial activity is the most impactful cause of food spoilage in terms of financial losses and health concerns. It has been estimated that microbial spoilage alone accounts for the loss of a quarter of the global food supply, reaching 50% of losses for non-grain staples, fruits, and vegetables [1, 2]. As reported by Schaefer et al. [3], it has been found that more than 6 million cases of foodborne diseases occur every year in the United States and 60 cases every 100.000 inhabitants in Spain.

The factors affecting microbial spoilage can be classified into extrinsic, intrinsic, implicit, and modes of processing and preservation. Intrinsic parameters are related to inherent chemical, structural, and physical properties of the food itself such as pH, chemical composition, redox potential, natural antimicrobial substances, nutrients, and water content. Extrinsic parameters refer to external conditions in which food is stored and include temperature, humidity, and

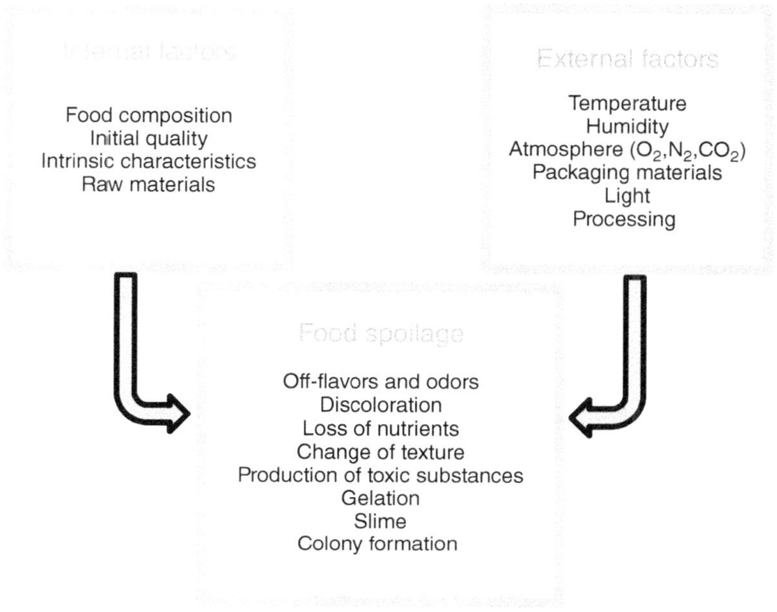

Figure 14.1 External and internal factors and sensory manifestations of food spoilage.

atmospheric composition. The methods followed in the production, processing, and preservation and the associated physical and chemical treatments induce changes in the characteristics of a food product determining a microflora, which can widely differ because of the specific effects of such practices. Implicit parameters are defined as the synergistic or antagonistic influences among microorganisms [3, 4].

The range of spoilage microorganisms is wide and comprises bacteria, yeasts, and fungi. Bacteria can be both Gram-negative such as *Pseudomonas, Escherichia coli,* and *Salmonella* and Gram-positive such as *Clostridium, Micrococcus, Bacillus,* and lactic acid bacteria [1]. Bacteria induce a rapid and evident spoilage of protein-rich foods such as meat, fish, shellfish, milk, and some dairy products [4]. Yeasts and molds grow more slowly compared to bacteria but they can proliferate on a variety of substrates such as pectin and other carbohydrates, organic acids, proteins, and lipids, and they tolerate more extreme conditions of pH, water activity, temperature, and the presence of preservatives. It should be noted that yeasts can utilize food ingredients, such as organic acids like lactic, citric, and acetic acids that are generally considered to have an inhibitory effect on the growth of many microorganisms [4]. For instance, *Zygosaccharomyces* spoil high sugar and salt products, and *Dekkera* and *Brettanomyces* are primary spoilers of alcoholic beverages. With regard to fungi, *Zygomycetes, Penicillia, Aspergilli,* and *Fusaria* are the most common species and they can grow on all kinds of food including cereals, meat, milk, fruits, vegetables, and nuts. Until recently, fungi were generally considered to cause only unaesthetic spoilage of food. After 1960, when the famous "TurkeyX" disease killed 100 000 turkey poults in Great Britain, it became evident that common spoilage molds could produce dangerous toxins [5]. Today more than 400 mycotoxins exhibiting diverse toxicity levels and effects have been identified, with aflatoxins being the best known [6].

14.1.3 Physical and Chemical Spoilage

Food spoilage is caused not only by the action of enzymes or microorganisms but also by a wide range of physical and chemical processes which may be triggered by light, metal ions, or excessive heat during processing or storage such as oxidation and lipolysis, leading to changes in color, flavor, viscosity, or to gelation and sedimentation phenomena. Oxidation reactions are among the most common causes of food deterioration and typically occur in food containing pigments and polyphenols, such as fruits, and unsaturated fats, like oils, and can be initiated by photosensitizers as well as by naturally occurring enzymes (e.g. lipoxygenase and peroxidases). Fats can also undergo an enzymatic hydrolysis by lipases responsible for the common off-flavor, referred to as rancidity [4, 7].

14.1.4 Factors Determining Food Spoilage

Food deterioration can occur at any stage from production to use by the final consumer. According to the United Nations, one third of all edible food (approximately 1.3 billion metric tons) is wasted annually due to poor manufacturing practices such as insanitary agricultural or processing conditions, inappropriate storage, defective packaging, incorrect handling, and even intentional adulteration by profit-oriented producers [3, 8, 9]. Moreover, if, on one hand, the global economy imposes the circulation of competitively priced food products worldwide, on the other, the lack of harmonized international regulations and standardization procedures regarding food safety and permitted additives renders food quality control operations complex and often subject to a country-to-country variability. Finally, it is anticipated that another major factor when addressing food safety is the limited availability of technological solutions that enable the early and real-time detection of spoilage processes both by specialized personnel that operate in the food industry and by the final consumers.

14.2 Food Spoilage Detection

14.2.1 Conventional Methods and Technologies for the Detection of Food Spoilage

Traditionally, spoilage detection has relied on sensory evaluation by highly trained panels that render this practice costly and subject to interpretation. In fact, the degree of freshness of food samples is attributed assigning numeric scores on the basis of the perceived intensity of the aroma, taste, aftertaste, and appearance. However, this method provides a simple indication of the food quality and it cannot be used to determine the spoilage causes, the pathogens concentration, and any relevant information that could lead to the optimization of in-plant quality assurance and contamination control protocols. Furthermore, it cannot serve for shelf life prediction and is therefore incompatible with routine applications in the food industry and consumers needs related to product safety information [10].

Another common approach consists in the identification of foodborne pathogens and in the tracking of bacterial metabolites, toxins, contaminants, biogenic gases, and other markers (Table 14.1) of organoleptic changes in food products based on biological, chemical, and enzymatic assays, such as bioluminescence, bacterial count, flow cytometry, and polymerase chain reaction and chromatographic techniques. However, these strategies are not suitable for the food shelf life evaluation and for quality control operations in food processing industries, due to their complex design, the time consuming procedures required for the extraction and purification of the target molecules and for the isolation and culturing of the microorganisms, and the long detection times. On top of that, specialized personnel are needed for the operation, costly reactants, growth media, and instrumental equipment are required and such analytical methods are easily susceptible to chemical and physical interferences and cross-contaminations [10–13]. For instance, the most common way of assessing the microbiological quality is to determine the Standard Plate Count (SPC) or the Psychrotrophic Bacteria Count (PBC), which require a minimum of 2 and 10 days, respectively [10, 14]. Headspace chromatography is a very powerful instrumental technique for both the qualitative and quantitative assessment of volatile compounds generated during the deterioration of food like milk and wine but it typically involves long analyte extraction and pre-concentration steps to ensure an adequate signal-to-noise ratio and a sample purity suitable for the analysis [14–17].

Therefore, the development and implementation of rapid, scalable, cost-effective, and univocally interpretable detection technologies applicable in all the phases of food production, storage, and distribution is of foremost importance. To this aim, much effort has been devoted to the possibility to incorporate sensing devices within packaging materials and an overview of the most recent advances in this field will be presented in Section 14.2.2.

14.2.2 On Package and on Site Sensing Technologies: A New Strategy for Food Spoilage Detection

The increasing demand, from both the food industry and end-consumers, for cost-effective and highly sensitive detection technologies to prevent microbial risk and allow a real-time and practical monitoring of the food state, has led to the recent development of engineered solutions that allow integration of portable and disposable sensing platforms within the packaging body, comprehensively defined as smart packaging. According to the legal definition of the EU 450/2009, smart packaging contains a component that enables the monitoring of the condition of packaged food or the environment surrounding the food during transport and storage [13].

Table 14.1 Overview of common target analytes used for sensing in food safety and quality monitoring from Refs. [10, 11].

Analyte type	Microbial agent/molecule	Food product
Bacteria, viruses, parasites	*Escherichia coli*	Meat and dairy products
	Salmonella groups B, D, and E	Range of foods
	Bacillus cereus	Milk
	Psychrobacter spp.	Refrigerated milk and raw meat
	Pseudomonas spp.	Refrigerated milk and raw meat
	Listeria monocytogenes	Processed meat and dairy products
	Clostridium perfringens	Raw meat
	Toxoplasma gondii	Swine
	Vibrio spp.	Seafood
	Staphylococcus aureus	Meat products
	Campylobacter spp.	Poultry and raw milk
	Shigella spp.	Range of foods
	Norovirus	Range of foods
Chemical or biological molecules	Oxygen	Range of foods
	Carbon dioxide	Range of foods
	Gaseous amines	Fish
	Lactate oxidase	Wine and yogurt
	Acetyl cholinesterase	Range of foods
	Lead	Range of foods
	Mercury	Fish
	Volatile compounds	Range of foods
	Gaseous acids	Range of foods
Yeast, mold, mycotoxins	*Aspergillus versicolor*	Yogurt, cheese
	Penicillium chrysogenum	Yogurt, cheese
	Geotrichum candidum	Milk
	Candida lipolyticum	Milk
	Mucor circinelloides	Yogurt
	Rhodotorula mucilaginosa	Yogurt
	Cladosporium cladosporioides	Cheese

Smart packaging is thus a system that provides to consumers reliable and correct information on the conditions of the food and/or of the packaging integrity based on its ability to sense, detect, or record changes in the product or its environment. In addition, the portable sensing technologies used in smart packaging can contribute to the on-site identification of health hazards and practices that strongly affect the food attributes in production and processing plants, as well

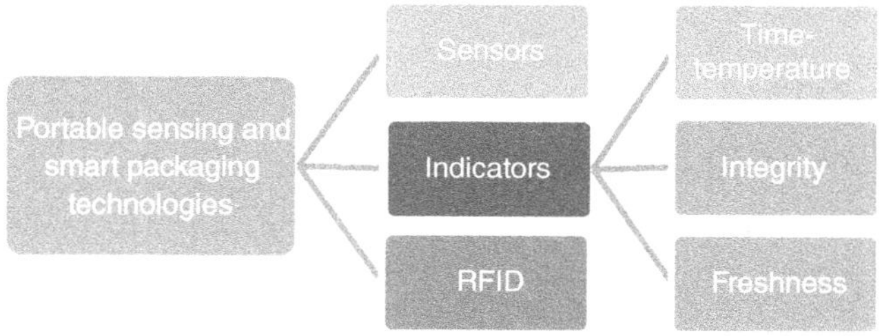

Figure 14.2 Overview of the different sensing portable technologies for on-site and on-package detection of food spoilage.

as to the establishment of strategies to reduce their occurrence and to efficiently improve the quality of the final product throughout the production process and supply chain operations [11].

Owing to their tremendous potential, over the next decade the global demand for sensing technologies in packaging applications is expected to increase with an annual growth rate of 7.4% reaching US$1.5 billion globally by 2023 [3] and $ 3600 million in the US market, $ 2360 million in Japan, $ 1390 million in Australia, followed by European countries by 2026 [18, 19].

To date, the major portable on-site and on-package sensing technologies for food spoilage detection rely on sensors, indicators, and on radiofrequency identification (RFID) devices (Figure 14.2). The adopted materials have many forms such as nanoparticles, hydrogels, supramolecular assemblies, films, and porous/fibrous substrates and many functionalities relying on their nature, which can be polymers, responsive dyes and pigments, metal nanoparticles, and carbon nanotubes [13, 20–23]. In these systems, the sensing mechanism is exerted through the response of an active component to a stimulus present in food or in its environment that triggers changes in the steric conformation, conductance, charge, physical structure (shrinking/swelling), hydrophobicity, or chemical functionality of the sensory platform. Such physical and chemical changes are in turn transduced into a measurable signal (e.g. electrochemical, optical, colorimetric) that enables the detection of the target compounds. Stimuli can be contaminants, pathogenic microorganisms or their metabolites that negatively affect the food freshness, safety and quality, or factors such as moisture, pH, gases, and temperature [13].

In sensors, the response is converted into an output signal that is proportional to the amount of analyte detected, providing, thus, a quantitative estimation of its concentration. Typical analytes include oxygen, biogenic headspace gases such as hydrogen sulfide, carbon dioxide, volatile amines, acid vapors, and volatile organic compounds. Biosensors operate in a similar manner but the receptors are organic or biological materials such as enzymes, hormones, nucleic acids, antigens, and microbes [11, 13, 20, 22].

Indicators are substances that can signal the presence or the absence of one or more target analytes providing immediate visual and qualitative information about the packaged food by means of a mechanical deformation, a color change, an increase in color intensity, or diffusion of a dye along a straight path [11, 24].

In most of the cases, the basic requirement of an indicator is that the changes should be irreversible, in order to avoid false information. The following types of indicators can be distinguished:

- Time-temperature indicator
- Freshness indicators
- Integrity indicators

Temperature is one of the most important environmental factors for determining the kinetics of physical, chemical, and microbial spoilage in food products. According to EC/450/2009, time temperature indicators (TTIs) provide information on whether a threshold temperature has been exceeded over time and the length of time the targeted food was exposed to such temperature (time–temperature history) during distribution and storage [13]. TTIs are thermoresponsive materials and their operation is based on mechanical, chemical, electrochemical, diffusion, enzymatic, or microbiological time- and temperature-dependent changes, usually expressed as a visible response (e.g. a color change). TTIs are particularly useful for warning of temperature abuse for chilled or frozen food products, thus ensuring food safety [25]. Currently available commercial TTIs (Table 14.2) include diffusion-based Freshness Check® of 3 M Company, OnVu™ (Ciba Speciality Chemicals, Inc.) which contains benzopyridines, organic pigments that change color with time at rates determined by temperature, and Tempix (Tempix AB) based on a barcode printed with fading inks that disappear due to temperature abuse [13].

Freshness indicators provide direct product visual quality information resulting from microbial growth or chemical changes within a food product. Freshness indicators can also be used to estimate the remaining shelf life of perishable products. The main limitations in these indicators are related to false positive cases in which color changes occur in products free from any significant sensory or quality deterioration [24]. Among commercially available freshness indicators, FreshTag® by Cox technologies is a colorimetric label able to react with volatile amines to detect spoilage in fish and seafood products. The FreshTag indicator is attached to the outside and monitors the freshness of seafood products via a barb on the backside of the tag that penetrates the packaging film, allowing any volatile amines to interact with the indicator turning it from yellow to blue [26]. SensorQ® developed by Food Quality Sensor International Inc. (FQSI) is a stick-on label that detects the freshness/spoilage level of packed meat and poultry. This indicator consists of an orange-colored pH-sensitive "Q" comprising anthocyanin dyes, such as betanin or flavonoid, which turns dark green if the meat is spoiled [27].

Integrity indicators allow to monitor the package integrity, which is an essential requirement for the maintenance of quality and safety of food products. Such systems usually provide qualitative or semi-quantitative information through visual colorimetric changes or through comparison with standard references. Much work on the development of integrity detection (e.g. leaking seals) for packaged foods has focused on visual gas oxygen and carbon dioxide indicators. These indicators have to be in contact with the gaseous environment inside the package and hence are in direct contact with the food. For instance, oxygen indicators

Table 14.2 Commercially available smart packaging systems.

Trade name	Manufacturer	Type
Ageless Eye	Mitsubishi Gas Chemical Inc.	Integrity indicator
Freshtag®	COX Technologies	Freshness indicator
Sensorq®	DSM NV and Food Quality Sensor International	Freshness indicator
Timestrip Complete®	Timestrip UK Ltd.	Time temperature indicator
Timestrip® PLUS Duo	Timestrip UK Ltd.	Time temperature indicator
Freshness Check	3 M	Time temperature indicator
Fresh-Check	Temptime Corporation	Time temperature indicator
Checkpoint®	Vitsab	Time temperature indicator
OnVu	Chiba Specialty Chemicals and Freshpont	Time temperature indicator
Colour-Therm	Colour-Therm	Time temperature indicator
Tempix	Tempix AB	Time temperature indicator
Thermax	Thermographic Measurements Ltd.	Time temperature indicator
Timestrip	Timestrip Ltd.	Integrity indicator
Novas®	Insignia Technologies Ltd.	Integrity indicator
Easy2log®	CAEN RFID Srl	RFID
Intelligent Box	Mondi Plc	RFID
CS8304	Convergence Systems Ltd.	RFID
Food Sentinel System	SIRA Technologies	RFID

Source: Adapted from Biji et al. 2015 [13].

based on redox dyes that respond colorimetrically to changes in oxygen concentration have been reported [26]. Ageless Eye® by Mitsubishi Gas Chemical company (Table 14.2) is a commercial oxygen indicator tablet which monitors the oxygen levels by a change in color to pink when the oxygen concentration is lower than 0.01% and to blue when it reaches 0.5% or higher levels [28]. CO_2 indicators typically function on the principle of Severinghaus CO_2 electrode, which involves the diffusion of CO_2 from the test medium (gaseous or aqueous) through a thin gas permeable membrane, and a quick establishment of equilibrium with the entrapped aqueous layer, which contains a pH-sensitive dye. Therefore, the detection occurs upon the interaction of the gas to be tested with a chemical reagent/dye of the indicator which results in a color change/shift of the system [29, 30]. However, these technologies still present some drawbacks, as indicators containing pH-sensitive dyes generally suffer from physical (e.g. humidity and temperature) and chemical interferences (e.g. other volatile acid or basic gases produced during spoilage).

RFID systems use wireless sensors for identification, automatization, antitheft prevention or counterfeit protection purposes, and for monitoring the temperature, relative humidity, light exposure, pressure, and pH of the products. A RFID is based on tags and readers that communicate through radio waves. Tags store some sort of identification number used by readers to recognize the tag and retrieve the information encoded in the tag memory [11]. By integrating RFID systems with materials that react to food analytes, a detectable signal can be acquired based on the change in electronic potential of the RFID. Fiddes et al. [31] exploited this approach for the detection of gaseous volatile amines. Volatile amines swell the conductive polymer film combined in the RFID, inducing a shift to a lower range in the optimal broadcasting frequency of the system. A number of RFIDs able to detect the possible interruptions of cold chain which are harmful to the food quality and safety is also available on the market [20]. For example, SIRA Technologies (Table 14.2) developed Food Sentinel System®, a RFID for food packaging systems based on a biosensor, which alerts consumer of the presence of contaminating bacteria. Food pathogens will be recognized by specific antibodies attached to the membrane forming part of the barcode causing localized dark bar formation and rendering the barcode unreadable upon scanning [32].

The most recent technological solutions in the sensors domain are based on biological recognition, antibody immobilization techniques, printed electronics, carbon nanotechnology, silicon photonics, molecularly imprinted polymers, metallic and chromogenic sensor arrays, and responsive organic dyes. Among them, systems that allow an optical and colorimetric detection are the most promising for the development of commercial portable devices and smart packaging applications related to food spoilage. This is because such systems are highly sensitive, their operation does not require electrical power and the output signal can be read with a variety of hand-held equipment such as Raman, infrared, or ultraviolet spectrometers and allows for direct naked eye assessment easily interpretable by non-specialized personnel and general consumers.

In addition to the need for excellent technical performance, practical design, and intuitive utilization, another major criterion in the evaluation of the suitability of sensing platforms for the detection of food spoilage concerns safety. In fact, since the sensing surface needs to be in contact with the food to detect the targeted changes, it cannot be toxic, it should not release any harmful substances, or alter the food's chemical, structural, and organoleptic characteristics, and it must comply with food contact regulations established by organizations like FDA and EFSA. The possible toxicity and health hazards disqualify the use of many effective materials such as metal nanoparticles and carbon nanomaterials as well as the majority of synthetic pH-responsive dyes such as bromophenol blue, bromothymol blue, chlorophenol red, cresol red, bromocresol green, bromocresol purple, phenolphthalein, xylenol blue, *p*-naphtholbenzein, and hexamethoxy red [33, 34]. Moreover, the demand for increased sustainability has been raising environmental concerns imposing a thorough investigation of the impact of sensory components integrated in packaging on the recycling, disposal, and compostability of the final materials [13].

Section 14.3 will address the topic of safety and sustainability by describing the most relevant natural and biocompatible materials reported in the literature that hold promise for the industrial production and commercialization of the next generation of eco-friendly, safe, efficient and cost-effective optical, and colorimetric indicators for the detection of food spoilage processes.

14.3 Natural and Biocompatible Optical Indicators for Food Spoilage

14.3.1 Optical and Colorimetric Detection

In recent years, there has been a growing interest in the development of optical and colorimetric sensors for real-time monitoring of the freshness of food products because of their advantages such as small size suitable for integration on-site and on-package, high sensitivity, low cost, and the possibility to obtain direct information just by optical/visual analysis.

Optical detection is based on the change in the optical properties, namely spectral absorption or fluorescence, of a sensory material that arises from a modification in its electronic or molecular structure in response to one or more analytes. Such changes are often accompanied by a chromic transition visible by naked eye. The most common optochemosensing mechanisms involve complex formation, supramolecular interactions, aggregation, and bond formation/cleavage.

$$\Delta E = \sqrt{(L_1 - L_2)^2 + (a_1 - a_2)^2 + (b_1 - b_2)^2} \tag{14.1}$$

Colorimetric detection typically relies on color analysis through the CIELab color space in which a given color is represented by the tristimulus values L (lightness), a (red/green), and b (yellow/blue) to mimic the human perception of colors. A color variation elicited by an indicator response can be calculated as the quantitative difference between a color and a reference, and is expressed as the Euclidean distance (ΔE) between their tristimulus values calculated according to Eq. (14.1) [35].

14.3.2 Natural and Biocompatible Indicators

Optical and colorimetric sensing platforms typically consist of a polymeric support in which the sensory functional component is embedded. The research efforts toward the development of greener and nontoxic sensing materials as optical and colorimetric indicators for food spoilage have focused, on one hand, on the utilization of renewable natural substrates, especially bio-based polymers such as cellulose, starch, gelatin, silk, alginate, and chitosan [23, 33–36] due to their availability, low cost, biocompatibility, and good processability, and, on the other hand, on exploring the sensing potential of natural dyes and pigments extracted from fruits and plants such as curcumin and anthocyanins because they are biocompatible and environment friendly [33, 36]. However, the use of natural dyes is challenging due to reduced stability toward light, oxidation, temperature, and pH. Among them, anthocyanins from red cabbage (*Brassica oleracea* L.) were found to possess superior properties in terms of chemical, thermal and color stability, water solubility and color spectrum at extended pH range [37], and many indicators that will be reviewed in the following paragraphs have been fabricated exploiting the anthocyanins.

With regard to the indicator fabrication technology, physical adsorption on films or paper strips are mostly used due to their ready availability and preparation, but engineered sensing technologies comprising the incorporation of anthocyanins into hydrogels, micelles, foams, and fibrous and nanofibrous mats are now starting to emerge [23, 33–44]. The latter provide several advantages such as faster response times and higher sensitivity due to increased porosity and surface area, tunable response based on the morphological features of the polymer substrate, and limited dye leaching due to its more efficient encapsulation into the polymer matrix.

In Sections 14.3.3–14.3.6, the most recent advances in the optical detection of food spoilage by means of natural and biocompatible sensing platforms will be presented for the major food spoilage markers. Much attention has been devoted to systems that can signal changes in pH and the presence of volatile and nonvolatile amines and acids [23, 33–42]. However, indicators of headspace oxygen and carbon dioxide and bacteria are scarce [43–45], and no report has yet specifically addressed the detection of volatile organic vapors such as ethylene, ketones, short chain esters, aldehydes, and alcohols.

14.3.3 Detection of pH, Acids, and Amines

Changes in pH provide the most immediate indication of ongoing degradation of food materials and are typically associated with the development of toxic volatile and nonvolatile biogenic amines and acids that accumulate both in the food and in the headspace of the package.

Biogenic amines are low molecular weight organic bases that possess biological activity. They include tyramine, trimethylamine, 2-phenylethylamine, histamine, putrescine, cadaverine, spermine, spermidine, tryptamine, and agmatine. Their presence in foods serves as a useful indicator of spoilage as they are generated from the decarboxylation of amino acids in meat, fish, and milk by bacteria [46, 47].

Organic acids are another kind of leading indicators of acidic food spoilage for a number of different products such as canned foods, packed meat, milk and milk products, beer, and wine [23]. For example, H. Korkeala et al. [48] demonstrated that lactic acid and pH could be used as indicators of spoilage for vacuum-packed cooked ring sausage. Milk spoilage results in the generation of lactic acid or free fatty acids by lipolytic enzymes. A number of bacterial species can degrade poultry meat with the evolution of volatile sulfur-containing compounds, such as hydrogen sulfide (H_2S), which is a breakdown product of cysteine with intense off-flavors and low threshold levels [10].

Zia et al. [49] developed a universal and reversible colorimetric indicator by combining polyvinyl alcohol and cellulose with a pH-responsive red cabbage anthocyanin to obtain a porous foam. Cellulose was found to play an important role in enhancing the porosity of the material, providing in turn a fast response. The indicator was effective when exposed to solutions in the 2–14 pH range and to volatile acids and bases. The colorimetric response, ascribed to the protonation/deprotonation of the anthocyanin molecules, was clearly visible by naked eye inspection as well as through UV–Vis spectroscopy (Figure 14.3). In an acidic environment (pH 2–6), the foams appeared red or purple and

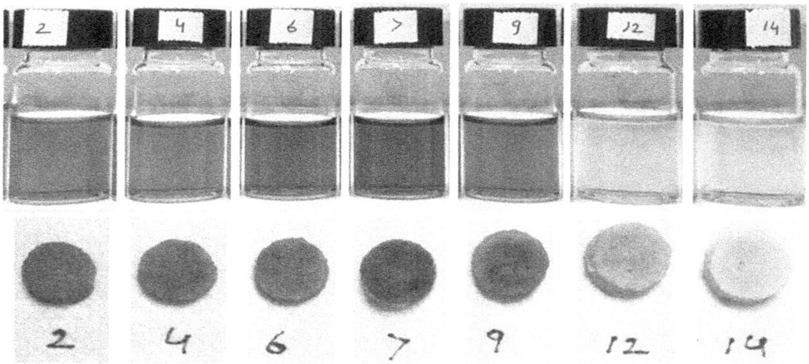

Figure 14.3 Colorimetric response to a pH range of 2–14 of (a) anthocyanin in solution (b) anthocyanin treated polyvinyl alcohol/cellulose foams. Source: From Zia et al. [49]. (See online version for color figure).

exhibited an absorption maximum centered at 527–592 nm. At neutral pH, the foams were purple–blue and their absorption maximum was centered at around 607 nm. In alkaline media (pH 9–14), the foams were green/yellow with the main absorbance peak at 355–384 and a shoulder at 610 nm. The response to volatile acids and bases was assessed upon exposure to vapors of hydrochloric acid and of ammonium hydroxide, and the foams changed color from purple to red and green respectively with a response time of c. 10 seconds. The indicator was able to detect the spoilage of chicken and could thus be used to evaluate the freshness of food products. Specifically, the foam was attached to a closed petri dish containing fresh chicken as shown in Figure 14.4. In the beginning of the experiment, the color of the foam was purple, corresponding to a neutral environment, while, after three days, the indicator turned greenish as a consequence of the generation of volatile amines upon spoilage.

A similar colorimetric bio-based pH-responsive system was developed by Yoshida et al. [38] by mixing natural compounds such as chitosan and anthocyanin with potential use as an indicator of metabolites derived from microbial growth such as *n*-butyrate, L-lactic acid, D-lactate, and acetic acid. The initial color of the dried film was dark violet. The color varied from pink in acid pH to bluish-green in neutral pH and to violet in basic pH. Another anthocyanin-based system was proposed by Silva-Pereira et al. [39] that combined chitosan, corn starch, and red cabbage extracts to obtain a hydrogel film cross-linked using a sodium tri-polyphosphate solution. The color variation after activation in different pH range was measured with the CIELab methodology. In order to validate the use of this system as a fish spoilage sensor, application tests were conducted by inserting the film in a package containing fish fillets. At room temperature, for 12 hours no color change was visible in the indicator film, after 16 hours the color began to change to blue, indicating a pH increase and initial spoilage, and after 72 hours the color completely changed to yellow thus indicating complete fish spoilage. At refrigeration temperature, the color changed to blue after 72 hours, and to yellow after seven days. Similarly, Zhang

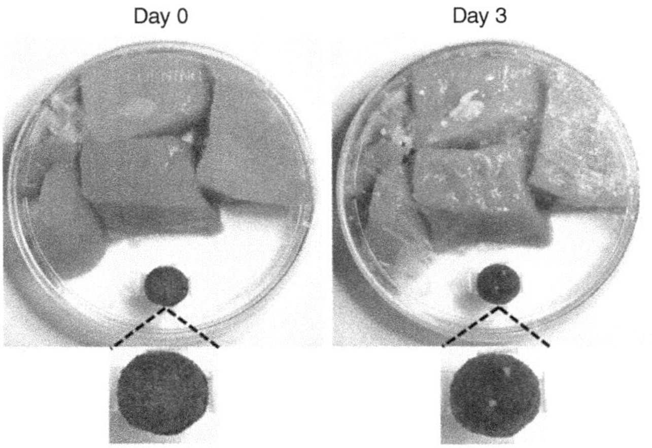

Figure 14.4 Application of the anthocyanin treated polyvinyl alcohol/cellulose indicator foam as a colorimetric sticker to monitor chicken freshness. Source: From Zia et al. [49]. (See online version for color figure).

et al. [33] extracted anthocyanin dyes from *Bauhinia Blakeana Dunn* and then immobilized them onto chitosan to monitor pork and fish deterioration in ambient conditions. The color change of the pH could be observed by naked eye in the 2.2–9.0 pH range. In pork samples, the prepared sensing film gave a significant color change from purple to brown after approximately 12 hours and from brown to green after 24 hours when the pH was between 6.0 and 7.0. A similar phenomenon could be observed during fish spoilage: the sensing film color changed steadily from purple to brown after approximately 12 hours, and then from brown to green after 16 hours.

Wu et al. [50] prepared a chitosan/agarose/anthocyanin film from the electrochemical patterning of a hydrogel by exploiting the change in pH of the anthocyanin. After drying, the resulting film was placed in the headspace of glass bottles containing crucian fish samples stored at room temperature (25 °C) for 20 hours. The film presented a red color at the beginning, turned into a pink color after 10 hours and finally reached a purple color after 20 hours. The observed color change was in agreement with the rejection limit for total nitrogen content that was reached after 20 hours. The results demonstrate that the film was capable of indicating real-time freshness of crucian fish.

Zhai et al. [36] fabricated a colorimetric sensor based on starch/polyvinyl alcohol films doped with anthocyanins extracted from flowers of roselle (*Hibiscus sabdariffa* L.) using a casting/solvent evaporation method. It was observed that the colors of the developed films were stable at refrigeration and at room temperature up to 14 days. As an example, the authors subjected this sensor for freshness evaluation of silver carp (*Hypophthalmichthys molitrix*) at refrigeration temperature. The sensor exhibited color changes over time due to changes in the pH attributed to the evolution of total volatile amines from the fish spoilage.

In another study [51], beads of alginate, a nontoxic, renewable, and linear copolymer consisting of α-L-guluronic acid (G) and β-D-mannuronic acid (M) residues linearly linked by 1,4-glycosidic linkages obtained from the brown algae *Laminaria*, were used as a colorimetric sensor to monitor the pH of packaged rainbow trout. The sensor was developed through the encapsulation of pH-responsive red cabbage extract in alginate beads. The beads were prepared by dissolving alginate in water and cross-linking it with calcium chloride to form an interconnected open pore network. The color change of the beads during fish storage in closed petri dishes at room temperature for 64 hours and at 4 °C for 12 days was measured using the CIELab method. Specifically, the alginate beads exhibited a pink color, which turned to purple after 12 hours in samples stored at room temperature and on day 3 in samples stored at 4 °C. The color of the beads further changed from purple to yellow upon fish storage for 64 hours and 12 days at room temperature and at 4 °C, respectively. The chromic variation of alginate beads during the storage at room and refrigerator temperature had a strong relationship with the pH increase of fish samples ascribed to the microbial production of volatile amine compounds. A comparison of the visual and colorimetric response of the beads revealed that not only was the colorimetric sensor response proportional to the levels of the total volatile basic nitrogen and was well correlated with variations in bacterial populations, but the onset of color change was also correlated with the rejection level of the product which was reached after 12 hours of storage at room temperature and approximately after three days of storage under refrigeration.

Kuswandi et al. [52] proposed an on-package sticker sensor based on curcumin for the real-time monitoring of shrimp spoilage through the detection of volatile amines. The authors extracted the yellow curcumin pigment, a commonly used spice derived from the rhizome of the plant *Curcuma longa*, and immobilized it on bacterial cellulose membrane via physical adsorption. The curcumin/bacterial cellulose nanocomposite exhibited color changes from yellow to orange to reddish orange as a result of its interactions with the volatile amines produced gradually in the package headspace due to shrimp spoilage processes. Such changes were assessed through visual inspection and color analysis via Photoshop software, which allowed for the quantitative determination of the total volatile amines. The same authors also developed a similar on-package colorimetric indicator for real-time visual monitoring of the freshness of packaged guava fruits (*Psidium guajava* L.) [53]. During ripening, guava acidity increases due to the production of organic volatile acids (e.g. acetic acid) by enzymatic oxidation of ethanol. The proposed indicator was fabricated by the immobilization of a bromophenol blue (BPB) dye onto a bacterial cellulose membrane via adsorption. The accumulation of volatile acidic compounds in the package headspace induced a color change in the material from blue to green at 28–30 °C and such change was clearly detectable even by naked eye. Another bromophenol blue on-package indicator was developed by Shukla et al. [54] for real-time monitoring of meat quality. The strip type indicator-sensor was fabricated by coating the dye onto a filter paper via centrifugation and by a successive coating with porous low-density polyethylene. The indicator was placed inside a red meat pack paper and reacted with the volatile bases released from stored meat, changing its color from yellow to blue during the storage period of 24 hours at ambient temperature. The relationship drawn between the color change of indicator-sensor and meat quality parameters indicated that the developed system can be successfully utilized to predict degree of spoilage of buffalo meat just by observing the color of indicator-sensor by naked eyes.

He et al. [55] fabricated a multifunctional film incorporating the natural aggregation-induced emission (AIE) luminogen quercetin via physically mixing quercetin and polyvinyl alcohol. Quercetin from Sophora japonica is a phenolic compound, which has antibacterial and antioxidant properties, and a good biocompatibility. The as-prepared film was transparent and showed good AIE enhancement at 540 nm and a color change from blue to green when exposed to biogenic amines, produced during spoilage of packaged salmon at room temperature (25 °C) for up to 8 hours. The AIE of the indicator was noticeably enhanced due to the coordination bond or hydrogen bond of the amines with phenolic hydroxyl moieties, which limit the motion of aromatic rings promoting the formation of aggregates that gives rise to the fluorescence enhancement. The system was found to be very sensitive in the detection of volatile amines with a detection limit of 0.06 ppm

Xiaowei et al. [56] designed a simple, cheap, and convenient colorimetric sensor array by printing nine natural chromoresponsive pigments on a hydrophobic nanoporous film for the evaluation of meat freshness. The pigment showed a high sensitivity in detecting biogenic amines evolved upon deterioration of Yao pork meat.

Schaude et al. [57] reported the production of a sensing platform produced by covalent immobilization of synthetic dyes onto cellulose microparticles with subsequent embedding into food-grade silicone. The sensor had an intuitive color code, changing from green, indicating that the food was edible, to red, indicating that the food was no longer edible, allowing for easy visual inspection from the outside of transparent food packaging. Since the dye was not in direct contact with food due to the silicon coating, the material exhibited a good cell biocompatibility. The spoilage of food stored at room temperature triggered full color changes from green to red within one day for chicken and two days for pork. However, the sensor was not sensitive enough to detect low amine concentrations resulting from microbial decay at refrigeration temperatures.

Regarding volatile acid detection, M. Smolander et al. [58] proposed a freshness indicator for modified atmosphere packed poultry meat based on the detection of H_2S using myoglobin immobilized on agarose. The indicator was heat-sealed between two polyethylene films and subsequently Inserted into packages containing unmarinated broiler stored at 5 °C. The indicator changed color from brown to bright red and exhibited a detection limit of 0.1 mg/l. The colorimetric variation correlated well with the microbiological and sensory quality of broiler samples and the onset of color change took place concurrently with the sensory rejection of the odor. Moreover, the indicator was selective for H_2S as the color change was not affected by the presence of an N_2 and CO_2 mixture in the package atmosphere.

Genovese et al. developed *on demand* acidoresponsive composite films and fibrous mats that could serve as a low-cost in situ food quality indicator, enabling both the optical and visual colorimetric detection of acidic vapors and, thus, ideal to signal acidic spoilage of food products [23, 35, 42, 59]. Such systems are based on the incorporation of photochromic spiropyran (SP) dyes in biocompatible and natural polymer substrates. SP molecules can be reversibly converted into their active zwitterionic merocyanine (MC) isomers upon photoactivation through UV irradiation and be restored through irradiation with visible light [59]. The MC species in presence of acidic vapors undergo protonation giving rise to a color change from purple to yellow, visible by naked eye, accompanied by a spectrally detectable blueshift of the absorption maximum from ~550 to ~420 nm and quenching of the MC fluorescence [35, 59]. Depending on the surface area and morphology of the specific substrate used to embed the functional SP, the response times could be tuned from 3.5 to 30 seconds using nanofibers and fibers and from 200 seconds to 10 minutes using films [23, 35, 42, 59]. For example, the authors reported that the incorporation of SP in silk fibroin and polyethylene oxide nanofibers enabled the reversible detection of gaseous acetic acid in 30 seconds through the abovementioned colorimetric change and a quenching in the optical emission at 633 nm. Moreover, the response was reversible so that upon acid vapor desorption the pristine MC could be regenerated with a recovery of the spectral emission, allowing to replicate the detection 20 times without significant signal losses [35]. Although SPs are generally not recognized as safe in cellular media due to hydrolysis and other degradation processes, the SP-doped silk polyethylene nanofibers exhibited a good biocompatibility due to the immobilization and stabilization of the SP molecules in the matrix, as assessed through cytotoxicity assays and confocal imaging of adherent fibroblast cells seeded on the nanofibers [35].

In a recent study [23], the same authors demonstrated the applicability of a photochromic composite for the optochemical and colorimetric detection of acidic spoilage in milk and wine. The developed sensing system was fabricated through the impregnation of filter paper with a photochromic acrylate polymer modified with appended spiropyran moieties. The composite paper disks were integrated in the caps of vials containing milk and wine to be in contact with the headspace of the food product. In particular, ultrahigh temperature (UHT) and pasteurized milk and three Sangiovese-based red wines were used in the study (Figure 14.5). By monitoring the color and the absorbance of the caps every day for 10 days immediately after UV irradiation with a portable lamp, the acidic vapors (e.g. lactic acid and acetic acid), generated during the samples deterioration, could be detected optically, through the blue shift in the absorption maximum from 545 to 415 nm of the photogenerated MC upon protonation, and visually by naked eye, through the progressive transition of the substrates from pink, characteristic of the photogenerated MC, to pale yellow-white, characteristic of the protonated MC species (MCH). Specifically, the evolution of acidic vapors progressively converted all the MC moieties into MCH so that the pink color was no longer detectable after five days in both milk samples and after only one day in all the red wine samples. More importantly, the optical response of the composite films could be linearly correlated with changes in the chemical environment (pH and total acidity) of the food samples indicative of spoilage processes. In addition, the indicators exhibited a high selectivity toward volatile acids as they did

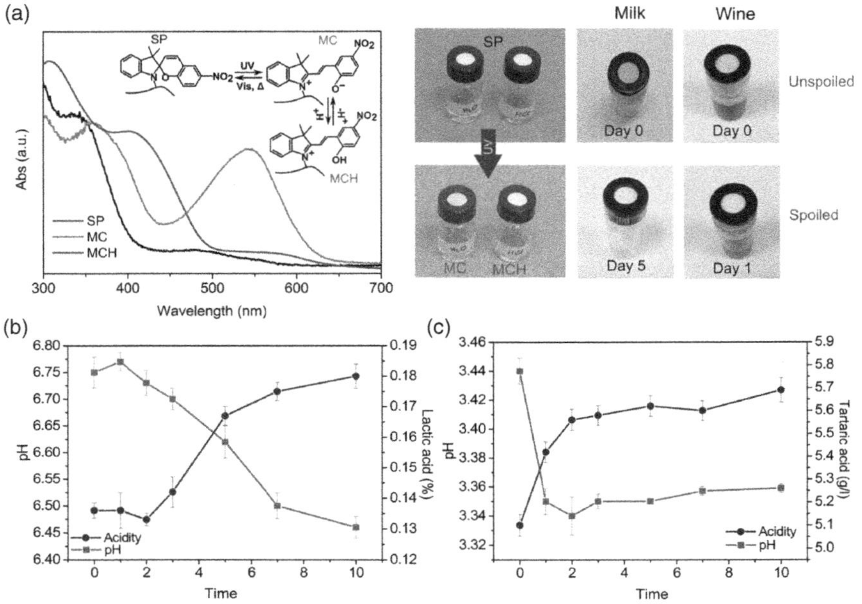

Figure 14.5 (a) Absorption spectra and photographs of paper films impregnated with a photochromic polymer (https://pubs.acs.org/doi/10.1021/acsomega.8b02570). Further permissions related to the material excerpted should be directed to the ACS. The spectra show the colorless SP, purple MC, and pale yellow-white protonated MC (MCH) species. After the UV irradiation for 90 seconds ($\lambda = 320$–375 nm, power density = 4.15 mW/cm^2) or for 5 seconds ($\lambda = 365$ nm, power density = 5 mW/cm^2), in the absence of acidic vapors (e.g. water vapors), the samples inserted in hollow screw caps become purple due to the formation of MC from the photoisomerization of SP, whereas upon exposure to hydrochloric acid (HCl) vapor they appear pale yellow-white due to the protonation of MC and subsequent formation of MCH. Similar changes are observed when the samples are in contact with headspace volatile acids generated during spoilage of milk and wine. Variation of pH and total acidity of UHT milk (b) and wine (c) samples as a function of time. Source: Genovese et al. [23]. 2018, © American Chemical Society. (See online version for color figure).

not respond to other volatile spoilage markers such as acetaldehyde, ethanol, and ethyl acetate [23].

14.3.4 Detection of Oxygen

The presence of oxygen in a package accelerates the oxidative deterioration of food as it facilitates the growth of aerobic microbes, off flavor and odor development, color changes, and nutritional losses [11]. Sensory materials to detect oxygen with potential for food spoilage monitoring were developed by Vu and Won [43, 60] by dip-coating a solution of a redox dye (thionine, azure A, or metyhelene blue), a sacrificial electron donor (glycerol), and a UV-absorbing semiconducting photocatalyst (TiO_2) on a nylon/polyethyelene film. The obtained composite films were then dipped in a polysaccharide solution of alginate or carrageenan. The encapsulation of the dyes within these natural polymers reduced their leaching in water down to ~2% with carrageenan and ~6% with alginate. Upon UV irradiation, the dyes underwent a photobleaching reduction due to the electrons generated from the sacrificial electron donor and regained their original color upon oxidation when exposed to atmospheric oxygen. The simple dye immobilization technique and the colorimetric response of the films are promising for applications of such indicators in smart packaging. In contrast to UV-activated oxygen indicators, Won et al. [61] designed an in-package activated oxygen indicator exploiting enzymatic oxidation. They incorporated a naturally occurring and nontoxic biocatalyst with an oxygen-dependent activity (laccase), an organic substrate forming colored oxidation products upon enzymatic reaction (guaiacol), and a biological reducing agent (cysteine) on a membrane of polyethylene terephthalate and polyethylene. The components were physically separated into two compartments by a fragile barrier. When the barrier was broken, the components mixed and started functioning as an oxygen indicator exhibiting different response times and color differences depending on their concentrations and with a reaction rate proportional to the oxygen concentration. It was shown that even the presence of 1% oxygen in the package could be detected easily within two hours.

14.3.5 Detection of Carbon Dioxide

Carbon dioxide can be added to the packaging environment to suppress the microbial growth in certain products such as fresh meat, poultry, fish, cheese, and backed goods [62]. However, carbon dioxide produced in meat products during storage is also an indication of food spoilage.

J. Jung et al. [44] proposed a chitosan-based CO_2 indicator to signal the onset of kimchi ripening, exploiting the pH-dependent solubility of chitosan. The indicator was developed by pouring a chitosan solution at pH 7.0 within low-density polyethylene (film sachets) and placing the obtained sensing platform in the headspace of kimchi packages. Due to the scarce solubility of chitosan at neutral pH, the solution appeared turbid. When the CO_2 produced by fermentation upon ripening of kimchi decreased the pH of the solution, the solution became transparent due to the dissolution of chitosan in acidic environment. A remarkable change from below 30% to nearly 100% was observed in the degree of transparency of the indicators at the end of the testing period (two weeks). The chitosan-based CO_2 indicator system offers a great potential as a convenient, easy-to-use technique for recognition of unripened packaged kimchi and/or onset of optimal ripening period. In another study [63], methylcellulose and hydroxypropyl cellulose and synthetic dyes were enclosed into polyethylene trays and then heat-sealed with a nylon/polyethylene film in order to monitor the CO_2 levels in skinless chicken breast packages stored at 4 and 10 °C. The indicators responded colorimetrically with a transition from green to orange-yellow due to the permeation of carbon dioxide through the polymer membrane and subsequent dissolution in the cellulose solution where it protonated the dyes. However, such type of indicator is in liquid state and thus it requires careful handling, which means inconvenience and safety problems for commercial products.

14.3.6 Detection of Bacteria

Pires et al. [45] developed a colorimetric biosensor for the detection of foodborne bacteria in apple juice based on cellulose strips filled with polydiacetylenic vesicles of *N*-[(2-tetradecanamide)-ethyl]-ribonamide/10,12-pentacosadyinoic acid (TIDER/PCDA). The sensor became red in presence of common food pathogens such as *Staphylococcus aureus* and *E. coli*.

14.4 Concluding Remarks and Future Perspectives

Food spoilage caused by physical or chemical agents or by the action of enzymes or microorganisms still constitutes a major unresolved issue for both the food industry and the final consumers due to the incurred financial losses and health concerns. The primary factors associated with food spoilage are related to the intrinsic perishable nature of food and contamination during harvesting, processing, storage, and distribution operations that render quality control practices complex and hardly standardized due to the variability in the food characteristics and manufacturing methods and in socio-economic aspects such as geographical location and regulations.

The conventional methods used for the identification of deterioration processes and foodborne threats in food commodities based on sensory evaluation, on one hand, and on biological and chromatographic assays, on the other, are limited by costly and time-consuming laboratory operations and do not address the need for a reliable, real-time on-site and on-package detection accessible at the supply chain and end-consumer level. In this regard, fully natural and biocompatible and/or edible spoilage indicators have been emerging as a new branch of smart packaging and portable sensing materials that will offer exciting opportunities for enhancing food safety, quality, shelf life prediction, traceability and convenience, overpassing conventional techniques that require breakage of package integrity and are not applicable for real-time, on-line control, or large-scale usage. These sensing technologies are based on blends of natural, eco-friendly, and food contact-safe polymers and active molecules that can be easily integrated in the packaging, to interact with substances indicative of food deterioration and to signal their presence by colorimetric and optical changes, enabling the efficient monitoring of packaged food products by direct naked eye visualization and wireless detection.

Although very few commercial sensing platforms are currently available, the smart packaging market is predicted to grow steadily over the next years owing to the recent advances in the development of natural optical indicators combined with the increasing demand for transparency and sustainability from consumers and regulatory agencies. We foresee that the use of natural and biocompatible food spoilage indicators will have a tremendous impact not only by directly addressing the needs of the food industry for retail and to better control the food production chain, reducing food waste, and limiting foodborne illness outbreaks, but at the same time by providing a greener and safer alternative to existing packaging technologies based on nonrecyclable and nonfood contact sensors and an improved transparency toward consumers.

Currently most of the reported results are focused on the detection of spoilage of meat and fish and the number and variety of sensing strategies is relatively limited to the response to pH, acids, and amines. Future developments should involve the detection of other volatile spoilage markers such as ethylene, ketones and alcohols, and bacteria. To this aim, the use of materials structured with a high surface area and porosity such as foams, fibrous films, and nanofibers as well as the use of natural active molecules or enzymes will be beneficial to improve both sensitivity and detection time.

Moreover, in order to address the present imbalance between potential and realization of smart sustainable packaging concepts, a number of research gaps need to be filled through a transversally integrated multidisciplinary approach that combines packaging science, food science, biotechnology, sensor science, information technology, and nanotechnology. Further research is necessary to overcome the limitations of the present natural optical and colorimetric indicators in terms of sensitivity limits, reproducibility, costs, scalability, selectivity, recyclability, etc. The functionality of the indicators must be stable in time and be robust against the possible interaction with external factors such as temperature, humidity, and presence of other gases. More exact correlations need to be established between target metabolites, product type and organoleptic quality and safety. Further modeling of the interactions between foods and microbes and their metabolites under dynamic storage conditions is necessary. A stronger focus has to be put on better understanding the behavior of smart packaging devices into high-volume packaging processes and at all points of the storage and distribution chain. Finally, both technical and marketing issues relating to sensitivity (including over-sensitivity), to the implementation of the signal read-out through a user-friendly interface and to the acceptance and cost-effectiveness for industry and consumers need to be addressed.

References

1 De W. Blackburn, C. (ed.) (2006). *Food Spoilage Microorganisms*. Cambridge, UK: Woodhead Publishing.
2 Lewis, D.C. and Shibamoto, T. (1986). Shelf life of fruits. In: *Handbook of Food and Beverage Stability* (ed. G. Charamlabous), 353–391. Orlando: Academic Press.
3 Schaefer, D. and Cheung, W.M. (2018). Smart packaging: opportunities and challenges. *Procedia CIRP* 72: 1022–1027.
4 Jay, J.M., Loessner, M.J., and Golden, D.A. (2005). Intrinsic and extrinsic parameters of foods that affect microbial growth. In: *Modern Food Microbiology*, Food Science Text Series, 39–59. Boston: Springer.
5 In't Veld, J.H.H. (1996). Microbial and biochemical spoilage of foods: an overview. *International Journal of Food Microbiology* 33 (1): 1–18.
6 Pitt, J.I. and Hocking, A.D. (2009). Introduction. In: *Fungi and Food Spoilage*, 1–2. Boston: Springer.

7 Filtenborg, O., Frisvad, J.C., and Thrane, U. (1996). Moulds in food spoilage. *International Journal of Food Microbiology* 33 (1): 85–102.

8 Pavase, T.R., Lin, H., Shaikh, Q. et al. (2018). Recent advances of conjugated polymer (CP) nanocomposite-based chemical sensors and their applications in food spoilage detection: a comprehensive review. *Sensors and Actuators B* 273: 1113–1138.

9 UN World Health Organization (2017). WHO Model lists of essential medicines. http://www.who.int/medicines/publications/essentialmedicines/en/index.html ().

10 Dainty, R.H. (1996). Chemical/biochemical detection of spoilage. *International Journal of Food Microbiology* 33 (1): 19–33.

11 Brockgreitens, J. and Abbas, A. (2016). Responsive food packaging: recent progress and technological prospects. *Comprehensive Reviews in Food Science and Food Safety* 15 (1): 3–15.

12 Pierami, R.M. and Stevenson, K.E. (1976). Detection of metabolites produced by psychrotrophic bacteria growing in milk. *Journal of Dairy Science* 59 (6): 1010–1015.

13 Biji, K.B., Ravishankar, C.N., Mohan, C.O. et al. (2015). Smart packaging systems for food applications: a review. *Journal of Food Science and Technology* 52 (10): 6125–6135.

14 Martin, N.H., Ranieri, M.L., Murphy, S.C. et al. (2011). Results from raw milk microbiological tests do not predict the shelf-life performance of commercially pasteurized fluid milk. *Journal of Dairy Science* 94 (3): 1211–1222.

15 Ortega-Heras, M., González-SanJosé, M.L., and Beltrán, S. (2002). Aroma composition of wine studied by different extraction methods. *Analytica Chimica Acta* 458 (1): 85–93.

16 Lehotay, S.J. and Hajšlová, J. (2002). Application of gas chromatography in food analysis. *TrAC Trends in Analytical Chemistry* 21 (9–10): 686–697.

17 Magan, N., Pavlou, A., and Chrysanthakis, I. (2001). Milk-sense: a volatile sensing system recognises spoilage bacteria and yeasts in milk. *Sensors and Actuators B: Chemical* 72 (1): 28–34.

18 De Jong, X.A.R., Boumans, H., Slaghek, T. et al. (2005). Active and intelligent packaging for food: is it the future? *Food Additives and Contaminants* 22 (10): 975–979.

19 Fuertes, G., Soto, I., Carrasco, R. et al. (2016). Intelligent packaging systems: sensors and nanosensors to monitor food quality and safety. *Journal of Sensors* 2016: 4046061.

20 Vanderroost, M., Ragaert, P., Devlieghere, F. et al. (2014). Intelligent food packaging: the next generation. *Trends in Food Science and Technology* 39 (1): 47–62.

21 Hu, J. and Liu, S. (2010). Responsive polymers for detection and sensing applications: current status and future developments. *Macromolecules* 43 (20): 8315–8330.

22 Zelzer, M., Todd, S.J., Hirst, A.R. et al. (2013). Enzyme responsive materials: design strategies and future developments. *Biomaterials Science* 1 (1): 11–39.

23 Genovese, M.E., Abraham, S., Caputo, G. et al. (2018). Photochromic paper indicators for acidic food spoilage detection. *ACS Omega* 3 (10): 13484–13493.

24 Hogan, S.A. and Kerry, J.P. (2008). Smart packaging of meat and poultry products. In: *Smart Packaging Technologies for Fast Moving Consumer Goods* (eds. J. Kerry and P. Butler), 33–59. West Sussex, UK: Wiley.

25 Wanihsuksombat, C., Hongtrakul, V., and Suppakul, P. (2010). Development and characterization of a prototype of a lactic acid-based time–temperature indicator for monitoring food product quality. *Journal of Food Engineering* 100 (3): 427–434.

26 Kerry, J.P., O'Grady, M.N., and Hogan, S.A. (2006). Past, current and potential utilisation of active and intelligent packaging systems for meat and muscle-based products: a review. *Meat Science* 74 (1): 113–130.

27 Williams, J. R., Myers, K. E., Owens, M. M. and Bonne, M. A. (2005). Food quality indicator. US Patent 11,225,410, filed 13 September 2005 and issued 16 March 2006. Food quality sensor international, Inc.

28 Mitsubishi Gas Chemical (2014). AGELESS EYE oxygen indicator. http://wwwmgccojp/eng/products/abc/ageless/eyehtml ().

29 Nopwinyuwong, A., Trevanich, S., and Suppakul, P. (2010). Development of a novel colorimetric indicator label for monitoring freshness of intermediate-moisture dessert spoilage. *Talanta* 81 (3): 1126–1132.

30 Puligundla, P., Jung, J., and Ko, S. (2012). Carbon dioxide sensors for intelligent food packaging applications. *Food Control* 25 (1): 328–333.

31 Fiddes, L.K., Chang, J., and Yan, N. (2014). Electrochemical detection of biogenic amines during food spoilage using an integrated sensing RFID tag. *Sensors and Actuators B: Chemical* 202: 1298–1304.

32 Yam, K.L., Takhistov, P.T., and Miltz, J. (2005). Intelligent packaging: concepts and applications. *Journal of Food Science* 70 (1): R1–R10.

33 Zhang, X., Lu, S., and Chen, X. (2014). A visual pH sensing film using natural dyes from Bauhinia blakeana Dunn. *Sensors and Actuators B: Chemical* 198: 268–273.

34 Musso, Y.S., Salgado, P.R., and Mauri, A.N. (2016). Gelatin based films capable of modifying its color against environmental pH change. *Food Hydrocolloids* 61: 523–530.

35 Genovese, M.E., Caputo, G., Nanni, G. et al. (2017). Light responsive silk nanofibers: an optochemical platform for environmental applications. *ACS Applied Materials & Interfaces* 9 (46): 40707–40715.

36 Zhai, X., Shi, J., Zou, X. et al. (2017). Novel colorimetric films based on starch/polyvinyl alcohol incorporated with roselle anthocyanins for fish freshness monitoring. *Food Hydrocolloids* 69: 308–317.

37 Devarayan, K. and Kim, B.S. (2015). Reversible and universal pH sensing cellulose nanofibers for health monitor. *Sensors and Actuators B: Chemical* 209: 281–286.

38 Yoshida, C.M.P., Maciel, V.B.V., and Mendonça, M.E.D. (2014). Chitosan biobased and intelligent films: Monitoring pH variations. *LWT - Food Science and Technology* 55 (1): 83–89.

39 Silva-Pereira, M.C., Teixeira, J.A., Pereira-Junior, V.A. et al. (2015). Chitosan/corn starch blend films with extract from *Brassica oleraceae* (red cabbage) as a visual indicator of fish deterioration. *LWT - Food Science and Technology* 61 (1): 258–262.

40 Kuswandi, B., Jayus, R., Oktaviana, A. et al. (2014). A novel on-package sticker sensor based on methyl red for real-time monitoring of broiler chicken cut freshness. *Packaging Technology and Science* 27 (1): 69–81.

41 Zhong, H., Liu, C., Ge, W. et al. (2017). Self-assembled conjugated polymer/chitosan-graft-oleic acid micelles for fast visible detection of aliphatic biogenic amines by "Turn-On" FRE. *ACS Applied Materials & Interfaces* 9 (27): 22875–22884.

42 Genovese, M.E., Colusso, E., Colombo, M. et al. (2017). Acidochromic fibrous polymer composites for rapid gas detection. *Journal of Materials Chemistry A* 5 (1): 339–348.

43 Vu, C.H.T. and Won, K. (2013). Novel water-resistant UV-activated oxygen indicator for intelligent food packaging. *Food Chemistry* 140 (1): 52–56.

44 Jung, J., Lee, K., Puligundla, P. et al. (2013). Chitosan-based carbon dioxide indicator to communicate the onset of kimchi ripening. *LWT - Food Science and Technology* 54 (1): 101–106.

45 Dos Santos Pires, A.C., Soares, N.D.F.F., Da Silva, L.H.M. et al. (2011). A colorimetric biosensor for the detection of foodborne bacteria. *Sensors and Actuators B: Chemical* 153 (1): 17–23.

46 Rokka, M., Eerola, S., Smolander, M. et al. (2004). Monitoring of the quality of modified atmosphere packaged broiler chicken cuts stored in different temperature conditions: B. Biogenic amines as quality-indicating metabolites. *Food Control* 15 (8): 601–607.

47 Ruiz-Capillas, C. and Jimenez-Colmenero, F. (2005). Biogenic amines in meat and meat products. *Critical Reviews in Food Science and Nutrition* 44 (7–8): 489–599.

48 Korkeala, H., Alanko, T., Makela, P. et al. (1990). Lactic acid and pH as indicators of spoilage for vacuum-packed cooked ring sausages. *International Journal of Food Microbiology* 10 (3–4): 245–253.

49 Zia, J., Mancini, G., Bustreo, M. et al. (2021). Porous pH natural indicators for acidic and basic vapor sensing. *Chemical Engineering Journal* 403: 126373.

50 Wu, S., Wang, W., Yan, K. et al. (2018). Electrochemical writing on edible polysaccharide films for intelligent food packaging. *Carbohydrate Polymers* 186: 236–242.

51 Majdinasab, M., Hosseini, S.M.H., Sepidname, M. et al. (2018). Development of a novel colorimetric sensor based on alginate beads for monitoring rainbow trout spoilage. *Journal of Food Science and Technology* 55 (5): 1695–1704.

52 Kuswandi, B., Larasati, T.S., Abdullah, A. et al. (2012). Real-time monitoring of shrimp spoilage using on-package sticker sensor based on natural dye of curcumin. *Food Analytical Methods* 5 (4): 881–889.

53 Kuswandi, B., Maryska, C., Abdullah, A. et al. (2013). Real time on-package freshness indicator for guavas packaging. *Journal of Food Measurement and Characterization* 7 (1): 29–39.

54 Shukla, V., Kandeepan, G., and Vishnuraj, M.R. (2015). Development of on package indicator sensor for real-time monitoring of meat quality. *Veterinary World* 8 (3): 393–397.
55 He, T., Wang, H., Chen, Z. et al. (2018). Natural Quercetin AIEgen Composite film with antibacterial and antioxidant properties for in situ sensing of Al^{3+} residues in food, detecting food spoilage, and extending food storage times. *ACS Applied Bio Materials* 1 (3): 636–642.
56 Xiaowei, H., Xiaobo, Z., Jiewen, Z. et al. (2015). Monitoring the biogenic amines in Chinese traditional salted pork in jelly (Yao-meat) by colorimetric sensor array based on nine natural pigments. *International Journal of Food Science & Technology* 50 (1): 203–209.
57 Schaude, C., Meindl, C., Fröhlich, E. et al. (2017). Developing a sensor layer for the optical detection of amines during food spoilage. *Talanta* 170: 481–487.
58 Smolander, M., Hurme, E., Latva-Kala, K. et al. (2002). Myoglobin-based indicators for the evaluation of freshness of unmarinated broiler cuts. *Innovative Food Science & Emerging Technologies* 3 (3): 279–288.
59 Genovese, M.E., Athanassiou, A., and Fragouli, D. (2015). Photoactivated acidochromic elastomeric films for on demand acidic vapor sensing. *Journal of Materials Chemistry A* 3 (44): 22441–22447.
60 Vu, C.H.T. and Won, K. (2014). Leaching-resistant carrageenan-based colorimetric oxygen indicator films for intelligent food packaging. *Journal of Agricultural and Food Chemistry* 62 (29): 7263–7267.
61 Won, K., Jang, N.Y., and Jeon, J. (2016). A Natural component-based oxygen indicator with in-pack activation for intelligent food packaging. *Journal of Agricultural and Food Chemistry* 64 (51): 9675–9679.
62 Lopez-Rubio, A., Almenar, E., Hernandez-Munoz, P. et al. (2004). Overview of active polymer based packaging technologies for food applications. *Food Reviews International* 20 (4): 357–387.
63 Rukchon, C., Nopwinyuwong, A., Trevanich, S. et al. (2014). Development of a food spoilage indicator for monitoring freshness of skinless chicken breast. *Talanta* 130: 547–554.

Part V

Technological Developments in the Engineering of Biocomposite Materials for Food Packaging Applications

15

Biopolymers in Multilayer Films for Long-Lasting Protective Food Packaging: A Review

Ilker S. Bayer

Smart Materials, Istituto Italiano di Tecnologia, Via Morego 30, Genova 16163, Italy

15.1 Introduction

Multilayer, co-extruded laminate films have become increasingly important for applications in the food industry. Laminated flexible films are mainly applied in the packaging of products such as fresh pasta, meats, and cut vegetables to prolong their shelf life [1]. Commercial multilayer films are generally made up of several layers, ranging from 3 to 9, containing different polymers with/without interlayer aluminum [2]. The outer layers consist of common, water barrier polymers with good mechanical properties such as polyethylene (PE) and polystyrene (PS), and the inner layers consist of other synthetic polymers that offer good gas-barrier properties such as polyvinyl chloride (PVC) and polyethylene terephthalate (PET). These layers are generally bound by various types of adhesives [3]. For effective food packaging applications, high oxygen and water vapor barriers are the foremost requirements for preserving the quality of foods throughout their whole life cycle [4]. Today, some biopolymer films are also being adapted in tandem with oil-derived polymers or aluminum to address sustainable food packaging trends and reducing recycling [5]. Conventional plastics, manufactured from fossil fuels not only consume nonrenewable and finite resources, but also impact heavily on waste disposal. In order to substitute these nonrenewable materials, current research efforts are focused on the development of sustainable flexible laminate structures, while maintaining the functional properties of the resulting packaging materials [6–8]. A standard multilayer liquid food container structure, for instance, is given in Figure 15.1. This is built on paper board as support with several layers of polyethylene and aluminum bonded with adhesives [9].

Proteins and biopolymers as either composites or laminates can be implemented in food packaging formulations, and as such, they can replace some major parts of nonrenewable multilayer packing materials with the eventual aim for 100% bio-based multilayer food packaging [10, 11]. Introduction of polysaccharides, plant/animal-based proteins, and other biodegradable resins into food packaging has started with applying them as coatings or laminates on conventional flexible food packaging polymers such as polypropylene (PP), polyethylene, and polystyrene [12–21]. For instance, whey protein isolate

Sustainable Food Packaging Technology, First Edition. Edited by Athanassia Athanassiou.

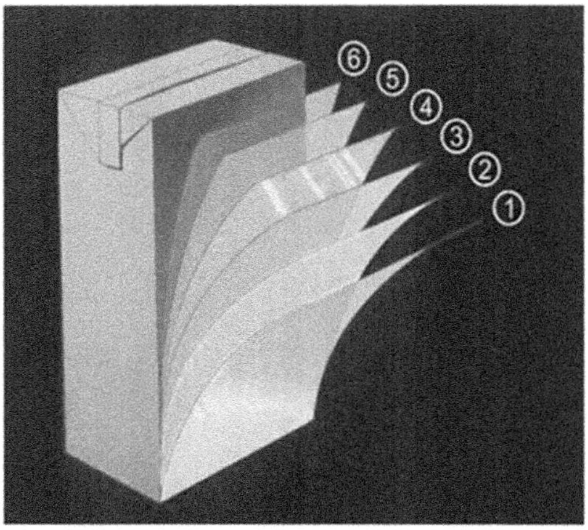

Figure 15.1 Schematic representation of a liquid food product (milk) packaging comprising of six laminated layers. Source: Modified from Arvanitoyannis et al. 1998 [9].

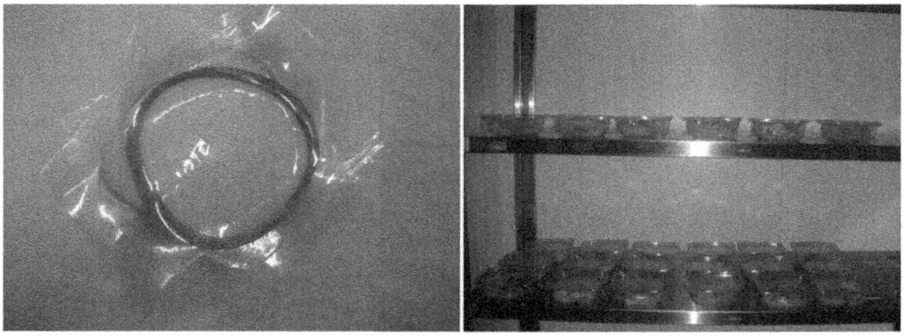

(a) (b)

Figure 15.2 (a) Fresh-cut apples packed in whey protein isolate film and (b) place of storage of the samples. Source: Gällstedt et al. [22]. © 2001, John Wiley and Sons, Inc.

(WPI) and its blends have been extensively utilized as biodegradable films in combination with conventional plastics as shown in Figure 15.2 [22].

Similarly, various polysaccharides, such as starches, chitin, chitosan, alginates, and carrageenan (CR) have been used in tandem with conventional petroleum-based flexible films [10, 23]. Figure 15.3 summarizes naturally occurring biopolymers originating from different sources such as agro-waste, marine industry, meat/animal processing, and microbial/fermentation-based sectors [24].

This chapter provides an introduction to laminated food packaging materials and requirements, and highlights current state-of-the-art on the use of bio-based polymers as films or coatings. It also aims to demonstrate that bio-based food packaging materials need to have tailored mechanical properties such as low brittleness and adequate tensile strength, a sufficient barrier for oxygen, CO_2,

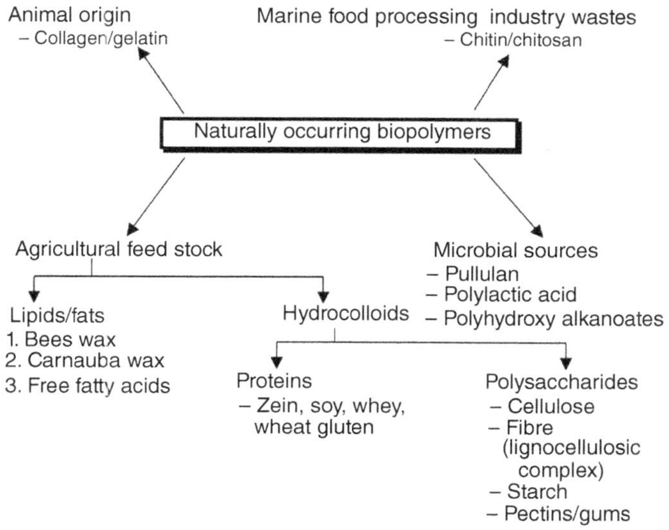

Figure 15.3 Naturally occurring biopolymers that can be used in the design, fabrication, and testing of bio-based food packaging materials including flexible multilayer systems. Source: Tharanathan [24]. © 2003, Elsevier.

and aromatic flavors, high UV-resistance, and high water retention-capacity to contain food moisture, and prevent humidity-induced degradation. As such, various bio- and natural polymers are considered as promising materials to compete with oil-derived plastics for food packaging market. Nevertheless, several short-comings related to material performance and economics of their synthesis and purification should still be addressed toward broad implementation for food packaging. This chapter also discusses advantages and drawbacks of these biopolymers as food packaging coatings or multilayer constituents, and demonstrates how desired properties can be improved by the design of novel composite laminates using different innovative approaches such as surface modification, no adhesive use, crystallization control, and other natural additives.

15.2 Biopolymer Coatings and Laminates on Common Oil-Derived Packaging Polymers

In this section, we will review published studies on biopolymer-coated oil-derived polymer films such as PE, PET, PS, PP, and PVC to name a few. Coated biopolymer films are envisioned to function as gas and solute barriers and complement oil-derived polymer films by minimizing food spoilage and extending the shelf life of foods. In particular, proteins derived from various animal and plant sources have been applied as films and/or coatings. Such bio-based films can be further utilized as smart coatings as they can be fortified with various additives such as antimicrobials, antioxidants, and other nutrients to name a few [15]. According to Guilbert et al. [25], the notion of using edible films as protective coatings for foodstuffs is not new at all. The earliest documented applications appeared in the early 1800s. Commercial patents on edible films for better and extended food shelf life were filed in the 1950s that targeted film production for frozen meat, poultry, and sea-food using alginates, fats, gums, and starches [15]. The highly cited work of Guilbert et al. [25] also presented an overview of the gas barrier properties of a range of edible and synthetic films in order to illustrate the influence of the film composition as shown in Table 15.1. They indicated that gas barrier properties depend on the relative humidity (RH) and conditions of the measurements, and as the RH increases the gas barrier (particularly oxygen gas) tend to decrease/worsen. The addition of lipid components (beeswax, glyceride esters) to wheat gluten (WG)based films caused a marked decrease in permeability. This may be attributed to the hydrophobicity of the lipids as they can reduce water uptake of the biopolymers, thereby preventing swelling of the polymers [25].

Lukowiak et al. [26] demonstrated polyglycerol (PG)coated PP surfaces for protein and bacteria resistance. PG-coated PP films were produced by a two-step approach using plasma bromination of PP and subsequently by grafting hyperbranched polyglycerols having amino functionalities. They investigated the effect of molecular weights and density of reactive linkers on the grafting efficiency, the protein adsorption, and bacteria attachment of the PG-grafted PP. PG with low amine content was shown to prevent bacteria attachment as shown in Figure 15.4.

Hong and Krochta studied oxygen permeation characteristics of WPI coatings on PP films [20]. Heat-denatured aqueous solutions of WPI including common plasticizers such as glycerol, sorbitol, sucrose, propylene glycol, and polyethylene glycol were coated on PP films previously treated with corona discharge to increase surface adhesion. Sucrose-plasticized WPI coatings had the best oxygen barrier property. Oxygen permeability (OP) of the resulting WPI-coated films increased significantly with temperature, in agreement with the Arrhenius model. Oxygen permeability of the coated films increased (worsened) exponentially with relative humidity. Similarly, Lee et al. [15] prepared three different protein coatings (corn zein, WPI, and soy protein isolate, SPI) with various plasticizers on common PP films. Protein and plasticizer types exerted significant influence on the physical properties of the resulting coated films. They observed no compatibility between corn zein and PP, however, both WPI and SPI formed satisfactory coatings once the proper plasticizer was applied. They also indicated that WPI or SPI coatings can be loaded with bioactive compounds for active food packaging systems. For instance, they have investigated antimicrobial effects of the nisin-impregnated protein-coated PP films on typical food spoilage and pathogen bacteria [15]. Schmid et al. [16] presented an introduction and review of food packaging studies on the use of whey-based coatings, their barriers properties, and discussed key properties of innovative aspects of multilayer packaging containing whey-based proteins. They demonstrated that certain whey protein formulations had barrier properties comparable to ethylene-vinyl acetate copolymers (EVA) with an oxygen barrier (oxygen transmission rate, OTR) levels less than 2 cm^3/m^2 d bar as shown in Figure 15.5.

Theapsak et al. [17] modified PE film surfaces with dielectric barrier discharge (DBD) plasma under medium vacuum pressure in air gas. The hydrophilization

Table 15.1 Several natural polymer barrier properties and in certain cases in tandem with other polymers as bilayers.

Film type	Oxygen permeability pO_2 ml mm/(m^2 d atm)	Carbon dioxide permeability pCO_2 ml mm/(m^2 d atm)	Film selectivity (pCO_2/pO_2)	Relative humidity (%)
Pectin	57.5	—	—	87
Pectin	258.8	4132	16	96
Chitosan	91.4	1553	17	93
Pullulan	3.3	14	4.24	30
Pullulan/arabic gum	3.05	10	3.27	36
(Wheat) gluten	190	4750	25	91
(Wheat) gluten	250	7100	28.4	94.5
Fish proteins	56	—	—	86
Fish proteins	169	2156	12.75	92
Na caseinate	77	462	6	77
Gluten–DATEM	153	1705	11.14	94.5
Gluten–beeswax	133	1282	9.64	91
Na caseinate/Myvacet	83	154	1.85	48
MC/HPMC/fatty acids	46.6	180	3.86	52
MC and beeswax (bilayer)	4	27	6.75	42
Gluten–DATEM and beeswax (bilayer)	< 3	15	>5	56
Gluten–beeswax and beeswax (bilayer)	< 3	13	>5	56
PET[a]	1	—	—	100
Ethylene/polyvinylic alcohol[a]	4	—	—	100
Polyamide 6	9.8	—	—	100
Oriented polypropylene[a]	44	—	—	100
Cellophane[b]	55	—	—	100
Methylcellulose–palmitic acid[c]	78.8	—	—	100

AMs, acetylated monoglycerides; DATEMs, diacetylated tartaric ester of monoglycerides; HPMC, hydroxypropyl-methylcellulose; MC, methylcellulose [25].
Values indicated by a, b and c: Not measured by authors but taken from literature.
Source: Guilbert et al. [25]. © 1996, Elsevier.

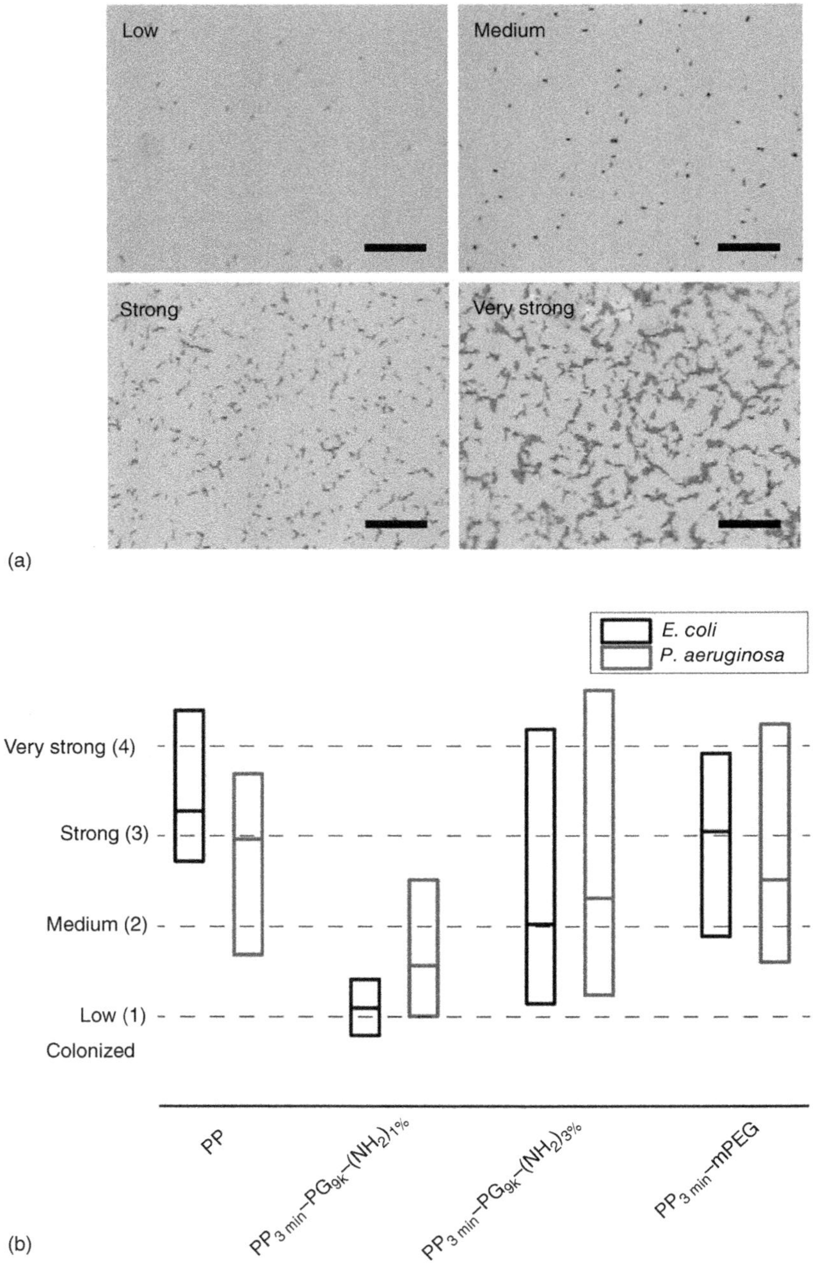

Figure 15.4 (a) Four different classes of bacteria colonization of PG-coated PP surfaces (low, medium, strong, and very strong colonization). Scale bar: 25 μm. (b) The pillar diagram shows range of bacteria attachment on the three minutes plasma brominated polymer coated PP surfaces. Top end of pillars: upper quartile (Q3), middle lines: median, lower end of pillars: lower quartile (Q1). Median and quartiles were estimated by interpolation according to index values 1–4. Source: Lukowiak et al. [26]. © 2015, The Royal Society of Chemistry.

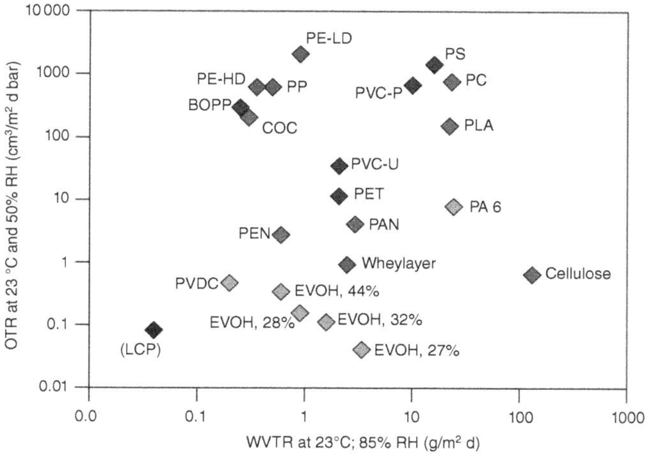

Figure 15.5 Barrier properties of whey-based layer compared to other plastics commonly used in the packaging industry. Source: Schmid et al. 2012 [16]. Licensed under CC-BY-4.0.

of PE surfaces was controlled by water contact angle measurements and mechanical properties were also tested to investigate the optimum plasma time. They immersed the plasma-treated films in chitosan acetate solutions with different concentrations of the chitosan and subsequently evaluated the antibacterial property of the plasma-treated chitosan-coated PE films against gram-negative *Escherichia coli* and gram-positive *Staphylococcus aureus* bacteria [17]. Similarly, Zemljič et al. [27] showed that chitosan-based films formed on PET surface demonstrated sufficient and effective antimicrobial and flexible compatibility with PET. They indicated that chitosan was bonded to the surface of PET plastic films due to its protonated amino groups and it successfully inhibited *E. coli* and *Listeria monocytogenes*, as well as fungi *Candida albicans* proliferation. They immersed the double layer films in a model acidic medium and observed time-dependent proportional decrease of the amino group amount due to chitosan desorption. They proposed this method to reversibly bind chitosan on the surface of the PET on demand.

Carneiro-da-Cunha et al. [28] reported preparation of multilayer films having nanoscale coatings; they characterized surface properties, water vapor permeability (WVP), and thermal and mechanical properties of such nanolayered films. The films were made from two polysaccharides with opposite charges, namely, chitosan and sodium alginate (see Figure 15.6), deposited on to functionalized PET surfaces. Contact angle measurements showed differences in the films with a successively higher number of layers and the films had a WVP of 0.85×10^{-11} g/(m s Pa), while the hardness increased to 0.245 GPa levels. Kurek et al. [18] demonstrated that chitosan coatings on polyethylene films enhanced their gas barrier properties at low relative humidity. In dry conditions, the coated films were up to 1000 times less permeable than the PE film such that the permeation followed the order $P(CO_2) > P(O_2) > P(\text{air})$. They indicated that due to inherent hydrophilicity of chitosan, water vapor barrier properties of the

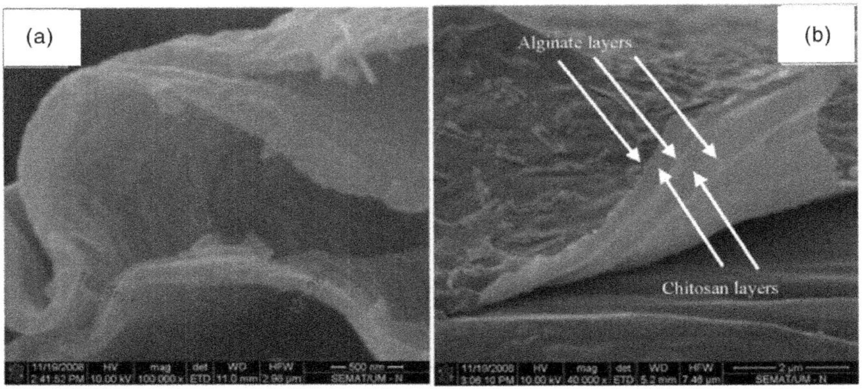

Figure 15.6 Scanning electron microscopy images of surface morphology of the alginate/chitosan films on PET. (a) Thickness of the outermost alginate layer and of five polysaccharide layers and (b) image of nanolayered film cross-section. Source: Carneiro-da-Cunha et al. [28]. © 2010, Elsevier.

PE films were slightly diminished but statistically no significant difference was observed [18].

Ozcalik and Tihminlioglu studied corn zein nanocomposite coatings (CZNC) as a barrier layer on PP films [6]. They measured the effect of layered silicate content in the CZNC coatings on the barrier and surface hydrophobicity of the modified PP films and showed that incorporation of organomodified nanoclays (Cloisite 10A or CLO10A) by solution intercalation into zein matrix significantly enhanced oxygen and water vapor barrier of the films. Particularly, the CZNC coating of PP films reduced the oxygen permeability about four times, while WVP reduced by 30% with 5 wt% OMMT in zein films having a thickness of about 6 μm (see Figure 15.7 for more details). Han et al. [29] fabricated and investigated packaging films consisting of cast polyvinyl alcohol (PVOH) layered PP films with rhubarb ethanolic extracts (REEs) and cinnamon essential oil (CEO) in order to improve fresh beef packaging quality. Fresh beefsteaks were packed into these antimicrobial double-layer films containing different concentrations of REE and CEO. They measured various beef characteristics such as pH, total viable counts, color, weight loss, total volatile base nitrogen (TVB-N), and sensory quality. They showed that all experimental films significantly inhibited bacterial growth with acceptable pH levels and efficiently maintained the quality of fresh beefsteaks during storage. Georgelos et al. [30], in their patent, describe multilayer polyolefin or nylon films laminated to various rigid or soft biopolymer resins such as polylactic polymers or polyhydroxyalkanoates (PHAs) and polycaprolactone (PCL) to name a few. They used various adhesives to maintain continuous bonding among the layers. However, adhesive bonding is an additional step that can increase the complexity and time of fabrication. It can also increase capital and production costs, and reduce production speeds. Finally, Reščček et al. [31] modified PE films as double-layer polyethylene/polycaprolactone (PE/PCL) film. Magnetite and casein were added to the PCL-coating film to improve barrier properties and thermal stability. Significant improvements were measured with regard to mechanical (tensile strength, elongation at break) and thermal properties, while

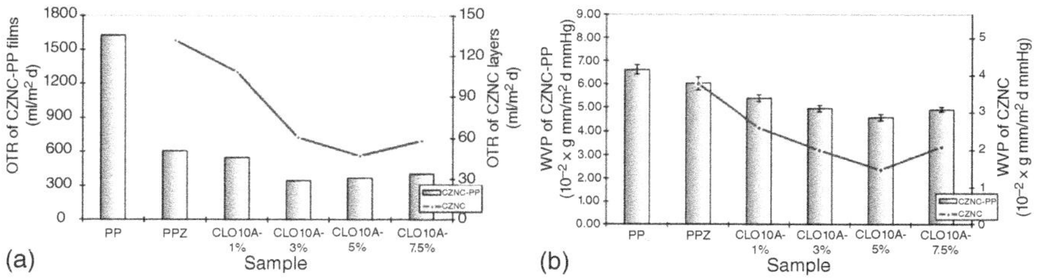

Figure 15.7 (a) Oxygen transmission rates (OTR) values of CZNC–PP films (in columns – left *y*-axis) and single CZNC layers (curve – right *y*-axis). (b) Water vapor permeability (WVP) values of CZNC–PP films (in columns – left *y*-axis) and single CZNC layers (curve – right *y*-axis). Films were labeled according to their clay content as CLO10A-1% to CLO10A-7.5%; PP and PPZ were used for the commercial polypropylene film and plasticized corn zein-coated PP films without nanoclay, respectively. Source: Ozcalik and Tihminlioglu [6]. © 2013, Elsevier.

oxygen barrier permeability properties were somewhat improved. The overall acetic acid migration values were found to be lower (from 1.0 to 4.6 mg/d m^2) than the upper limit set by the general requirements. Specific migration of iron was also measured in the bilayer films and was found to be at levels of μg/l, an order of magnitude lower than mg/kg.

15.3 Multilayer Films Based on Proteins

Proteins have long been implemented in the fabrication of biodegradable, renewable, and edible films. Several cereal and vegetable proteins such as corn zein, wheat gluten, and soy proteins and animal proteins such as milk proteins, collagen, gelatin, keratin, and myofibrillar proteins are commonly used for forming naturally hydrophobic and gas barrier films. In general, two standard approaches are followed: the "wet (or solvent) process" based on dispersion or solubilization of proteins in a solvent medium, and the "dry process" based on the thermoplastic forming and casting. In both cases, several plasticizers have been used since pure protein films tend to be very rigid and brittle. In their review article, Gómez-Estaca et al. [32] indicated that food packaging materials made from proteins can be tuned in many different ways since they can be modified enzymatically, chemically, or physically, giving rise to materials with improved physicochemical properties that are adjustable to each specific application. They reviewed [32] such advances in the use of proteins for the manufacture of packaging materials. They demonstrated main types of protein sources that have been used (soy, whey proteins, gelatins, zein, collagens, milk proteins, wool keratin, wheat proteins, etc.); manufacturing methods (coating on conventional materials, thermoforming, extrusion), with special emphasis on materials made by extrusion; possible strategies for improving the physicochemical properties of the materials and the main future prospects and limitations of plant and animal based proteins. They also stressed that an important application of proteins could be in the design of active packaging due to their characteristic water sensitivity, which allows a greater release of active agents when packaging foods with medium–high water activity.

Recently, Coltelli et al. [33] published an extensive review on the state of the art in the development and properties of protein-based films and coatings and their applicability particularly to cellulose based products. They stressed that innovations to current multilayer (or bilayer) systems can be made by following two different routes, namely, by replacing the two polymers with compostable ones or by replacing the adhesive bonding within multilayer plastic or paper laminates with a protein-based material, which can also improve the barrier properties. Cinelli et al. [34] investigated whey protein layers on the oxygen barrier properties of a commercial compostable plastic film, under the trade name Bio-FlexR F 2110 (Previously, Bio-FlexR 467F) produced by FKuR Kunststoff GmbH, Germany, with a thickness of 58 μm. The BIO-FLEX® trade name indicates blends of co-polyester and poly(lactic acid) (PLA). They showed that oxygen barrier properties were significantly improved by the whey protein layer. They also assessed the biodegradability of the whey protein layer itself since natural polymers would not be degradable if cross-linked or blended with certain additives even if they are derived naturally. The material based on denatured whey protein and plasticizer biodegraded readily after application on the commercial film. Guillaume et al. [35] studied how the structural, surface, and water vapor and gas barrier properties of WG-coated paper could be influenced by the features of paper. They employed a surface-treated paper (TP) and an untreated paper (UTP) as substrate layers. Penetration of WG-coating into the paper bulk was more in UTP than TP. They argued that WP-infused UTP formed a significant junction zone resulting in an interpenetrating network of WG–UTP, a composite-like structure, while a bilayer film was obtained for WG–TP. Differences in WG penetration were related to the hydrophilicity, roughness and porosity of the paper used. Even though the extent of penetration did not greatly affect the surface properties (water and oil resistance), the transfer properties (water vapor, O_2 and CO_2) of WG-coated papers were significantly improved when WG-coating penetrated into the bulk, while the WG–TP bilayer behaved as a micro-perforated material. Schmid et al. [36] evaluated the influence of RH on the oxygen permeability and water vapor transmission rate (WVTR) of whey protein monolayer films and whey protein-coated PET. In addition, they also measured the activation energies for the permeability of oxygen, carbon dioxide and nitrogen as well as the permselectivities (a term used to define the preferential permeation of certain species through membranes) under different set of temperatures. Their results showed that the permeability values through the whey protein coating and whey protein film increased with increasing RH. The WVP measured at 50% RH of $(9.3 \pm 0.6)^{-10}$ cm^3 (STP) cm/(cm^2 s Pa) increased to $(16.2 \pm 0.9)^{-10}$ cm^3 (STP) cm/(cm^2 s Pa) at 85% RH. An increase in temperature showed an expected increase of the permeability in both uncoated PET and the whey protein monolayer. The permselectivity of whey protein film was determined and the calculated ratios of the permselectivity ($P(N_2)/P(O_2)/P(CO_2)$): 1/(4–6/(34–35)) somewhat differed from the simplified ratios given in the literature ($P(N_2)/P(O_2)/P(CO_2)$: 1/4/16).

Martucci and Ruseckaite [37] produced three-layer PLA-based films with a gelatin interfacial layer (PLA/Ge/PLA) by using sequential thermal compression. They conducted comprehensive biodegradation experiments on this

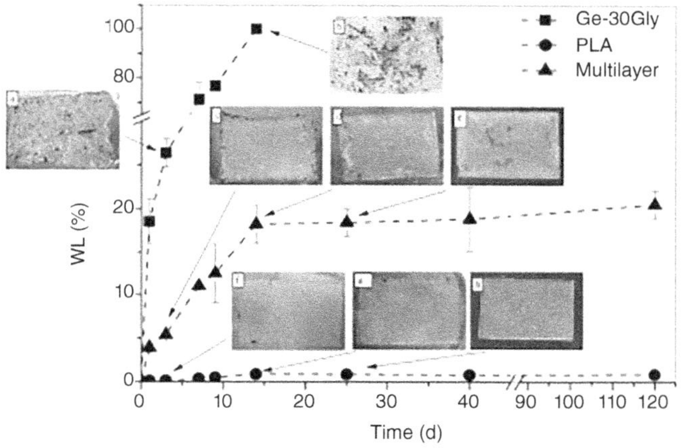

Figure 15.8 Soil burial weight loss (WL) curves of 30 wt% glycerol plasticized gelatin (Ge-30Gly), neat PLA and three-layer sheet against exposure time. Variations of visual aspect are also included: Ge-30Gly (inset a, b), PLA (inset c, d, e) and three-layer sheet (inset f, g, h). Source: Martucci et al. [37]. © 2015, Elsevier.

multilayer bio-based film. The soil burial tests showed that the degradation of the multilayer in Pinocha-type soil environment was sluggish. The multilayer sheet only lost 20% of weight after 120 days of indoor soil burial, while control PLA was practically unaltered during the same time period. They found that the gelatin inner layer enhanced water accessibility around the three-layer sheet soil micro-environment. They argued that hydrolysis of gelatin (chemical and enzymatic) prompts water penetration and consequently the abiotic hydrolysis of PLA. The products of the abiotic PLA hydrolysis possibly stimulate the action of more active soil microorganisms against PLA. Their results indicated that multilayer sheets need much longer time for complete biodegradation under realistic indoor soil conditions, where the average temperature never exceeded $20 \pm 2\,°C$ and soil humidity content was kept at 40% (see Figure 15.8).

Wang et al. [38] demonstrated design of multilayered film with directional water barrier properties based on water-soluble (kafirin) and water-insoluble (gelatin) proteins by sequential solvent casting using acetic acid solutions. Multilayered films showed good UV light protection property, and Kafirin layer improved the transparency of multilayered film. They argued that although kafirin side was more hydrophilic than gelatin, the water vapor penetration capacity was remarkably lower when moisture transfer took place from the kafirin side to the gelatin side of the multilayer. They reported the moisture absorption induced compact arrangement of the kafirin layer, hence reducing further moisture penetration and migration (see Figure 15.9). They suggested that the multilayer film must be used in such a way that the kafirin layer should face the airside and the gelatin layer the package's inner compartment to obtain the optimum moisture resistance against environment and moisture retaining property within the package. They argued that such multilayer films can lead to applications in fresh fruits and vegetables packaging.

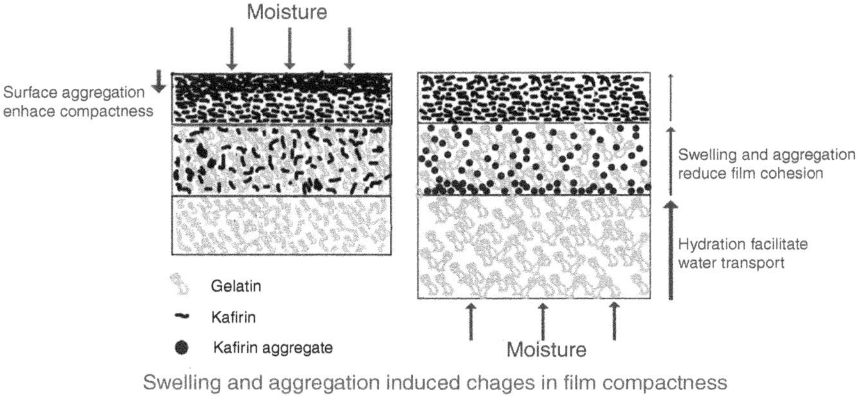

Figure 15.9 Schematic illustration of the mechanism for one-way water barrier property of multilayered film. Source: Wang et al. [38]. © 2018, Elsevier.

Figueroa-Lopez et al. [39] evaluated the effect of electrospun PCL/black pepper oleoresin (OR) extract blends as a barrier coating on the morphology, thermal, mechanical, antimicrobial, oxygen, and water vapor barrier properties of solvent-cast gelatin films. The antimicrobial activity of the multilayer system was tested against *S. aureus* strains for 10 days that indicated no bacterial growth or contamination. The results showed that the multilayer system containing PCL and OR increased the thermal resistance, elongated the gelatin film, and significantly reduced water vapor transport properties.

Hosseini et al. [40] prepared multilayer and optically clear films from PLA and fish gelatin (FG), by solvent-casting technique. PLA was kept as the outer layer. The scanning electron microscopy (SEM) images of multilayer films showed that the outer PLA layers were fused into the inner FG layer forming a continuous film. The measured oxygen permeability of the multilayer film ($5.02\,cm^3/m^2$ d bar) decreased more than 8-fold compared to pure PLA film, and the WVP of the multilayer film ($0.125\,g\,mm/kPa\,h\,m^2$) also decreased 11-fold compared to pure FG film. They asserted that the PLA lamination profoundly increased the water resistance of the bare gelatin film. The tensile strength of the multilayer films (~27 MPa) were shown to be better than pure FG films (~9 MPa). Multilayer films also featured very uniform surface morphology and the root-mean-square roughness value of the PLA/FG/PLA multilayer film (see Figure 15.10) was not markedly different from virgin PLA and FG indicating that the PLA layers were evenly coated over the FG film.

Xia et al. [41] fabricated gradient multilayer films of zein/gelatin with one-way water barrier and prolonged antioxidant release (tea polyphenol) properties using a layer-by-layer solvent-casting approach. Multilayer films had better UV-protection and higher transparency than neat gelatin and zein films, respectively. They showed that the mechanical response of multilayer films can be regulated by the zein/gelatin ratio in middle layer. Although the zein side of multilayer films was more hydrophilic than the gelatin side, WVP value declined when the moisture transported from the zein to gelatin side. Tea polyphenol

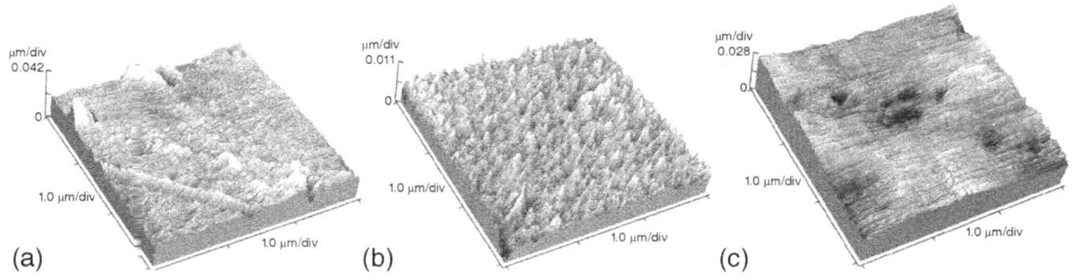

Figure 15.10 3D AFM images of (a) (PLA), (b) (FG), and (c) multilayer films. Source: Hosseini et al. [40]. © 2016 Elsevier.

loaded multilayer films were effective in controlling water loss of freshly cut fruits and reduce microbial growth. Rhim et al. [42] made multilayer films from a SPI inner layer and PLA outer layer using solvent casting. The tensile strength of the multilayer film increased more than fivefold compared with that of the SPI film. The mechanical properties of the multilayer film were comparable to those of low-density polyethylene (LDPE) or high-density polyethylene (HDPE) films. The lamination of PLA layers on SPI film enabled good gas-barrier properties combining low WVP of PLA and low OP of SPI. The WVP of the multilayer film (6.6×10^{-4} kg m/m^2 s Pa) decreased 40-fold compared with that of the SPI film, and the OP of the multilayer film (2.4×10^{-18} m^3 m/m^2 s Pa) decreased more than 26-fold compared with neat PLA film. A very recent work by Li et al. [43] showed that the development of a WG-based PLA-WG-PLA multilayer film highly enhanced the barrier properties of PLA. The most efficient packaging thickness was found to be 60 µm, with a WG coating of 20 µm. It clearly amplified the barrier properties of neat PLA (60 µm) to oxygen more than 20 times and approximately 20% to water molecules. They measured the transfer rates in realistic conditions for food packaging application (25 °C, 50% RH) and not only in dry conditions. They used hot-pressing and high-pressure homogenization to compatibilize WG and PLA layers in the construction of the multilayer films.

A review by Fabra et al. [44] concluded that when biopolymers or proteins need to be used in food packaging, multilayering and (nano)-composite layering will be indispensable. They call this approach biolayering. The biolayering refers to biopolymer-based monolayer coatings, multilayer structures, layer-by-layer nanoassemblies, and electrospun intermediate nanostructured layers and coatings. They determined that biolayering has been used as a highly efficient alternative to nanocomposite films for improving physical and barrier properties of bio-based materials without significant changes in mechanical performance.

15.4 Multilayer Films Based on Polysaccharides

Biodegradable films based on polysaccharides are known to be an effective barrier to gas transfer such as O_2 and CO_2, but at the same time, these materials are generally very hydrophilic with poor water vapor barrier properties. However, alginate and carrageenan can be applied as relatively thick films on foodstuff surfaces in order to absorb water, resulting in a temporary moisture protection against moisture loss from the food. As such, foodstuff does not lose significant moisture until the coating gets dehydrated. The review by Cazón et al. [21] summarizes the advances in polysaccharide-based films and coatings for food packaging. The polysaccharide materials for biodegradable packaging films and coatings are mainly cellulose, chitosan, starch, pectin, and alginate. These polysaccharides are known to form films and coatings with good barrier properties against the transport of gases. On the other hand, they are rigid and brittle polymers with high sensitivity to humidity. For example, tensile strength values of films based on high amylose starch or chitosan are comparable to those found for HDPE films [21]. Flexibility and elongation are the main concerns, which are far from the desirable values found in synthetic polymers. However, they can be improved by combining polysaccharides with other biopolymers. Current research is focused on finding the correct combination between the polymers and food products applied. Efficiency and functional properties of polysaccharide films depend on many factors such as use of additives, plasticizers, solvents for casting, pH levels during processing, relative humidity, etc. Similarly, Ferreira et al. [45] review these issues with emphasis on polysaccharide-based food packaging materials. These reviews have clearly shown that significant progress have been made with polysaccharides, particularly in multilayer packaging and multilayer composite packaging as Section 15.4 will review.

For instance, Sanyang et al. [46] developed environmentally friendly bilayer films from sugar palm starch (SPS) and PLA by sequential solvent casting from water and chloroform. They reported physical, mechanical, thermal, and water barrier properties of the SPS–PLA bilayer films. Addition of 50% PLA layer onto 50% SPS layer (SPS50-PLA50) increased the tensile strength of neat SPS film from 7.74 to 13.65 MPa but reduced their elongation at break from 46.66% to 15.53%. Existence of PLA layer in the system significantly reduced WVP as well as the water uptake and solubility of bilayer films. Furthermore, SEM image of SPS50-PLA50 revealed some problems related to interfacial adhesion between SPS and PLA that would require some surface treatment for PLA before deposition of the SPS layer.

Wang et al. [47] published a comprehensive review on advances made in chitosan-based films for food packaging [47]. They also reviewed chitosan coatings since chitosan is intrinsically antimicrobial in nature. Ferreira et al. [48] prepared bilayer films of a microbial polymer FucoPol, an exopolysaccharide that is rich in fucose and produced by the bacterium Enterobacter A47, and chitosan, and characterized various properties such as optical, morphologic, hygroscopic, mechanical, and barrier for assessing their suitability in food packaging. They used sequential solvent casting to prepare the bilayer films. Films still demonstrated high swelling degree in water (263.3%) and a high WVP (0.75×10^{-11} mol/m s Pa). However, they had low permeability to O_2 and CO_2 (0.47×10^{-16} mol m/m^2 s Pa and 5.8×10^{-16} mol m/m^2 s Pa, respectively). Tensile tests indicated relatively flexible and resistant films with an elongation at break of 38% and an elastic modulus of 137 MPa. Gu et al. [49] used a layer-by-layer (LBL) approach (see Figure 15.11a) to produce alternating layers of sodium alginate (ALG)/polyethyleneimine (PEI) on biaxially oriented poly(lactic acid) (BOPLA) films in order to produce multilayer bio-based films with low gas permeability.

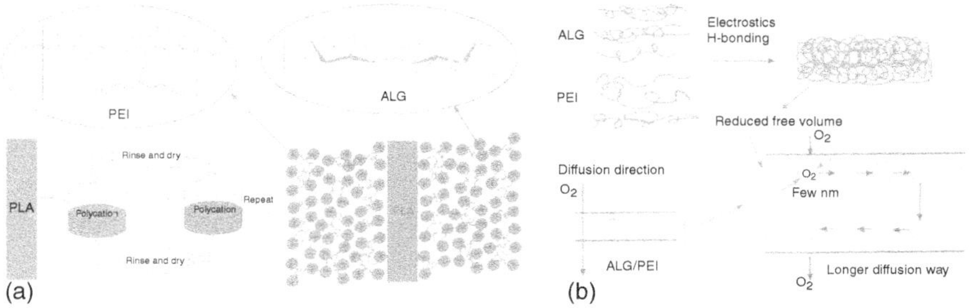

Material	Oxygen permeability (cm³ µm/m² day kPa)
$(PEI/NFC)_{50}$ assembly	0.34
$(PEI/CMC)_{50}$ assembly	0.71
NFC (carboxymethylated)[15]	0.52
Poly(vinyl alcohol)[48]	0.21
Ethylene vinyl alcohol[48]	0.30
High-density polyethylene[2]	427
Amylopectin (glycerol 40 wt%)[48]	14
Amylose (glycerol 40 wt%)[48]	7
O-Acetyl-galactoglucomannan[14]	2.0

(c)

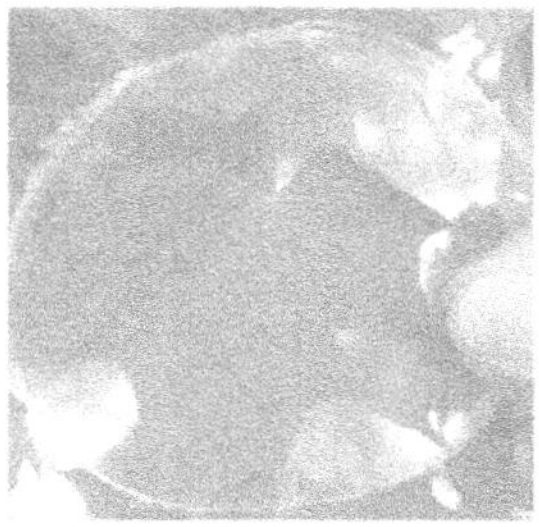

(d)

Figure 15.11 (a) Schematic of the LBL deposition process and a simplified image of the resulting multilayer ALG/PEI structure deposited on the BOPLA films. (b) Schematic of reduced free volume in highly interdiffused ALG/PEI layers and O_2 diffusion path. Source: (a, b) Gu et al. [49]. © 2013, Elsevier. (c) Oxygen permeability data at 23 °C and 50% RH for layer-by-layer assemblies of PEI/NFC and PEI/CMC and other published data on some renewable and synthetic polymers. Source: (c) Aulin et al. [50]. © 2013, American Chemical Society. (d) Multilayer nanofibrillated cellulose (NFC) or carboxymethyl cellulose (CMC) with guar gum films showing good transparency. Please refer to Ref. [49] for superscripts in the Table in (c).

The number of layers of both ALG and PEI polymers increased up to 30 layers with a final thickness of about 4 µm. After 30 ALG/PEI layers were deposited, the final film had an OTR of $1.22\,cm^3/(m^2\,d\,atm)$. When normalized by thickness, the resulting OP was found to be less than $3.8 \times 10^{-17}\,cm^3\,cm/cm^2\,s\,Pa$. This was about 3 orders of magnitude lower than that of uncoated BOPLA film ($1.8 \times 10^{-14}\,cm^3\,cm/cm^2\,s\,Pa$). They also showed that the multilayer-coated BOPLA films had high optical clarity and satisfactory tensile properties. A schematic illustration of reduced free volume in interpenetrating ALG/PEI layers and O_2 diffusion path are shown in Figure 15.11b. Figure 15.11b shows the tangling/interpenetration of the two polymer networks. The circles in the lower part denote reduced free volume due to the interpenetrating network. They argued that the reduced free volume results in longer time for oxygen molecules to travel through, causing better barrier effect in these films [49].

Likewise, layer-by-layer (LbL) deposition method was used by Aulin et al. [50] to form alternating layers of nanofibrillated cellulose (NFC) or carboxymethyl cellulose (CMC) with a branched, cationic polyelectrolyte, PEI on flexible PLA substrates. They produced optically transparent nanocellulosic films with tunable gas barrier properties. Up to 50 composite layers of PEI/NFC and PEI/CMC were deposited on PLA demonstrating OP values of 0.34 and $0.71\,cm^3\,\mu m/m^2\,d\,kPa$ at

23 °C and 50% relative humidity, respectively, matching the performance values of polymers such as PVOH and ethylene-vinyl acetate (see Figure 15.11c). The oxygen permeability of these multilayer nanocomposites outperforms those of pure NFC films prepared by solvent casting. The PEI/NFC composite layers showed a highly entangled network of nanofibrils. Flexible and transparent all-polysaccharides multilayer films were prepared by Dai et al. [51] using cationic guar gum (CGG) and TEMPO-mediated oxidized cellulose nanofibrils (TOCNs) based on LbL casting (see Figure 15.11c). The multilayer CGG/TOCNs films had good barrier and solvent resistance properties. For instance, films consisting of eight layers of CGG/TOCNs had low WVTR (68.52 g/(m^2 d atm)) and OTR (5.56 cm^3/(m^2 d atm)). Moreover, the films resisted oil penetration, and they did not dissolve in organic solvents even after 48 hours of soaking.

Sipahi et al. [52] used multilayered antimicrobial alginate-based edible coatings to extend the shelf life of fresh-cut watermelon (*Citrullus lanatus*). A set of solutions comprising sodium alginate (0.5, 1, 2 g/100 g), beta-cyclodextrin and microencapsulated *trans*-cinnamaldehyde (natural antimicrobial agent), pectin, and calcium lactate were made into a multilayer coating (see Figure 15.12a). The samples were coated using the layer-by-layer dipping technique and stored at 4 °C for 15 days. They monitored texture, color, weight loss, Brix (sugar content of an aqueous solution), pH, and growth of total coliforms, yeasts and molds, aerobics, and psychrotrophs every three days throughout storage. Controls were uncoated washed fruits. They also conducted a real consumer acceptance study with high acceptance ($P < 0.05$) of the coated samples. They indicated that certain coatings extended the shelf life of fresh-cut watermelon from 7 (control) to 12–15 days. Similarly, Martiñon et al. [53] developed a multilayered edible coating with an antimicrobial agent to extend the shelf life of fresh-cut cantaloupe kept at 4 °C. The films were made of layers of chitosan and pectin with encapsulated *trans*-cinnamaldehyde as an antimicrobial agent (see Figure 15.12b, c). They also measured changes in fruits texture, color, moisture, acidity, and pH. Sensory testing was carried out to support the measurements. Their multilayer films extended the shelf life of fresh-cut cantaloupe for up to nine days.

In sustainable food packaging, another important aspect is to develop biodegradable multilayer materials that also resist migration of oil, in other words a good barrier against not only gases but also oils. Chitosan in this respect is considered promising. Ham-Pichavant et al. [54] aimed at potentially replacing fluorocarbon-treatment of paper-based materials by chitosan coating to produce oil barrier packaging. As a model, lipid interactions between oleic acid and chitosan were particularly investigated. The ability of chitosan to interact with lipids was shown to be pH sensitive. They proved that in acidic conditions, the polycationic chitosan has adsorbed the lipid from fatty acid emulsions, which had been attributed to its capacity to bind with anionic lipid molecules. The fat barrier properties of paper and chitosan-coated paper were also compared in control experiments. They argued, however, that chitosan coating cost remained high compared with fluorinated resins. Also Despond et al. [55] made a natural multilayer material based on parchment paper, chitosan, and carnauba wax for food packaging applications. They used a lab-scale roll coating apparatus to apply sequential layers of chitosan and wax from aqueous solutions and emulsions,

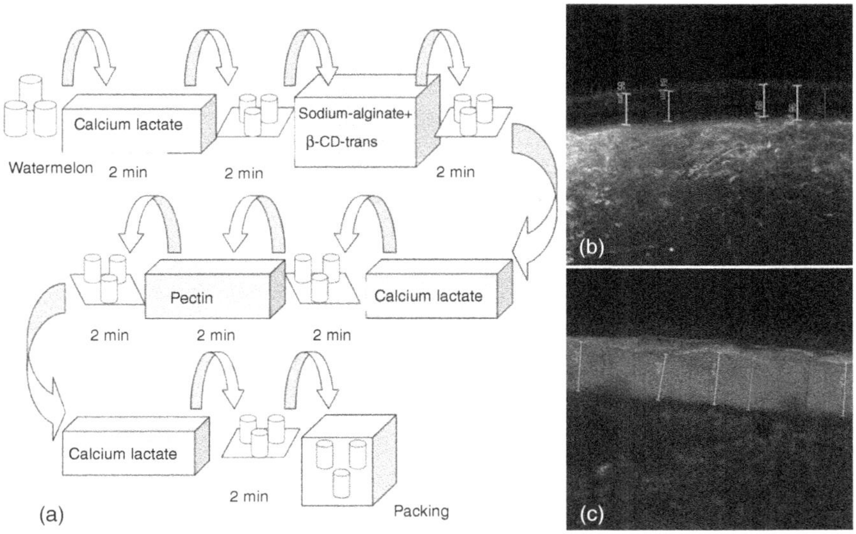

Figure 15.12 (a) Layer-by-layer coating manufacturing process. Source: (a) Sipahi et al. [52]. © 2013, Elsevier. (b) Microscopic image of cross-section of multilayer edible coated (0.5 g/100 g chitosan, 1 g/100 g pectin, 2 g/100 g antimicrobial) cantaloupe showing thickness of coating. (c) Cross-section of multilayer edible coated (2 g/100 g chitosan, 2 g/100 g pectin, 2 g/100 g antimicrobial) cantaloupe showing thickness of coating. Source: (b, c) Mauricio [53]. © 2014, Elsevier.

respectively. Chitosan layer controlled the level of gas barrier properties in the anhydrous state. A decrease in CO_2 permeability in the anhydrous state from 1635 barrer (permeability of neat parchmentized paper) to 0.08 barrer for the chitosan–paper bilayer was measured (1 barrer = 3.348×10^{-16} mol m/(m^2 s Pa)). Due to the hydrophilic character of chitosan, they did not make gas barrier measurements under the hydrated state. A layer of carnauba wax, a hydrophobic material, was then coated on the chitosan side of the bilayer. For a coating of 2 g of wax per square meter, gas P values lower than 0.5 barrer were obtained in the hydrated state.

Thermoplastic starch has also been utilized in multilayer bio-laminates [44]. The work of Ortega-Toro et al. [56] demonstrated compression molded bilayer films consisting of one layer of PCL with either pure thermoplastic starch (S) or thermoplastic starch with 5% PCL (S95). Before compression, aqueous solutions of ascorbic acid or potassium sorbate were sprayed onto the S or S95 layers in order to plasticize them and enable adhesion. S95–PCL bilayer films had good mechanical performance, especially when potassium sorbate was added at the interface. These films also had improved barrier properties to water vapor (up to 96% compared to net starch films) and oxygen (up to 99% compared to neat PCL). Fabra et al. [57] studied the effect of adding different biopolyester electrospun coatings fabricated from PCL, PLA, and polyhydroxybutyrate (PHB) on oxygen and water vapor barrier properties of a thermoplastic corn starch (TPCS) film. Presence of these electrospun biopolyester coatings led to an exponential oxygen and WVP drop as the amount of the electrospun coating quantity and thickness increased. Again Fabra et al. [58] constructed antioxidant multilayer packaging film based on TPCS and electrospinning. They first evaluated the effect of adding electrospun outer layers made of PCL on the microstructure and barrier properties of a TPCS film. Secondly, β-carotene was encapsulated by PCL during electrospinning and the fibers were directly electrospun as coating layers onto both sides of TPCS films. Barrier properties of the TPCS-based films were improved by the the electrospun coatings. However, they found that β-carotene addition did not improve barrier properties compared to plain PCL nanofiber coatings but both permeability values were 75% lower than pure TPCS. They attributed this to the fact that stability of the β-carotene was greatly affected by the annealing step (used to improve the adhesion between layers) even when it was encapsulated into the electrospun PCL layer.

Lustig and Schuetz [59] disclosed a patented multilayer packing that can be used to cook meat *in situ*. It is a meat product package comprising a starch-based multilayer film and an *in situ* aqueous medium-cooked meat product type. The meat was placed in a flexible tube formed of this film, hermetically sealed, and *in situ* cooked in an aqueous (soup-like) medium. They indicated that to produce multilayer packaging films based on TPS that can be used as *in situ* cooking, there are numerous requirements including: delamination resistance, low oxygen permeability, and high temperature strength. Another necessary characteristic for meat cook-in films or pouches is adherence to the meat outer surface, thereby preventing "cook-out," which is the collection of juices between the meat outer surface and film inner surface. They have prepared various co-extruded films of ethylene-vinyl acetate (EvOH) with thermoplastic starch inner layer. Their films were mostly three layers. One layer was a composite of EvOH with dispersed starch microparticles. The inner layer could be thermoplastic starch and the outer layer could be EvOH or plasticized PVC. Further details can be found in their patent [59]. In a more comprehensive study, Martin et al. [60] made multilayer films based on plasticized wheat starch (PWS) and various biodegradable aliphatic polyesters via flat film co-extrusion and compression molding. PLA, polyesteramide (PEA), PCL, poly(butylene succinate adipate) (PBSA), and poly(hydroxybutyrate-co-valerate) (PHBV) were chosen as the outer layers of the "polyester/PWS/polyester" multilayer films. They used the polyester layer in order to improve mechanical performance and moisture resistance. They indicated that no compatibilizers, ties, or adhesive layers were used and the properties of multilayer films depended on the compatibility between the polymers only. They reported on the effects of plasticizing glycerol content on PWS, polyester type, and film composition on the mechanical properties and adhesion strength of multilayers. In particular, PEA had the best adhesion to the PWS layer, due to its polar amide groups. The PCL and PBSA showed medium adhesion values, and PLA or PHBV were the least compatible polyesters. They reported a similar trend in the magnitude of adhesion strength between co-extruded and hot-pressed multilayers. They also observed a gradual decrease of peel strength as the glycerol content in PWS was increased.

Laufer et al. [61] produced multilayered thin films of chitosan (CH), CR, and montmorillonite (MMT) clay by employing layer-by-layer technique for low oxygen permeability. Several layers of CH/MMT/CR (<40 nm thick) deposited on a model PET film resulted in promising oxygen permeability values (1.76×10^{-15} cm^3 cm/cm^2 s Pa). They also showed that by adding an additional chitosan over the trilayer sequence, a "quadlayer" film of CH/CR/CH/MMT (<60 nm thick) was made that could reduce oxygen permeability of PET by 2 orders of magnitude under the same conditions. They correlated this to a strong association among the oppositely charged polysaccharides.

As can be seen so far, polysaccharides are being actively explored in sustainable flexible multilayer food packaging applications. There are several other noticeable works that utilize different forms of thermoplastic starch in combination with biodegradable polyesters [62–64], proteins [65] and alginates to name a few [66]. In general, polysaccharides are really attractive for multilayer food packaging [67, 68] but there is still not a clear indication as to what part of these bio-based multilayer packaging materials should be made from them (outer layer facing air, inner layer food side, intermediate layers, etc.). Finally, it is important to state that xylans and mannans are also attractive and interesting polysaccharides for food packaging. Xylan- and mannan-based films and coatings show low oxygen and grease permeability with good tensile strength [69]. They can be blended with other biopolymers to obtain new generation composites for multilayer lamination.

15.5 Coatings on Biopolyesters

Herein, we will review multilayer films that are made up of different types of biopolyesters available today such as PCLs, polylactic acids, PHBs and PHAs, etc. A review of literature indicates that many works have so far focused on their blends [70, 71] or with other agro-based polymers [72] rather than coating or laminating them with one another. However, several different effective coating strategies have been demonstrated on biopolyesters with the main aim of reducing their OTRs and rendering them antioxidant or antimicrobial for preserving food quality. For instance, Iotti et al. [73] demonstrated organic–inorganic hybrid coatings based on PCL, poly(ethylene oxide), or PLA and silica from tetraethoxysilane precursor as inorganic phase using the sol–gel approach. The coatings were developed to modify PLA films for food packaging to enhance barrier properties toward oxygen and water vapor. They reported that all coatings were dense and homogeneous with good adhesion to the substrate. Gas permeation performance of the coating layers were an order of magnitude better than that of the uncoated PLA film. Due to the hydrophilic nature of the coatings, water vapor barrier performances were not improved as much. González et al. [74] reported preparation and characterization of biodegradable isolated soy protein (SPI) covered PLA films. Both films were prepared by sequential solvent casting using an alkaline aqueous solution for the protein and chloroform for PLA. The films had good transparency and strong adhesion between layers obtained without chemically modifying the film surfaces. Presence of PLA improved the mechanical properties of the films made of pure SPI. The biodegradation

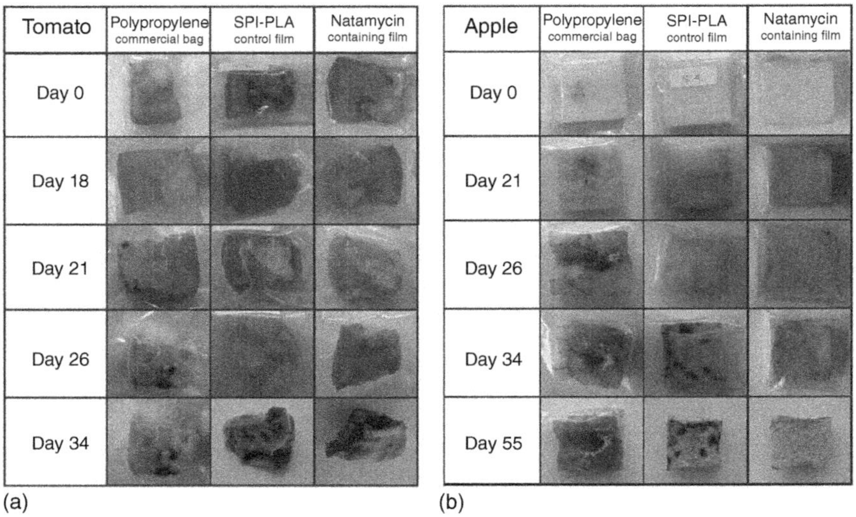

Figure 15.13 Visual analysis of (a) tomato and (b) apple chunks packed in a commercial polypropylene, SPI–PLA 60/40 film without active agent and SPI–PLA 60/40 film containing natamycin. Source: González and Igarzabal [74]. © 2013, Elsevier.

under soil burial conditions was done by measuring weight loss as a function of time. The films were transformed into active packaging by incorporation of an antifungal and an antibacterial agent (natamycin and thymol, respectively) to the SPI layer, indicating noticeable growth inhibition of mold (see Figure 15.13), yeast and two strains of bacteria by *in vitro* microbiological assays.

Lantano et al. [75] prepared natamycin based sol–gel antimicrobial coatings on PLA films for food packaging. Coatings employed tetraethoxysilane as a precursor of the inorganic phase and PVOH as the organic component, and incorporating natamycin as the active agent. The antifungal properties of the coated PLA films were investigated against mold growth on the surface of commercial semi-soft cheese. The release of natamycin from the films was measured and the maximum release was about 0.105 mg/d m^2, which was in compliance with allowed concentrations by legislation. Rocca-Smith et al. [76] prepared a PLA-WG-PLA multilayer film with highly enhanced oxygen barrier properties compared to neat PLA. The thickness of the multilayers was 60 μm, with a WG coating of 20 μm. More specifically, the oxygen barrier properties of neat PLA (60 μm) increased more than 20 times (or 2000%) and approximately 20% to water molecules. The reduction of the transfer rate of oxygen was due to the WG layer, while the reduction of the water vapor rate was due to PLA. Test conditions were 25 °C, 50% RH in addition to dry conditions. In a rare study, Cunha et al. [77] produced bilayer biopolyesters. They used PHBV a biodegradable polymer. They indicated that by itself it is a difficult film-forming polymer. They prepared 70/30 wt% blends of PHBV with a biodegradable polybutylene succinate (PBS) based composite (PBSebT), or with poly(butylene adipate-*co*-terephtalate) (PBAT) and tested for extrusion

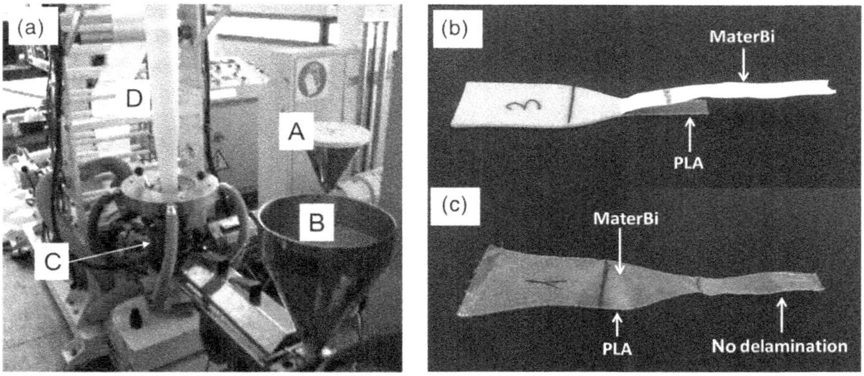

Figure 15.14 (a) Film blowing process of co-extruded layers of PHBV and PBAT. (A) extruder feeding the co-extrusion tubular die (C) with the external layer. (B) Extruder feeding the die C with the internal layer. (D) Bilayered bubble. Photographs illustrating bilayer systems after tensile test: (b) bilayer sheet and (c) bilayer film by co-extrusion. Source: Hilliou et al. [77]. © 2015, John Wiley and Sons, Inc.

film blowing. Both blends had similar rheological form at 175 °C, which was the maximum processing temperature before thermal degradation. To eliminate some issues related to thermal effects during extrusion related to polymer blending, PHBV/PBAT films were co-extruded (see Figure 15.14a). The double layers had enhanced functional properties when compared with films blown from blends. The mechanical properties of bilayered films matched those of films blown from commercial PBAT designed for food packaging.

Similarly, Scaffaro et al. [78] demonstrated fabrication of biodegradable bilayer films from PLA and MaterBi®. The prepared films were characterized using a number of different approaches such as morphological, mechanical, thermal, chemical, optical, and water contact angle analyses. Furthermore, the hydrolytic degradation of monolayer and bilayer films was investigated by performing tests at three different pH levels. The bilayer film had a higher stiffness than that of MaterBi and a larger elongation at break compared to neat PLA (see Figure 15.14b). Their degradation tests showed that weight loss was faster at basic pH levels for all the systems and at pH 10, the bilayer film had the fasted degradation rate. No barrier property measurements have been reported. In a very recent work, Quiles-Carrillo et al. [79] prepared bi- and multilayer PLA films by forming electrospun nanostructured PLA coatings and interlayers containing gallic acid (GA) antioxidant at 40 wt% onto extruded PLA films. The multilayers were made by coating the cast PLA films with electrospun PLA and then sandwiching with a 10-μm PLA film on the other side, and the resultant multilayer was thermally annealed at 150 °C without pressure. They found that GA delayed thermal degradation of PLA for approximately 10 °C, and all films were thermally stable up to 345 °C. PLA multilayers had very good sustained release of GA, having the capacity to deliver the antioxidant for over 1000 hours. Thellen et al. [80] made multilayer films from PVOH as the core layer and PHA as the outer skin layers by co-extrusion of the polymers. To improve interfacial adhesion, PHA was grafted with maleic anhydride. OTR measurements indicated that the multilayer films had a value of 27 cc/m^2 d at 0% RH and values of 41 and 52 cc/m^2 d at 60% and 90% RH, respectively. These results were better than monolayer PVOH that had an OTR of 60 cc/m^2 d at 0% RH and 999 cc/m^2 d at 60% RH. Fabra et al. [81, 82] published two rare works on multilayers of PHA polymers with zein either as solid zein film or electrospun fiber networks. They focused more on thermomechanical and physical properties of the multilayers rather than their barrier properties and indicated that such PHA–zein multilayer structures could be potentially used in food packaging applications with further improvements. As can be seen, biopolyesters have the advantage of thermal and melt processability at higher temperatures compared with many polysaccharides and proteins that might experience irreversible thermal degradation issues. Coating or laminates on biopolyesters appear to be an emerging and reliable approach. Current inclination in the food packaging industry is introducing more bio-based polymers with lighter weight for reduction of raw material use, transportation costs, and minimizing the amount of waste. Interest in sustainable materials combined with barrier-improving fillers and multilayer laminates will continue to grow. This chapter shows that single layer biopolymer films rarely meet the competitive properties of oil-derived polymers and hence the combination of individual layers to multilayer structures is unavoidable [11].

15.6 Summary and Outlook

Flexible multilayer plastic packaging constitutes more than 75% of food packaging materials today. Most of these laminated films are neither recyclable nor compostable. PLA appears to be commonly used in many types of flexible bio-packaging today [83]. For efficient gas barrier requirements, a coating or a metallization surface treatment of PLA films is needed. However, this review chapter showed that using a proper natural barrier polymer might essentially eliminate this need. Ideally, bio-based multilayers should offer increased melt strength, better processability, good transparency (not always needed, i.e. Tetra Pak containers), and good adhesion during sealing. Mechanical properties should also be carefully tailored. Moreover, high carbon footprint production methods should be reduced or avoided to make multilayers more sustainable. The use of new biodegradable or recyclable materials reduces the consumption of water and CO_2 emissions in their manufacturing processes. The production costs can also be reduced by making more efficient composite multilayer bio-based packaging materials [84]. Table 15.2 presents a summary of the bio-based or natural polymers used in manufacturing flexible multilayer packaging films in the form of laminates or coatings. The first column represents both bio-based and oil-derived thermoplastic polymers and the first row represents natural origin polymers. Note that therein "chitosan" refers to a class of chitosan polymers including acetylated chitosan; similarly, "starch" or "wax" refer to the class of starch-based polymers including thermoplastic starch and different types of waxes, respectively. The term "n/a" does not necessarily indicate that

Table 15.2 Summary of bi- or multilayer food packaging systems made by combining bio- or oil-derived thermoplastic polymers with natural polymers.

Polymers	Starch	Chitosan	Alginate	Cellulose derivatives	Zein	Gelatin	Whey proteins	Soy protein	Gluten	Wax	Polyglycerol
PLA	ML/C	C	C	ML	C	ML	C	ML	ML	n/a	C
PCL	ML	C	C	ML	C	n/a	n/a	n/a	n/a	n/a	n/a
PHB	ML	C	C	n/a	ML	n/a	n/a	n/a	n/a	n/a	n/a
PHA	ML	C	C	n/a	ML	n/a	n/a	n/a	n/a	n/a	n/a
PHVB	ML	C	C	n/a	n/a	n/a	n/a	n/a	n/a	n/a	n/a
PE	ML	C	C	ML/C	ML/C	ML	ML/C	ML/C	C	n/a	n/a
PP	ML	C	C	C	C	C	ML/C	C	C	n/a	n/a
PET	ML	C	C	ML/C	ML/C	C	C	C	ML/C	n/a	n/a

ML stands for melt lamination and C stands for coating (usually by solvent processing).
n/a indicates insufficient literature data to draw conclusions on food packaging applications.

there exist no studies on such combinations but merely suggests that there are either very few publications or mentions in patents that would not allow one to draw reliable conclusions as to their suitability for food packaging. For instance, some of them have been published as biomedical layers or surgical systems or in gel form (i.e. PCL–cellulose derivatives, etc.).

The review of the table indicates that proteins are utilized both as coatings or laminated layers, since they can be both processed by solvent casting or by extrusion. The marine origin polysaccharides, i.e. chitosan appear to be more suitable as coatings carrying and releasing active agents such as antimicrobials and antioxidants. Cellulose derivatives have been utilized both as coating or as laminated layers with other polymers. However, cellulose is mainly utilized as nano-cellulose reinforcing agent with biopolymers rather than as a laminate layer. Some edible polymers such as polyglycerols have not yet been completely explored. They are biocompatible, edible, and reactive polyether polyols that can have significant new applications in multilayer biopolymer packaging. Some bi- or multilayer systems among biopolyesters are not identified in this table but the reader can refer to the last part of Section 15.5 on such systems. Moreover, very recent advances on nanocomposites and nanocoatings for food packaging that are beyond the scope of this chapter can be found in the review of Vasile [85].

In closing, it is important to stress that both consumer and environmental protection demands are driving research and development for alternatives to oil-derived food packaging materials including those with recyclable or edible properties, as well as plastics made from renewable/sustainable agricultural products. Use of biopolymers in multilayer laminates is gaining momentum since both researchers and consumers have clearly seen that bio-based, edible, or biopolymer packaging not only enhance the quality or safety of the food, but also slow deterioration, extend the shelf life, and in some cases, impart desirable characteristics (color, flavor, etc.) to the food. Multilayer laminates can "hide" some unwanted properties of natural polymers while exposing their desired properties due to their construction. However, as with all new or novel packaging developments destined for consumers, cost, reliable fabrication methods, consumer preference, toxicological, safety, and regulatory considerations must be addressed if bio-based multilayer flexible packaging technologies are to be adopted and implemented by the food industry.

References

1 Kaiser, K., Schmid, M., and Schlummer, M. (2018). Recycling of polymer-based multilayer packaging: a review. *Recycling* 3 (1): 1.
2 Kopacic, S., Walzl, A., Zankel, A. et al. (2018). Alginate and chitosan as a functional barrier for paper-based packaging materials. *Coatings* 8 (7): 235.
3 Bekhta, P., Lyutyy, P., Hiziroglu, S., and Ortynska, G. (2016). Properties of composite panels made from Tetra-Pak and polyethylene waste material. *Journal of Polymers and the Environment* 24 (2): 159–165.

4 Aznar, M., Vera, P., Canellas, E. et al. (2011). Composition of the adhesives used in food packaging multilayer materials and migration studies from packaging to food. *Journal of Materials Chemistry* 21 (12): 4358–4370.
5 Chung, D., Papadakis, S.E., and Yam, K.L. (2003). Simple models for evaluating effects of small leaks on the gas barrier properties of food packages. *Packaging Technology and Science: An International Journal* 16 (2): 77–86.
6 Ozcalik, O. and Tihminlioglu, F. (2013). Barrier properties of corn zein nanocomposite coated polypropylene films for food packaging applications. *Journal of Food Engineering* 114 (4): 505–513.
7 Siracusa, V., Rocculi, P., Romani, S., and Dalla Rosa, M. (2008). Biodegradable polymers for food packaging: a review. *Trends in Food Science & Technology* 19 (12): 634–643.
8 Padgett, T., Han, I.Y., and Dawson, P.L. (1998). Incorporation of food-grade antimicrobial compounds into biodegradable packaging films. *Journal of Food Protection* 61 (10): 1330–1335.
9 Arvanitoyannis, I., Biliaderis, C.G., Ogawa, H., and Kawasaki, N. (1998). Biodegradable films made from low-density polyethylene (LDPE), rice starch and potato starch for food packaging applications: Part 1. *Carbohydrate Polymers* 36 (2–3): 89–104.
10 Jia, F., Wu, Z., and Gosling, J. (2019). Tetra Pak: sustainable initiatives in China. In: *Managing Sustainable Business* (eds. G. Lenssen and N. Smith), 63–82. Dordrecht: Springer.
11 Vartiainen, J., Vähä-Nissi, M., and Harlin, A. (2014). Biopolymer films and coatings in packaging applications—a review of recent developments. *Materials Sciences and Applications* 5 (10): 708.
12 Youssef, A.M. and El-Sayed, S.M. (2018). Bionanocomposites materials for food packaging applications: concepts and future outlook. *Carbohydrate Polymers* 193: 19–27.
13 Hong, S.I., Lee, J.W., and Son, S.M. (2005). Properties of polysaccharide-coated polypropylene films as affected by biopolymer and plasticizer types. *Packaging Technology and Science: An International Journal* 18 (1): 1–9.
14 Kurek, M., Brachais, C.H., Nguimjeu, C.M. et al. (2012). Structure and thermal properties of a chitosan coated polyethylene bilayer film. *Polymer Degradation and Stability* 97 (8): 1232–1240.
15 Lee, J.W., Son, S.M., and Hong, S.I. (2008). Characterization of protein-coated polypropylene films as a novel composite structure for active food packaging application. *Journal of Food Engineering* 86 (4): 484–493.
16 Schmid, M., Dallmann, K., Bugnicourt, E. et al. (2012). Properties of whey-protein-coated films and laminates as novel recyclable food packaging materials with excellent barrier properties. *International Journal of Polymer Science* 2012: 1–7.
17 Theapsak, S., Watthanaphanit, A., and Rujiravanit, R. (2012). Preparation of chitosan-coated polyethylene packaging films by DBD plasma treatment. *ACS Applied Materials & Interfaces* 4 (5): 2474–2482.
18 Kurek, M., Ščetar, M., Voilley, A. et al. (2012). Barrier properties of chitosan coated polyethylene. *Journal of Membrane Science* 403: 162–168.

19 Tihminlioglu, F., Atik, İ.D., and Özen, B. (2010). Water vapor and oxygen-barrier performance of corn–zein coated polypropylene films. *Journal of Food Engineering* 96 (3): 342–347.
20 Hong, S.I. and Krochta, J.M. (2003). Oxygen barrier properties of whey protein isolate coatings on polypropylene films. *Journal of Food Science* 68 (1): 224–228.
21 Cazón, P., Velazquez, G., Ramírez, J.A., and Vázquez, M. (2017). Polysaccharide-based films and coatings for food packaging: a review. *Food Hydrocolloids* 68: 136–148.
22 Gällstedt, M., Törnqvist, J., and Hedenqvist, M.S. (2001). Properties of nitrocellulose-coated and polyethylene-laminated chitosan and whey films. *Journal of Polymer Science Part B: Polymer Physics* 39 (10): 985–992.
23 Azevedo, V.M., Dias, M.V., de Siqueira Elias, H.H. et al. (2018). Effect of whey protein isolate films incorporated with montmorillonite and citric acid on the preservation of fresh-cut apples. *Food Research International* 107: 306–313.
24 Tharanathan, R.N. (2003). Biodegradable films and composite coatings: past, present and future. *Trends in Food Science & Technology* 14 (3): 71–78.
25 Guilbert, S., Gontard, N., and Gorris, L.G. (1996). Prolongation of the shelf-life of perishable food products using biodegradable films and coatings. *LWT Food Science and Technology* 29 (1–2): 10–17.
26 Lukowiak, M.C., Wettmarshausen, S., Hidde, G. et al. (2015). Polyglycerol coated polypropylene surfaces for protein and bacteria resistance. *Polymer Chemistry* 6 (8): 1350–1359.
27 Zemljič, L.F., Tkavc, T., Vesel, A., and Šauperl, O. (2013). Chitosan coatings onto polyethylene terephthalate for the development of potential active packaging material. *Applied Surface Science* 265: 697–703.
28 Carneiro-da-Cunha, M.G., Cerqueira, M.A., Souza, B.W. et al. (2010). Physical and thermal properties of a chitosan/alginate nanolayered PET film. *Carbohydrate Polymers* 82 (1): 153–159.
29 Han, C., Wang, J., Li, Y. et al. (2014). Antimicrobial-coated polypropylene films with polyvinyl alcohol in packaging of fresh beef. *Meat Science* 96 (2): 901–907.
30 Georgelos, P., Montefusco, P., Skaggs, J.M., and Sanfilippo, J.J. (2016). Film for thermal adhesion to a rigid biopolymer structure. US Patent No. 9,248,947. Washington, DC: U.S. Patent and Trademark Office.
31 Rešček, A., Ščetar, M., Hrnjak-Murgić, Z. et al. (2016). Polyethylene/polycaprolactone nanocomposite films for food packaging modified with magnetite and casein: oxygen barrier, mechanical, and thermal properties. *Polymer-Plastics Technology and Engineering* 55 (14): 1450–1459.
32 Gómez-Estaca, J., Gavara, R., Catalá, R., and Hernández-Muñoz, P. (2016). The potential of proteins for producing food packaging materials: a review. *Packaging Technology and Science* 29 (4–5): 203–224.
33 Coltelli, M.B., Wild, F., Bugnicourt, E. et al. (2016). State of the art in the development and properties of protein-based films and coatings and their applicability to cellulose based products: an extensive review. *Coatings* 6 (1): 1.

34 Cinelli, P., Schmid, M., Bugnicourt, E. et al. (2014). Whey protein layer applied on biodegradable packaging film to improve barrier properties while maintaining biodegradability. *Polymer Degradation and Stability* 108: 151–157.
35 Guillaume, C., Pinte, J., Gontard, N., and Gastaldi, E. (2010). Wheat gluten-coated papers for bio-based food packaging: structure, surface and transfer properties. *Food Research International* 43 (5): 1395–1401.
36 Schmid, M., Zillinger, W., Müller, K., and Sängerlaub, S. (2015). Permeation of water vapour, nitrogen, oxygen and carbon dioxide through whey protein isolate based films and coatings—permselectivity and activation energy. *Food Packaging and Shelf Life* 6: 21–29.
37 Martucci, J.F. and Ruseckaite, R.A. (2015). Biodegradation behavior of three-layer sheets based on gelatin and poly(lactic acid) buried under indoor soil conditions. *Polymer Degradation and Stability* 116: 36–44.
38 Wang, W., Xiao, J., Chen, X. et al. (2018). Fabrication and characterization of multilayered kafirin/gelatin film with one-way water barrier property. *Food Hydrocolloids* 81: 159–168.
39 Figueroa-Lopez, K., Castro-Mayorga, J., Andrade-Mahecha, M. et al. (2018). Antibacterial and barrier properties of gelatin coated by electrospun polycaprolactone ultrathin fibers containing black pepper oleoresin of interest in active food biopackaging applications. *Nanomaterials* 8 (4): 199.
40 Hosseini, S.F., Javidi, Z., and Rezaei, M. (2016). Efficient gas barrier properties of multi-layer films based on poly(lactic acid) and fish gelatin. *International Journal of Biological Macromolecules* 92: 1205–1214.
41 Xia, C., Wang, W., Wang, L. et al. (2019). Multilayer zein/gelatin films with tunable water barrier property and prolonged antioxidant activity. *Food Packaging and Shelf Life* 19: 76–85.
42 Rhim, J.W., Mohanty, K.A., Singh, S.P., and Ng, P.K. (2006). Preparation and properties of biodegradable multilayer films based on soy protein isolate and poly(lactide). *Industrial & Engineering Chemistry Research* 45 (9): 3059–3066.
43 Li, Y., Mi, J., Fu, H. et al. (2019). Nanocellular foaming behaviors of chain-extended poly(lactic acid) induced by isothermal crystallization. *ACS Omega* 4 (7): 12512–12523.
44 Fabra, M.J., Busolo, M.A., Lopez-Rubio, A., and Lagaron, J.M. (2013). Nanostructured biolayers in food packaging. *Trends in Food Science & Technology* 31 (1): 79–87.
45 Ferreira, A., Alves, V., and Coelhoso, I. (2016). Polysaccharide-based membranes in food packaging applications. *Membranes* 6 (2): 22.
46 Sanyang, M.L., Sapuan, S.M., Jawaid, M. et al. (2016). Development and characterization of sugar palm starch and poly(lactic acid) bilayer films. *Carbohydrate Polymers* 146: 36–45.
47 Wang, H., Qian, J., and Ding, F. (2018). Emerging chitosan-based films for food packaging applications. *Journal of Agricultural and Food Chemistry* 66 (2): 395–413.
48 Ferreira, A.R., Torres, C.A., Freitas, F. et al. (2016). Development and characterization of bilayer films of FucoPol and chitosan. *Carbohydrate Polymers* 147: 8–15.

49 Gu, C.H., Wang, J.J., Yu, Y. et al. (2013). Biodegradable multilayer barrier films based on alginate/polyethyleneimine and biaxially oriented poly(lactic acid). *Carbohydrate Polymers* 92 (2): 1579–1585.

50 Aulin, C., Karabulut, E., Tran, A. et al. (2013). Transparent nanocellulosic multilayer thin films on polylactic acid with tunable gas barrier properties. *ACS Applied Materials & Interfaces* 5 (15): 7352–7359.

51 Dai, L., Long, Z., Chen, J. et al. (2017). Robust guar gum/cellulose nanofibrils multilayer films with good barrier properties. *ACS Applied Materials & Interfaces* 9 (6): 5477–5485.

52 Sipahi, R.E., Castell-Perez, M.E., Moreira, R.G. et al. (2013). Improved multilayered antimicrobial alginate-based edible coating extends the shelf life of fresh-cut watermelon (*Citrullus lanatus*). *LWT Food Science and Technology* 51 (1): 9–15.

53 Martiñon, M.E., Moreira, R.G., Castell-Perez, M.E., and Gomes, C. (2014). Development of a multilayered antimicrobial edible coating for shelf-life extension of fresh-cut cantaloupe (*Cucumis melo* L.) stored at 4 °C. *LWT Food Science and Technology* 56 (2): 341–350.

54 Ham-Pichavant, F., Sèbe, G., Pardon, P., and Coma, V. (2005). Fat resistance properties of chitosan-based paper packaging for food applications. *Carbohydrate Polymers* 61 (3): 259–265.

55 Despond, S., Espuche, E., Cartier, N., and Domard, A. (2005). Barrier properties of paper–chitosan and paper–chitosan–carnauba wax films. *Journal of Applied Polymer Science* 98 (2): 704–710.

56 Ortega-Toro, R., Morey, I., Talens, P., and Chiralt, A. (2015). Active bilayer films of thermoplastic starch and polycaprolactone obtained by compression molding. *Carbohydrate Polymers* 127: 282–290.

57 Fabra, M.J., López-Rubio, A., Cabedo, L., and Lagaron, J.M. (2016). Tailoring barrier properties of thermoplastic corn starch-based films (TPCS) by means of a multilayer design. *Journal of Colloid and Interface Science* 483: 84–92.

58 Fabra, M.J., López-Rubio, A., Sentandreu, E., and Lagaron, J.M. (2016). Development of multilayer corn starch-based food packaging structures containing β-carotene by means of the electro-hydrodynamic processing. *Starch-Stärke* 68 (7–8): 603–610.

59 Lustig, S. and Schuetz, J.M. (1988). Meat product package containing an inner layer of starch/ethylene vinyl acetate. US Patent No. 4,784,863. Washington, DC: U.S. Patent and Trademark Office.

60 Martin, O., Schwach, E., Avérous, L., and Couturier, Y. (2001). Properties of biodegradable multilayer films based on plasticized wheat starch. *Starch-Stärke* 53 (8): 372–380.

61 Laufer, G., Kirkland, C., Cain, A.A., and Grunlan, J.C. (2013). Oxygen barrier of multilayer thin films comprised of polysaccharides and clay. *Carbohydrate Polymers* 95 (1): 299–302.

62 Alix, S., Mahieu, A., Terrie, C. et al. (2013). Active pseudo-multilayered films from polycaprolactone and starch based matrix for food-packaging applications. *European Polymer Journal* 49 (6): 1234–1242.

63 Zhou, X., Yang, R., Wang, B., and Chen, K. (2019). Development and characterization of bilayer films based on pea starch/polylactic acid and use

in the cherry tomatoes packaging. *Carbohydrate Polymers*, https://www.sciencedirect.com/science/article/pii/S0144861719305478.

64 Requena, R., Vargas, M., and Chiralt, A. (2018). Obtaining antimicrobial bilayer starch and polyester-blend films with carvacrol. *Food Hydrocolloids* 83: 118–133.

65 Jagannath, J.H., Nanjappa, C., Das Gupta, D.K., and Bawa, A.S. (2003). Mechanical and barrier properties of edible starch–protein-based films. *Journal of Applied Polymer Science* 88 (1): 64–71.

66 Şen, F., Uzunsoy, İ., Baştürk, E., and Kahraman, M.V. (2017). Antimicrobial agent-free hybrid cationic starch/sodium alginate polyelectrolyte films for food packaging materials. *Carbohydrate Polymers* 170: 264–270.

67 Rhim, J.W. (2013). Effect of PLA lamination on performance characteristics of agar/κ-carrageenan/clay bio-nanocomposite film. *Food Research International* 51 (2): 714–722.

68 Vidal, R., Martínez, P., Mulet, E. et al. (2007). Environmental assessment of biodegradable multilayer film derived from carbohydrate polymers. *Journal of Polymers and the Environment* 15 (3): 159–168.

69 Mikkonen, K.S. and Tenkanen, M. (2012). Sustainable food-packaging materials based on future biorefinery products: xylans and mannans. *Trends in Food Science & Technology* 28 (2): 90–102.

70 Plackett, D.V., Holm, V.K., Johansen, P. et al. (2006). Characterization of l-polylactide and l-polylactide–polycaprolactone co-polymer films for use in cheese-packaging applications. *Packaging Technology and Science: An International Journal* 19 (1): 1–24.

71 Razavi, S.M., Dadbin, S., and Frounchi, M. (2012). Oxygen-barrier properties of poly(lactic acid)/poly(vinyl acetate-co-vinyl alcohol) blends as biodegradable films. *Journal of Applied Polymer Science* 125 (S2): E20–E26.

72 Muller, J., González-Martínez, C., and Chiralt, A. (2017). Combination of poly(lactic) acid and starch for biodegradable food packaging. *Materials* 10 (8): 952.

73 Iotti, M., Fabbri, P., Messori, M. et al. (2009). Organic–inorganic hybrid coatings for the modification of barrier properties of poly(lactic acid) films for food packaging applications. *Journal of Polymers and the Environment* 17 (1): 10–19.

74 González, A. and Igarzabal, C.I.A. (2013). Soy protein–poly(lactic acid) bilayer films as biodegradable material for active food packaging. *Food Hydrocolloids* 33 (2): 289–296.

75 Lantano, C., Alfieri, I., Cavazza, A. et al. (2014). Natamycin based sol–gel antimicrobial coatings on polylactic acid films for food packaging. *Food Chemistry* 165: 342–347.

76 Rocca-Smith, J.R., Pasquarelli, R., Lagorce-Tachon, A. et al. (2019). Toward sustainable PLA-based multilayer complexes with improved barrier properties. *ACS Sustainable Chemistry & Engineering* 7 (4): 3759–3771.

77 Cunha, M., Fernandes, B., Covas, J.A. et al. (2016). Film blowing of PHBV blends and PHBV-based multilayers for the production of biodegradable packages. *Journal of Applied Polymer Science* 133 (2).

78 Scaffaro, R., Sutera, F., and Botta, L. (2018). Biopolymeric bilayer films produced by co-extrusion film blowing. *Polymer Testing* 65: 35–43.
79 Quiles-Carrillo, L., Montanes, N., Lagaron, J.M. et al. (2019). Bioactive multilayer polylactide films with controlled release capacity of gallic acid accomplished by incorporating electrospun nanostructured coatings and interlayers. *Applied Sciences* 9 (3): 533.
80 Thellen, C., Cheney, S., and Ratto, J.A. (2013). Melt processing and characterization of polyvinyl alcohol and polyhydroxyalkanoate multilayer films. *Journal of Applied Polymer Science* 127 (3): 2314–2324.
81 Fabra, M.J., Sánchez, G., López-Rubio, A., and Lagaron, J.M. (2014). Microbiological and ageing performance of polyhydroxyalkanoate-based multilayer structures of interest in food packaging. *LWT Food Science and Technology* 59 (2): 760–767.
82 Fabra, M.J., Lopez-Rubio, A., and Lagaron, J.M. (2013). High barrier polyhydroxyalcanoate food packaging film by means of nanostructured electrospun interlayers of zein. *Food Hydrocolloids* 32 (1): 106–114.
83 Nagarajan, M., Prodpran, T., Benjakul, S., and Songtipya, P. (2017). Properties and characteristics of multi-layered films from tilapia skin gelatin and poly(lactic acid). *Food Biophysics* 12 (2): 222–233.
84 Galgano, F. (2015). Biodegradable packaging and edible coating for fresh-cut fruits and vegetables. *Italian Journal of Food Science* 27 (1): 1–20.
85 Vasile, C. (2018). Polymeric nanocomposites and nanocoatings for food packaging: a review. *Materials* 11 (10): 1834.

Index

a

acetyl-tri-n-butyl citrate (ATBC) 124
acid-soluble collagen (ASC) 37, 38
acidoresponsive composite films and fibrous mats 384
active antioxidant packaging 350
active coating
  antimicrobial agents 85
  antioxidant agents 85–86
active edible coatings 85, 86
active food packaging 10, 338, 349–353, 356, 359–363
active packaging (AP) systems
  active-releasing systems 325
  active scavenging systems 325
  antimicrobial packaging 326–328
  benefit 340
  definition 325
  using polymeric films 326
active packaging compositions
  and natural additives effect 358
Ageless Eye® 377
aggregation-induced emission (AIE) 384
aleuritic acid 169, 186, 187
alginate 63
alginate-based edible coatings 63, 412
alginate/chitosan films 404
aliphatic–aromatic copolymer 9
alkaline treatment with sodium hydroxide 241
alkene ketene dimer wax (AKD) 283
alkenyl succinic anhydride (ASA) 283
3-aminopropyltriethoxysilane (APTS) 117, 316
amorphous cellulose-based bioplastics 361
amorphous starch/zein blend films 215, 217
amylopectin 15, 16, 71, 137–141, 143–145, 154, 165, 294
amylose 15, 16, 71, 137–141, 143–145, 154, 165, 212, 213, 294, 410
anthocyanin treated polyvinyl/cellulose indicator foam 382
anthocyanins 44, 354, 376, 379–382
antibacterial properties, PLA 115–116, 123
antimicrobial agents 40, 58, 85, 116, 122, 201, 327–340, 412
antimicrobial food packaging systems 326, 328, 336
  natural antimicrobial agents 328–340
    bacteriocins 336–337
    chitosan 338–340
    enzymes 337–338
    organic acids, salts and anhydrides 335–336
    plant extracts 328–335
  types 327
antioxidant agents 42, 60, 85–87, 212
antioxidant food waste
  chocolate industry 360
  red beetroot 359
antioxidant natural extracts
  berry family 354
  fruit seed extract 356
  herbs and spices 354
  phenolic compounds 353

antioxidant natural extracts (*contd.*)
  tea 353
apple fruit 333
Aticelca test method 275–276
attenuated total reflectance (ATR-FTIR) spectroscopy
  cutin-inspired polyhydroxy esters 180, 181
  of polyaleuritate films on stainless steel 185, 186
Aureobasidium pullulans 71

b

bacteria colonization, of PG coated PP surfaces 402
bacterial cellulose nanocrystal (BNC) films 316
bacterial nanocellulose 307, 309
bacteriocins, in antimicrobial packaging systems 336
barrier properties 3, 15–17, 20–22, 40–41, 73–75, 88, 110–113, 115–116, 123–124, 146–150, 200–201, 231–232, 281–283, 285–291, 296–300, 305–307, 312–316, 352–354, 400–401, 403–404, 406–416, 418
beeswax 80, 82, 84, 86, 187–188, 400
biaxially oriented poly(lactic acid) (BOPLA) films 410, 411
bio-based, biodegradable polymers
  classification 164–165
  monomer units and applications 166–168
bio-based multilayer food packaging 395
bio-based polybutylene succinate (bio-PBS) 10
bio-based polyethylene (bio-PE) 11
bio-based polyethylene terephthalate (bio-PET) 6, 13–14
bio-based polymers 4–6, 12, 166, 188, 217, 227, 228, 237, 282, 379, 398, 418
bio-based polymer matrices, antimicrobial agents incorporation
  bacteriocins 332
  enzymes 333
  organic acids, salts and anhydrides 331
  plant extracts 329
biocircular economy 248
biocompatible cyanoacrylate monomers 283
biocomposite films 361
biocomposites, defined 227
biodegradability 123–124
biodegradable bilayer film fabrication 417
biodegradable biopolymers 21, 135, 236, 281
biodegradable films 21, 146, 201, 307, 335, 398, 409
biodegradable packaging 16, 17, 20, 23, 111, 133, 135, 151, 154, 340, 410
biodegradable polymers 4–6, 11, 22, 23, 111, 114, 133, 135, 136, 146, 153, 164–166, 218, 227, 228, 236, 249, 265, 293, 416
  classification 228
  definition 227
bio-elastomers
  antioxidant 360
  cocoa 360
BIO-FLEX® 406
biogenic amines (food spoilage indicators) 380, 384
biolayering approach 409
biomass carbohydrates 5
biomaterials 153, 220, 236, 353
  sustainability packaging 135–137
bioplastics 4–7, 22–23, 153, 161–166, 169, 171–172, 227–228, 247–248, 361
bioplastic films
  active
    containing tea and berry extracts 354–355
  incorporated with herbal extracts 356–357
bioplastics production 6, 22
  from biopolymers, advantages and drawbacks 163, 164

biopolyesters 8, 166, 171, 420
  advantages 418
  multilayer film coatings on 415–418
biopolymers 281, 282
  advantages 352
  classification of 4–6
  definition 227
  food packaging 3
    bio-based polyethylene 11–12
    bio-based polyethylene terephthalate 13–14
    poly(ε-caprolactone) 15
    poly(butylene adipate-*co*-terephtalate) 9–10
    polybutylene succinate 10–11
    polyethylene furanoate 14–15
    polyhydroxyalkanoates 8–9
    polylactide 7–8
    protein-based films 17–18
    thermoplastic starch 15–17
biopolymer coated oil-derived polymer films 399
biopolymer-tannic acid films 334
biorefinery 6
bisphenol A (BPA) resins 163, 184, 185
blown extrusion technique 201, 202
blueberry 71, 333, 354
bromophenol blue dye (BPB) 383

c

C16-based cutins 169
C18-based cutins 169
candelilla wax 80–82, 84
capstone treated cardboard 287, 288
Carbon fibershave 122
carboxymethyl cellulose (CMC) 67, 84, 86, 148, 171, 336, 338, 411
cardboard sheet
  capstone and ECA treatment 283
  chemical characterization 285
  FTIR spectra 285, 287
  mechanical characterization 291–292
  moisture content 289–291
  morphological characterization 283–284
  oil contact angle 285–288
  sample preparation 283
  SEM images of 284
  stereo-micro photograph of 283–284
  water contact angle 285, 288–290
  water drop deposition 285
  XPS analysis 285, 286
carnauba wax 80–82, 186, 187, 412, 413
carrageenans 63–67
casein 18, 21, 73–74, 136, 166, 282, 404
caseinates 21, 74
cast polyvinyl alcohol layered PP films with REE and CEO 404
cellulose 17, 67, 165, 281
  microfibres arrangement 237
  types 308
cellulose derivatives 17, 67, 68, 171, 420
  solubility properties 68
cellulose fibers 121, 148, 165, 237, 245, 264–267, 272, 273, 275, 277, 285, 289, 291, 312
  advantages 264
  as reinforcement, development of 264
  characteristics 264
cellulose nanocrystals (CNC) 147, 148, 308, 309, 312, 313, 315, 316
  for barrier applications 315
cellulose nanofibers (CNF) 115, 119, 309–311, 313–315, 332, 336
  for barrier applications 314–315
cellulose nanofibrils 309, 311
cellulose nanostructures 305
cellulose-based paper and cardboard 282
cellulosic packaging 263
chain extenders 9, 235
chemical fibre modification treatments 241
Chilean berry 354
chitin 36, 42–43, 69, 119, 147, 148, 165, 338, 352, 398
chitin extraction 42–43

chitosan 69–70, 165
  in antimicrobial packaging systems 338
  coating, on paper-based materials 412, 413
chitosan films and coatings
  characterization 43–44
  chitin 42–43
  food shelf-life extension 44–46
chitosan-based antimicrobial films 334
chitosan-based CO_2 indicator 387
chitosan-carrageenan-montmorillonite (CH/CR/MMT) layers 414–415
chitosan-LAE films 339, 340
chitosan-sodium montmorillonite films 339
chitosan/agarose/anthocyanin film 382
chitosan/methyl cellulose films 335
cinnamon essential oil (CEO) 328, 404
circular economy 8, 188, 237, 248, 264, 351
cloud point 203
CO_2 indicators 377, 387
coating process 313
cocoa bean shell waste
  bio-elastomers 360
  bio-filaments for 3D printing application 360
collagen 18
  extraction
    fish gelatin 37–42
  proteins 74–75
colorimetric bio based pH responsive system 381
composite edible coatings 83
compostable plastics 227, 228, 406
continuum of mobility model 233
copolymers 4, 8, 9, 11–13, 15, 63, 114, 214, 215, 230, 231, 242, 283, 382, 400
corn zein 18, 20–21, 77, 200, 201, 212–214, 400, 404, 405
corn zein nanocomposite coatings (CZNC), of PP films 404, 405
corn zein-soy protein isolate (CZ/SPI) bilayer film 213, 214
corn-zein-coated PP film 213
corn-zein-laminated linear low-density polyethylene (LLDPE) films 3, 212, 213
curcumin based on-package sticker sensor 383
cutins
  biodegradability 171–172
  C16 and C18 169
  characteristics 166
  vs. commercial plastics and bioplastics 169–172
  composition 169
  dihydroxylated C16 fatty acids 169
  mechanical properties of 171
  sustainable renewable resource 172–173
cutin-inspired coatings on metal substrates 183–186
cutin-inspired polyesters 173–183
  direct melt polycondensation 173
  polymerization 173
cutin-inspired polyhydroxyesters
  physical and chemical properties
    ^{13}C solid-state NMR 174, 176
    chemical structure 174
    monomers influence 173
    tensile stress-train curves 176, 178
    TGA and TGA-derivative profiles 174, 177
    transmission IR spectra 174, 175
cutin-inspired polyhydroyesters
  oxidation effect 177–180
  surface *vs.* bulf properties 180–183
β-cyclodextrine 336

d

degradation time, of fuel-based plastics 163, 164
degree of freshness (food samples) 372
dicumyl peroxide (DCP) 115, 242, 245
dip-coating 77, 283, 386
direct melt polycondensation 173

double-layer polyethylene/poly-caprolactone (PE/PCL) film 404
dried plant waste powders 359
drop-casting 283
dry process 43, 46, 59, 405

e
ECA treated cardboard 287
edible biopolymers
  active coating
    antimicrobial agents 85
    antioxidant agents 85–86
  edible composite materials 82–85
  lipids 78–80
    beewax 80
    candelilla wax 80–81
    carnauba wax 81
    shellac 81
  material selection 86–87
  polysaccharide
    alginate 63
    carrageenans 63–67
    cellulose 67–69
    chitosan 69–70
    pectin 70–71
    pullulan 71
    starch 71–72
  proteins 72–78
    casein 73–74
    collagen 74
    gelatin 74–75
    silk fibroin 76–77
    wheat gluten 75
    whey protein 75–76
    zein 77–78
edible composite materials 80, 82–85
edible films 58–60, 70, 72–73, 80, 85, 87, 333, 338, 400, 405
edible packaging 58, 60, 217, 218, 307
electrospun polycaprolactone (PCL)/black pepper oleoresin extract (OR) blends 408
enzyme-soluble collagen (ESC) 37, 38
enzymes, in antimicrobial packaging systems 337
epigallocatechin-3-gallate (EGCG) films 333
essential oils (EOs) 45, 85, 88, 217, 327, 328, 332
ethyl cyanoacrylate (ECA) 282–292
ethylene-vinyl acetate (EvOH)-thermoplastic starch films 4, 200, 281, 314, 414
ethylenediaminetetraacetic acid (EDTA) 37, 336, 337
extrinsic parameters (microbial food spoilage) 370
extrusion-hydro-extraction (EHE) 38

f
fatty hydroxyesters, chemical structure of 174
fibre purification/delignification methods 241
fibre-matrix compatibility 244
fish gelatin films and coating
  characterization of 39–40
  collagen extraction 37–39
  food shelf-life 40–42
fish production 36
fish spoilage sensor 381
fluorinated organic polymers 283
fluoropolymers 283, 288
food demand 349, 361
food lifetime 349, 352
food packaging
  bio-based polyethylene 11–12
  bio-based polyethylene terephthalate 6, 13–14
  biodegradable 9
  cellulose derivatives 17, 18, 67, 68, 171, 420
  chitosan films 42–46
  fish gelatin films 37–42
  materials 3
  poly(ε-caprolactone) 15
  poly(butylene adipate-*co*-terephtalate) 9–10
  polybutylene succinate 10–11
  polyethylene furanoate 14–15
  polyhydroxyalkanoates 8–9
  polylactic acid 111

food packaging (*contd.*)
  polylactide 7–8
  protein-based films
    corn zein 20–21
    gelatin 18
    milk proteins 21
    soy protein 20
    wheat gluten 18–20
  thermoplastic starch 15–17
food packaging sector, PHA-based composites 247–248
food safety and quality monitoring 374
Food Sentinel System® 378
food shelf-life 35, 40–42, 44–46, 87, 361, 362, 369, 373, 400
  chitosan films 44–46
  fish gelatin films 40–42
food spoilage
  biogenic amines 380–386
  carbon dioxide detection 387–388
  colorimetric detection 378, 379
  colorimetric indicator 380, 381, 383
  colorimetric sensor array 384
  description 370
  factors determining 372
  foodborne bacteria detection 388
  H_2S detection 384
  lipid oxidation 350
  manifestations 370
  microbial activity 370–372
  natural and biocompatible indicators 379–380
  on-package colorimetric indicator 383
  optical detection 378, 379
  organic acid indicator 380
  oxygen detection 386–387
  pH changes 380
  physical and chemical process 372
  synthetic dye indicators 384
food spoilage detection
  on package and on site sensing technologies 373–378
  sensory evaluation 372
food waste 3, 57, 87, 154, 164, 247, 264, 325, 327, 340, 356, 359, 360, 388
foodborne bacteria detection 388
freshness indicators (food spoilage detection) 376, 384
fruit seeds 356, 359
FucoPol bilayer films 410
fuel-based plastics, degradation time of 163, 164
2,5-furandicarboxylic acid (FDCA) 14, 15
fused deposition modelling (FDM) 276

g

gallic acid (GA) antioxidant 146, 356, 417
gas permeability, of nanocrystalline based films 311
gelatin 18, 37–42
  proteins 74–75
generally recognized as safe (GRAS) 7, 40, 60, 70, 80, 81, 328, 335, 336, 338
glass fibers (GF) 121
gliadins 18, 75
glutaraldehyde (GDA)-modified zein 205, 339
grease resistance test 282, 297
green packaging materials 200, 281
green polyethylene 11
green polymers 6

h

headspace chromatography 373
healthy diets 349
hemicelluloses 237–241
herbs and spices 354, 356
hexamethylene diisocyanate (HMDI) 247
high amylose potato starch (HAP) 144
high methoxy pectin (HMP) 70
high-amylose corn starch (HACS)-zein composite films 212, 213
hydrochloric acid-treated cellulose nanocrystals 316
hydrophilization of PE surfaces 403
hydroxypropyl methylcellulose (HPMC) 67–69, 82, 85, 336

i
implicit parameters (microbial food spoilage) 371
in corn gluten meal/starch extrudates 209, 210
injection moulding
  Poly-Paper composites 269
innovative active food packaging 349
integrity indicators (food spoilage detection) 376, 377
intrinsic parameters (microbial food spoilage) 370

j
Joncryl® 119, 247

k
kafirin-gelatin multilayer films 407
kit test 297
konjac glucomannan (KGM)/zein blend films 214

l
lactoferrin 337–338
lactoperoxidase system-whey protein isolate films 338
laminated flexible films 397
layer-by-layer (LbL) approach 313–314, 408, 409, 412–414
  deposition method 410, 411
lignin 163–165, 172, 237, 239–241, 339, 360
lignocellulosic fibres 236, 237
  mechanical properties of 239
  in polymer matrix composite materials
    advantages and disadvantages 240
    biological modification 241
    chemical modification 241
    fibre-matrix chemical anchor 242
    physical modification 241
  properties of 239
  structure of 237–240
  thermal stability of 239
lipid oxidation 45, 70, 75–76, 84–85, 212, 350, 354, 356, 362
lipid-based edible coatings 78, 81
lipids 78
  beewax 80
  candelilla wax 80–81
  carnauba wax 81
  shellac 81
loads (filling materials) 240
low methoxy pectin (LMP) 70
lumen 237
lysozyme 337–338
lysozyme-chitosan composite films 337

m
macro-debris 163
Maillard reaction 40, 214
maleic anhydride (MA) 113, 118–119, 152, 242, 244, 418
mannan-based films 415
marine plastic debris 163
MaterBi® 417
matrix-fibre adhesion 245
McLafferty rearrangement 234
mechanical properties, PLA 114
merocyanine (MC) 385
metal-oxide composite, PLA 123
methylcellulose (MC) 67–68, 333, 336, 387
micro-debris 163
microbial food spoilage 370–372
microbial polyethylene 11
migration process, of molecules 312
milk proteins 21, 73, 75, 405
minimally processed food 58, 60, 337
modified atmosphere packaging (MAP) 58, 69–70, 314
moisture content (MC) 44, 144, 317
  of treated and untreated cardboard 289–291
montmorillonite (MMT) 20, 123, 146, 202, 414
multi-layer liquid food container structure 397
multilayer films 397
  based on polysaccharides 409–415
  based on proteins 405–409

multilayer films (*contd.*)
coatings on biopolyesters 415–418
one-way water barrier property 407–408

n
nanocellulose (NC) 305, 308
crystallinity of 310
effect on networking 310–311
in food packaging industry 307–308
need for packaging 310
O_2 barrier properties of 306
overview of 308
types 308–310
utilization of 306
water vapor transmission rate of 315
nanocellulose-based multi-dimensional structures 305–317
nanocrystalline cellulose 308–309, 316
nanocrystalline-based structures, processing routes of 312
coating process 313
layer-by-layer approach 313–314
solvent casting method 313
spinning 312–313
vacuum filtration method 313
nanofibrillar cellulose 309
nanofibrillated cellulose 309, 411
nanofillers 9, 114–116, 122, 147–149, 235
nanolayered films 403–404
natural biopolymers 136, 229, 338, 362
natural fiber-PLA 111, 116
natural fibers 17, 113, 114, 116–121, 266
advantages 264
natural plant-based antioxidants 352, 353
naturally occurring bioactive compounds 337
naturally occurring biopolymers 398, 399
nisin 336
nisin grafted CNF 336
nisin-impregnated protein coated PP films 400
non-biodegradable biological based polymers 228–229
non-migratory antimicrobial polymers 327
normal potato starch (NPS) 144, 146
nucleating agents 114, 119, 235, 243

o
odor-intensive volatiles 332
oil contact angle (OCA) 282, 285, 288
of untreated and treated cardboard 289
onsite and on-package sensing technologies, for food spoilage detection
biosensors 375
CO_2 indicators 377
freshness indicators 376
integrity indicators 376, 377
overview 375
oxygen indicators 376, 377
RFID systems 377
time temperature indicators 376
optical properties, PLA 113
optochemosensing mechanisms 379
oregano essential oil 45, 328
oregano extract 356
oregano-EO-whey protein isolate films 328
organic acids (food spoilage indicators) 380
organic acids, in antimicrobial packaging systems 335–336
oxidation effect
cutin-inspired polyhydroyesters 177–180
oxybisenzenesulphonylhydrazide (OBSH) expanding agent 270
oxygen indicators 376–377, 387
oxygen permeability
of CNF films 314
of nanocrystalline film 315
oxygen transmission rates (OTR)
of CNF films 314–315
for PLA nanocomposites 316

p

paper sheet
  capstone and ECA treatment 283
  sample preparation 283
paper surfaces
  grease resistance properties 298, 299
  morphological analysis 295
  oxygen barrier properties 299
  physical characteristics and mechanical properties 296–300
  SEM analysis 295
  water and oil resistance properties 296
paper-based packaging materials 212
pectins 61, 70–71, 82, 84, 136, 164, 166, 237, 239, 352, 371, 410, 412–413
pepper fruit cutin composition 170
pepsin-soluble collagen (PSC) 38
peptization 203
peracetic acid treatment 241
permselectivity of whey protein film 406
petroleum-based plastics 161
  advantages 161–163
  drawbacks 162, 163
petroleum-based polymers 6, 8, 281
  biodegradation rates 171
pH (food spoilage indicator) 380
PHA-based composites, in food packaging industry 247–248
PHBV/TPU blends 245
PHBV/PBAT films 417
phenolic compounds 46, 165–166, 328, 332, 353, 356, 384
photochromic composite 385
physical aging 232–234
physical fibre modification methods 241
PLA-gelatin based films (PLA/Ge/PLA) 406
PLA-WG-PLA multilayer film 409, 416
PLA/clay nanocomposites 122–123
plant cuticle-like film composites 186–188
plant cuticles
  chemical composition and location 166, 169
  description 166
plant extracts, in antimicrobial packaging systems 328–335
plant-based food packaging materials, antioxidant capacity of 352–361
plant-based monoethylene glycol (bio-MEG) 13
plasticizers 16–17, 20–21, 39, 44, 58, 60, 74, 113, 114, 117, 124, 141–146, 153–154, 200, 212, 214–215, 217, 218, 235, 244, 264, 354, 356, 400, 405–406, 410
plastics production (2004–2017) 162
plastics
  for packing materials 351
  threat 134
  wastes 163, 361
pMDI compatibilizer 245
poly (ethyl cyanoacrylate) (PECA) 283–285, 288, 291
poly 3-hydroxybutyrate (PHB) 8, 9, 229
  agave compounds 243
  cellulose composites
    achievements and limitations 242–245
    trends 245, 247
  opportunities 235–236
  strengths 231–232
  threats 236
  weakness 232–235
poly(ε-caprolactone)(PCL) 15
poly(butylene adipate-*co*-terephtalate) (PBAT) 9–10, 119, 150–151, 245, 335, 416–417
poly(butylene succinate-*co*-adipate) (PBSA) 11, 150, 414
poly(butylene succinate-*co*-terephthalate (PBST) 11
poly(ethylene glycol) (PEG) 114

poly(lactic acid) films for food packaging
  biodegradable isolated soy protein coatings 415
  biodegradable isolated soy protein coatings on 415
  natamycin based sol–gel antimicrobial coatings on 416
  PLA-WG-PLA multilayer film 416
poly(lactic acid)-fish gelatin (PLA-FG) multilayer films 408
poly(vinyl alcohol)-chito-oligosaccharide (COS) films 339
poly-3-hydroxy-butyrate-co-3-hydroxy-valerate (PHBV)
  agave compounds 243
  -cellulose composites
    achievements and limitations 242–245
    trends 245, 247
  opportunities 235–236
  strengths 231–232
  threats 236
  weakness 232–235
Poly-Paper 265
  3D printed samples 272
  adhesive property results of 274
  Aticelca method 275, 277
  end of life 275–277
  self-healing by water 273–275
  shape memory forming of 272–273
  water dissolution kinetics 275, 276
Poly-Paper composites 269
  cellulose fiber reinforcement material 266
  expansion tests 270–271
  extrusion process 267
  formulations 266
  injection moulding 269–270
  PVA matrix material 265
  thermoforming 270
polyamides (PAs) 4
polybutylene adipate co terephtalate (PBAT) 9–10, 119, 150–151, 245, 335, 416–417
polybutylene succinate (PBS) 10–11, 416
polybutylene succinate adipate (PBSA) 11, 150, 414
polycaprolactone (PCL)/zein blend films 215
polycondensation 10, 14–15, 110, 173–175, 177–178, 186, 188
polydimethylsiloxane (PDMS)-based water-repellent layer 293
polyester amide (PEA) 150, 414
polyesters 4, 10, 14–15, 109–110, 149–151, 153, 166, 174, 177–178, 188, 248, 314, 406, 414
polyethylene (PE) 11–12, 109, 113, 134, 143, 161, 171, 281, 292, 384–385, 387, 397, 403
polyethylene furanoate (PEF) 14–15
polyethylene oxide (PEO) 151, 385
polyethyleneimine (PEI)-nanofibrillated cellulose (NFC) composite layers 412
polyglycerol-coated polypropylene films 400, 402
polyhydroxyalkanoates (PHA) 8, 229
  characteristics 231
  classification 231
  description 229–231
  growth rate 236
  production process 231
  structure 230
  -zein multilayer structures 418
polyhydroxybutyrate-*co*-valerate (PHBV) 8, 151, 227–249, 282, 414, 416–417
polylactic acid (PLA) 109
  antibacterial properties 115–116
  barrier properties 111, 113
  functional fillers
    biodegradability 123–124
    clay nanocomposites 122–123
    metal-oxide composite 123
  mechanical properties 114
  natural fiber 116–121
  optical properties 113
  properties 110
  structural formula 110
  synthetic fibers 121–122
  thermal properties 114–115

polylactide (PLA) 5, 7
polymer treated cardboard 285–287
polyolefins 3, 8, 200–201, 404
polysaccharides 59, 147
  alginate 63–65
  carrageenans 63, 66, 67
  cellulose 67–69
  chitosan 69–70
  pectin 70–71
  pullulan 71
  starch 71–78
polysaccharide-based edible coatings 64
polysaccharide-based films and coatings 410
  alginate-based edible coatings 412
  chitosan 412, 413
  chitosan based films 410
  chitosan-carrageenan-montmorillonite layers 414, 415
  efficiency and functional properties 410
  EvOH-thermoplastic starch based films 414
  PEI-NFC composite layers 412
  SPS-PLA bilayer films 410
  TPCS based films 413
polyterpenes 165
potassium sorbate 335, 413
propionic acid (PA) 110, 335–336
proteins 72
  casein 73–74
  collagen 74
  gelatin 74–75
  silk fibroin 76–77
  wheat gluten 75
  whey protein 75–76
  zein 77–78
protein-based edible coatings 79
protein-based films
  corn zein 20–21
  gelatin 18
  milk proteins 21
  soy protein 20
  wheat gluten 18, 20
protein-based food packaging 200
protein-derived coatings 59
Psychrotrophic Bacteria Count (PBC) determination 373
pullulan 61, 71, 228, 338
PVA/cellulose micro-fibers composites 266–268
  mechanical tensile properties of 268, 269

q
quercetin 356, 384

r
radiofrequency identification (RFID) systems 375, 377–378
rancidity 61, 76, 84–85, 214, 360, 372
raspberry extract 354
reactive compatibilization in PHA/fibre composites 244–246
reactive processing 152, 242
recycled newspaper cellulose fiber (RNCF) 121
red beetroot 359–360
reinforcement materials 240, 268, 308
rhubarb ethanolic extracts (REE) 404
ring-opening polymerization (ROP) 7, 14–15, 110–111
rosemary extract 356
rosin 283

s
salt-soluble collagen (SSC) 37, 38
secondary crystallization 151, 232, 234–235
self-healing effects, Poly-Paper 273
self-polycondensation reaction 174
semicrystalline aliphatic polyester 10
SensorQ® 376
shape memory forming, Poly-Paper 272–273
shellac 80, 81, 136, 169
silicone oil 282, 294
silicone oil-functionalized starch/PDMS treatment on paper 294
silk fibroin 73, 76–78, 88, 385
smart packaging systems 373, 374, 377
sodium alginate (ALG) 62, 63, 82, 187, 328, 403, 410–412

sodium alginate (ALG)/polyethyleneimine (PEI) coated BOPLA films 410, 411
sodium dodecyl sulfate–polyacrylamide gel electrophoresis (SDS-PAGE) 208
  solvent-cast zein films 208
soil burial weight loss curves 407
solvent cast zein films
  dimethyl sulfoxide (DMSO) 204
  mechanical properties 205
solvent casting method 10, 143, 202, 313, 353, 361
solvent-cast zein films
  chemical characteristics 206
  Fourier Transfor-Raman spectroscopy 206, 207
  solid-state 13C-NMR spectra 208, 209
  surface elemental compositions 208
  surface morphology 206
soy protein (SP) 18, 20, 86, 136, 166, 214, 219, 331, 354, 356, 400, 405, 415
spices and herbs 352, 354
spinning process 312–313
spiropyran (SP) dyes 385
Standard Plate Count (SPC) determination 373
starch 71, 165, 294
  gelatinization 140
  granules 140
starch/PDMS treatment, on paper 298
starch/polyvinyl alcohol (PVA)/citric acid blend films 335
sugar palm starch (SPS)-PLA bilayer films 410
surface hydrophobicity, of fatty polyhydroxyesters 182
sustainability 46, 133–137, 145, 153, 184, 214, 248, 264, 349, 352, 361, 378, 388
sustainability packaging 135
  biomaterials 135
  plastic threat 134
sustainable packaging technology 363
SWOT analysis 248, 249
  of PHA's 236
synthetic dyes, covalent immobilization of 384
synthetic fibers 116, 121–122, 264

t

tea extracts 41, 353, 354
TEMPO-oxidized CNFs 314–315, 336
terpenes 164, 165
thermal degradation 14, 116, 140, 149, 187, 209, 219, 234–235, 239, 241, 359, 417–418
  of PHA's 234
thermal properties, PLA 114–115
thermal stability 15, 115, 124, 140, 145–146, 149, 152–153, 175, 215, 239–240, 242–244, 249
  PHB and PHBV 234, 235
thermoforming process 234
  Poly-Paper composites 270–271
thermoplastic corn starch (TPCS) based films 413–414
thermoplastic starch (TPS) 5, 15, 16, 114, 141, 413
  chemical modification 152
  commercial packaging materials 152–153
  fillers composites 146–147
  packaging materials
    technical and legal 145–146
  physico-chemical properties 137
  polyesters 149–151
  polysaccharides composites 147–149
  processing of 143–145
  sustainability packaging 135
    biomaterials 135
    plastic threat 134
  plasticizers 142–143
thiobarbituric acid reacting substances (TBARS) 46
third generation feedstock 6, 22
three-phase model 232–233

time temperature indicators (TTI) (food spoilage detection) 376
tomato fruit cutin
composition 170
mechanical properties 171
tomato pomace 172–173, 187–188
treated cardboard 284–289, 291
treated paper surface 294
triallyl isocyanurate (TAIC) 114–115
trifluoroacetic acid 361
triglycidyl isocianurate (TGIC) 247
type I cellulose 238

u
untreated cardboard
chemical characterization 285
FTIR spectra 285, 287
mechanical characterization 291–292
moisture content 289–291
morphological characterization 284
oil contact angle 285
SEM images of 284
stereo-micro photograph of 283
water contact angle 285, 288–290
water drop deposition 285
XPS analysis 285, 286
untreated paper surfaces
morphological analysis 295
oxygen barrier properties 299
physical characteristics and mechanical properties 296–300
SEM analysis 295
water and oil resistance properties 296
UV-vis blocking capacity, of polyhydroxyester films 182–183, 188

v
vacuum filtration approach 313, 316
vanillin 334–335
vegetable fibres 236–237

w
water contact angle (W-CA)
of fatty polyhydroxyester 182
of treated and untreated cardboard 285, 288–290
water vapor 15, 20, 59–61, 63, 67, 70–75, 82, 111, 135, 145–149, 153–154, 182, 213, 215, 281, 306, 308, 310–311, 315, 353–354, 359, 386, 397, 403–404, 406–409, 415–416
water vapor permeability (WVP) 17, 21, 44, 59, 61, 73, 76–77, 80, 113, 134, 146–148, 150, 182, 205, 213, 219, 316, 403–406, 408, 413
of PLA 409–410
water vapor transmission rate (WVTR) 213, 315, 412
of whey protein monolayer films 406
water-soluble protein 18
wet/solvent process 405
wheat gluten (WG) 18, 20, 73, 75, 136, 201, 400
coated paper 406
wheat gluten coated surface treated paper (WG–TP) 406
wheat gluten coated untreated paper (WG–UTP) 406
whey protein 18, 21, 73, 75–76, 214, 328, 356, 397–398, 400, 405–406
whey protein isolate (WPI) 21
biodegradable films 398
coatings
barrier properties 400, 401
on PP films, oxygen permeation characteristics of 400

x
xylan-based films 415

z
Z19 zein 199, 200
zein 20–21, 77
aging effect in aqueous ethanol solutions 204
biodegradation 201
biopolymer coatings 200
blend films 214–217

zein (*contd.*)
  chemical amino acid composition
    analysis 199
  extrusion processing 209–212
  extrusion techniques 201
  molecular structure 199
  nanoparticles/nanofibers 219
  plasticized 211–212
  solubility behavior 203, 204
  types 199
zein film forming 201
zein microfluidic devices 205
zein/F127 composite films 215, 216
zein/gelatin multilayer films 408
zein/poly(vinyl alcohol) blends films
    214–215